Biomechanics: Bones and Joints

Biomechanics: Bones and Joints

Edited by Randall Calloway

hayle
medical

New York

Hayle Medical,
750 Third Avenue, 9th Floor,
New York, NY 10017, USA

Visit us on the World Wide Web at:
www.haylemedical.com

ISBN: 978-1-63241-450-2

The publisher's policy is to use permanent paper from mills that operate a sustainable forestry policy. Furthermore, the publisher ensures that the text paper and cover boards used have met acceptable environmental accreditation standards.

Trademark Notice: Registered trademark of products or corporate names are used only for explanation and identification without intent to infringe.

Printed in the United States of America.

Cataloging-in-Publication Data

Biomechanics : bones and joints / edited by Randall Calloway.
    p. cm.
Includes bibliographical references and index.
ISBN 978-1-63241-450-2
1. Biomechanics. 2. Bones--Mechanical properties. 3. Joints--Mechanical properties.
I. Calloway, Randall.
QH513 .B56 2017
571.43--dc23

Table of Contents

Preface

It is often said that books are a boon to mankind. They document every progress and pass on the knowledge from one generation to the other. They play a crucial role in our lives. Thus I was both excited and nervous while editing this book. I was pleased by the thought of being able to make a mark but I was also nervous to do it right because the future of students depends upon it. Hence, I took a few months to research further into the discipline, revise my knowledge and also explore some more aspects. Post this process, I began with the editing of this book.

Biomechanics is the study and research of biological systems. This book on biomechanics discusses the mechanisms under work in the musculoskeletal system when faced with various stimuli. Biomechanics is applied in the fields of sports and exercise physiology to study movement and muscle development. Procedures in this field are also applied in engineering. The topics covered in this book offer the readers new insights in the field of biomechanics. It covers in detail some existent theories and innovative concepts revolving around this discipline. The various advancements in biomechanics are glanced at and their applications as well as ramifications are discussed. This book will prove to be immensely beneficial to students and researchers in this field.

I thank my publisher with all my heart for considering me worthy of this unparalleled opportunity and for showing unwavering faith in my skills. I would also like to thank the editorial team who worked closely with me at every step and contributed immensely towards the successful completion of this book. Last but not the least, I wish to thank my friends and colleagues for their support.

Editor

Non-invasive Predictors of Human Cortical Bone Mechanical Properties: T_2-Discriminated ^{1}H NMR Compared with High Resolution X-ray

R. Adam Horch[1,2], Daniel F. Gochberg[2,3], Jeffry S. Nyman[1,4,5,6]*, Mark D. Does[1,2,3,7]*

1 Department of Biomedical Engineering, Vanderbilt University, Nashville, Tennessee, United States of America, **2** Institute of Imaging Science, Vanderbilt University, Nashville, Tennessee, United States of America, **3** Department of Radiology and Radiological Sciences, Vanderbilt University, Nashville, Tennessee, United States of America, **4** VA Tennessee Valley Healthcare System, Nashville, Tennessee, United States of America, **5** Department of Orthopaedics and Rehabilitation, Vanderbilt University Medical Center, Nashville, Tennessee, United States of America, **6** Center for Bone Biology, Vanderbilt University Medical Center, Nashville, Tennessee, United States of America, **7** Department of Electrical Engineering, Vanderbilt University, Nashville, Tennessee, United States of America

Abstract

Recent advancements in magnetic resonance imaging (MRI) have enabled clinical imaging of human cortical bone, providing a potentially powerful new means for assessing bone health with molecular-scale sensitivities unavailable to conventional X-ray-based diagnostics. To this end, ^{1}H nuclear magnetic resonance (NMR) and high-resolution X-ray signals from human cortical bone samples were correlated with mechanical properties of bone. Results showed that ^{1}H NMR signals were better predictors of yield stress, peak stress, and pre-yield toughness than were the X-ray derived signals. These ^{1}H NMR signals can, in principle, be extracted from clinical MRI, thus offering the potential for improved clinical assessment of fracture risk.

Editor: Stuart Phillips, McMaster University, Canada

Funding: The authors would like to acknowledge financial support from NIH Grant # EB001744, #EB001452, the Vanderbilt Discovery Grant program, and NSF Grant #0448915. The funders had no role in study design, data collection and analysis, decision to publish, or preparation of the manuscript.

Competing Interests: The authors have filed a provisional patent on inventions related to the findings presented in this manuscript.

* E-mail: mark.does@vanderbilt.edu (MDD); jeffry.s.nyman@vanderbilt.edu (JSN)

Introduction

Current bone diagnostics are incomplete. The estimate of areal bone mineral density (BMD) by dual energy x-ray absorptiometry (DXA) does not fully predict fracture risk: for a given DXA score, there is an unexplained increase in fracture risk with age [1,2], as well as with progression of various disease states, such as diabetes [3]. The limitations of DXA related to BMD depending on bone size [4] may be somewhat overcome by quantitative computed tomography imaging, but, ultimately, any X-ray based diagnostic is only sensitive to the mineral portion of the bone, which accounts for only $\approx 43\%$ of bone by volume. The remaining soft-tissue components of bone, including collagen and collagen-bound water, are essentially invisible to DXA and quantitative computed tomography. In contrast, clinical magnetic resonance imaging (MRI), which is based on the ^{1}H NMR signal, cannot directly detect bone mineral but is sensitive to the soft tissue of bone. Further, a recent study has demonstrated that ^{1}H NMR transverse relaxation time constants (T_2) distinguishes proton signals from collagen, collagen-bound water, and pore water [5]. With this technology and the idea that the presence and hydration-state of collagen play a critical role in dissipating energy in bone [6], we hypothesized that ^{1}H NMR can report on the material strength of bone, and we present here compelling experimental observations of ^{1}H NMR, X-ray CT and mechanical tests of cadaveric bone samples which indicate that MRI has the potential to better diagnose fracture risk than DXA.

Results

Figure 1 shows the mean (and standard deviation and range) spectrum of ^{1}H NMR transverse relaxation time constants (T_2 spectrum) from 40 cadaveric bone samples. In this mean spectrum and in each individual sample spectrum, signals from three distinct domains of T_2 were readily identified, as previously found [5]: 50 µs$<T_2<$150 µs, defined as pool A, due primarily to collagen methylene protons; 150 µs$<T_2<$1 ms, pool B, due primarily to collagen-bound water protons; and 1 ms$<T_2<$1 s, pool C, due to water protons in pores in lipid protons. From these three signals, six parameters were extracted: 3 signal amplitudes (S_A, S_B, S_C, in absolutes units of mole ^{1}H per liter bone) and 3 corresponding mean relaxation rate constants ($R_{2,A}$, $R_{2,B}$, $R_{2,C}$ in s^{-1}). Note that while the signal amplitudes are computed in absolute units of concentration, the correspondence between signal amplitudes, S_A, S_B, and S_C, and actual concentrations of collagen methylene protons, bound water protons, and pore-water or lipid protons, respectively, is potentially affected by a number of factors, including the line shape of the methylene protons, the magnetization exchange rate between bound and methylene protons, and overlap of T_2 components from different sources.

Each of the three NMR signal amplitudes (S_A, S_B, S_C) was found to linearly correlate ($r^2 = 0.34$, 0.68, 0.61, $p<0.05$) with peak stress (Fig. 2), but note that the sum of all three signals did not ($r^2 = 0.06$, $p>0.05$). Similar pair-wise linear correlations (and lack thereof) also existed between NMR signal amplitudes and the other three

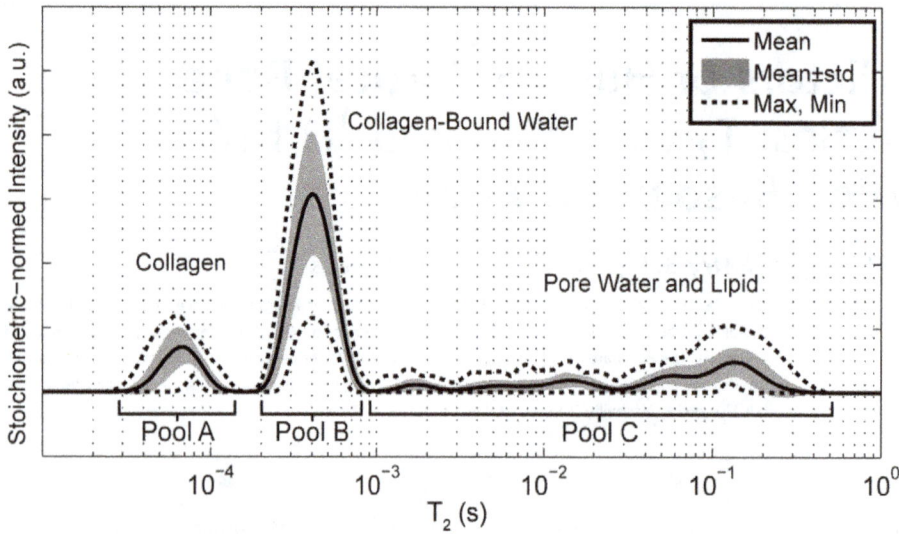

Figure 1. Summary of T₂ spectra measured from 40 human cortical bone samples. All spectra exhibited a short-T_2 component ($T_2 \approx 60$ μs), derived primarily from collagen protons, an intermediate T_2 components ($T_2 \approx 400$ μs), derived primarily from collagen-bound water protons, and a broad distribution of long-T_2 components (1 ms$<T_2<$1 s), derived from a combination of pore water and lipid protons.

measured mechanical properties. These findings indicate that peak cortical bone stress, and the other measured mechanical properties, are directly related to the amount of collagen and collagen-bound water in bone, and inversely related to the bone pore volume. Micro-computed tomography (μCT)-derived measures of bone porosity and the apparent volumetric bone mineral density (avBMD, akin to DXA) were also found to linearly correlate with mechanical properties, but S_A and S_B were better predictors (i.e., higher r^2 values) than μCT-porosity for three of four mechanical properties (flexural modulus being the exception), and better predictors than avBMD (i.e., DXA) for all four mechanical properties. Table 1 summarizes the pairwise linear correlations between imaging measure (^{1}H NMR and X-ray) and the four mechanical properties.

Note that without the two apparent outlier data (peak stress $\approx$ 100 MPa), the predictive power of S_B and S_C decreased to r^2

values of 0.52 and 0.49, respectively, but the r^2 of avBMD with peak stress decreased to a greater extent (to 0.16). That is, the relative predictive power of S_B and S_C compared with avBMD *increased* without these two data points. Also note that multiple linear regression analysis told a similar story: combination of NMR signal parameters (R_B and S_B) best predicted of three of four mechanical properties (adjusted R^2: 0.56-0.70, again, flexural modulus was the exception), and better predicted all four mechanical properties than did avBMD.

Discussion

As a surrogate to radiation-based CT, MRI has been developed to characterize trabecular volume and architecture as a means to assess fracture risk [7,8]. For example, such MRI-derived measurements of bone volume fraction and trabecular thickness

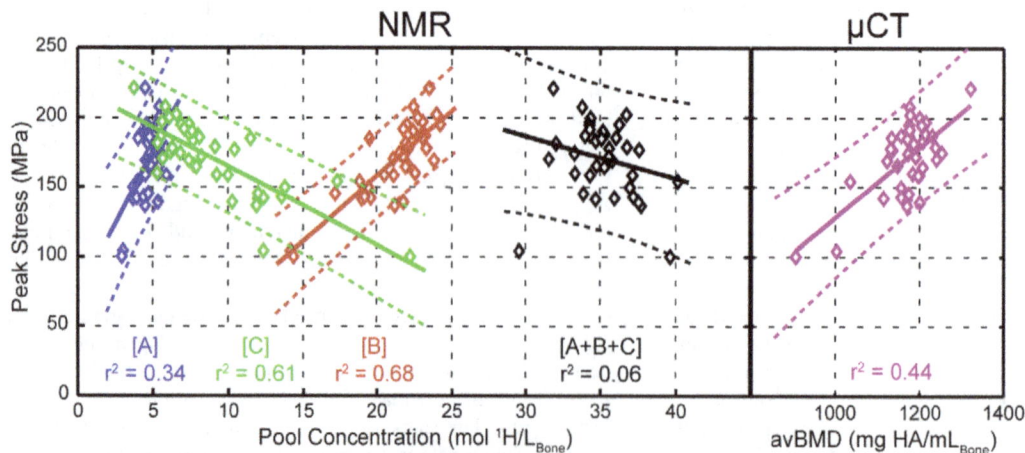

Figure 2. Correlations of measured peak stress and T₂ spectral component amplitudes (NMR, left) and avBMD measured by μCT (right). Blue, red, and green data show integrated amplitudes (S_A, S_B, and S_C) of the T_2-discriminated signals from pools A, B, and C, respectively. The black data show the total ^{1}H NMR signal ($S_A+S_B+S_C$), and the purple data are derived from μCT-based measures of avBMD. Each of the NMR signals amplitudes shows a significant linear correlation with peak stress and both S_B and S_C correlate more strongly with peak stress than does avBMD. Note that the total ^{1}H NMR signal does not correlate well with peak stress.

Table 1. A summary of Pearson's r^2 for pairwise correlations between imaging measures (^{1}H NMR and X-ray) and mechanical properties.

	Yield Stress	Peak Stress	Flexural Modulus	Pre-Yield Toughness
$R_{2,A}$	0.10	0.12	0.04*	0.12
$R_{2,B}$	0.19	0.22	0.12	0.19
$R_{2,C}$	0.00*	0.01*	0.01*	0.00*
S_A	0.41	0.34	0.39	0.34
S_B	**0.62**	**0.68**	0.48	**0.57**
S_C	0.57	0.61	0.49	0.49
$S_A+S_B+S_C$	0.05*	0.06*	0.06*	0.03*
AVBMD	0.43	0.44	0.46	0.33
POROSITY	0.58	0.60	**0.59**	0.46

All correlations were significant ($p<0.05$) *except* those indicated with *. The imaging measure that was most predictive (highest r^2) of each mechanical measure is indicated with boldface type.

correlated with the compressive strength of human trabecular bone, although the correlations were not as strong as that between CT-derived BMD and strength [9]. These MRI techniques do not assess the inherent quality of the bone tissue, and this is a significant shortcoming given the importance of ultrastructural characteristics of the extracellular matrix (e.g., collagen integrity) to the fracture resistance of bone [10]. From ex vivo studies of bone, various quantifications of water by NMR have been correlated with the mechanical competence of bone. In a rabbit model of diet-induced hypomineralization, a ^{1}H NMR-derived measurement of water content was directly related to the bending strength of cortical bone [11]; however, in a study of ovariectomized and treated mice, only group-mean total water ^{1}H NMR signal correlated with mechanical properties—no correlation was found across pooled data from 60 bones, which may be explained by the findings of total ^{1}H signal shown here (Fig. 2). Also, an NMR technique known as "decay from diffusion in an internal field" (DDIF) found an inverse correlation between this NMR-derived pore water parameter and the yield stress of bovine trabecular bone in compression [12], in rough agreement with the present observations of pore-water. Prior to the present study though, only one study attempted to correlate NMR measurements of both pore water and water bound to the

extracellular matrix to the mechanical properties of human bone [13]. That study used T_2^*-discriminated rather than T_2-discriminated (used herein) ^{1}H NMR signals at low static magnetic field, and while a direct relationship existed between the so-called T_2^*-defined bound water and peak stress, it described a much lower fraction of the peak stress variance ($r^2 = 0.36$, compared to 0.68, above). Also, the translation of T_2^* based discrimination to clinical imaging may be problematic due to the presence of lipid in bone [5,11], and the inability of T_2^* to discriminate bone ^{1}H pools at clinical field strengths (no discrimination was found at 4.7T [5] and no discrimination has been reported at clinical fields strengths (≥1.5 T)).

Current uTE protocols on human MRI systems use echo times <100 µs [14] (and references therein), more than short enough to capture the majority of the bound water signal and some of the collagen proton signal, but the translation of the present findings to clinical MRI will require practical imaging methods of distinguishing these short-T_2 signals from the longer-T_2 pore water and lipid signals. There are numerous strategies for integrating T_2-selective magnetization preparation into a clinically practical uTE-type sequence [15,16,17], and the optimal approach for bone imaging has not yet been determined. However, Fig. 3 shows two T_2 spectra from one bone specimen. The solid line shows the normal T_2 spectrum, as used in the above analysis, while the dotted line shows the spectrum that results following the complex average of two CPMG signals, with and without the preceding hyperbolic secant radiofrequency (RF) pulse. This RF pulse effectively inverts only the long T_2 signals while largely saturating the collagen proton and bound-water signal, so the complex average cancels only the long T_2 signals and results in a net NMR signal that is $\approx95\%$ derived from protons with $T_2<1$ ms. This result demonstrates in principle that a simple RF pre-pulse, which can be readily integrated into a standard uTE pulse sequence, can distinguish pore water from collagen protons and collagen bound water protons in bone. Once implemented on clinical scanners, such an MRI method can then assess both the contribution of structure to whole bone strength as well as the contributions of collagen integrity and porosity, thus proving a more complete assessment of fracture risk than current X-ray based methods.

Materials and Methods

Human cortical bone processing

The Musculoskeletal Tissue Foundation (Edison, NJ), a non-profit tissue allograft bank, and the Vanderbilt Donor Program

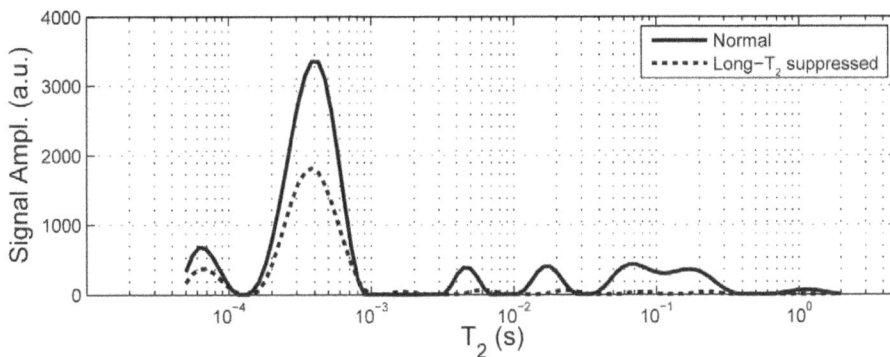

Figure 3. Solid line shows a the T_2 spectrum from a typical bone sample, and the dotted line shows the spectrum that results following the complex average of two signals, with and without an adiabatic full passage magnetization preparation. The total integrated signal from this long-T_2 suppressed spectrum is 95% from signals with $T_2<1$ ms, thereby demonstrating in principle a simple and practical method for generating a MRI contrast dominated by S_A+S_B.

(Nashville, TN) supplied human femurs from 40 cadaveric donors (26 male, 14 female, aged 21-105 years old, mean ± standard deviation: 67±24 years) under instruction to not provide tissue from donors who had tested positive for a blood borne pathogen (e.g., HIV or Hepatitis C). One human cortical bone sample per donor was extracted from the medial quadrant of the mid-shaft and was machined to $70 \times 5 \times 2$ mm^3 dimensions via end mill. During dimensioning, care was taken to remove endosteal and periosteal surfaces such that the final specimens were pure cortical bone. From each milled sample, three specimens were extracted for NMR, μCT, and mechanical testing (Fig. 4). Specimens were stored in phosphate-buffered saline at $-20°C$ then thawed at $4°C$ for approximately 18 hours prior to NMR measurements. (No more than three freeze-thaw cycles were involved for a given specimen, and separate experiments found that up to six freeze-thaw cycles had negligible impact on the NMR properties.) Final specimen dimensions were measured with digital caliper for volume determination.

NMR

From one of the three specimens per donor sample, ^{1}H NMR transverse relaxation (T_2) characteristics were measured and reduced to three independent signal components, which we have recently identified as being primarily derived from collagen methylene protons, collagen-bound water protons, and water protons in pores [5]. All NMR measurements were performed in a Varian/Magnex 4.7 T horizontal bore magnet with a Direct Drive Receiver. An in-house loop-gap style RF coil with Teflon structural support was used (similar to the coil described in [18]), which provided $90°/180°$ RF pulses of ≈ 8 μs/16 μs duration and contributed negligible background ^{1}H signal ($\approx 1\%$ of net HCB signal).

Carr-Purcell-Meiboom-Gill (CPMG) measurements with a total of 10000 echoes were collected at 100 μs echo spacing, which was empirically determined to be a suitable minimum threshold for both maximizing the range of T_2 detection while minimizing spin-locking effects. Echo magnitudes were fitted to a sum of 128 decaying exponential functions (with time constants log-spaced between 20 μs and 10 sec) in a non-negative least-squares sense, subject to a minimum curvature constraint, which produced a so-called T_2 spectrum [19]. In order to quantitatively compare the absolute signal amplitudes of T_2 components across days, a reference sample with long T_2 (≈ 2 s) and known proton content was included in each CPMG measurement. The presence of the reference sample, together with the known specimen volumes, enabled the calculation of proton concentrations in the bulk bone specimens for each CPMG relaxation component by comparing integrated areas of each T_2 spectral component to the area of the marker. As a simple demonstration of the potential for acquiring signal from a specific T_2 domain without the full CPMG acquisition, from one bone specimen, an additional CPMG measurement was acquired with a preceding a 10-ms duration, 3500 Hz bandwidth hyperbolic secant inversion pulse [20], so chosen to selectively invert the long-T_2 ^{1}H signal.

μCT

The second specimen from each donor sample ($\sim$ volume of 40 mm^3) was studied at high resolution (6 μm), with low noise micro-CT (μCT) to quantify apparent volumetric bone mineral density (avBMD) and intracortical porosity (for pores ≥ 6 μm in diameter). Note that for a given specimen size avBMD is a volumetric analog to areal BMD as measured by DXA, and intracortical porosity at this resolution is not readily determined from clinical radiographs or QCT including high-resolution peripheral QCT scanners (which obtain resolutions of 80–100 μm) [21]. The specimen was scanned by acquiring 1000 projections per $180°$ at 70 keV using a Scanco, model μCT-40. From an hydroxyapatite (HA) phantom image (acquired weekly), linear attenuation coefficients derived from the μCT images were equated to volumetric bone mineral density (vBMD) in units of mg-HA/cm^3. Using the Scanco software, the outer perimeter of the sample was defined to determine the total bone volume. The avBMD was defined as the mean of vBMD for all voxels within the total bone volume. The bone tissue volume was segmented from air or soft tissue at a threshold of 800 mg-HA/cm^3 to determine the porosity (= 1 minus bone tissue volume per total bone volume) (Fig. 5).

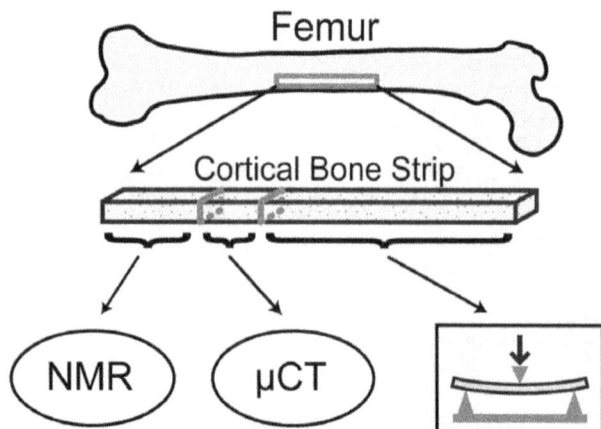

Figure 4. From each cadaveric bone studied, one strip of cortical bone was extracted, three separate pieces of which were used for NMR, μCT, and mechanical testing.

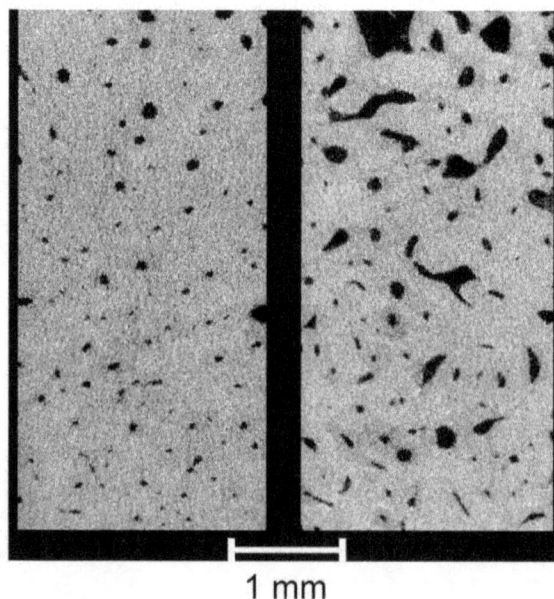

Figure 5. Axial μCT images are shown for cortical bone specimens from a 48 y.o. male donor (left) and an 82 y.o. male donor (right). For the 48 and 82 y.o. donors, respectively, avBMD was 1222 and 1135 mg-HA/cm^3, and porosity was 4% and 11.3%.

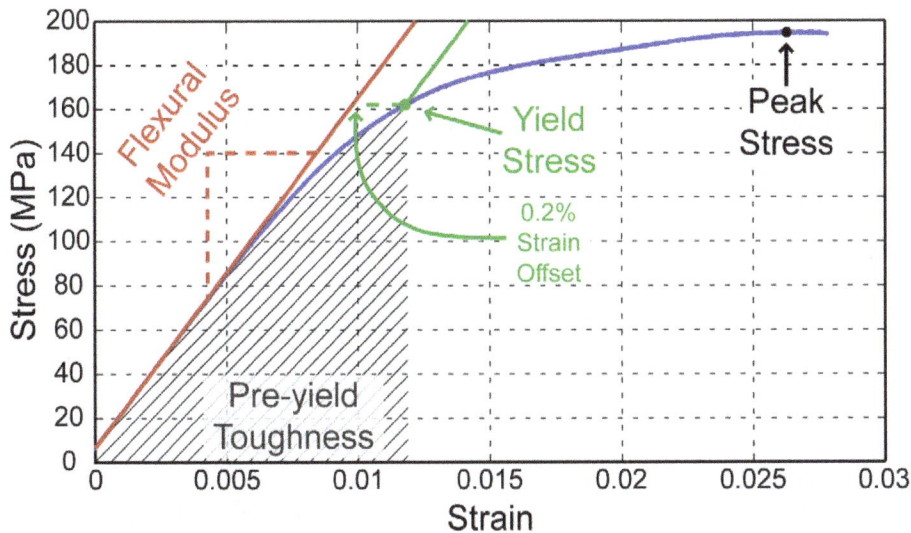

Figure 6. A representative stress vs. strain curve for cortical bone is shown (blue) along with graphical depictions of mechanical parameters. Flexural modulus is the slope of the initial linear mechanical response, yield stress is defined at 0.2% offset from the flexural modulus line, and peak stress is the maximum observed stress. Pre-yield toughness (see text for definition) is proportional to the area under the curve, up to the yield stress.

Mechanical

Finally, we subjected the third, parallelpiped specimen (nominal dimensions of 2 mm×5 mm×40 mm) from each donor sample to a three point bending test, and measured four mechanical properties relevant to fracture risk in bone: yield stress, peak stress, flexural modulus, and pre-yield or elastic toughness. A material testing system (Dynamight 8841, Instron, Canton, OH) recorded the force-displacement data (Fig. 6) from a 100 N load cell and the linear variable differential transformer, respectively, at 50 Hz, as the hydrated bone was loaded to failure at 5 mm/min. The span was 35 mm, and all tests were performed at room temperature. Applying the flexure formula to the yield force, as determined by the 0.2% offset, or to the peak force endured by the

bone specimen, and applying the deflection equation to the slope of the linear section of the force-displacement curve provided the material properties, yield stress, peak stress, and flexural modulus, respectively [6]. Pre-yield or elastic toughness was the area under the force-displacement curve from zero displacement to the yield displacement divided by the cross-sectional area of the bone sample to account for slight differences in specimen dimensions.

Author Contributions

Conceived and designed the experiments: RAH DFG JSN MDD. Performed the experiments: RAH. Analyzed the data: RAH. Contributed reagents/materials/analysis tools: RAH DFG JSN MDD. Wrote the paper: JSN MDD. Supervised the project: JSN MDD.

References

1. Kanis JA, Johnell O, Oden A, Dawson A, De Laet C, et al. (2001) Ten year probabilities of osteoporotic fractures according to BMD and diagnostic thresholds. Osteoporos Int 12: 989–995.
2. Johnell O, Kanis JA, Oden A, Johansson H, De Laet C, et al. (2005) Predictive value of BMD for hip and other fractures. J Bone Miner Res 20: 1185–1194.
3. Vestergaard P (2007) Discrepancies in bone mineral density and fracture risk in patients with type 1 and type 2 diabetes–a meta-analysis. Osteoporos Int 18: 427–444.
4. Carter DR, Bouxsein ML, Marcus R (1992) New approaches for interpreting projected bone densitometry data. J Bone Miner Res 7: 137–145.
5. Horch RA, Nyman JS, Gochberg DF, Dortch RD, Does MD (2010) Characterization of 1H NMR Signal in Human Cortical Bone for Magnetic Resonance Imaging. Magnetic Resonance in Medicine 64: 680–687.
6. Nyman JS, Roy A, Shen XM, Acuna RL, Tyler JH, et al. (2006) The influence of water removal on the strength and toughness of cortical bone. Journal of Biomechanics 39: 931–938.
7. Wehrli FW (2007) Structural and functional assessment of trabecular and cortical bone by micro magnetic resonance imaging. J Magn Reson Imaging 25: 390–409.
8. Majumdar S (2008) Magnetic resonance imaging for osteoporosis. Skeletal Radiol 37: 95–97.
9. Majumdar S, Kothari M, Augat P, Newitt DC, Link TM, et al. (1998) High-resolution magnetic resonance imaging: three-dimensional trabecular bone architecture and biomechanical properties. Bone 22: 445–454.
10. Nyman JS, Reyes M, Wang X (2005) Effect of ultrastructural changes on the toughness of bone. Micron 36: 566–582.
11. Fernandez-Seara MA, Wehrli SL, Takahashi M, Wehrli FW (2004) Water content measured by proton-deuteron exchange NMR predicts bone mineral density and mechanical properties. J Bone Miner Res 19: 289–296.

12. Sigmund EE, Cho H, Chen P, Byrnes S, Song YQ, et al. (2008) Diffusion-based MR methods for bone structure and evolution. Magn Reson Med 59: 28–39.
13. Nyman JS, Ni Q, Nicolella DP, Wang X (2008) Measurements of mobile and bound water by nuclear magnetic resonance correlate with mechanical properties of bone. Bone 42: 193–199.
14. Robson MD, Bydder GM (2006) Clinical ultrashort echo time imaging of bone and other connective tissues. NMR in Biomedicine 19: 765–780.
15. Larson PEZ, Conolly SM, Pauly JM, Nishimura DG (2007) Using adiabatic inversion pulses for long-T-2 suppression in ultrashort echo time (UTE) imaging. Magnetic Resonance in Medicine 58: 952–961.
16. Rahmer J, Blume U, Bornert P (2007) Selective 3D ultrashort TE imaging: comparison of "dual-echo" acquisition and magnetization preparation for improving short-T-2 contrast. Magnetic Resonance Materials in Physics Biology and Medicine 20: 83–92.
17. Du J, Takahashi AM, Bae WC, Chung CB, Bydder GM (2010) Dual Inversion Recovery, Ultrashort Echo Time (DIR UTE) Imaging: Creating High Contrast for Short-T-2 Species. Magnetic Resonance in Medicine 63: 447–455.
18. Horch RA, Wilkens K, Gochberg DF, Does MD (2010) RF coil considerations for short T2 MRI. Magnetic Resonance in Medicine 64: 1652–1657.
19. Whittall KP, Mackay AL (1989) Quantitative interpretation of NMR Relaxation Data. Journal of Magnetic Resonance 84: 134–152.
20. Silver MS, Joseph RI, Hoult DI (1984) Highly Selective Pi/2 and Pi-Pulse Generation. Journal of Magnetic Resonance 59: 347–351.
21. Burghardt AJ, Kazakia GJ, Ramachandran S, Link TM, Majumdar S (2010) Age- and Gender-Related Differences in the Geometric Properties and Biomechanical Significance of Intracortical Porosity in the Distal Radius and Tibia. Journal of Bone and Mineral Research 25: 983–993.

Enhanced Th17-Cell Responses Render CCR2-Deficient Mice More Susceptible for Autoimmune Arthritis

Rishi R. Rampersad[1], Teresa K. Tarrant[1], Christopher T. Vallanat[1], Tatiana Quintero-Matthews[1], Michael F. Weeks[1], Denise A. Esserman[2,3], Jennifer Clark[3], Franco Di Padova[4], Dhavalkumar D. Patel[1,4], Alan M. Fong[1], Peng Liu[1]*

1 Department of Medicine, Thurston Arthritis Research Center, University of North Carolina at Chapel Hill, Chapel Hill, North Carolina, United States of America, 2 Division of General Medicine and Clinical Epidemiology, Department of Medicine, University of North Carolina at Chapel Hill, Chapel Hill, North Carolina, United States of America, 3 Department of Biostatistics, University of North Carolina at Chapel Hill, Chapel Hill, North Carolina, United States of America, 4 Novartis Institutes for BioMedical Research, Basel, Switzerland

Abstract

CCR2 is considered a proinflammatory mediator in many inflammatory diseases such as rheumatoid arthritis. However, mice lacking CCR2 develop exacerbated collagen-induced arthritis. To explore the underlying mechanism, we investigated whether autoimmune-associated Th17 cells were involved in the pathogenesis of the severe phenotype of autoimmune arthritis. We found that Th17 cells were expanded approximately 3-fold in the draining lymph nodes of immunized CCR2$^{-/-}$ mice compared to WT controls (p = 0.017), whereas the number of Th1 cells and regulatory T cells are similar between these two groups of mice. Consistently, levels of the Th17 cell cytokine IL-17A and Th17 cell-associated cytokines, IL-6 and IL-1β were approximately 2–6-fold elevated in the serum and 22–28-fold increased in the arthritic joints in CCR2$^{-/-}$ mice compared to WT mice (p = 0.04, 0.0004, and 0.01 for IL-17, IL-6, and IL-1β, respectively, in the serum and p = 0.009, 0.02, and 0.02 in the joints). Furthermore, type II collagen-specific antibodies were significantly increased, which was accompanied by B cell and neutrophil expansion in CCR2$^{-/-}$ mice. Finally, treatment with an anti-IL-17A antibody modestly reduced the disease severity in CCR2$^{-/-}$ mice. Therefore, we conclude that while we detect markedly enhanced Th17-cell responses in collagen-induced arthritis in CCR2-deficient mice and IL-17A blockade does have an ameliorating effect, factors additional to Th17 cells and IL-17A also contribute to the severe autoimmune arthritis seen in CCR2 deficiency. CCR2 may have a protective role in the pathogenesis of autoimmune arthritis. Our data that monocytes were missing from the spleen while remained abundant in the bone marrow and joints of immunized CCR2$^{-/-}$ mice suggest that there is a potential link between CCR2-expressing monocytes and Th17 cells during autoimmunity.

Editor: Raffaella Bonecchi, Università degli Studi di Milano, Italy

Funding: This work was supported by an Arthritis Foundation Postdoctoral Fellowship, a pilot grant from the North Carolina Translational and Clinical Science (NC TraCS) Institute, home of the UNC-Chapel Hill Clinical and Translational Science Award (CTSA), and NIH grant HL077406. The funders had no role in study design, data collection and analysis, decision to publish, or preparation of the manuscript.

Competing Interests: Drs. Patel, DD and Di Padova, F are employees of Novartis Institutes for BioMedical Research. All other authors report no potential conflicts.

* E-mail: liupz@med.unc.edu

Introduction

Th17 cells are IL-17-producing T helper effector cells that are distinct from Th1 and Th2 cells, and from regulatory T (Treg) cells. Th17 cells have been suggested to mediate inflammation and play a key role in the pathogenesis of tissue-specific autoimmune diseases including experimental autoimmune encephalomyelitis (EAE), collagen-induced arthritis (CIA), and psoriasis [1,2]. Specifically, studies found that mice lacking a Th17 cell-promoting cytokine IL-23 were resistant to EAE or CIA [3], whereas mice deficient for the classic Th1 cell cytokine IL-12 or IFN-γwere more susceptible to these diseases [4,5]. Adoptive transfer of Th17 cells was shown to be required for EAE [6]. Furthermore, gene targeted deletion of IL-17 or treatment with a neutralizing anti-IL-17 antibody results in CIA resistance [7,8]. These results suggest that Th17 cells are potent inducers of autoimmune disorders. Since the IL-17 receptor is expressed on epithelial and parenchymal cells, Th17 cells are thought to promote tissue

inflammation by producing IL-17 to stimulate the production of IL-6, IL-1, tumor necrosis factor (TNF), and other proinflammatory factors [9]. In humans, anti-IL-17A antibodies have shown positive clinical responses and good safety in patients with active rheumatoid arthritis (RA) in randomized, double-blinded proof-of-concept trials [10,11], suggesting that targeting of IL-17A is a promising therapeutic approach against human autoimmune disease, such as RA.

CCR2 is a chemokine receptor for monocyte chemoattractant protein-1 (MCP-1 or CCL2) and important for monocyte trafficking toward sites of inflammation, a process that is critical for autoimmune diseases like rheumatoid arthritis (RA) and CIA [12]. Since CCR2 is highly expressed on joint infiltrated monocytes/macrophages in RA patients and CIA animals [13], targeted inhibition of CCR2 was initially thought to be a promising therapeutic strategy for the treatment of RA. However, clinical trials using CCR2 or CCL2 neutralizing antibodies did not show efficacy [14,15]. Application of anti-CCR2 during disease progression or deletion of Ccr2 genes ($CCR2^{-/-}$) unexpectedly induce exacerbated CIA in mice [13,16]. Although our previous study and others show that inhibiting CCR2 is beneficial to other inflammatory diseases, such as vascular inflammations [17,18], the mechanism underlying the CCR2 paradox in autoimmune inflammation is not completely understood.

Autoimmune inflammation uniquely involves antigen-specific activation of T cells that leads to subsequent B cell activation and autoantibody formation. Given the specific proinflammatory role of Th17 cells in autoimmune diseases, we hypothesized that skewing of Th17 cells and Th17-cell responses may account for the exacerbated arthritis in $CCR2^{-/-}$ mice. To test this hypothesis, we examined Th17 cells, Th1 cells, Treg cells, and Th17 cell-associated events, such as cytokine profile, autoantibody production, and neutrophil activities, in $CCR2^{-/-}$ mice induced with collagen-induced arthritis. We also treated these animals with an IL-17A neutralizing antibody. Our results demonstrate that both Th17 cells and Th17-cell associated responses are markedly enhanced in immunized $CCR2^{-/-}$ mice and neutralizing IL-17A has an ameliorating effect on the severe autoimmune arthritis seen in CCR2 deficiency. Our data that monocytes decrease in the spleen, but not in the bone marrow and arthritic joints in $CCR2^{-/-}$ mice suggest a potential biological link between CCR2-expressing monocytes and Th17 cells.

Methods

CIA induction and evaluations

$CCR2^{-/-}$ mice on DBA/1J background were generated by backcrossing $CCR2^{-/-}$ C57BL/6J mice with wildtype DBA/1J mice (Jackson Laboratory, Bar Harbor, Maine) for 12 generations. 8–10 week-old male $CCR2^{-/-}$ DBA/1J mice ($CCR2^{-/-}$ mice) were used in the experiments and the age- and gender-matched littermates of $CCR2^{+/+}$ DBA/1J mice (WT mice) were used as controls. All mice were bred and maintained in a barrier facility and fed with Prolab RMH-3000 (PMI Nutrition International, Richmond, IN), a normal rodent chow diet. All experimental protocols were in compliance with IACUC (Institutional Animal Care and Use Committee) guidelines and were approved by the IACUC at the University of North Carolina at Chapel Hill (ID: 09-344).

CIA is typically performed with either one or two injections of heterologous type II collagen on a susceptible congenic mouse strain. The main purpose of a second booster is to increase the overall disease severity so that the difference between control mice and gene-deficient or treatment animals that have less disease can be apparent. In our case, CCR2-deficient mice have exacerbated disease that starts from day 14 after one injection, which was prior to the second booster at day 21. Therefore, we chose to utilize single injection to examine effect of CCR2 deficiency on the pathogenesis of CIA. In this study, CIA was induced by immunization with type II bovine collagen (CII, Chondrex, Redmond, WA) emulsified with complete Freund's adjuvant (CFA, Sigma). The immunization and disease evaluation methods were performed according to protocols validated by our group and others with minor revisions [19,20]. Briefly, CII was dissolved overnight in 0.1 M acetic acid and emulsified in an equal volume of CFA. Mice were immunized with a subcutaneous injection of 100 µl of emulsion containing 100 µg of CII at approximately 1 cm from the base of the tail at day 0. The clinical disease of arthritis was evaluated as follows: 0 = normal; 1 = swelling in 1 joint; 2 = swelling in >1 joint; 3 = swelling of the entire paw. Paw swelling was measured in mm range with a Dial Thickness gauge (Geneva Gage, Albany, OR). Both arthritis scoring and measurement were performed in a blinded manner.

Flow cytometry

Single cell suspension of the spleen or draining lymph nodes was prepared using a manual tissue homogenizer. Red blood cells (RBC) were lysed with a lysis buffer (0.14 M NH_4Cl, 0.017 M Tris-HCl adjust to pH 7.2). Cells were washed with RPMI containing 2 mM Hepes and 1% bovine serum albumin and passed through a 70 µm nylon cell strainer. Isolation of synovial cells was performed as described previously [21] with minor modifications. Briefly, the hind paws were dissected out and incubated with Collagenase D (Sigma) at 2 mg/ml in PBS at 37°C for 30 minutes. The digestion was stopped by adding 10 mM EDTA and incubated for an additional 5 minutes. Synovial cells were isolated and passed through a 70 µm cell strainer. A CyAn flow cytometer and Summit software (CyAn, Dako) were used to analyze cell surface markers that were stained with fluorophore-conjugated primary and isotype control antibodies (eBioscience and Biolegend). Antibodies used in the experiments included anti-CD3, CD4, CD8, CD45R/B220, CD11b, CD11c, F4/80, Ly6C, Ly6G and their isotype controls that were conjugated with FITC, PE, PerCP, Pacific Blue, Pacific Orange, allophycocyanin (APC), and Alexa Fluor-APC. For intracellular cytokine staining, single cell suspension of draining lymph nodes were cultured in petri dishes pre-coated with anti-CD3 in the presence of anti-CD28 for 6 hours, and Golgi Stop (BD Bioscience) was added for the last two hours. Cells then were harvested and stained with anti-CD4 on the surface, fixed and permeabilized with BD Cytofix/ Cytoperm (BD Bioscience), and stained with anti-IL-17A and anti-IFN-γ intracellularly. For staining of Treg cells, anti-CD4/ CD25 surface stained cells were fixed and permeabilized overnight at 4°C. The following day the cells were blocked for 15 minutes with anti-CD16/CD32, and then stained with anti-FoxP3-PE (eBioscience, San Diego, CA) at 4°C for 30 minutes.

Analysis of systemic cytokine profile

Systemic cytokine profiles of IL-1β, IL-6, IL-17A, and IFN-γ were determined by Luminex using blood sera collected through submandibular bleeding of $CCR2^{-/-}$ and WT mice with CIA. Sera cytokine levels were measured with a Premixed Beadlyte Kit (Millipore, Billerica, MA) by the Luminex 100 dual laser system and XY Platform (Luminex Corp., Austin, TX), which was controlled by the MasterPlex CT 1.2 software (MiraiBio, Alameda, CA). MasterPlex QT 4.0 system (MiraiBio, Alameda, CA) was used for data analysis and the five-parameter regression formula was used to calculate cytokine concentrations from the standard curves. All luminex assays were performed by the UNC Clinical Proteomics Laboratory. BAFF was detected using a Quantikine Mouse BAFF ELISA Kit from R&D System (Minneapolis, MN) according to manufacture instruction.

Production of antigen-specific antibodies

The production of antigen-specific antibodies was measured by ELISA using the Mouse Anti-Bovine Type II Collagen IgG Assay Kit (Chondrex, Redmond, WA). Briefly, collagen pre-coated wells were washed and blocked with the blocking buffer and incubated with mouse sera samples with a 1:10,000 dilution, incubated with an HRP-labeled polyclonal secondary antibody, and developed with the OPD solution. OD values were read at 490 nm on an Emax Precision Microplate Reader and analyzed with SOFTmax PRO, 3.0 (Molecular Devices).

Quantitative real-time PCR

The expression of IL-17A in the joint tissues was measured by quantitative real-time PCR. Briefly, total RNA was isolated from CIA mouse paws by Trizol followed by a RNA cleaning step using the Qiagen RNeasy Mini Kit (Qiagen, Valencia, CA), and cDNA was generated using the First-Strand cDNA Synthesis Superscript II RT (Invitrogen, Carlsbad, CA). Primers used for the amplification of murine IL-17A, IFN-γ, IL-6, IL-1β, CXCL1, CXCL2, RANKL and 18s rRNA were as follows: IL-17A, 5'-TCTCTGATGCTGTTGCTGCT-3'(forward) and 5'-CTCCA-GAAGGCCCTCAGACTAC-3'(reverse); IFN-γ, 5'-ACTGG-CAAAAGGATGGTGAC-3' (forward) and 5'-ACCTGTGGG-TTGTTGACCTC-3' (reverse); IL-6, 5'-TTCCATCCAGTT-GCCTTCTT-3' (forward) and 5'-CAGAATTGCCATTGCAC-AAC-3' (reverse); IL-1β, 5'-GGTCAAAGGTTTGGAAGCAG-3' (forward) and 5'-TGTGAAATGCCACCTTTTGA-3' (reverse); CXCL1, 5'-CAATGAGCT GCGCTGTCAGTG-3' (forward) and 5'-CTTGGGGAC ACCTTT TAGCAT C-3' (reverse); CXCL2, 5'-CCAAGGGTTGACTTCAAGAAC-3' (forward) and 5'-AGCGAGGCACATCAGGTACG-3' (reverse); RANKL, 5'-TGTACTTTCGAGCGCAGATG-3' (forward) and 5'-AGGCTTGTTTCATCCTCCTG-3' (reverse) and 18s rRNA, 5'-GACCATAAACGATGCCGACT-3' (forward) and 5'-GTGAGGTTTCCCGTGTTGAG-3' (reverse). Quantitative real-time PCR was performed in a SYBR Green Master Mix with primers listed above in an iCycler instrument (BioRad Laboratories Hercules, CA). The $2^{-\Delta\Delta Ct}$ method [22] was used for data analysis.

Anti-IL-17A antibody treatment

CCR2$^{-/-}$ mice were immunized for CIA and divided into two groups of treatment. At day 14 post-immunization, each mouse was given 200 μg of either a murine anti-IL-17A antibody (Novartis, Basel, Switzerland) or an isotype control IgG1 (Biolegend, San Diego, CA) through intraperitoneal (i.p.) injection. Then additional injections were given twice a week for two weeks.

Statistical analysis

Numerical data presented in the text and figures were expressed as mean ± SEM. Student's unpaired two-tail t test was utilized to compare average numbers of cells or percentages between experimental groups. For clinical disease assessment and antibody treatment experiments, arthritis severities were analyzed using a Mixed Effect Model to determine significances over time in arthritis scores and paw swelling. The overall group effect was assessed using a likelihood ratio test (LRT). Analysis was conducted using SAS, v. 9.2 statistical software package (Cary, NC). In all cases, p<0.05 was considered significant.

Results

Th17 cells are expanded in the draining lymph nodes and joints of collagen immunized CCR2$^{-/-}$ mice

Both CCR2 and its ligand CCL2 are up-regulated in the synovium of patients with RA, and thus are considered to be proinflammatory in autoimmune inflammation [12]. However, more severe arthritis was developed in CCR2$^{-/-}$ mice that were induced by collagen immunization [13]. Indeed, significant differences were identified over time for CCR2$^{-/-}$ mice in both the clinical arthritis scores (LRT = 26.4, df = 4, p<0.0001) and the paw swelling (LRT-37.5, df = 4, p<0.0001) compared to WT controls (Figure 1A). Disease in CCR2$^{-/-}$ mice occurred 14 days earlier than that in WT controls after one immunization. Between days 42 and 49 after immunization, 100% of CCR2$^{-/-}$ mice and

approximately 40% of WT mice received maximal disease, which proceeded to the recovery and resolution phase after this period of time. The severity of arthritis was approximately three times worse during the follow-up time period in CCR2$^{-/-}$ mice with deteriorated histological changes of leukocyte infiltration, bone erosion, and joint destruction (Figure 1 B). These results confirm the previous finding that the absence of CCR2 does not ameliorate but rather aggravates autoimmune arthritis.

Given that autoimmune arthritis is a T cell-mediated disease; and Th17 cells are potent inducers of tissue inflammation in this disease [23], we hypothesized that CCR2 deficiency promotes autoimmune arthritis by facilitating Th17 cell activities. To test this hypothesis we examined Th17 cells in immunized CCR2$^{-/-}$ and wild type mice. We also tested the abundance of Th1 cells and Treg cells. Our data showed that the number of Th17 cells was increased approximately 3-fold in the draining lymph nodes of CCR2$^{-/-}$ mice prior to disease onset (p = 0.017), whereas Th1 cells remained similar between CCR2$^{-/-}$ and WT mice (Figure 1C and D). There was no significant difference of Treg cells in peripheral lymphoid tissues between immunized CCR2$^{-/-}$ and wild type mice (Figure 1E).

Consistently, levels of IL-17A, but not IFN-γ in the serum of CCR2$^{-/-}$ mice were significantly increased (Figure 2A) in the early phase of arthritis. In the meantime, serum IL-6 and IL-1β, which are closely related to Th17 cell differentiation and downstream Th17 cell-mediated proinflammatory responses, were also increased in CCR2$^{-/-}$ mice (Figure 2A). However, such significant elevation of sera cytokines in CCR2$^{-/-}$ mice became less obvious when robust joint disease developed (Figure 2B). Then we went on to test these cytokines in the arthritic joints and found that the expression levels of IL-17A, IL-6, and IL-1β were up-regulated in CCR2$^{-/-}$ mice compared to WT mice (Figure 2C and D). These results suggest that CCR2 deficiency induces increased Th17 cell activities in the generation and development of autoimmune arthritis.

Increased autoantibody production in CCR2$^{-/-}$ mice

IL-17 has been shown to act in synergy with B cell-activating factor (BAFF) to influence B cell biology in systemic lupus erythematosus [24]. Therefore we analyzed B cell abundance and their ability to produce type II collagen (CII)-specific antibodies in immunized CCR2$^{-/-}$ mice. We found that CCR2$^{-/-}$ mice have approximately 26% more B lymphocytes in the draining lymph nodes than that in WT mice (Figure 3A). In agreement with this result, significantly higher levels of anti-CII IgG were detected in the serum of these mice (Figure 3B). However, such increased B cell activities were BAFF-independent as we did not find significant BAFF elevation in CCR2$^{-/-}$ mice (Figure 3C). Thus, the expansion of Th17 cells in CCR2$^{-/-}$ mice appears to be associated with enhanced humoral immune responsiveness to type II collagen.

Upregulated neutrophils in CCR2$^{-/-}$ mice

IL-17 promotes inflammation through stimulation of granulopoiesis in vivo [25]. To examine activities of neutrophils in immunized CCR2$^{-/-}$ mice, we investigated the abundance of Ly6-G$^+$Ly6C$^+$CD11b$^+$ neutrophils in the spleen and the joints. In the spleen, we found substantially more neutrophils in CCR2$^{-/-}$ mice than in WT mice (Figure 4A and B). Correspondingly, enhanced neutrophil infiltration into the joints was detected in these mice (Figure 4A and B). The expansion of neutrophils in CCR2$^{-/-}$ mice was also observed in the histological joint sections with the H&E stain (Figure 1B). As neutrophil recruitment is driven by neutrophil-tropic chemokines, CXCL1 and CXCL2, we examined joint expressions of these cytokines. Indeed, these cytokines increased substantially in the

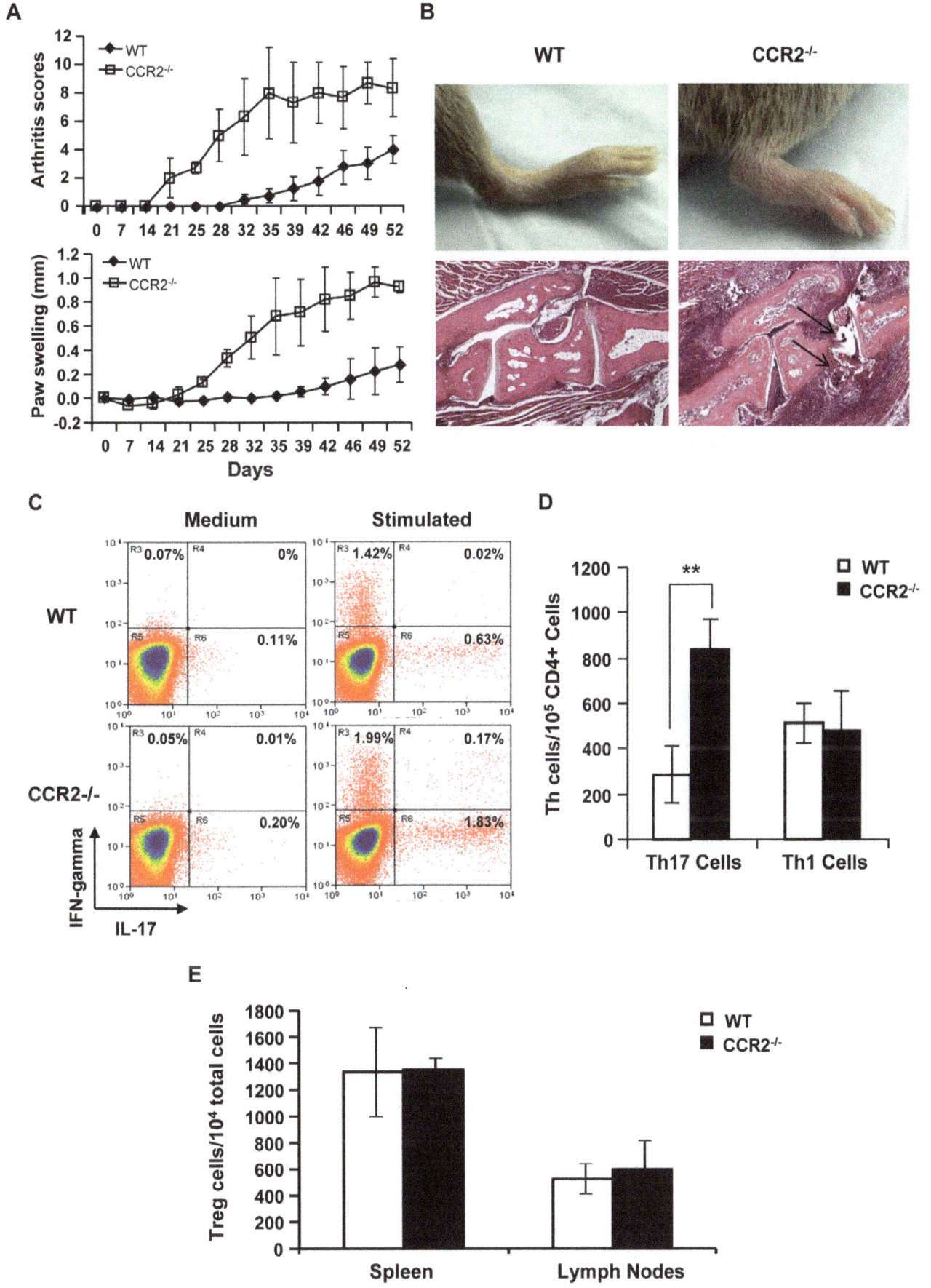

Figure 1. Th17 cells expand in the draining lymph nodes of immunized CCR2$^{-/-}$ mice. A. CCR2$^{-/-}$ and WT mice were immunized with bovine type II collagen in CFA at day 0. Arthritis was measured and recorded in arthritis scores and paw swelling over time. These results are representative of one of three experiments with a total of 25 mice. **B.** Representative photos of hind paws (day 26) and their histological sections with the H&E stain (day 48) from WT and CCR2$^{-/-}$ mice are shown. Arrows indicate the extensive inflammation and bone erosion in the joint section of CCR2$^{-/-}$ mice with CIA. **C–D.** Single cell suspension of the draining lymph nodes of immunized mouse was cultured in medium with or without anti-CD3 and anti-CD28 antibody stimulation for 6 hours. They were then harvested and stained with anti-CD4 antibody on the surface and anti-IL-17A and anti-IFN-γantibodies intracellularly. C shows the dot plots of cells gated on CD4^{+} cells in flow cytometry, and D demonstrates the average numbers of Th17 cells versus Th1 cells in the draining lymph nodes of CCR2$^{-/-}$ and WT mice 14 days after immunization, each group contains 5 animals, **p = 0.017. **E.** Treg cells were identified from the spleen and lymph nodes of immunized mice as CD4^{+}CD25^{+}FoxP3^{+} cells. These data are results of at least three separate experiments.

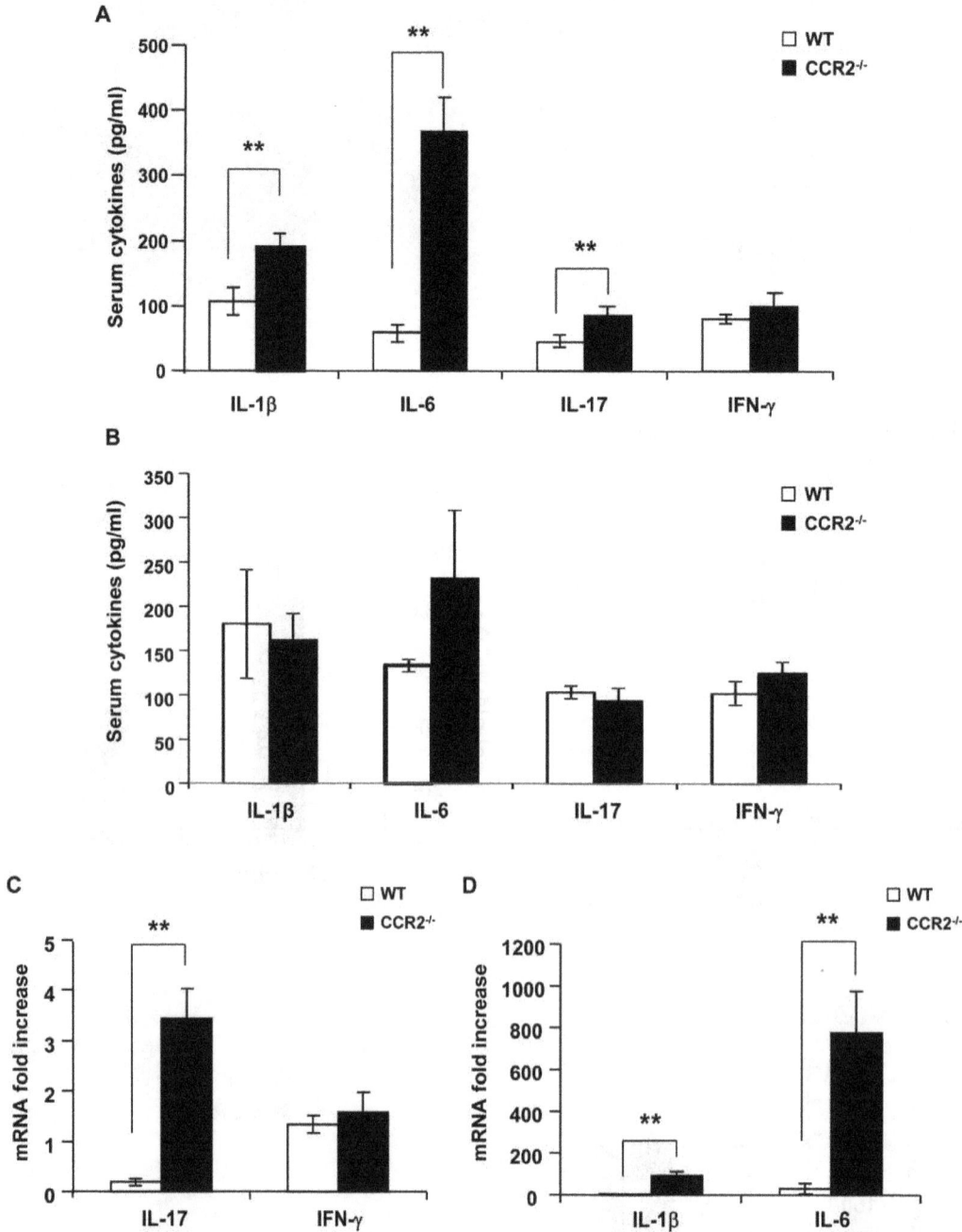

Figure 2. Serum and joint cytokine profiles favor Th17 cell expansion in CCR2$^{-/-}$ mice. A. Blood sera were collected at day 26 after immunization from WT and CCR2$^{-/-}$ mice. Levels of cytokines were measured by Luminex assays. Significant differences indicated as asterisks (**) were found in IL-1β (p = 0.01), IL-6 (p = 0.0004), and IL-17 (p = 0.04). Eight animals in each group were used. B. Sera levels of cytokines were measured at day 48 after immunization by Luminex. Four animals in each group were used. **C.** Joint tissues were isolated from the hind paws of mice at day 48 after immunization. The expression of IL-17 and IFN-γ mRNA was measured by quantitative real time PCR in WT (n = 5) versus CCR2$^{-/-}$ (n = 4) mice, **p = 0.009. **D.** Joint tissues were isolated from the hind paws of mice at day 48 after immunization. The expression of IL-1β and IL-6 mRNA was measured by real time PCR in WT (n = 5) versus CCR2$^{-/-}$ (n = 4) mice, **p ≤ 0.02.

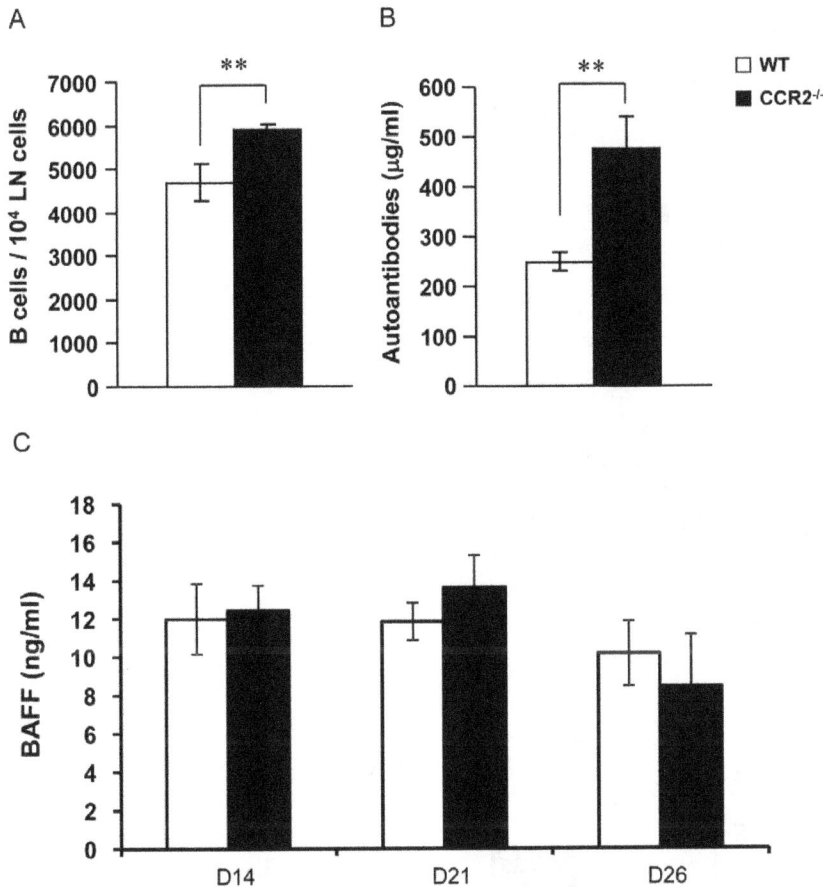

Figure 3. Antigen-specific antibodies increase with B cell expansion in CCR2$^{-/-}$ mice. A. Single cell suspensions were isolated from the draining lymph nodes of mice at day 14 after immunization. B cells were identified as CD45R/B220$^+$ cells by flow cytometry. Data shown are average numbers of B cells from WT (n = 6) and CCR2$^{-/-}$ mice (n = 4), **p = 0.0004. **B.** Blood sera were collected from WT (n = 7) and CCR2$^{-/-}$ mice (n = 8) at day 26 after immunization, and levels of anti-type II collagen antibodies (anti-CII) were determined by ELISA, **p = 0.005. **C.** ELISA results of serum levels of BAFF of WT and CCR2$^{-/-}$ mice at day 14 (n = 4 vs. 5), day 21 (n = 7 vs. 8), and day 26 (n = 7 vs. n = 5) after immunization are shown.

joints of immunized CCR2$^{-/-}$ mice (Figure 4C). In addition, increased expression of receptor activator of nuclear factor kappa-B ligand (RANKL) was also observed in the joint of CCR2$^{-/-}$ mice (Figure 4C). These results indicate that Th17 cell expansion induced by CCR2 deficiency is associated with cellular responses in collagen-induced arthritis.

Anti-IL-17A antibody treatment ameliorates CIA in CCR2$^{-/-}$ mice

To confirm a role of IL-17 in the pathogenesis of CCR2-mediated exacerbation of arthritis, we examined the effect of a murine anti-IL-17A antibody on the development of CIA in CCR2$^{-/-}$ mice. As shown in Figure 5, CCR2$^{-/-}$ mice that were treated with the anti-IL-17A antibody developed milder disease than CCR2$^{-/-}$ mice that were treated with an isotype control antibody. In addition, the onset of CIA in the anti-IL-17A treated group was also delayed. These results indicate that IL-17A contributes, at least in part, to the exacerbation of collagen-induced arthritis in CCR2$^{-/-}$ mice.

Monocytes decrease in the spleen, but are abundant in the bone marrow and arthritic joints in immunized CCR2$^{-/-}$ mice

Given that CCR2 is highly expressed on monocytes [26] that have been recently identified as T cell suppressors under

autoimmune conditions [27], we examined monocytes in immunized CCR2$^{-/-}$ mice. We found that Ly6C^{high}CD11b$^+$ CCR2-expressing monocytes were almost absent from the spleen. In contrast, these monocytes were more abundant in the bone marrow and the inflamed joints in CCR2$^{-/-}$ mice compared to WT mice. Together, these results indicate that CCR2 mediates monocyte immigration from the bone marrow while serving no role in monocyte recruitment to the arthritic joints (Figures 6A–C). These data also suggest that the lack of CCR2-expressing monocytes in CCR2$^{-/-}$ mice in the spleen after immunization may promote Th17 cell polarization and proliferation in the priming phase of collagen induced arthritis.

Discussion

CCR2 is a chemokine receptor predominantly expressed on monocytes and is considered proinflammatory in response to inflammation. However, CCR2 deficiency unexpectedly induces severe autoimmune arthritis with accelerated disease onset; and the underlying mechanism is not completely understood. In this study, we show that Th17 cells are selectively expanded prior to disease onset in the draining lymph nodes in collagen-immunized CCR2$^{-/-}$ mice compared to WT controls. Consistently, augmented IL-17A, IL-6, and IL-1β levels are observed in the blood and in the inflamed joints of CCR2$^{-/-}$ mice, which are

A

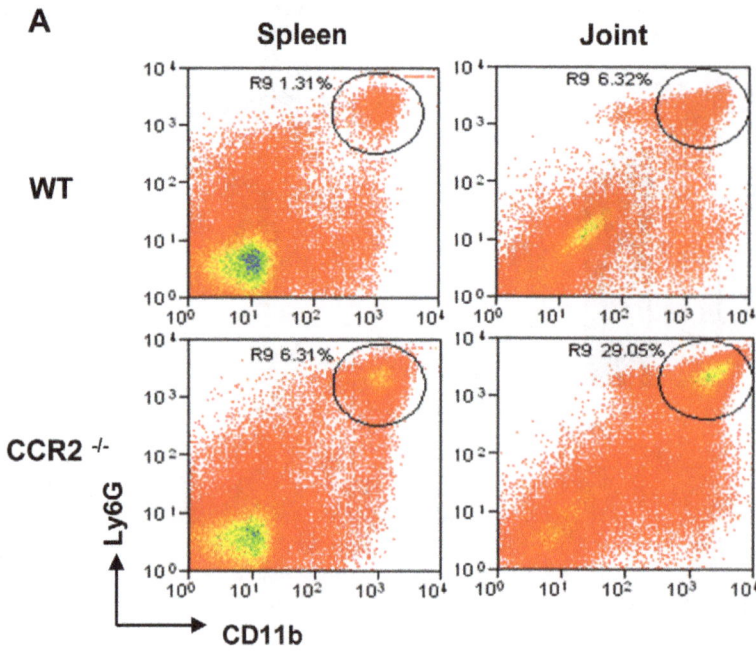

B

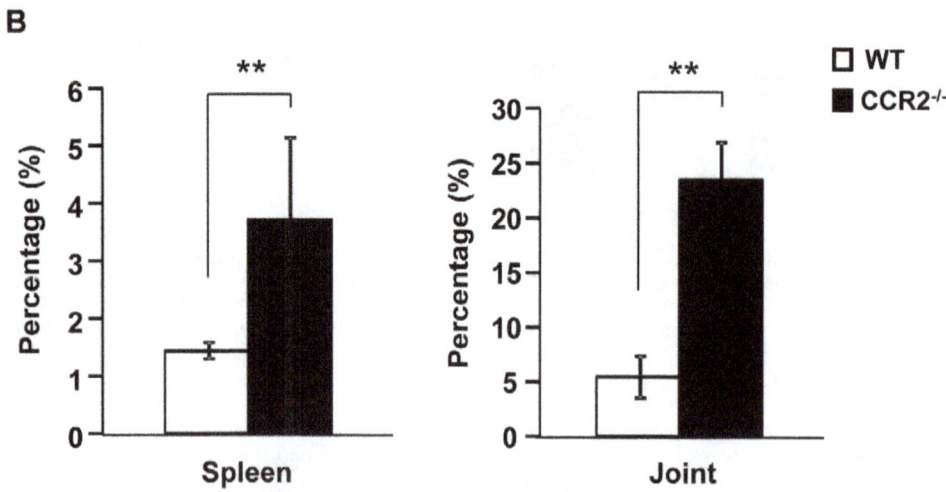

C

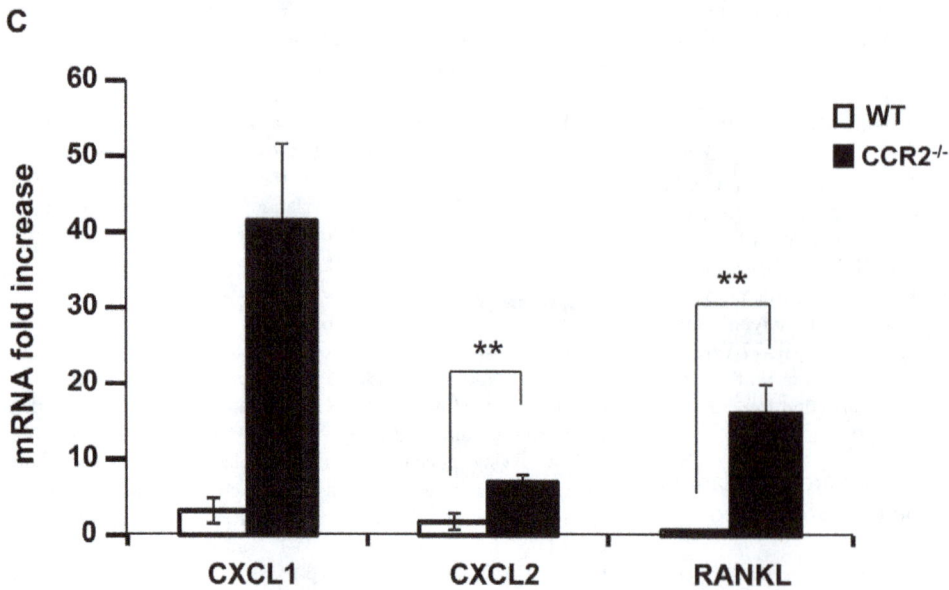

Figure 4. Neutrophil infiltration increases in the spleen and the joints of CCR2$^{-/-}$ mice. Cells were isolated from the spleen and the hind paws of mice at day 48 after immunization and were stained for flow cytometry analysis. Ly6G$^+$CD11b$^+$ (Ly6C$^+$) cells were gated as neutrophils. Shown are representative dot plots (**A**) and quantification of the number of neutrophils (**B**) of three experiments, **p = 0.02 in the spleen and 0.002 in the joints. **C.** Joint tissues were isolated from the hind paws of mice at day 48 after immunization. The expression of CXCL1, CXCL2, and RANKL mRNA was measured by real time PCR in WT (n = 5) versus CCR2$^{-/-}$ (n = 4) mice, p = 0.059, 0.016, 0.026, respectively.

accompanied by enhanced autoantibody production and neutrophil infiltration. Neutralizing IL-17A has an ameliorating effect on the severe arthritis observed in CCR2$^{-/-}$ mice. Our results suggest a previously unrecognized mechanism that CCR2 may be important in maintaining immunological homeostasis and protecting against collagen-induced arthritis via regulation of Th17 cell responses. Given that monocytes suppress T cell function during autoimmunity, our finding that monocytes are substantially decreased in the spleen but not in the bone marrow and the arthritic joints in CCR2$^{-/-}$ mice may provide a biological link between CCR2 and Th17 cells during the pathogenesis of collagen-induced arthritis.

Accumulated evidence has implicated an important role for Th17 cells in autoimmune disease such as RA [28,29,30]. Th17 cell differentiation from naïve CD4+ T cells depends on IL-6 and TGF-β production. Given that TGF-β is also required for Treg cell generation and IL-6 can reverse Treg cell-mediated suppression of autoreactive T cells [31], IL-6 is an essential differentiation factor for Th17 polarization. On the other hand,

IL-17 activates cells in synovium, such as synovial fibroblasts and monocytes/macrophages, to produce IL-6, IL-1 and TNF-α and thus sustain joint inflammation [32,33,34]. Consistent with these findings, our data show that early Th17 cell expansion in CCR2$^{-/-}$ mice is accompanied by elevated IL-6 levels in the blood. Increased levels of IL-17A expression are also observed together with augmented IL-6 and IL-1β abundance in the arthritic joints of CCR2$^{-/-}$ mice. Additionally, RANKL expression is significantly up-regulated in CCR2-deficient joints, which is consistent with findings that joint IL-17A stimulates osteoclasts to express RANKL that promotes bone erosion [35,36]. Together, our results support the hypothesis that a positive feedback loop between Th17 cells and IL-6 is a key factor involved in tolerance breakdown and tissue injury in autoimmune arthritis. Our data further suggest a potential beneficial role of CCR2 in this process.

Th17 cells promote autoimmune pathology through effects on cellular and humoral immune responses. Indeed, increased Th17 cells are not only accompanied by a substantial expansion of neutrophils in both the spleen and joints but also B cell up-regulation with a consequent elevation of type II collagen-specific antibodies in immunized CCR2$^{-/-}$ mice. IL-17A is known to stimulate granulopoiesis *in vivo* [25] and stimulates neutrophil-specific chemokines such as MIP-2 and IL-8 [37]. Blockade of IL-17A in IFN-γ receptor knockout mice with CIA ameliorates joint inflammation and bone erosion by significant reductions of the splenic expansion and joint influx of neutrophils [38], suggesting a critical role of IL-17-stimulated neutrophils in autoimmune pathology. IL-17 receptor is expressed on B cells, and IL-17 has been shown to work together with B cell-activating factor to control the survival and differentiation of B cells into antibody producing cells in autoimmunity [24,39]. In addition, IL-17 is found to drive autoimmune responses by promoting the formation of spontaneous germinal centers [40], and IL-17A-deficient mice have decreased autoantibodies in CIA and EAE [7,41]. These results provide rationales for our data that increased Th17-cell responses contribute to the exacerbated arthritis in CCR2$^{-/-}$ mice.

Recently, Fuji et al. have published a well-designed study showing that ablation of the Ccr2 gene exacerbated polyarthritis in IL-1 receptor antagonist-deficient (Il1rn$^{-/-}$) mice [42]. Increased neutrophil accumulation and IL-I7 production were observed in the inflamed joints of those double knockout mice. Neutralizing CXCR2, a neutrophil chemokine receptor, reduced arthritis severity in Il1rn$^{-/-}$Ccr2$^{-/-}$ mice. Since our collagen-induced arthritis model is different from the Il1rn-dependent model in which polyarthritis develops spontaneously [43], it is difficult to compare directly which therapeutic method, anti-IL-17A or anti-CXCR2 is more effective. It is possible that the augmented neutrophil infiltration into the joints in CCR2$^{-/-}$ mice contribute to the aggravated disease process through an additional Th17-independent mechanism. However, since neutrophils are key innate cells against infection, targeting such cells for the treatment of chronic disease like RA needs to be done cautiously.

Monocytes traffic to tissues and differentiate into macrophages or dendritic cells in response to inflammation [44,45]. Recent evidence shows that CCR2 does not mediate monocyte trafficking into the inflamed tissues [46,47], but is rather involved in the

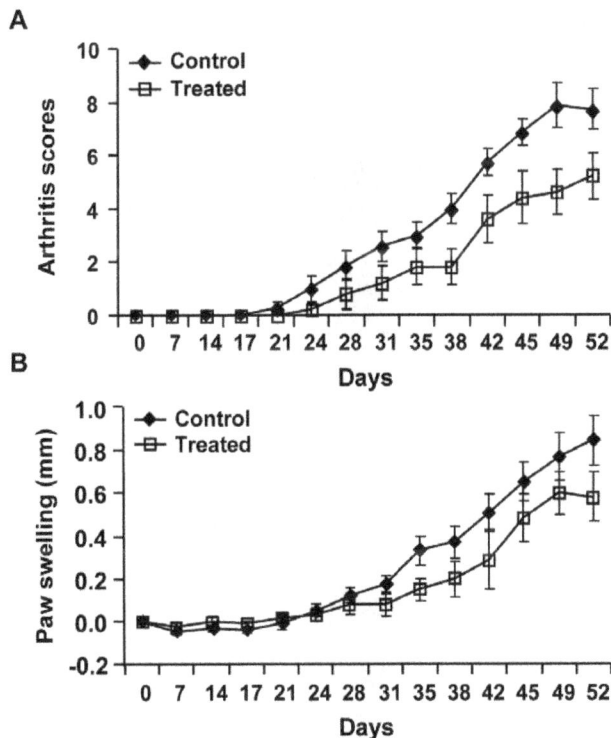

Figure 5. Treatment with anti-IL-17A antibodies reduces CIA in CCR2$^{-/-}$ mice. CCR2$^{-/-}$ mice were immunized with bovine type II collagen at day 0 and were treated 5 times in 2.5 weeks with either an anti-IL-17A antibody (Treated, n = 5) or an isotype control IgG1 (Control, n = 7) starting from day 14 after immunization. Arthritis score (A) and paw swelling (B) were recorded over time (Days). Data represent one of the two independent experiments performed. Differences of both arthritis scores and paw swelling during the course of disease between treated and control groups are significant (p<0.01).

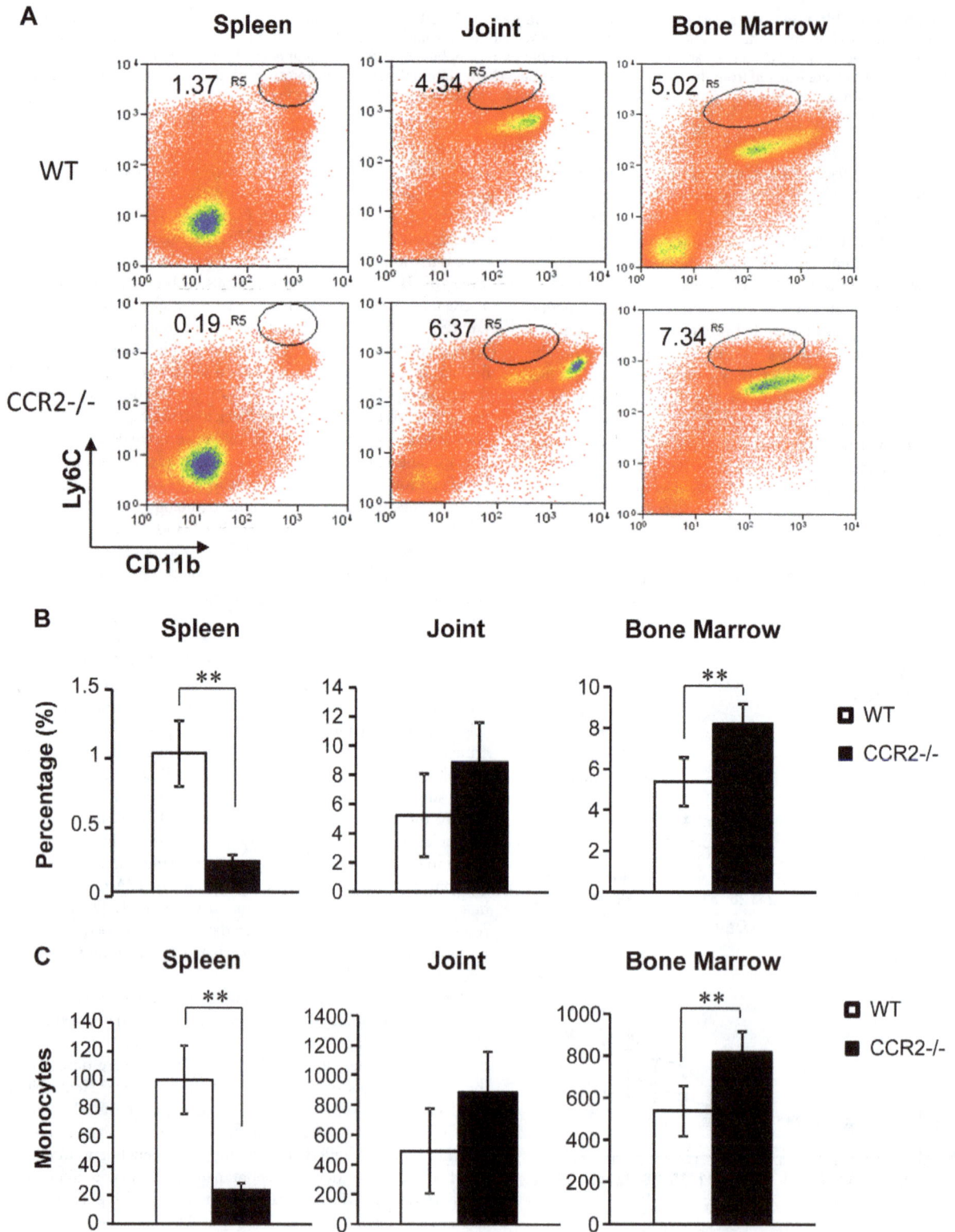

Figure 6. Monocytes decrease in the spleen, but are abundant in the bone marrow and arthritic joints in immunized CCR2$^{-/-}$ mice.
Single cell suspensions of the spleen, joints, and bone marrow from collagen-immunized mice were labeled with corresponding antibodies. The monocyte populations were recognized by flow cytometry as L6C^{high}CD11b$^+$Ly6G$^-$ cells (A). The abundance of monocytes in each compartment was quantified as percentage (B) and number of monocytes per 10^4 of total gated live cells (C) in immunized CCR2$^{-/-}$ and WT mice. These data are results of at least three separated experiments, **$p < 0.05$.

egress of monocytes from the bone marrow [46,48]. Indeed, monocytes are almost absent from the spleen, but are abundant in the bone marrow and inflamed joints in immunized CCR2$^{-/-}$ mice. Intriguingly, CCR2-expressing monocytes have been recently shown to have an immune regulatory role via suppression of T cell activities under tumor and autoimmune conditions [27,49]. It is conceivable that CCR2 deficiency alters T-cell suppressive monocytes, which results in Th17 cell expansion. CCR2 is also expressed on a subset of Treg cells that inhibit T cell proliferation *in vitro* [50]. However, we find similar numbers of Treg cells between immunized CCR2$^{-/-}$ and WT mice, suggesting that upregulated Th17 cells in CCR2$^{-/-}$ mice are Treg-cell-independent. But our data cannot exclude the possibility that a small subset of Treg cells is affected by CCR2 deficiency.

Although CCR2 is a drug target for the treatment of some inflammatory conditions, blocking CCR2 or deleting Ccr2 gene worsens autoimmune arthritis. Our data show that Th17 cell dysregulation is responsible, at least in part, for the exacerbated phenotype in CCR2-deficient mice. CCR2 may have a protective

role against autoimmune arthritis. Neutralization of IL-17 is a promising therapeutic approach against autoimmune arthritis possibly by inhibiting the excessive joint recruitment of neutrophils and monocytes as local injection and expression of IL-17 promote migration of both cell types to the joints [51,52].

Acknowledgments

The authors would like to thank Marcus McGinnis for technical assistance in the arthritis disease models, Dr. Roland Tisch for helpful discussions about the project, Dr. Michael Wu for suggestions on statistical analysis and Drs. Dave Fitzhugh and Maureen Su for proofreading the manuscript.

Author Contributions

Conceived and designed the experiments: PL. Performed the experiments: PL RRR CTV TQ MFW. Analyzed the data: PL RRR TKT DAE DDP. Contributed reagents/materials/analysis tools: PL AMF DDP FDP DAE JC. Wrote the paper: PL RRR.

References

1. Bettelli E, Oukka M, Kuchroo VK (2007) T(H)-17 cells in the circle of immunity and autoimmunity. Nat Immunol 8: 345–350.
2. McGeachy MJ, Cua DJ (2008) Th17 cell differentiation: the long and winding road. Immunity 28: 445–453.
3. Cua DJ, Sherlock J, Chen Y, Murphy CA, Joyce B, et al. (2003) Interleukin-23 rather than interleukin-12 is the critical cytokine for autoimmune inflammation of the brain. Nature 421: 744–748.
4. Gran B, Zhang GX, Yu S, Li J, Chen XH, et al. (2002) IL-12p35-deficient mice are susceptible to experimental autoimmune encephalomyelitis: evidence for redundancy in the IL-12 system in the induction of central nervous system autoimmune demyelination. J Immunol 169: 7104–7110.
5. Ferber IA, Brocke S, Taylor-Edwards C, Ridgway W, Dinisco C, et al. (1996) Mice with a disrupted IFN-gamma gene are susceptible to the induction of experimental autoimmune encephalomyelitis (EAE). J Immunol 156: 5–7.
6. Langrish CL, Chen Y, Blumenschein WM, Mattson J, Basham B, et al. (2005) IL-23 drives a pathogenic T cell population that induces autoimmune inflammation. J Exp Med 201: 233–240.
7. Nakae S, Nambu A, Sudo K, Iwakura Y (2003) Suppression of immune induction of collagen-induced arthritis in IL-17-deficient mice. J Immunol 171: 6173–6177.
8. Lubberts E, Koenders MI, Oppers-Walgreen B, van den Bersselaar L, Coenen-de Roo CJ, et al. (2004) Treatment with a neutralizing anti-murine interleukin-17 antibody after the onset of collagen-induced arthritis reduces joint inflammation, cartilage destruction, and bone erosion. Arthritis Rheum 50: 650–659.
9. Awasthi A, Kuchroo VK (2009) IL-17A directly inhibits TH1 cells and thereby suppresses development of intestinal inflammation. Nat Immunol 10: 568–570.
10. Genovese MC, Van den Bosch F, Roberson SA, Bojin S, Biagini IM, et al. LY2439821, a humanized anti-interleukin-17 monoclonal antibody, in the treatment of patients with rheumatoid arthritis: A phase I randomized, double-blind, placebo-controlled, proof-of-concept study. Arthritis Rheum 62: 929–939.
11. Hueber W, Patel DD, Dryja T, Wright AM, Koroleva I, et al. Effects of AIN457, a fully human antibody to interleukin-17A, on psoriasis, rheumatoid arthritis, and uveitis. Sci Transl Med 2: 52ra72.
12. Tarrant TK, Patel DD (2006) Chemokines and leukocyte trafficking in rheumatoid arthritis. Pathophysiology 13: 1–14.
13. Quinones MP, Ahuja SK, Jimenez F, Schaefer J, Garavito E, et al. (2004) Experimental arthritis in CC chemokine receptor 2-null mice closely mimics severe human rheumatoid arthritis. J Clin Invest 113: 856–866.
14. Haringman JJ, Gerlag DM, Smeets TJ, Baeten D, van den Bosch F, et al. (2006) A randomized controlled trial with an anti-CCL2 (anti-monocyte chemotactic protein 1) monoclonal antibody in patients with rheumatoid arthritis. Arthritis Rheum 54: 2387–2392.
15. Vergunst CE, Gerlag DM, Lopatinskaya L, Klareskog L, Smith MD, et al. (2008) Modulation of CCR2 in rheumatoid arthritis: A double-blind, randomized, placebo-controlled clinical trial. Arthritis Rheum 58: 1931–1939.
16. Bruhl H, Wagner K, Kellner H, Schattenkirchner M, Schlondorff D, et al. (2001) Surface expression of CC- and CXC-chemokine receptors on leucocyte subsets in inflammatory joint diseases. Clin Exp Immunol 126: 551–559.
17. Boring L, Gosling J, Cleary M, Charo IF (1998) Decreased lesion formation in CCR2−/− mice reveals a role for chemokines in the initiation of atherosclerosis. Nature 394: 894–897.
18. Jerath MR, Liu P, Struthers M, Demartino JA, Peng R, et al. Dual targeting of CCR2 and CX3CR1 in an arterial injury model of vascular inflammation. Thromb J 8: 14.

19. Rampersad RR, Esserman D, McGinnis MW, Lee DM, Patel DD, et al. (2009) S100A9 is not essential for disease expression in an acute (K/BxN) or chronic (CIA) model of inflammatory arthritis. Scand J Rheumatol 38: 445–449.
20. Wooley PH, Luthra HS, Stuart JM, David CS (1981) Type II collagen-induced arthritis in mice. I. Major histocompatibility complex (I region) linkage and antibody correlates. J Exp Med 154: 688–700.
21. Bruhl H, Cihak J, Plachy J, Kunz-Schughart L, Niedermeier M, et al. (2007) Targeting of Gr-1+,CCR2+ monocytes in collagen-induced arthritis. Arthritis Rheum 56: 2975–2985.
22. Livak KJ, Schmittgen TD (2001) Analysis of relative gene expression data using real-time quantitative PCR and the 2(−Delta Delta C(T)) Method. Methods 25: 402–408.
23. Annunziato F, Cosmi L, Liotta F, Maggi E, Romagnani S (2009) Type 17 T helper cells-origins, features and possible roles in rheumatic disease. Nat Rev Rheumatol 5: 325–331.
24. Doreau A, Belot A, Bastid J, Riche B, Trescol-Biemont MC, et al. (2009) Interleukin 17 acts in synergy with B cell-activating factor to influence B cell biology and the pathophysiology of systemic lupus erythematosus. Nat Immunol 10: 778–785.
25. Schwarzenberger P, La Russa V, Miller A, Ye P, Huang W, et al. (1998) IL-17 stimulates granulopoiesis in mice: use of an alternate, novel gene therapy-derived method for in vivo evaluation of cytokines. J Immunol 161: 6383–6389.
26. Geissmann F, Jung S, Littman DR (2003) Blood monocytes consist of two principal subsets with distinct migratory properties. Immunity 19: 71–82.
27. Zhu B, Bando Y, Xiao S, Yang K, Anderson AC, et al. (2007) CD11b+Ly-6C(hi) suppressive monocytes in experimental autoimmune encephalomyelitis. J Immunol 179: 5228–5237.
28. Lubberts E (2008) IL-17/Th17 targeting: on the road to prevent chronic destructive arthritis? Cytokine 41: 84–91.
29. van den Berg WB, Miossec P (2009) IL-17 as a future therapeutic target for rheumatoid arthritis. Nat Rev Rheumatol 5: 549–553.
30. Kato H, Fox DA. Are Th17 cells an appropriate new target in the treatment of rheumatoid arthritis? Clin Transl Sci 3: 319–326.
31. Pasare C, Medzhitov R (2003) Toll pathway-dependent blockade of CD4+CD25+ T cell-mediated suppression by dendritic cells. Science 299: 1033–1036.
32. Chabaud M, Miossec P (2001) The combination of tumor necrosis factor alpha blockade with interleukin-1 and interleukin-17 blockade is more effective for controlling synovial inflammation and bone resorption in an ex vivo model. Arthritis Rheum 44: 1293–1303.
33. Jovanovic DV, Di Battista JA, Martel-Pelletier J, Jolicoeur FC, He Y, et al. (1998) IL-17 stimulates the production and expression of proinflammatory cytokines, IL-beta and TNF-alpha, by human macrophages. J Immunol 160: 3513–3521.
34. Katz Y, Nadiv O, Beer Y (2001) Interleukin-17 enhances tumor necrosis factor alpha-induced synthesis of interleukins 1,6, and 8 in skin and synovial fibroblasts: a possible role as a "fine-tuning cytokine" in inflammation processes. Arthritis Rheum 44: 2176–2184.
35. Page G, Miossec P (2005) RANK and RANKL expression as markers of dendritic cell-T cell interactions in paired samples of rheumatoid synovium and lymph nodes. Arthritis Rheum 52: 2307–2312.
36. Sato K, Suematsu A, Okamoto K, Yamaguchi A, Morishita Y, et al. (2006) Th17 functions as an osteoclastogenic helper T cell subset that links T cell activation and bone destruction. J Exp Med 203: 2673–2682.
37. Kolls JK, Linden A (2004) Interleukin-17 family members and inflammation. Immunity 21: 467–476.

38. Kelchtermans H, Schurgers E, Geboes L, Mitera T, Van Damme J, et al. (2009) Effector mechanisms of interleukin-17 in collagen-induced arthritis in the absence of interferon-gamma and counteraction by interferon-gamma. Arthritis Res Ther 11: R122.

39. Lai Kwan Lam Q, King Hung Ko O, Zheng BJ, Lu L (2008) Local BAFF gene silencing suppresses Th17-cell generation and ameliorates autoimmune arthritis. Proc Natl Acad Sci U S A 105: 14993–14998.

40. Hsu HC, Yang P, Wang J, Wu Q, Myers R, et al. (2008) Interleukin 17-producing T helper cells and interleukin 17 orchestrate autoreactive germinal center development in autoimmune BXD2 mice. Nat Immunol 9: 166–175.

41. Komiyama Y, Nakae S, Matsuki T, Nambu A, Ishigame H, et al. (2006) IL-17 plays an important role in the development of experimental autoimmune encephalomyelitis. J Immunol 177: 566–573.

42. Fujii H, Baba T, Ishida Y, Kondo T, Yamagishi M, et al. Ablation of the Ccr2 gene exacerbates polyarthritis in interleukin-1 receptor antagonist-deficient mice. Arthritis Rheum 63: 96–106.

43. Horai R, Saijo S, Tanioka H, Nakae S, Sudo K, et al. (2000) Development of chronic inflammatory arthropathy resembling rheumatoid arthritis in interleukin 1 receptor antagonist-deficient mice. J Exp Med 191: 313–320.

44. Tacke F, Randolph GJ (2006) Migratory fate and differentiation of blood monocyte subsets. Immunobiology 211: 609–618.

45. Liu P, Yu YR, Spencer JA, Johnson AE, Vallanat CT, et al. (2008) CX3CR1 deficiency impairs dendritic cell accumulation in arterial intima and reduces atherosclerotic burden. Arterioscler Thromb Vasc Biol 28: 243–250.

46. Serbina NV, Pamer EG (2006) Monocyte emigration from bone marrow during bacterial infection requires signals mediated by chemokine receptor CCR2. Nat Immunol 7: 311–317.

47. Tacke F, Alvarez D, Kaplan TJ, Jakubzick C, Spanbroek R, et al. (2007) Monocyte subsets differentially employ CCR2, CCR5, and CX3CR1 to accumulate within atherosclerotic plaques. J Clin Invest 117: 185–194.

48. Tsou CL, Peters W, Si Y, Slaymaker S, Aslanian AM, et al. (2007) Critical roles for CCR2 and MCP-3 in monocyte mobilization from bone marrow and recruitment to inflammatory sites. J Clin Invest 117: 902–909.

49. Gabrilovich DI, Nagaraj S (2009) Myeloid-derived suppressor cells as regulators of the immune system. Nat Rev Immunol 9: 162–174.

50. Bruhl H, Cihak J, Schneider MA, Plachy J, Rupp T, et al. (2004) Dual role of CCR2 during initiation and progression of collagen-induced arthritis: evidence for regulatory activity of CCR2+ T cells. J Immunol 172: 890–898.

51. Lubberts E, Joosten LA, Oppers B, van den Bersselaar L, Coenen-de Roo CJ, et al. (2001) IL-1-independent role of IL-17 in synovial inflammation and joint destruction during collagen-induced arthritis. J Immunol 167: 1004–1013.

52. Shahrara S, Pickens SR, Mandelin AM, 2nd, Karpus WJ, Huang Q, et al. IL-17-mediated monocyte migration occurs partially through CC chemokine ligand 2/monocyte chemoattractant protein-1 induction. J Immunol 184: 4479–4487.

3

Optimal Principle of Bone Structure

Yifang Fan[1]*, Yubo Fan[2]*, Zhiyu Li[3], Mushtaq Loan[4], Changsheng Lv[1], Zhang Bo[1]

1 Center for Scientific Research, Guangzhou Institute of Physical Education, Guangzhou, People's Republic of China, 2 Key Laboratory for Biomechanics and Mechanobiology of Ministry of Education, School of Biological Science and Medical Engineering, Beihang University, Beijing, People's Republic of China, 3 College of Foreign Studies, Jinan University, Guangzhou, People's Republic of China, 4 International School, Jinan University, Guangzhou, People's Republic of China

Abstract

Bone modeling and remodeling is an optimization process where no agreement has been reached regarding a unified theory or model. We measured 384 pieces of bone *in vivo* by 64-slice CT and discovered that the bone's center of mass approximately superposes its centroid of shape. This phenomenon indicates that the optimization process of non-homogeneous materials such as bone follows the same law of superposition of center of mass and centroid of shape as that of homogeneous materials. Based upon this principle, an index revealing the relationship between the center of mass and centroid of shape of the compact bone is proposed. Another index revealing the relationship between tissue density and distribution radius is followed. Applying these indexes to evaluate the strength of bone, we have some new findings.

Editor: Sudha Agarwal, Ohio State University, United States of America

Funding: This project was funded by National Natural Science Foundation of China (http://www.nsfc.gov.cn) under the grant numbers of 10772053, 10925208, 10972061 and 11172073 and by Guangdong Natural Science Foundation (http://gdsf.gdstc.gov.cn/) under the grant numbers of S2011010001829. The funders had no role in study design, data collection and analysis, decision to publish, or preparation of the manuscript.

Competing Interests: The authors have declared that no competing interests exist.

* E-mail: tfyf@gipe.edu.cn (Yifang Fan); yubofan@buaa.edu.cn (Yubo Fan)

Introduction

The optimization of bone's size, shape and structure is a physical process [1,2,3,4] and the process is an adaptive response [3,5,6]. The adaptive responses of bone tissue generated by activities such as bone modeling and remodeling maximize its bearing load [7]. However, it remains uncertain what principles of mechanics these adaptive changes of bone follow.

Wolff's law [8] on bone's adaptive changes served as a prelude to the study of bone modeling and remodeling. Wolff's law was refined by Frost who promoted his Mechanostat theory [9], describing the bone's transformation on the tissue level. An ideal description of its mechanism should be studied from the perspectives of cell, molecule or gene [10] though no matter from which perspective, no agreement on a unified theory or model has been reached [11,12]. What's more, the complexity of bone's loading has brought difficulties (such as the target function or constraint equation involved in the target optimization analysis) in defining when the minimal material can sustain the maximal loading [3,13,14].

We assume that the optimization process of the non-homogeneous bone follows the same law of superposition of its center of mass (COM) and centroid of shape (COS) of the homogeneous material. A spiral CT scanning with an accuracy of sub-millimeter is conducted to 32 feet *in vivo*. An analysis to the positional relationship between the COM and COS of 384 pieces of foot bone (12 pieces from each foot) verifies our assumption. According to the principle of superposition between the bone's COM and COS, an evaluation method is put forward to evaluate the bone strength. The result from our evaluation indexes is different from those derived from other evaluation methods such as the BMD (bone mineral density).

Materials and Methods

Equipment

The test equipment was Brilliance 64-slice Scanner by Philips, Netherlands, provided by Image Processing Center of Zhujiang Hospital. Scan settings were: frame bone tissue; power: 120kv; pixel size: 0.50 mm; layer distance: 0.50 mm. The scanning was conducted along both feet transect, from top to bottom.

Software

Software applied included a free trial of SMSolver (The Structural Mechanics Solver for Windows, Version 2.5. http://www.civil.edu.cn/sms/). The three-dimensional model was constructed by Mimics (Version 10) and the statistical analysis was performed by SPSS (Version 12) (provided by the Key Laboratory of Biomechanics and Mechanobiology of Ministry of Education).

Materials

Altogether, we collected data of 384 pieces of bone - both from the volleyballers (with average height, weight and age of 183.94 ± 3.90 cm, 69.80 ± 5.20 kg and 21.88 ± 0.99 yrs, respectively) and wrestlers (with average height, weight and age of 168.00 ± 5.68 cm, 65.52 ± 5.16 kg and 21.00 ± 2.78 yrs, respectively), i.e. 32 pieces of 12 types of bones: calcaneus, talus, navicular, cuboid, lateral cuneiform, intermediate cuneiform, medial cuneiform, first metatarsal, second metatarsal, third metatarsal, fourth metatarsal and the fifth metatarsal.

The subjects were male volleyball players from our institute and male wrestlers from Provincial Sports School. It was confirmed before the test that every subject had been trained as a professional player for more than five years. Before the test, each subject's medical history was inquired and all the subjects were x-rayed to exclude subjects with diseases such as foot pathological change, deformity or injury to make sure that their physical conditions meet the requirements of the test.

Definition of the concept

Consider the volume element's (VE) position coordinates (x,y,z) with respect to equipment coordinate system. g stands for VE's gray value, N the number of VE of the bone, M the number of VE of the cross-sectional image. With the help of the following equation, bone's physical quantities such as the COM or COS are defined by the following equation.

The bone's density is defined as

$$\rho = \frac{\sum_1^N \rho_i}{N}, \qquad (1)$$

where $\rho_i = \frac{g_i}{g_w}$, g_i stands for the gray value of the i-th VE, g_w stands for the gray value of water. The equipment has been calibrated, the gray value of the air is set to 0 and that of the water is 1024.

The bone's COS is defined as

$$(x_s, y_s, z_s) = \left(\frac{\sum_1^N x_i}{N}, \frac{\sum_1^N y_i}{N}, \frac{\sum_1^N z_i}{N} \right). \qquad (2)$$

The bone's COM is defined as

$$(x_c, y_c, z_c) = \left(\frac{\sum_1^N x_i \rho_i}{\sum_1^N \rho_i}, \frac{\sum_1^N y_i \rho_i}{\sum_1^N \rho_i}, \frac{\sum_1^N z_i \rho_i}{\sum_1^N \rho_i} \right). \qquad (3)$$

The distance between the bone's COS and COM is

$$d_{cs} = \sqrt{(x_s - x_c)^2 + (y_s - y_c)^2 + (z_s - z_c)^2}. \qquad (4)$$

To the CT data of bone, let's set $z_i = j$. When j is set as a constant value, then $(x_{(j)i}, y_{(j)i})$ stands for the collection of the j-th

cross-sectional VE, $\rho_{(j)i}$ for the density of VE, M_j the number of cross-sectional VE. Calculate the cross-sectional image COS by $\left(\frac{\sum_1^{M_j} x_{(j)i}}{M_j}, \frac{\sum_1^{M_j} y_{(j)i}}{M_j} \right)$, its COM by $\left(\frac{\sum_1^{M_j} x_{(j)i} \rho_i}{\sum_1^{M_j} \rho_{(j)i}}, \frac{\sum_1^{M_j} y_{(j)i} \rho_{(j)i}}{\sum_1^{M_j} \rho_{(j)i}} \right)$ and the distance between the two points $\sqrt{\left(\frac{\sum_1^{M_j} x_{(j)i}}{M_j} - \frac{\sum_1^{M_j} x_{(j)i} \rho_{(j)i}}{\sum_1^{M_j} \rho_{(j)i}} \right)^2 + \left(\frac{\sum_1^{M_j} y_{(j)i}}{M_j} - \frac{\sum_1^{M_j} y_{(j)i} \rho_{(j)i}}{\sum_1^{M_j} \rho_{(j)i}} \right)^2}.$

The bone tissue's radius is

$$r = \frac{1}{N} \sum_1^N \sqrt{(x_i - x_s)^2 + (y_i - y_s)^2 + (z_i - z_s)^2}. \qquad (5)$$

The same density tissue radius is

$$r_k = \frac{1}{Q} \sum_1^Q \sqrt{(x_{(k)i} - x_s)^2 + (y_{(k)i} - y_s)^2 + (z_{(k)i} - z_s)^2}, \qquad (6)$$

where $k = \rho_i$. When k is set as a constant value, it refers to the same density tissue of the bone. For example, when $k = 1.1$ (g/cm^3), $(x_{(1.1)i}, y_{(1.1)i}, z_{(1.1)i})$, it indicates the VE's coordinates of a density of 1.1. Q is the number of VEs when the density is 1.1.

Ethics Statement

The study received approval from the Ethical Committee of Guangzhou Institute of Physical Education. The subjects provided fully informed consent to participate in this study by signing a written consent form.

Results and Discussion

Following [15,16], we separated foot bone to calculate the volume, surface area and BMD. The results for the extracted measurements are shown in Table 1.

BMD is an important index to analyze bone strength. Table 1 shows that no significant difference exists in the foot bone of both groups of athletics. Is that true?

Table 1. Foot bone volume, surface area and bone density (Mean ±SD).

Item	Wrestler			Volleyballer		
	Volume	Area	Density	Volume	Area	Density
Calcaneus	71.01±8.46	107.39±8.83	1.47±0.04	83.94±6.05	120.70±5.56	1.49±0.05
Talus	38.30±4.33	71.38±5.41	1.63±0.04	43.87±3.33	80.11±5.97	1.65±0.04
Navicular	11.45±1.39	31.21±2.73	1.56±0.04	13.44±1.51	34.78±3.00	1.58±0.05
Cuboid	13.87±1.61	33.14±2.77	1.46±0.04	15.09±2.69	35.24±4.78	1.47±0.05
Lateral cuneiform	5.91±0.69	19.09±1.56	1.51±0.04	6.79±0.61	20.99±1.28	1.53±0.06
Intermediate cuneiform	4.43±0.66	15.69±1.56	1.59±0.04	5.20±0.44	17.56±1.00	1.64±0.06
Medial cuneiform	10.76±1.48	28.60±2.73	1.52±0.03	12.20±1.04	31.02±1.91	1.58±0.05
First metatarsal	16.94±2.23	44.90±3.89	1.62±0.05	20.93±2.25	51.94±3.47	1.65±0.05
Second metatarsal	9.01±1.29	33.72±3.25	1.73±0.07	11.65±0.77	40.56±2.08	1.76±0.08
Third metatarsal	7.72±0.58	30.23±1.60	1.70±0.05	8.99±1.07	34.50±2.61	1.68±0.07
Fourth metatarsal	7.47±0.78	28.80±2.19	1.66±0.04	8.88±0.92	32.97±2.05	1.66±0.05
Fifth metatarsal	8.83±1.09	30.92±2.64	1.72±0.05	9.65±1.07	33.73±2.49	1.71±0.05

Volume is cm^3, area is cm^2 and density is g/cm^3.

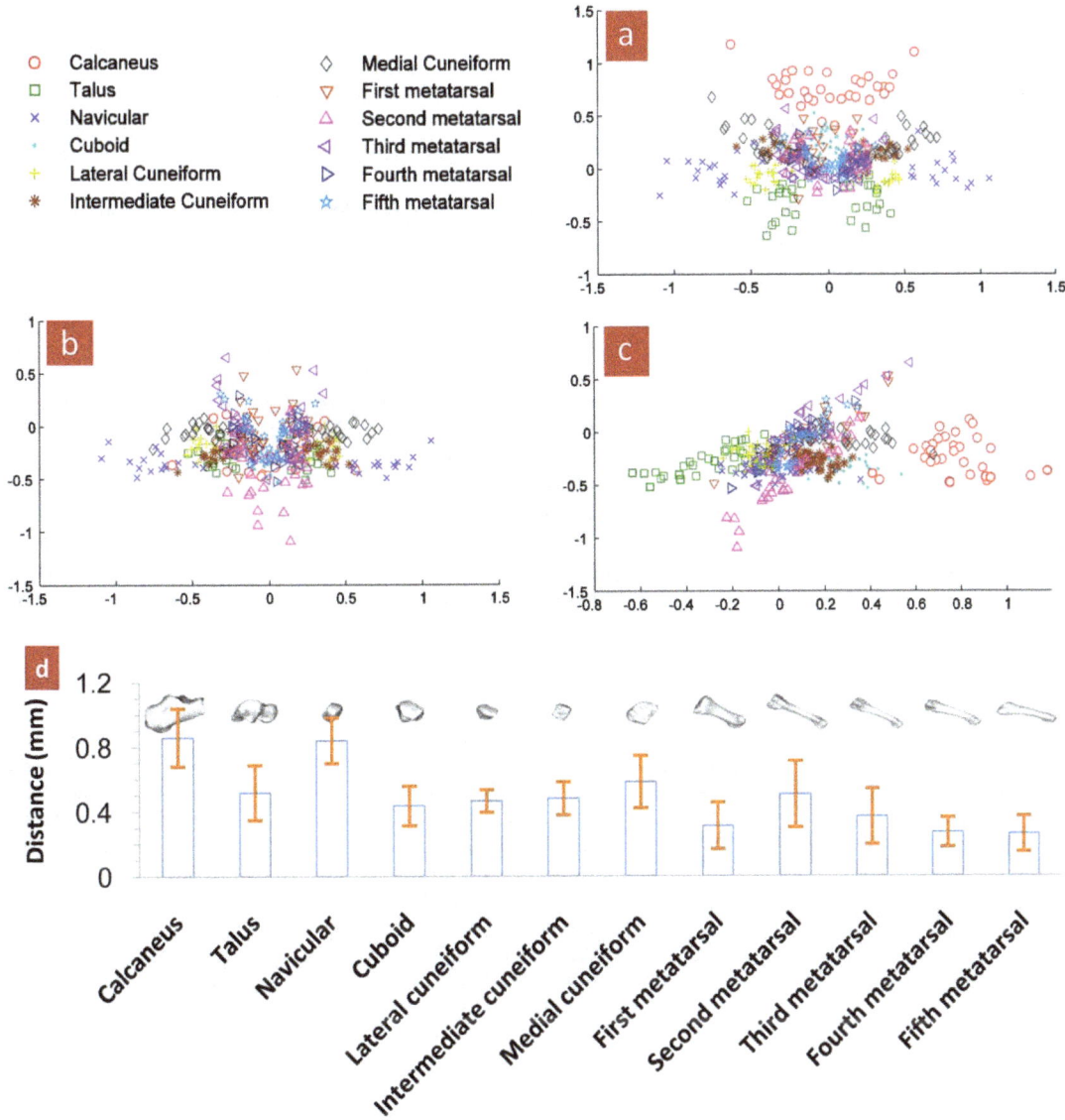

Figure 1. Positional relationship between a bone's COM and COS. Fig. 1a Positional relationship between COM and COS on x-y plane; Fig. 1b Positional relationship between COM and COS on x-z plane; Fig. 1c Positional relationship between COM and COS on y-z plane; Fig. 1d Distance between bone's COM and COS. The bones' COS and COM are derived from the calculation of Eqs. (2) and (3). When choosing coordinate system with origin at COM, the coordinates of COS relative to COM can be derived as $(x_s - x_c, y_s - y_c, z_s - z_c)$. By using $(x_s - x_c, y_s - y_c)$, $(y_s - y_c, z_s - z_c)$ and $(x_s - x_c, z_s - z_c)$, 384 pieces' bone coordinates of COS with respect to COM can be located on x-y, y-z and x-z planes. See Fig. 1a, 1b and 1c (unit is mm). Through Eq. (4), the distance of these 384 pieces of bones' COS to the COM can be calculated, resulting in Fig. 1d.

The COM and COS of homogeneous materials superpose exactly one another while those of non-homogeneous materials do not. Bone is a typical non-homogeneous material [17,18,19]. Using CT scanning, bone will be separated into a collection of finite VE. The coordinates of each VE and gray value can be provided [16,20]. This makes it easy to calculate the bone's COM and COS. Setting the bone's COS as the coordinate origin, the positional relationship between the bone's COM and COS can be established. See Fig. 1.

It is known that significant difference exists in the size, density and shape of the navicular and calcaneus. However, Fig. 1 shows that there is no significant difference (p>0.05) in the positional superposition of the COM and COS of both. Shape similarity does exist between the first and second metatarsal, but there is significant difference (p<0.01) in the positional superposition of

the COM and COS of both. Therefore, within the range of measurement accuracy, the phenomenon of superposition does exist in the positions of the COM and COS of non-homogeneous bone. It is furthermore unaffected by such different factors as bone size, density or shape.

When the cross section passes through the COS of a symmetrical geometry, the COS of the cross section and the COS of the geometry are in the same position. Setting the coordinate origin as the bone's COS, the relationship between the COM and COS of the cross-sectional image through the coordinate origin is set up. See Fig. 2.

Fig. 2a suggests that the COM and COS of the cross-sectional image through the COS of the bone also superpose. Fig. 2b shows difference in the COS position of the cross section and that of the whole bone. Fig. 1 and 2 show that superposition of COM and

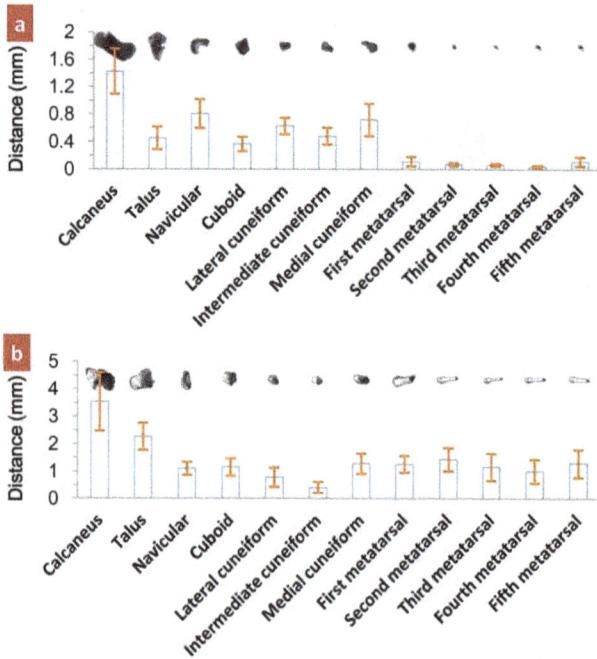

Figure 2. Relationship between the COM and COS of the cross-sectional image. Fig. 2a Positional relationship between the COM and COS of the cross-sectional image through the coordinate origin; Fig. 2b Positional relationship between the COS of the cross section and the COS of the bone. When the position value of the cross-sectional VE at z-axis is approximately equal to the bone's COS, i.e. $z_i \approx z_s$, the cross section is the tomography that goes through the bone's COS. Calculate the bone's cross-sectional COM and COS, and then calculate the distance between the two points by using the plane distance formula. See Fig. 2a. Fig. 2b is the distance between the cross-sectional COS and the COS of bone (x_s, y_s) on x-y plane calculated by the plane distance formula.

COS does not only exist in the whole bone, but also in the cross section. Attention should be paid to the fact that it is risky to determine the bone's COS by the cross section's COS since the bone's shape is asymmetric [21].

The bone is then simplified to a truss structure, which is composed of an external square and an internal one. The external square refers to the cortical bone and the internal one to the cancellous bone. When the load and constraint remain the same, the structure strength changes when the position of the internal square changes. See Fig. 3.

Fig. 3 shows that bearing the same constraint and load, the structure where the COS of the internal square superposes with the COS of the external square is superior in the load-carrying capacity of shear and moment to the structure where there is not such a superposition. Therefore, when the force action line of the balance forces passes the COM of an object, the carrying capacity of the structure reaches its maximum.

Though the shape and structure of bone are more complicated than the truss in Fig. 3, the constraints and loads born by the bone *in vivo* are the same as the structure in Fig. 3 – they are both acted upon by out-of-balance forces. It can thus be assumed that when the COS (determined by the bone's shape) is in the same position as the COM (determined by the bone's density distribution), the bone's structure has optimal strength.

It can be concluded that to meet its functional requirements [22], the bone's size, shape [23] and density [22] all produce adaptive changes [3,5,6]. In this process, the principle of optimal structure where the COM superposes with COS is always followed. This holds the same idea that function determines the structure as that of the maximal strength with minimal materials [24], or mechanic stability theory [23] or the bone adaptation as an optimization process [6] and Wolf's law [8] (i.e. law of bone transformation) while the superposition of COM and COS is a quantitative description.

Why is the superposition of COM and COS a quantitative description? The following relationship can be established based upon the fact that the strength of compact bone is many times greater than that of the spongy bone [25,26], that the density

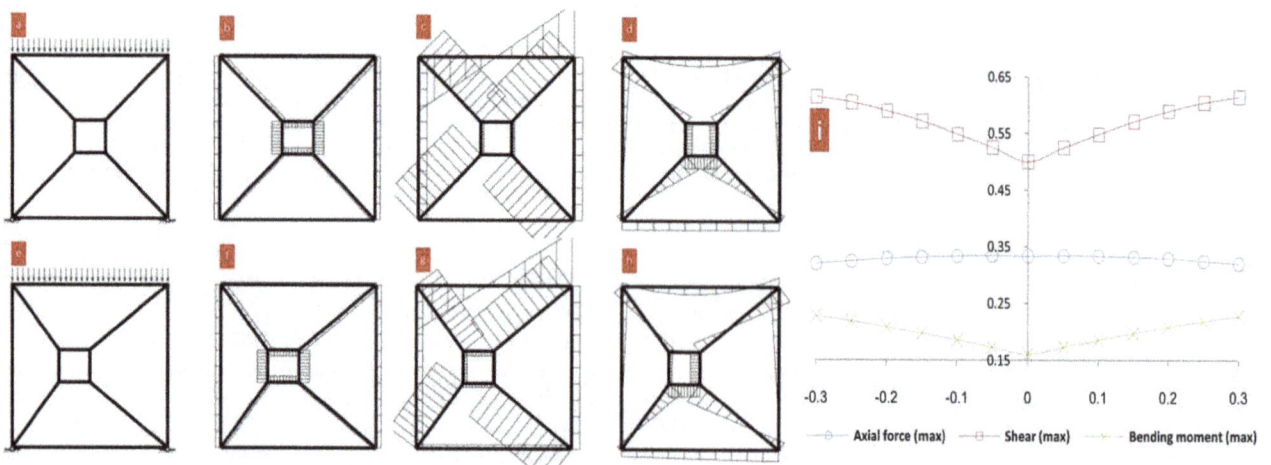

Figure 3. COM and COS of the truss. Fig. 3a and 3e Structure by constraints and loads; Fig. 3b and 3f Axial force distribution in the structure; Fig. 3c and 3g Shear distribution in the structure; Fig. 3d and 3h Bending moment distribution in the structure. Fig. 3i Relationship between internal square position and strength. The rods in the structure are all rigid and the connections between the rods are rigid also. Two squares are drawn with a side length of 1 and 0.2 respectively. Connect the vertices of the two squares and a simple structural mechanics model is forged. Set the two bottom vertices of the bigger square to connect with the hinge bearing on the ground. The top of the bigger square is subjected to distributed load (size is 1). The vertical coordinate of the smaller square COS superposes the bigger square. Change the horizontal coordinate from −0.3 to +0.3. By using the software of **SMSolver**, the calculation results are shown in Fig. 3a–3i.

distribution (relative to the bone's COS) of bone tissue is related to the bone's strength: 1) the relationship between the COS of the compact bone and the COS of the bone where the distance from the compact bone's COS to the bone's COS is standardized by the bone tissue's radius; 2) the relationship between the bone tissue's density and the distribution radius (relative to that of the bone's COS) where the same density tissue radius is standardized by the bone tissue's radius. See Fig. 4.

In Fig. 4a, the distance of the volleyballers' calcaneus compact bone's COS to the bone's COS is shorter than that of the wrestlers and it has a significant difference (p<0.05), which is in contrast with estimates in Table 1 where the no significant difference is observed. A similar trend is observed for the distance of fifth metatarsal compact bone's COS to the bones' COS. From Table 1, we can see that in the similar morphological first to fifth metatarsal, the lowest density goes to the first metatarsal, which does not sound very reasonable, suggesting the limitation of bone density assessment index, i.e. factors such as volume and joint

segmental area might have affected bone density. In Fig. 4a, the distance of both athletic groups' first metatarsal compact bone's COS to the bone's COS is the shortest.

When a volleyballer takes off to spike, the braking movement has a great impact on the calcaneus. In Fig. 4b, the distribution radius of the volleyballers' calcaneous begins to become larger than that of the wrestlers from the density of compact bone on; especially when comparing this with the results from the marrow and spongy bone tissues (when density $\rho < 1.14$, it is the marrow; when $1.14 \leq \rho \leq 1.65$, the spongy bone and when $\rho > 1.65$, the compact bone), this difference is outstanding. The wrestlers' fierce body combats carry great strength to their fifth metatarsal from the front, rear, left and right. The distribution radius of the wrestlers' fifth metatarsal begins to become bigger from the density of compact bone on than that of the volleyballers.

Fig. 4 shows that according to the superposition principle of the bone's COM and COS, the establishment of relationship between the compact bone's COS and the bone's COS and the relationship

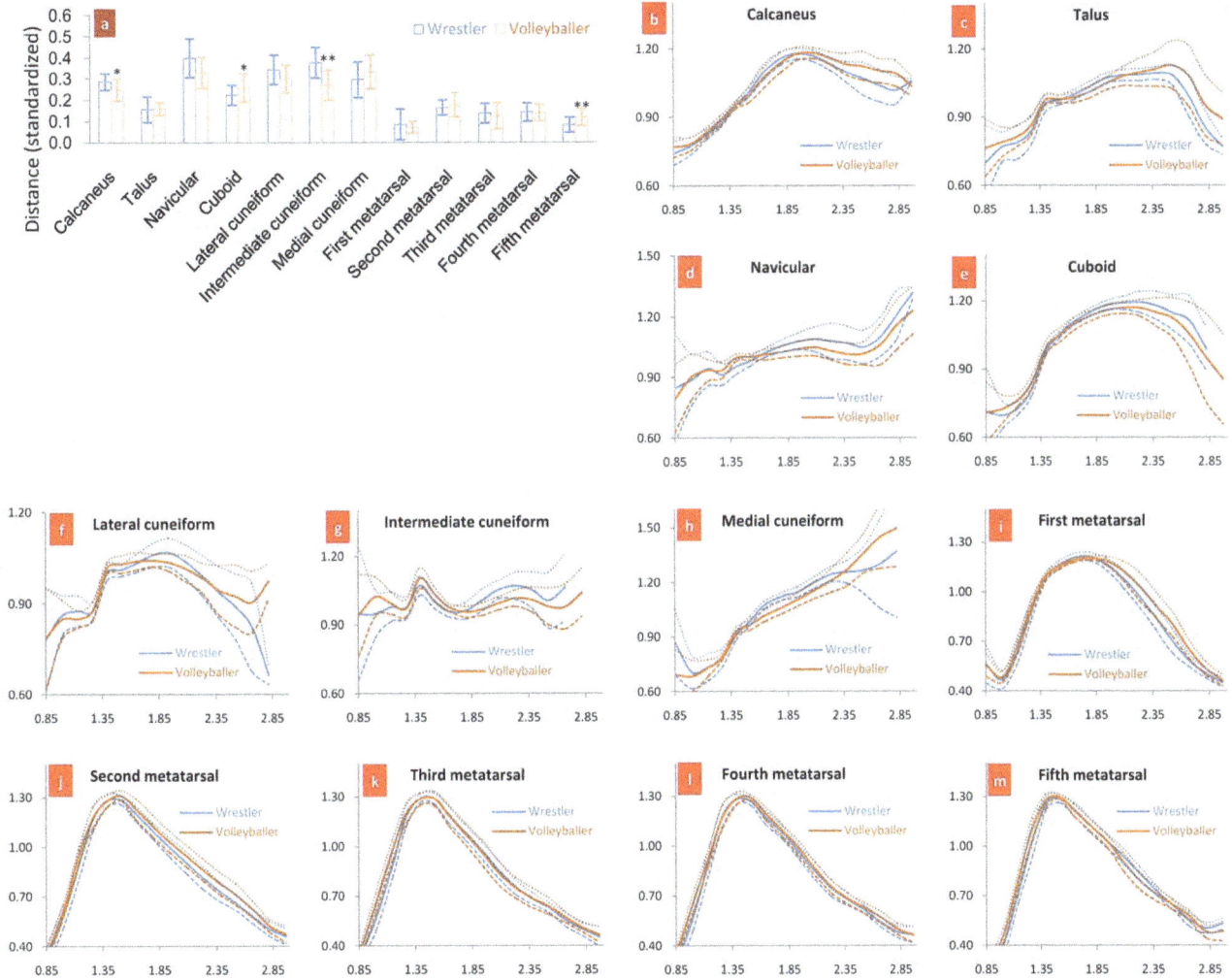

Figure 4. Application of the superposition principle of the bone's COM and COS. Fig. 4a Positional relationship between the COS of the compact bone and the COS of the bone; Fig. 4b–m Relationship between the bone tissue's density and distribution radius, where axis x stands for the tissue's density and axis y for the standardized mean distribution radius of the tissue. The data were collected from 192 pieces of foot bone of the wrestlers and 192 ones of the footballers. *p<0.05, **p<0.01. When $\rho > 1.65$, the bone tissue is defined as compact bone. Eqs (2) and (3) are used to calculate the compact bone's COM and COS while Eq. (4) the distance between the two points and Eq. (5) the distribution radius of bone tissue. Fig. 4a is the result of the distance between the compact bone's COM and COS standardized by the bone tissue's radius. Eq. (6) is applied to calculate same density tissue radius. Then standardize it by the bone tissue's radius. See Fig. 4b–4m.

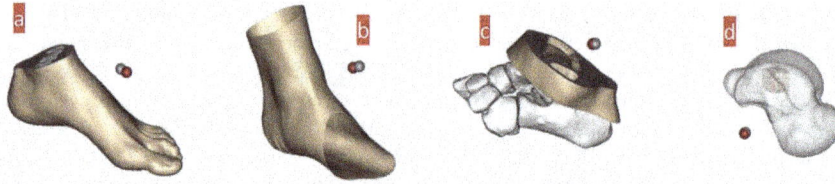

Figure 5. COM and COS of non-bone tissues. Fig. 5a Positional relationship between the whole foot's COM and COS; Fig. 5b Positional relationship between the COM and COS of ankle skin; Fig. 5c Positional relationship between the COM and COS of non-bone tissues (a group of cross sections selected around the ankle joint); Fig. 5d Positional relationship between the COM and COS when a ROI (region of interest) of 1 cm^3 is established around the COS of the talus. The grey ball stands for the COM position and the red ball for the COS position. The radius of the ball is 0.5 *mm*. According to the definition of non-bone tissue density, Eqs. (2) and (3) are used to calculate COM and COS of non-bone tissue. The three-dimensional model is constructed by the software of **Mimics**.

between the tissue's density and distribution radius has offered a new approach to study the bone's strength.

What insight will this phenomenon of superposition between the COM and COS bring to biomechanical research? Using the CT data of bone, we analyze the COM and COS of the other foot non-bone tissues. See Fig. 5.

Fig. 5 shows that the COM and COS of the whole foot, of its ankle skin, of the non-bone tissues around the ankle joint and of the ROI established around the whole bone's COS superpose highly. Further subdivisions tell us that if the COM and COS of the cell also follow this principle of high superposition, then a new method of dynamics can be set up to study activities such as cell growth and division.

The COS of a continuous closed geometry superposes with that of its surface (shape). The COS of the cell can thus be obtained through the numerical model of the cell surface. According to the dynamic principle of COM (i.e. the internal force cannot change the motion of the system's COM), if the forces acting on the cell are known, the cell's kinematic characteristics can be obtained. On the other hand, we can use the kinematic characteristics of the cell to analyze the characteristics of external mechanical signals. When the cell shape is asymmetrical, its geometric transformation invariance and the uniqueness of the principal moments of inertia axes [27,28] can be applied to study issues such as the rotational dynamics of the cell.

Conclusion

The physiological activities of the bone are a process of optimization. In this adaptive changing process, what remains unchanged is the optimal structure principle of superposition of

COM and COS. The mechanical significance of following the optimal structure principle is to use the optimal structure to bear the external load.

We propose the concept of distance between the tissue's COS and its bone's COS and discover the relationship between the distance (of the compact bone's COS and its bone's COS) and the loading type. This relationship is represented by the phenomenon that the impact strength has made the compact bone's COS move towards the bone's COS. This movement symbolizes a functional adaptation of bone in its structure. The physiological activity of the middle aged and seniors is mostly a reconstruction [29]. When their bone masses are gradually decreasing, it is essential to look into the possibility of whether physical exercises can diminish the bone loss and change the movement's direction. This is meaningful and worthwhile research.

With the advances of three-dimensional imaging technology [30,31], if this phenomenon of superposition of COM and COS also happens in cell, this will play a significant role in the study of cytokinetics.

Acknowledgments

The authors would like to acknowledge the support from the subjects.

Author Contributions

Conceived and designed the experiments: Yifang Fan Yubo Fan. Performed the experiments: Yifang Fan CL ZB. Analyzed the data: Yifang Fan Yubo Fan ZL ML. Contributed reagents/materials/analysis tools: Yifang Fan CL ZB. Wrote the paper: Yifang Fan ZL ML.

References

1. Harrigan TP, Mann RW (1984) Characterization of microstructural anisotropy in orthotropic materials using a second rank tensor. J Mater Sci 19: 761–767.
2. Turner CH (1992) Functional determinants of bone structure: beyond Wolff's Law of bone transformation. Bone 13: 403–409.
3. Huiskes R, Hollister SJ (1993) From structure to process, from organ to cell - recent developments of FE-analysis in orthopedic biomechanics. J Biomech Eng 115: 520–527.
4. Rusconi M, Zaikin A, Marwan N, Kurths J (2008) Effect of stochastic resonance on bone loss in osteopenic conditions. Phys Rev Lett 100: 128101.
5. Odgaard A (1997) Three-dimensional methods for quantification of cancellous bone architecture. Bone 20: 315–328.
6. Bagge M (2000) A model of bone adaptation as an optimization process. J Biomech 33: 1349–1357.
7. Burger EH, Klein-Nulend J (1999) Mechanotransduction in bone-role of the lacuno-canalicular network. Faseb J 13: S101–S112.
8. Wolff J (1892) Das Gesetz der Transformation der Knochen. Berlin: A. Hirchwild. [Macquet P, Furlong R, translators. The law of bone remodeling. Springer Berlin (1986)].
9. Frost HM (1987) Bone 'mass' and the 'mechanostat': a proposal. Anat Rec 219: 1–9.
10. Salter DM, Robb JE, Wright MO (1997) Electrophysiological responses of human bone cells to mechanical stimulation: Evidence for specific integrin function in mechanotransduction. J Bone Miner Res 12: 1133–1141.
11. Wang YL, Cai SX (1999) Biomechanics and bone tissue engineering. Advances in mechanics 29: 232–244. [in chinese].
12. ZhangY, Tao ZL (2000) The mechanocytobiological mechanisms of loading-induced bone formation. Advances in mechanics 30: 433–445. [in chinese].
13. Turner CH (1992) On Wolff's law of trabecular architecture. J Biomech 25: 1–9.
14. Harrigan TP, Hamilton JJ (1994) Bone remodeling and structural optimization. J Biomech 27: 323–328.
15. Ciarelli MJ, Goldstein SA, Kuhn JL, Cody DD, Brown MB (1991) Evaluation of orthogonal mechanical-properties and density of human trabecular bone from the major metaphyseal regions with materials testing and computed-tomography. J Orthop R 9: 674–682.
16. Rho JY, Hobatho MC, Ashman RB (1995) Relations of mechanical-properties to density and CT numbers in human bone. Med Eng Phys 17: 347–355.
17. Carter DR, Spengler DM (1978) Mechanical properties and composition of cortical bone. Clin Orthop Rel Res 135: 192–214.
18. Roesler H (1987) The history of some fundamental-concepts in bone biomechanics. J Biomech 20: 1025–1034.

19. Turner CH (1999) Toward a mathematical description of bone biology: The principle of cellular accommodation. Calcif Tissue Int 65: 466–471.
20. Ciarelli MJ, Goldstein S, Kuhn J, Cody D, Brown M (1991) Evaluation of orthogonal mechanical-properties and density of human trabecular bone from the major metaphyseal regions with materials testing and computed-tomography. J Orthop R 9: 674–682.
21. Aspden RM (2003) Mechanical testing of bone ex vivo. Bone Research Protocols (Edited by: Helfrich MH, Ralston SH). TotowaNew Jersey: Human Press Inc. pp 369–379.
22. Huiskes R, Weinans H, Grootenboer HJ, Dalstra M, Fudala B, et al. (1987) Adaptive bone-remodeling theory applied to prosthetic design analysis. J Biomech 20: 1135–1150.
23. Frost HM, Jee WSS (1992) On the rat model of human osteopenias and osteoporoses. Bone Miner 18: 227–236.
24. Sherwood RJ (1999) Pneumatic processes in the temporal bone of chimpanzee (Pan troglodytes) and gorilla (Gorilla gorilla). J Morphol 241: 127–137.
25. Evans FG, Lebow M (1957) Strength of human compact bone under repetitive loading. J Appl Physiol 10: 127–130.
26. Currey JD (1999) What determines the bending strength of compact bone? J Exp Biol 202: 2495–2503.
27. Coburn JC, Upal MA, Crisco JJ (2007) Coordinate systems for the carpal bones of the wrist. J Biomech 40: 203–209.
28. Fan YF, Fan YB, Li ZY, Lv CS (2010) Bone *in vivo*: Surface mapping technique. arXiv:1010.0617.
29. Taylor D, Hazenberg JG, Lee TC (2007) Living with cracks: damage and repair in human bone. Nat Mater 6: 263–268.
30. Frey TG, Mannella CA (2000) The internal structure of mitochondria. Trends Biochem Sci 25: 319–324.
31. Boehm B, Westerberg H, Lesnicar-Pucko G, Raja S, Rautschka M, et al. (2010) The Role of Spatially Controlled Cell Proliferation in Limb Bud Morphogenesis. PLoS Biol 8: e1000420.

4

Role of Calcitonin Gene-Related Peptide in Bone Repair after Cyclic Fatigue Loading

Susannah J. Sample, Zhengling Hao, Aliya P. Wilson, Peter Muir*

Comparative Orthopaedic Research Laboratory, School of Veterinary Medicine, University of Wisconsin-Madison, Madison, Wisconsin, United States of America

Abstract

Background: Calcitonin gene related peptide (CGRP) is a neuropeptide that is abundant in the sensory neurons which innervate bone. The effects of CGRP on isolated bone cells have been widely studied, and CGRP is currently considered to be an osteoanabolic peptide that has effects on both osteoclasts and osteoblasts. However, relatively little is known about the physiological role of CGRP *in-vivo* in the skeletal responses to bone loading, particularly fatigue loading.

Methodology/Principal Findings: We used the rat ulna end-loading model to induce fatigue damage in the ulna unilaterally during cyclic loading. We postulated that CGRP would influence skeletal responses to cyclic fatigue loading. Rats were fatigue loaded and groups of rats were infused systemically with 0.9% saline, CGRP, or the receptor antagonist, $CGRP_{8-37}$, for a 10 day study period. Ten days after fatigue loading, bone and serum CGRP concentrations, serum tartrate-resistant acid phosphatase 5b (TRAP5b) concentrations, and fatigue-induced skeletal responses were quantified. We found that cyclic fatigue loading led to increased CGRP concentrations in both loaded and contralateral ulnae. Administration of $CGRP_{8-37}$ was associated with increased targeted remodeling in the fatigue-loaded ulna. Administration of CGRP or $CGRP_{8-37}$ both increased reparative bone formation over the study period. Plasma concentration of TRAP5b was not significantly influenced by either CGRP or $CGRP_{8-37}$ administration.

Conclusions: CGRP signaling modulates targeted remodeling of microdamage and reparative new bone formation after bone fatigue, and may be part of a neuronal signaling pathway which has regulatory effects on load-induced repair responses within the skeleton.

Editor: Alejandro Almarza, University of Pittsburgh, United States of America

Funding: This study was funded by a grant from the AO Research Fund of the AO Foundation Switzerland (Project S-06-9M). Susannah J. Sample also received support from National Institutes of Health (T32 RR17503 and T32 RR023916). The funders had no role in study design, data collection and analysis, decision to publish, or preparation of the manuscript.

Competing Interests: The authors have declared that no competing interests exist.

* E-mail: muirp@vetmed.wisc.edu

Introduction

The failure of repair responses to protect the skeleton from fracture is an important problem, but the physiological pathways that regulate skeletal responses to loading are not fully understood. Fractures resulting from osteoporosis were estimated to cost the U.S.A. $17 billion in 2005, a number that is expected to rise nearly 50% by 2025 [1]. In addition to osteoporotic fracture, stress fractures are also common in military recruits and other human, canine and equine athletes, where bone fatigue can overwhelm repair mechanisms, with associated accumulation of fatigue damage within bone and development of a stress fracture [2–4].

Functional adaptation of the skeleton is thought to consist primarily of two processes, modeling and remodeling [5]. Bone modeling changes the spatial distribution of bone, while remodeling is a process of bone removal and replacement [5]. In situations where fatigue damage is present within bone, repair responses include targeted remodeling of microdamage [6] and woven bone formation on adjacent bone surfaces. Load-induced skeletal responses are thought to be locally regulated by bone cells [7,8]. However, recent work also suggests that the sensory innervation of bone may have regulatory effects on skeletal responses to bone loading [9,10].

The nervous system plays a role in the regulation of skeletal metabolism [11]. The sensory innervation of bone also has an important role in nociception and development of bone pain [12]. However, little work to date has addressed whether or not the innervation of bone has a functional role in the physiological responses of bone to loading. Periosteum, endosteum, and bone tissue are all innervated by nerve fibers [13–15]. This innervation exhibits plasticity in response to mechanical loading, in that a single loading event results in persistent changes in neuropeptide concentrations in both loaded and distant long bones, as well as changes in the neural circuits between limbs [9,10]. Of the three compartments of long bones, the periosteum has a particularly dense innervation, which is arranged in a net-like meshwork optimized for the detection of mechanical distortion [12]. This innervation is primarily peptidergic, and contains both sensory and sympathetic fibers [12,16,17]. Individual bone cells are directly connected to the nervous system via unmyelinated sensory neurons [18]. Bone cells express a range of functional neurotransmitter receptors and transporters, including those for calcitonin gene related peptide (CGRP) [19–21].

The calcitonin family of peptides has been extensively studied in bone over the past few years because of their effects on bone cells and potential as future drug targets. CGRP has pleiotropic effects on bone cells; both osteoclasts and osteoblasts have functional receptors for CGRP [20–22]. Physiological actions of CGRP are mediated through a family of type II G-protein coupled receptors, the most important of which is the $CGRP_1$ receptor [20,23,24]. In vitro, CGRP inhibits maturation of osteoclasts [25] and bone resorption [26], and is anabolic to osteoblasts by stimulation of canonical Wnt signaling and by inhibition of osteoblast apoptosis [21,27]. CGRP has two isoforms, CGRP-α and CGRP-β. In the rat, CGRP-α and CGRP-β differ by only one amino acid, despite being derived from separate genes, *Calca* and *Calcb*, respectively [28]. CGRP-α increases the proliferation rate of osteoblasts [27,29], prevents bone loss when delivered systemically to ovariectomized rats [30], and increases bone mass in transgenic mice with an osteoblast-specific promoter that over-expresses CGRP-α [31]. CGRP-α knockout mice are osteopenic [32]. Collectively, these findings suggest that CGRP-α is a peptide with osteoanabolic activity *in-vivo*. Unlike CGRP-α, CGRP-β is not considered osteogenic [33].

In the present study, our goal was to determine whether bone CGRP concentrations were modulated by cyclic fatigue loading *in vivo*. An increase would be expected with an osteoanabolic effect in a bone formation model, although earlier work suggests that short periods of loading without fatigue are associated with reduced CGRP concentrations in bone [9]. Using the fatigue loading model, we also sought to determine whether manipulation of systemic CGRP signaling *in vivo* would limit osteoclast activation for remodeling and increase reparative osteogenesis.

Materials and Methods

Animals

A homogeneous group of 68 actively growing male Sprague-Dawley rats (body weight 292–305 g, aged 67 ± 14 days) was used for the study. Rats were provided with food and water ad libitum. All procedures were performed in strict accordance with the recommendations in the Guide for the Care and Use of Laboratory Animals of the National Institutes of Health and the American Veterinary Medical Association and with approval from the Animal Care Committee of the University of Wisconsin-Madison. Ulna loading was performed under isoflurane anesthesia with butorphanol analgesia. Humane euthanasia was performed under isoflurane anesthesia at the end of the experimental period, using an intracardiac injection of pentobarbitone.

Experimental design

To determine the effect of unilateral ulna fatigue loading on CGRP peptide concentrations in the thoracic limb long bones, 12 rats were fatigue loaded until 40% loss of stiffness was attained, using an initial peak strain of $-3,000$ $\mu\varepsilon$ (Fatigue group). An additional 24 rats were used as controls: 12 rats were sham controls (Sham group), and thus given the same treatment regimen as rats in the treatment groups, but without being subjected to any mechanical loading; 12 rats served as baseline controls (Baseline group). Rats were euthanatized 10 days after loading or sham loading.

To determine how manipulation of systemic CGRP signaling might influence skeletal repair responses to fatigue loading, 32 rats were assigned one of 3 treatment groups. The right ulna of all rats was fatigue loaded until 40% loss of stiffness was attained, using an initial peak strain of $-3,000$ $\mu\varepsilon$. Immediately after loading, an osmotic pump (Alzet Corporation, Cupertino, CA) was implanted subcutaneously dorsally between the scapulae to provide a continuous infusion of either saline, CGRP, or the $CGRP_1$

receptor antagonist $CGRP_{8-37}$ [23,34], for the duration of the study. The 8 rats assigned to the Saline group received 40 $\mu l/kg/$ day 0.9% saline. The 12 rats assigned to the CGRP group received 100 $\mu g/kg/day$ of CGRP [35] and the 12 rats assigned to the $CGRP_{8-37}$ group received 100 $\mu g/kg/day$ of $CGRP_{8-37}$ [36]. All rats were additionally given an intraperitoneal injection of calcein (7 mg/kg) immediately after fatigue loading and again 7 days later to label load-induced bone formation. Rats were euthanatized 10 days after loading.

In-vivo ulnar fatigue loading

In-vivo fatigue loading of the right ulna was performed under isoflurane-induced general anesthesia. The right antebrachium of each rat was placed horizontally between two loading cups, which were fixed to the loading platen and actuator of a materials testing machine (Model 8800 DynaMight; Instron, Canton, MA, USA) with a 250N load cell (Honeywell Sensotec, Canton, MA, USA). The right ulna then underwent cyclic loading by means of axial compression, which accentuates the pre-existing mediolateral curvature of the diaphysis of the rat ulna, translating most of the axial force into a bending moment (Fig. 1). Cyclic fatigue loading was performed at 4 Hz, and was initiated at $-16N$. To induce fatigue, the load applied to the ulna was incrementally increased until fatigue was initiated, as indicated by increasing displacement amplitude from a stable baseline. Loading was then terminated when 40% loss of stiffness was attained.

Quantification of plasma and bone CGRP and plasma TRAP5b

At the time of euthanasia, plasma blood samples were collected and left and right ulnae were dissected with surrounding tissue; all samples were stored at $-80°C$ until processing. Bones were placed in liquid nitrogen and pulverized using a ball-mill grinder (Mikro-Dismembrator S; Sartorius Stedim Biotech, Aubagne Cedex, France). After homogenization, the samples were centrifuged at 3,000 g for 15 min. The supernatants were freeze-dried and dissolved in ELISA buffer. Plasma and bone CGRP concentrations were determined using a rat-specific ELISA assay (Cayman

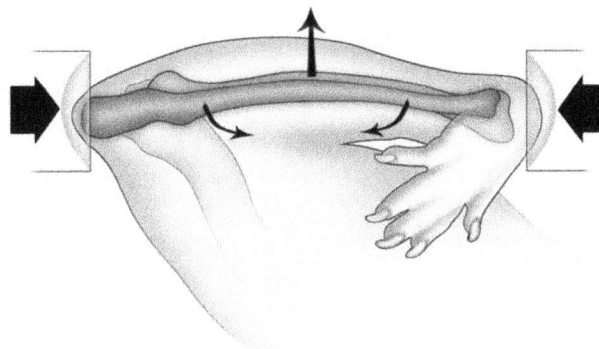

Figure 1. Schematic diagram of the rat ulna loading model. The antebrachium was placed horizontally in loading cups attached to a materials testing machine. The medio-lateral diaphyseal curvature of the rat ulna is accentuated through axial compression, most of which is translated into a bending moment, which is greatest at ~60% the total bone length measured from the proximal end of the ulna [37]. Ulnae underwent cyclic fatigue loading, initiated at $-3,000$ $\mu\varepsilon$, with incremental increases in load until fatigue was initiated. Loading was then terminated when 40% loss of stiffness was attained. Reproduced from [43] with permission from John Wiley & Sons.

A

B

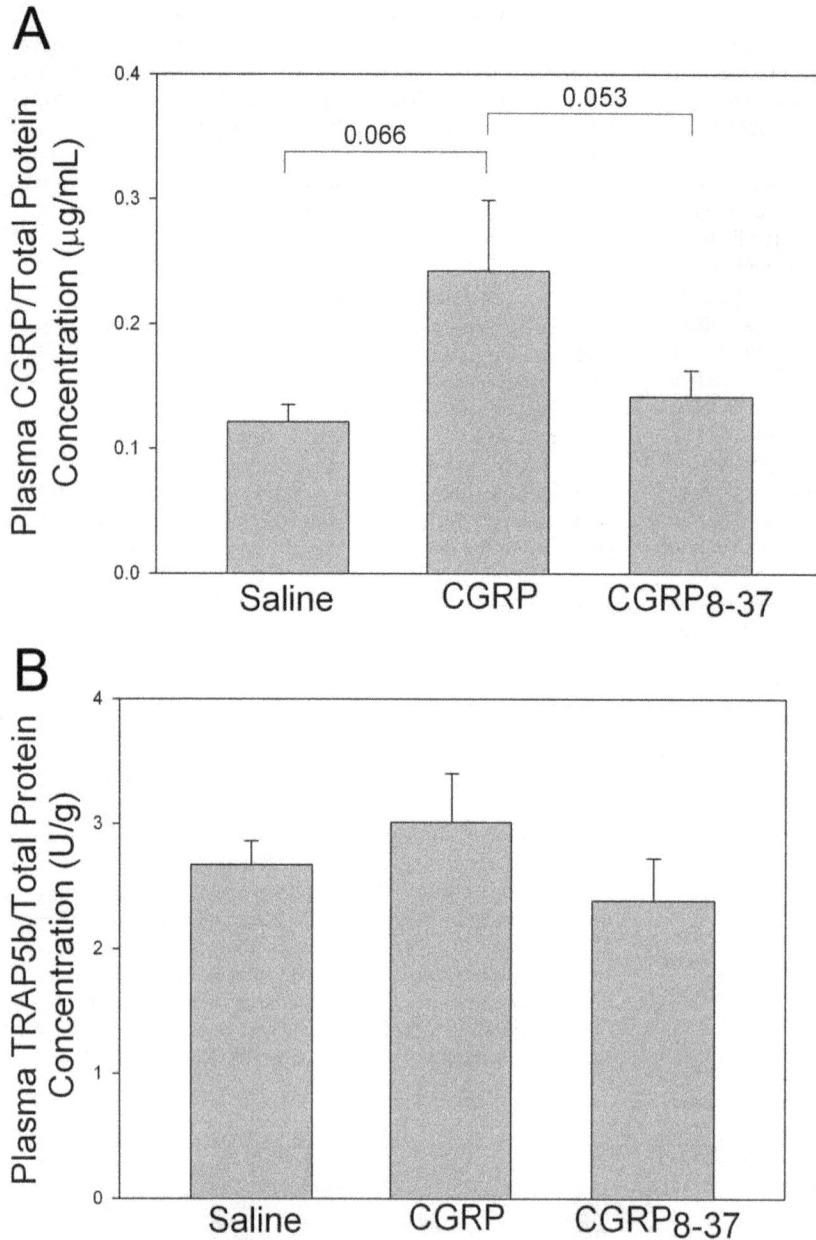

Figure 2. CGRP or CGRP$_{8-37}$ administration did not influence plasma TRAP5b in vivo. Plasma concentrations of CGRP and TRAP5b, normalized to plasma total protein, 10 days after fatigue loading. (**A**) Rats in the CGRP group had higher plasma CGRP concentrations when compared to rats in the Saline and CGRP$_{8-37}$ groups. No differences were seen between the Saline group and the CGRP$_{8-37}$ group. (**B**) Administration of CGRP or CGRP$_{8-37}$ did not have an effect on plasma TRAP5b levels. Error bars represent standard error of the mean. Saline group, n = 8; CGRP group n = 12; CGRP$_{8-37}$ group, n = 12.

Chemical, Ann Arbor, MI, USA). Total protein concentrations were also determined for both plasma and bone (BCA Protein assay, Thermo Scientific, Rockford, IL, USA). Plasma TRAP5b concentrations were determined using an ELISA kit validated for the rat (Immunodiagnostic Systems Ltd, Fountain Hills, AZ, USA). For each sample, CGRP and TRAP5b concentrations were normalized to the total protein concentration.

Bone Histomorphometry

The pairs of ulnae and humeri were dissected along with surrounding tissue. Fluorochrome labeled bones were dehydrated in a graded series of ethanol (70%, 100%), bulk stained with

Villanueva's solution for 3 days, and then embedded in methylmethacrylate. Transverse calcified sections, 125 µm thick, were made and mounted on standard microscope slides. Ulnae were sectioned at 60% of total bone length measured from the proximal end, where it has been shown maximal adaptation occurs with this model [37]. Humeri were sectioned at the mid-diaphysis (50% of total bone length). Confocal microscopy (MRC-1024 Laser Scanning Confocal Microscope; Bio-Rad, Hercules, CA, USA) was used to collect fluorescent images of each bone section. Periosteal and endosteal labeled bone areas (Ps.L.B.Ar and Es.L.B.Ar, %) were determined using a standard method [38], and total cortical bone area (Tt.L.B.Ar, %) was also determined by thresholding the image and pixel-counting (Image J; NIH) [9]. All measurements were made by a single observer (SJS). Data were normalized to the original cortical area to account for minor variations in rat size. Using the same sections, resorption space number density and microcrack surface density (Rs.Sp.Dn, $\#/mm^2$; Cr.S.Dn, $\mu m/mm^2$, respectively) were also determined in the left and right ulnae by assessment of Villanueva staining using bright-field microscopy.

Statistical Analysis

The Kolmogorov-Smirnov test was used to confirm that data were normally distributed. Limbs were treated as separate experiments. Group differences in normalized bone CGRP concentrations, normalized plasma CGRP and TRAP5b concentrations, and labeled adaptive bone formation were determined using a one-way ANOVA with a Dunnett post-hoc test; the Baseline group served as the control. Planned comparisons were used to determine differences between the CGRP and $CGRP_{8-37}$ groups. The Kruskal-Wallis ANOVA test and the Mann-Whitney U test were use to analyze Rs.Sp.Dn and Cr.S.Dn. Results were considered significant at $p<0.05$. Data are reported as mean ± standard error of the mean or median and range for non-parametric data.

Results

The number of cycles required to reach $47\pm15\%$ loss of stiffness in fatigue-loaded ulnae was $3,738\pm2,472$, with a final peak load of $-24.0\pm4.0N$. A minimally displaced intracortical stress fracture was noted in 31 of the 32 fatigue-loaded ulnae evaluated for microdamage; as expected, no intracortical fatigue damage was seen in the contralateral ulnae.

Infusion of CGRP elevated plasma CGRP concentrations but did not influence plasma TRAP5b

Plasma CGRP concentrations were measured to verify functional infusion of CGRP in the CGRP treatment group.

Plasma CGRP was elevated in the CGRP group compared with the $CGRP_{8-37}$ group ($p=0.05$), but not the Saline group ($p<0.07$) (Fig. 2A). Plasma TRAP5b concentrations were not altered as a result of CGRP or $CGRP_{8-37}$ administration ($p=0.41$) (Fig. 2B).

Cyclic fatigue loading increased bone CGRP concentrations in loaded and contralateral ulnae

To determine the effects of cyclic fatigue loading on bone CGRP concentrations, we quantified CGRP in both the loaded and contralateral ulnae of fatigue-loaded and control rats. The concentration of CGRP was increased in both the loaded right ulna ($p<0.05$) and the contralateral left ulna ($p<0.05$) of rats in the Fatigue group, when compared to the Baseline group (Fig. 3). No differences were seen between the Baseline group and the Sham group ($p=0.95$). Additionally, no differences in humeral CGRP concentrations were seen between groups.

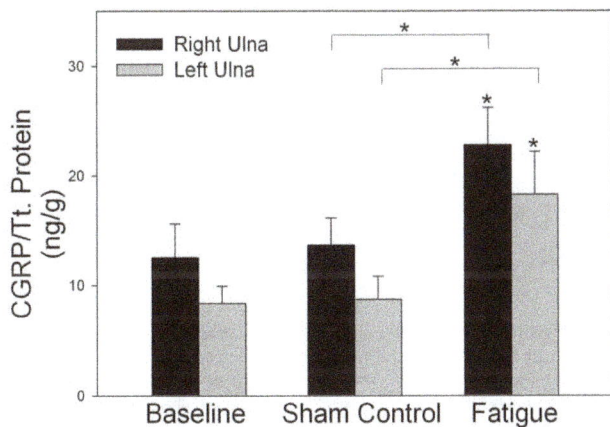

Figure 3. Bone CGRP is increased by mechanical loading. Cyclic fatigue loading of the right ulna resulted in increased CGRP concentrations in both the fatigue-loaded (right) ulna and the contralateral (left) ulna, when compared to the Baseline group. No differences in CGRP concentrations were seen between the Sham group and the Baseline group. The Fatigue group also had increased CGRP concentrations compared to the Sham group. * −p<0.05 versus the relevant baseline control bone. Error bars represent standard error of the mean. Baseline group n=12; Sham group n=12; Fatigue group n=12.

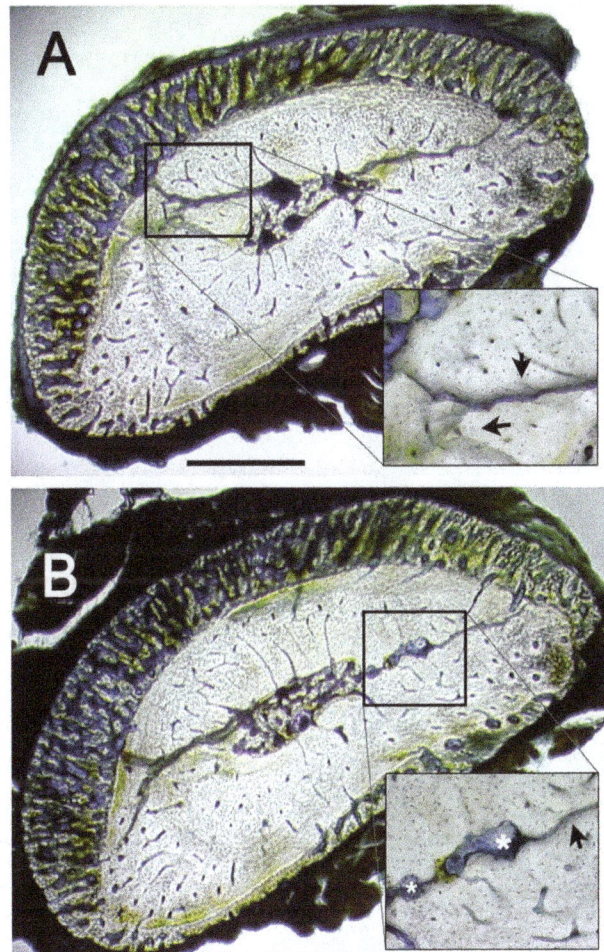

Figure 4. Targeted remodeling of bone microdamage. Photomicrographs of calcified transverse sections of ulna at 60% of bone length, from proximal to distal [37]. Fatigue loading induced microcrack formation and targeted remodeling. (**A**) Branching microcracks can be appreciated histologically in fatigue-loaded bones. (**B**) Targeted remodeling resulted in resorption space formation around the areas of microcracking. Bones were bulk-stained with Villanueva bone stain. Black arrows indicate fatigue damage; white asterisks are labeling resorption spaces. Bar = 0.5 mm.

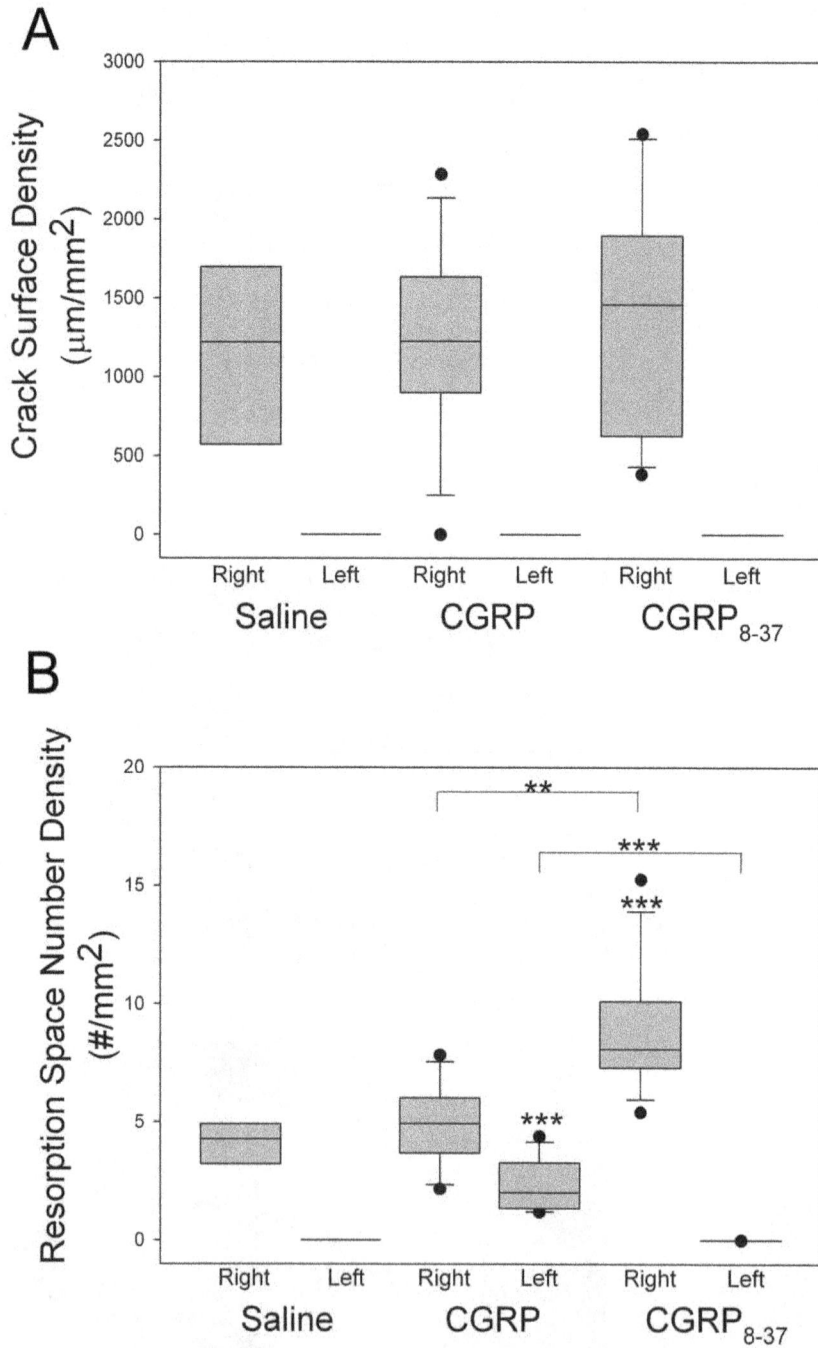

Figure 5. Remodeling of fatigue loaded bone was increased with administration of CGRP$_{8-37}$. Treatment with either saline, CGRP or CGRP$_{8-37}$ for 10 days after cyclic fatigue loading of the right ulna did not affect crack surface density (Cr.S.Dn), but resulted in altered resorption space density (Rs.Sp.Dn). **(A)** Cr.S.Dn was similar between groups. **(B)** Rs.Sp.Dn in the right ulna was increased in the CGRP$_{8-37}$ group compared to the CGRP group and the Saline group. In the left ulna, resorption space number density was increased in the CGRP group, compared to the Saline group and the CGRP$_{8-37}$ group. ** $-p<0.01$; *** $-p<0.001$; versus the relevant saline control bone. Error bars represent range. Saline group n = 8; CGRP group n = 12; CGRP$_{8-37}$ group n = 12.

Systemic administration of CGRP and CGRP$_{8-37}$ altered bone remodeling in response to unilateral cyclic fatigue loading

To quantify fatigue damage and associated osteoclast activation in rats treated with CGRP or CGRP$_{8-37}$, we measured Cr.S.Dn and Rs.Sp.Dn in both fatigue-loaded right ulnae and the contralateral left ulnae (Fig. 4). No differences in Cr.S.Dn were seen between groups (Fig. 5A). Rats in the CGRP$_{8-37}$ group had increased Rs.Sp.Dn in their right ulna 10 days after fatigue loading, compared to the CGRP group ($p<0.001$), and the Saline group ($p<0.01$) (Fig. 5B). In the contralateral left ulna, Rs.Sp.Dn was increased to a small extent in the CGRP group, compared to both the Saline group ($p<0.001$) and CGRP$_{8-37}$ group ($p<0.001$).

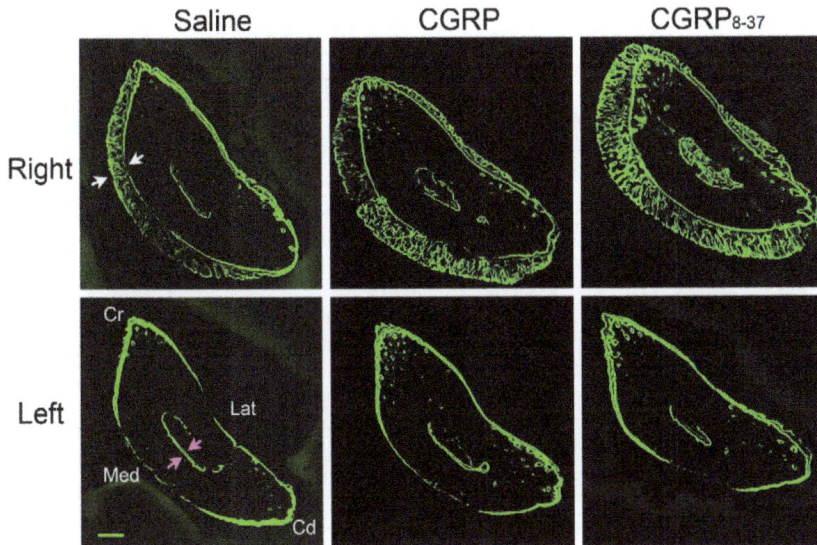

Figure 6. Reparative bone formation induced by fatigue loading was increased after treatment with CGRP or CGRP$_{8-37}$. Confocal photomicrographs of calcified transverse sections of ulna at 60% of bone length, from proximal to distal [37]. Administration of either CGRP or CGRP$_{8-37}$ for 10 days after cyclic fatigue loading of the right ulna increased reparative bone formation in the loaded ulna compared with saline-treated rats. Endosteal bone formation was particularly evident after CGRP$_{8-37}$ treatment. Rats treated with CGRP$_{8-37}$ also had greater bone formation in the contralateral (left) ulna, which was not loaded, when compared to the left ulna of the saline-treated rats. New bone formation was double labeled with calcein. White arrows indicate periosteal new woven bone formation; pink arrows indicate endosteal new bone. Bar = 250 μm. Cr, cranial; Cd, caudal; Med, medial; Lat, lateral. Saline group, n = 8; CGRP group n = 12; CGRP$_{8-37}$ group, n = 12.

Systemic administration of both CGRP and CGRP$_{8-37}$ increased reparative bone formation after unilateral cyclic fatigue loading

The effect of CGRP and CGRP$_{8-37}$ on load-induced bone formation after cyclic fatigue was evaluated in both the fatigue-loaded right ulnae and contralateral left ulnae. Rats treated with either CGRP or CGRP$_{8-37}$ for 10 days after cyclic fatigue loading had increased reparative bone formation, when compared to the rats that were treated with saline (Fig. 6). Ps.L.B.Ar was increased in the right ulnae of the CGRP ($p<0.001$) and the CGRP$_{8-37}$ ($p<0.001$) groups, and in the left ulna of the CGRP$_{8-37}$ group ($p<0.0001$), when compared to the Saline group; Ps.L.B.Ar was also increased in the left ulna of the CGRP$_{8-37}$ group when compared to the CGRP group (Fig. 7A, $p<0.05$). Similarly, Tt.L.B.Ar was increased in the right ulnae of the CGRP ($p<0.001$) and the CGRP$_{8-37}$ ($p<0.0001$) groups, and in the left ulna of the CGRP$_{8-37}$ group ($p<0.05$), when compared to the Saline group (Fig. 7B). Es.L.B.Ar was also increased in the right ulna of the CGRP ($p<0.05$) and the CGRP$_{8-37}$ ($p<0.001$) groups, when compared to the Saline group (Fig. 7C). No differences in humeral bone formation were seen between any of the groups.

Discussion

CGRP has been recognized as a neurotransmitter involved in the regulation of bone remodeling and fracture healing [39,40]. However, the potential effects of CGRP signaling on skeletal responses to bone loading and stress injury have not been investigated. In the present study, using the rat ulna end-loading model to induce cyclic fatigue in a single bone, we determined how CGRP and the CGRP$_1$ receptor antagonist, CGRP$_{8-37}$, influenced skeletal repair responses to stress injury, including bone remodeling and load-induced bone new formation. We also determined how cyclic fatigue loading influenced bone

CGRP concentrations both locally and at distant skeletal sites. We found that bone CGRP concentrations were increased after unilateral fatigue loading in both loaded and contralateral ulnae. We also found that both CGRP and CGRP$_{8-37}$ augmented new bone formation in response to fatigue but did not influence plasma TRAP5b concentrations. As expected, plasma CGRP concentrations were higher in rats given supplementary CGRP, as compared to rats that were treated with saline or CGRP$_{8-37}$.

TRAP5b is a marker of bone remodeling [41]. Both immature and mature osteoclasts express TRAP5b, and plasma concentrations of TRAP5b are proportional to osteoclast number [42]. Our data shows that systemic administration of CGRP or CGRP$_{8-37}$ did not affect TRAP5b concentrations in plasma, suggesting that 10 days after fatigue loading, overall numbers of activated osteoclasts in each group were not influenced by these treatments. Previous work from our laboratory has suggested that TRAP5b serum concentrations may be neuronally-regulated in fatigue-loaded male rats [43]. The data from this study suggest that neuronal signaling effects on plasma TRAP5b are not associated with CGRP signaling in sensory nerve fibers. It should be noted that TRAP5b levels were measured 10 days after fatigue loading; in future work it would be interesting to investigate markers of bone remodeling throughout a 10 day adaptive period, as the effect of CGRP on osteoclast activation may be more prominent earlier in the experimental period after bone loading.

CGRP-immunoreactive nerve fibers are widely distributed throughout bone [12,44,45]. During fracture healing, newly formed CGRP fibers can be found in areas with high bone formation rates [40]. The increase in CGRP concentrations 10 days after cyclic fatigue loading of the right ulna likely reflects altered release of CGRP from sensory nerve endings in loaded bone. Furthermore, the increase in CGRP concentrations in the contralateral ulna 10 days after loading supports the concept that mechanical loading of a single long bone may influence skeletal

A

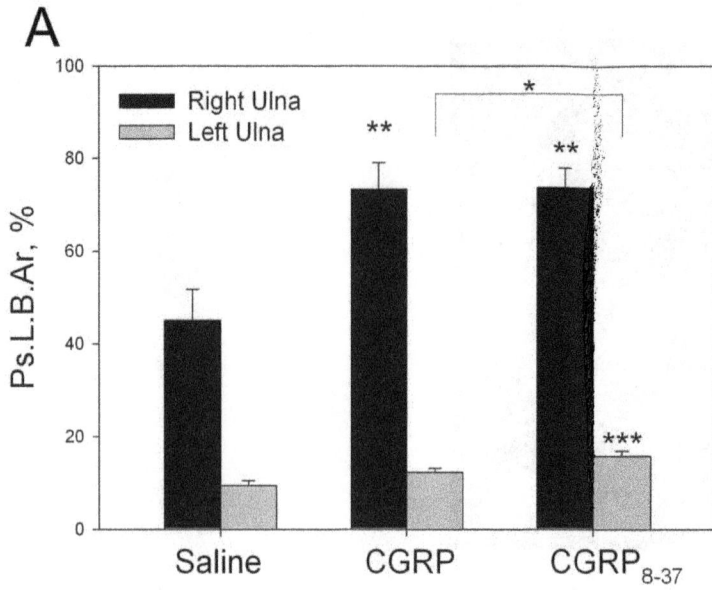

B

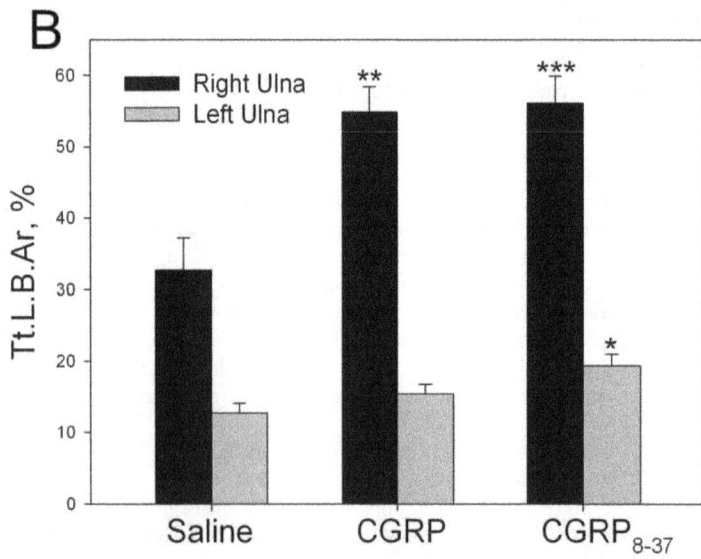

C

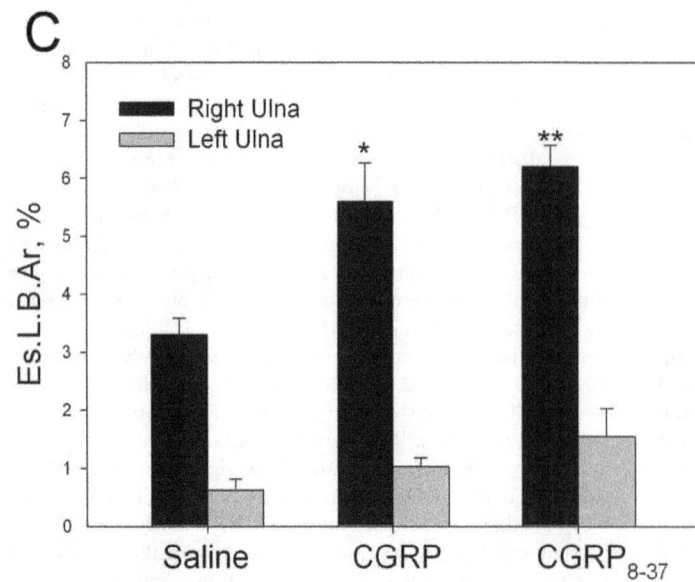

Figure 7. Calcein-labeled new bone formation in response to unilateral cyclic fatigue loading of the right ulna in rats. Treatment with either CGRP or CGRP$_{8-37}$ for 10 days after fatigue loading the right ulna resulted in increased reparative bone formation in the loaded ulna in both treatment groups; treatment with CGRP$_{8-37}$ also significantly influenced labeled bone formation in the contralateral ulna, when compared to saline treated rats. (**A**) Periosteal labeled bone area (Ps.L.B.Ar) was increased in the right ulnae of both treatment groups, and the left ulna of the CGRP$_{8-37}$ treatment group, when compared to the saline treated group. (**B**) Total labeled bone area (Tt.L.B.Ar) was also increased in the right ulnae of both treatment groups, and the left ulna of the CGRP$_{8-37}$ treatment group, when compared to the saline treated group. (**C**) Endosteal labeled bone area (Es.L.B.Ar) was increased in the right ulna of both the CGRP and CGRP$_{8-37}$ treatment groups, when compared to the saline treated group. . * $-p<0.05$; ** $-p<0.01$; *** $-p<0.001$ versus the relevant saline control bone. Error bars represent standard error of the mean. Saline group, n = 8; CGRP group n = 12; CGRP$_{8-37}$ group n = 12.

responses in distant bones through a neuronal mechanism [9]. We hypothesize that the CGRP in the contralateral bone is likely derived from the peptidergic sensory innervation of bone [12–15]. It is also possible that bone cells themselves may act as a source of CGRP, although expression in bone cells occurs at a low level [46]. It is well established that cyclic fatigue loading induces a repair response in the rat ulna loading model [37,47]. Previous work from our laboratory has suggested that loading a single bone may result in a systemic neurovascular response and increased bone formation in distant skeletal long bones, and that these changes are neuronally regulated [9,43,47]. CGRP may thus be part of a previously unrecognized neuronal mechanism that regulates skeletal responses to cyclic mechanical loading. In the first experiment in which we studied CGRP concentrations in bone after loading [9], we found decreased bone CGRP concentrations after loading without bone fatigue. Thus, it appears that the release of CGRP into bone tissue is modulated by the intensity and pattern of the applied mechanical signal. However, it is also possible that our results could be influenced by the methodology used to isolate neuropeptides from bone, as we used a different peptide isolation method in the previous study.

Interestingly, we found that systemic administration of CGRP and CGRP$_{8-37}$ over a 10 day period after unilateral fatigue loading increased bone formation in the periosteal, endosteal and intracortical envelopes. The increase in reparative bone formation with CGRP administration was anticipated; CGRP has been shown to be an inhibitor of bone resorption [48], and increases osteogenesis in a dose-dependent manner [49].

CGRP$_{8-37}$ principally antagonizes signaling via the CGRP$_1$ receptor [20,23,24]. This receptor is well defined as a heterodimer of the calcitonin receptor-like receptor (CRLR) and receptor activity modifying protein 1 (RAMP$_1$) [23,50]. In addition, CGRP is also thought to signal through a putative CGRP$_2$ receptor [23]. In rodents, CGRP binds to variants of both adrenomedullin (AM) (AM$_2$ consisting of CRLR and RAMP$_3$) and amylin (AMY) receptors (AMY$_{1(a)}$, AMY$_{3(a)}$) [23,51,52]. AMY$_{1(a)}$ consists of RAMP$_1$ and an insert negative form of the calcitonin receptor (CTR), while AMY$_{3(a)}$ consists of RAMP$_3$ and the same CTR variant [23,52]. Thus the 'CGRP$_2$ receptor' may in reality be an amalgamation of contributions from a variety of CGRP-activated receptors [23].

As CGRP has been shown to activate bone formation, it is not surprising that CGRP-α knockout mice are osteopenic [32]. The increase in bone formation in rats treated with CGRP$_{8-37}$, however, was an unexpected result. The heterogeneity of CGRP receptor signaling may explain this observation, and suggests that signaling via a CGRP receptor that is not CGRP$_1$ is responsible for the load-induced osteogenic response. However, CGRP also has actions on multiple organ systems, and it is also possible that the effect of CGRP on vascular tone may help explain this result [53–58]. Increased bone blood flow precedes bone repair in response to fatigue loading, and remodeling in response to decreased mechanical loading [47,58]. Systemic administration of CGRP also decreases blood pressure in a dose-dependent

manner [59]. Therefore, treatment with CGRP$_{8-37}$ may have increased intraosseus pressure, transcortical interstitial fluid flow, and associated bone formation [60,61], although such an effect appears less likely in our model since we did not detect significant changes in humeral bone formation with CGRP or CGRP$_{8-37}$ treatment.

As expected, 10 days after cyclic fatigue loading there were no differences in Cr.S.Dn between groups, indicating that rats in each group experienced a similar degree of fatigue damage during bone loading. However, we found that treatment with CGRP$_{8-37}$ did have a significant effect on osteoclast recruitment and activation for targeted remodeling in the fatigue-loaded ulna; targeted remodeling in the fatigue-loaded ulna was increased after treatment with CGRP$_{8-37}$. CGRP-immunoreactive nerve fibers have direct contact with osteoclasts [18]. CGRP is an inhibitor of bone resorption, possibly through interference with the action of the receptor activator of NF-kB (RANKL), and thus differentiation and recruitment of osteoclast precursors [48,62]. Our results suggest the CGRP$_1$ receptor is responsible for this in-vivo effect, such that in fatigue-loaded rats treated with the CGRP$_{8-37}$, the normal inhibition of CGRP on osteoclast activation for targeted remodeling appears diminished. Interestingly, this effect was detected in the face of increased bone CGRP concentrations in both loaded and contralateral ulnae at 10 days and a lack of difference in plasma TRAP5b between groups, suggesting that the signals regulating targeted remodeling of damaged bone are complex, and may involve other pathways that activate osteoclastic bone remodeling. CGRP signaling may have greater effects on osteoclastic activation, and a lesser effect on osteoclastic recruitment and proliferation in vivo. We also detected a small CGRP treatment effect on remodeling in the contralateral ulna; the biological significance of this observation is unclear.

A limitation to this study is that we analyzed bone CGRP concentrations and bone repair at a single time point, 10 days after fatigue loading. Additionally, it is not possible to fully isolate direct effects of CGRP signaling on bone cells from effects of CGRP on neuronal signaling, and possibly bone blood flow. As both CGRP and CGRP$_{8-37}$ augmented reparative bone formation, our data suggest that CGRP release from nerve endings during bone loading acts to modulate load-induced skeletal repair. Our results implicate the CGRP$_1$ receptor as responsible for effects on osteoclasts, and a CGRP receptor that is not CGRP$_1$ as responsible for the osteogenic effects we observed.

In conclusion, our data support the established hypothesis that CGRP is osteoanabolic to the skeleton in that it acts to increase osteogenesis and inhibit osteoclastic bone resorption in vivo in a bone fatigue model. The present study also suggests that CGRP has regulatory effects on skeletal responses to mechanical loading. Our results suggest that CGRP augments load-induced bone formation and inhibits osteoclastic remodeling through increased release of CGRP from the peptidergic sensory innervation of bone tissue that is being repaired. Future work should confirm whether or not mechanical loading of the skeleton leads to up-regulation of

CGRP in the sensory innervation of loaded bone, as is thought to occur during nociception [63], and also determine which CGRP receptors are responsible for these actions.

Author Contributions

Conceived and designed the experiments: SJS PM. Performed the experiments: SJS ZH APW PM. Analyzed the data: SJS ZH PM. Wrote the paper: SJS PM.

References

1. Burge R, Dawson-Hughes B, Solomon DH, Wong JB, King A, et al. (2007) Incidence and economic burden of osteoporosis-related fractures in the United States, 2005–2025. J Bone Miner Res 22: 465–467.

2. Burr DB, Forwood MR, Fyhrie DP, Martin RB, Schaffler MB, et al. (1997) Bone microdamage and skeletal fragility in osteoporotic and stress fractures. J Bone Miner Res 12: 6–15.

3. Muir P, Johnson KA, Ruaux-Mason C (1999) In vivo matrix microdamage in a naturally occurring canine fatigue fracture. Bone 25: 571–576.

4. Muir P, McCarthy J, Radtke CL, Markel MD, Santschi EM, et al. (2006) Role of endochondral ossification of articular cartilage and adaptation of the subchondral plate in the development of fatigue microcracking of joints. Bone 38: 342–349.

5. Frost HM (2001) From Wolff's Law to the Utah Paradigm: Insights about bone physiology and its clinical applications. Anat Rec 262: 398–419.

6. Burr DB (2002) Targeted and nontargeted remodeling. Bone 30: 2–4.

7. Chenu C (2004) Role of innervation in the control of bone remodeling. J Musculoskel Neuron Interact 4: 132–134.

8. Robling AG, Castillo AB, Turner CH (2006) Biomechanical and molecular regulation of bone remodeling. Annu Rev Biomed Eng 8: 455–498.

9. Sample SJ, Behan M, Smith L, Oldenhoff WE, Markel MD, et al. (2008) Functional adaptation to loading of a single bone is neuronally regulated and involves multiple bones. J Bone Miner Res 23: 1372–1381.

10. Wu Q, Sample SJ, Baker TA, Thomas CF, Behan M, et al. (2009) Mechanical loading of a long bone induces plasticity in sensory input to the central nervous system. Neurosci Lett 463: 254–257.

11. Elefteriou F (2008) Regulation of bone remodeling by the central and peripheral nervous system. Arch Biochem Biophys 473: 231–236.

12. Martin CD, Jimenez-Andrade JM, Ghilardi JR, Mantyh P (2007) Organization of a unique net-like meshwork of $CGRP^+$ sensory fibers in the mouse periosteum: Implications for the generation and maintenance of bone fracture pain. Neurosci Lett 427: 148–152.

13. Hill EL, Elde R (1991) Distribution of CGRP-, VIP-, D beta H-, SP-, and NPY-immunoreactive nerves in the periosteum of the rat. Cell Tissue Res 264: 469–480.

14. Hukkanen M, Konttinen YT, Rees RG, Gibson SJ, Santavirta S, et al. (1992) Innervation of bone from healthy and arthritic rats by substance P and calcitonin gene related peptide containing sensory fibers. J Rheumatol 19: 1252–1259.

15. Tabarowski Z, Gibson-Berry K, Felten SY (1996) Noradrenergic and peptidergic innervation of the mouse femur bone marrow. Acta Histochem 98: 453–457.

16. Hohmann EL, Elde RP, Rysavy JA, Einzig S, Gebhard RL (1986) Innervation of periosteum and bone by sympathetic vasoactive intestinal peptide-containing nerve fibers. Science 232: 868–871.

17. Konttinen Y, Imai S, Suda A (1996) Neuropeptides and the puzzle of bone remodeling. State of the art. Acta Orthop Scand 67: 632–639.

18. Imai S, Rauvala H, Konttinen YT, Tokunaga T, Maeda T, et al. (1997) Efferent targets of osseous CGRP-immunoreactive nerve fiber before and after their destruction in adjuvant arthritic rat: An ultramorphological study on their terminal-target relations. J Bone Miner Res 12: 1018–1027.

19. Spencer GJ, Hitchcock IS, Genever PG (2004) Emerging neuroskeletal signaling pathways: a review. FEBS Lett 559: 6–12.

20. Naot D, Cornish J (2008) The role of peptides and receptors of the calcitonin family in the regulation of bone metabolism. Bone 43: 813–818.

21. Mrak E, Guidobono F, Moro G, Fraschini G, Rinacci A, et al. (2010) Calcitonin gene-related peptide (CGRP) inhibits apoptosis in human osteoblasts by β-catenin stabilization. J Physiol 225: 701–708.

22. Lerner UH (2002) Neuropeptidergic regulation of bone resorption and bone formation. J Musculoskel Neuron Interact 2: 440–447.

23. Hay DL (2007) What makes a $CGRP_2$ receptor? Clin Exp Pharmacol Physiol 34: 963–971.

24. Lerner UH, Persson E (2008) Osteotropic effects by the neuropeptides calcitonin gene-related peptide, substance P and vasoactive intestinal peptide. J Musculoskelet Neuronal Interact 8: 154–165.

25. Owan I, Ibaraki K (1994) The role of calcitonin gene-related peptide (CGRP) in macrophages: the presence of functional receptors and effects on proliferation and differentiation into osteoclast-like cells. Bone Miner 24: 151–164.

26. Zaidi M, Chambers TJ, Gaines Das RE, Morris HR, MacIntyre I (1987) A direct action of human calcitonin gene-related peptide on isolated osteoclasts. J Endocrinol 115: 511–518.

27. Villa I, Dal Fiume C, Maestroni A, Rubinacci A, Ravasi F, et al. (2003) Human osteoblast-like cell proliferation induced by calcitonin-related peptides involves PKC activity. Am J Physiol Endocrinol Metab 284: E627–633.

28. Amara SG, Arriza JL, Leff SE, Swanson LW, Evans RM, et al. (1985) Expression in brain of messenger RNA encoding a novel neuropeptide homologous to calcitonin gene-related peptide. Science 229: 1094–1097.

29. Cornish J, Callon KE, Lin CQ, Xiao CL, Gamble GD, et al. (1999) Comparison of the effects of calcitonin gene-related peptide and amylin on osteoblasts. J Bone Miner Res 14: 1302–1309.

30. Valentijn K, Gutow AP, Troiano N, Gundberg C, Gilligan JP, et al. (1997) Effects of calcitonin gene-related peptide on bone turnover in ovariectomized rats. Bone 21: 269–274.

31. Ballica R, Valentijn K, Khachatryan A, Guerder S, Kapadia S, et al. (1999) Targeted expression of calcitonin gene related peptide to osteoblasts increases bone density in mice. J Bone Miner Res 14: 1067–1074.

32. Schinke T, Liese S, Priemel M, Haberland M, Schilling AF, et al. (2004) Decreased bone formation and osteopenia in mice lacking alpha-calcitonin gene-related peptide. J Bone Miner Res 19: 2049–2056.

33. Hirt D, Bernard GW (1997) CGRP-beta unlike CGRP-alpha has no osteogenic stimulatory effect in vitro. Peptides 18: 1461–1463.

34. Uzan B, Villemin A, Garel JM, Cressent M (2007) Adrenomedullin is anti-apoptotic in osteoblasts through CGRP1 receptors and MEK-ERK pathway. J Cell Physiol 215: 122–128.

35. Gangula PR, Wimalawansa SJ, Yallampalli C (2002) Sex steroid hormones enhance hypotensive effects of calcitonin gene-related peptide in aged female rats. Biol Reprod 67: 1881–1887.

36. Reinshagen M, Flämig G, Ernst S, Geerling I, Wong H, et al. (1998) Calcitonin gene-related peptide mediates the protective effect of sensory nerves in a model of colonic injury. J Pharmacol Exp Ther 286: 657–661.

37. Kotha SP, Hsieh YF, Strigel RM, Müller R, Silva MJ (2004) Experimental and finite element analysis of the rat ulnar loading model – correlations between strain and bone formation following fatigue loading. J Biomech 37: 541–548.

38. de Souza RL, Pitsillides AA, Lanyon LE, Skerry TM, Chenu C (2005) Sympathetic nervous system does not mediate the load-induced cortical new bone formation. J Bone Miner Res 20: 2159–2168.

39. Irie K, Hara-Irie F, Ozawa H, Yajima T (2002) Calcitonin gene-related peptide (CGRP)-containing nerve fibers in bone tissue and their involvement in bone remodeling. Microsc Res Tech 58: 85–90.

40. Li J, Kreicbergs A, Bergström J, Stark A, Ahmed M (2007) Site-specific CGRP innervation coincides with bone formation during fracture healing and modeling: a study in rat angulated tibia. J Orthop Res 25: 1204–1212.

41. Boyce BF, Xing L (2008) Functions of RANKL/RANK/OPG in bone modeling and remodeling. Arch Biochem Biophys 473: 139–146.

42. Henriksen K, Tanko LB, Qvist P, Delmas PD, Christiansen C, et al. (2007) Assessment of osteoclast number and function: application in the development of new and improved treatment modalities for bone diseases. Osteoporosis Int 18: 681–685.

43. Sample SJ, Behan M, Collins RJ, Wilson AP, Markel MD, et al. (2010) Systemic effects of ulna loading in young male rats during functional adaptation. J Bone Miner Res 25: 2016–2028.

44. Mach DB, Rogers SD, Sabino MC, Luger NM, Schwei MJ, et al. (2002) Origins of skeletal pain: sensory and sympathetic innervation of the mouse femur. Neuroscience 113: 155–166.

45. Ivanusic JJ (2009) Size, neurochemistry, and segmental distribution of sensory neurons innervating the rat tibia. J Comp Neurol 517: 276–283.

46. Drissi H, Hott M, Marie PJ, Lasmoles F (1997) Expression of the CT/CGRP gene and its regulation by dibutyryl cyclic adenosine monophosphate in human osteoblastic cells. J Bone Miner Res 12: 1805–1814.

47. Muir P, Sample SJ, Barrett JG, McCarthy J, Vanderby R, Jr., et al. (2007) Effect of fatigue loading and associated matrix microdamage on bone blood flow and interstitial fluid flow. Bone 40: 948–956.

48. Wang L, Shi X, Zhao R, Halloran BP, Clark DJ, et al. (2010) Calcitonin-gene related peptide stimulates stromal cell osteogenic differentiation and inhibits RANKL induced NF-κB activation, osteoclastogenesis and bone resorption. Bone 46: 1369–1379.

49. Bernard GW, Shih C (1990) The osteogenic stimulating effect of neuroactive calcitonin gene-related peptide. Peptides 11: 625–632.

50. Yu LC, Hou JF, Fu FH, Zhang YX (2009) Roles of calcitonin gene-related peptide and its receptors in pain-related behavioral responses in the central nervous system. Neurosci Biobehav Rev 33: 1185–1191.

51. Tilakaratne N, Christopoulos G, Zumpe ET, Foord SM, Sexton PM (2000) Amylin receptor phenotypes derived from human calcitonin receptor/RAMP coexpression exhibit pharmacological differences dependant upon receptor isoform and host cell environment. J Pharmacol Exp Ther 294: 61–72.

52. Hay DL, Christopoulos G, Christopoulos A, Poyner DR, Sexton PM (2005) Pharmacological discrimination of calcitonin receptor: receptor activity-modifying protein complexes. Mol Pharmacol 67: 1655–1665.

53. Han SP, Naes L, Westfall TC (1990) Calcitonin gene-related peptide is the endogenous mediator of nonadrenergic-noncholinergic vasodilation in rat mesentery. J Pharmacol Exp Ther 255: 423–428.

Role of Calcitonin Gene-Related Peptide in Bone Repair after Cyclic Fatigue Loading

54. Haegerstrand A, Dalsgaard CJ, Jonzon B, Larsson O, Nilsson J (1990) Calcitonin gene-related peptide stimulates proliferation of human endothelial cells. Proc Natl Acad Sci USA 87: 3299–3303.

55. Greenburg B, Rhoden K, Barnes P (1987) Calcitonin gene-related peptide (CGRP) is a potent non-endothelium-dependent inhibitor of coronary vasomotor tone. Br J Pharmac 92: 789–794.

56. Gangula PR, Zhao H, Supowit SC, Wimalawansa SJ, Dipette DJ, et al. (2000) Increased blood pressure in alpha-calcitonin gene-related peptide/calcitonin gene knockout mice. Hypertension 35: 470–475.

57. Kubota M, Moseley JM, Butera L, Dusting GJ, MacDonald PS, et al. (1985) Calcitonin gene-related peptide stimulates cyclic AMP formation in rat aortic smooth muscle cells. Biochem Biophys Res Commun 132: 88–94.

58. Gross TS, Damji AA, Judex S, Bray RC, Zernicke RF (1999) Bone hyperemia precedes disuse-induced intracortical bone resorption. J Appl Physiol 86: 230–235.

59. Wimalawansa SJ (1996) Calcitonin gene-related peptide and its receptors: molecular genetics, physiology, pathophysiology and therapeutic potentials. Endocr Rev 17: 533–585.

60. Kelly PJ, Bronk JT (1990) Venous pressure and bone formation. Microvasc Res 39: 364–375.

61. Bergula AP, Huang W, Frangos JA (1999) Femoral vein ligation increases bone mass in the hindlimb suspended rat. Bone 24: 171–177.

62. Ishizuka K, Hirukawa K, Nakamura H, Togari A (2005) Inhibitory effect of CGRP on osteoclast formation by mouse bone marrow cells treated with isoproterenol. Neurosci Lett 379: 47–51.

63. Yoneda T, Hata K, Nakanishi M, Nagae M, Nagayama T, et al. (2011) Involvement of acidic microenvironment in the pathophysiology of cancer-associated bone pain. Bone 48: 100–105.

5

Local Mechanical Stimuli Regulate Bone Formation and Resorption in Mice at the Tissue Level

Friederike A. Schulte, Davide Ruffoni, Floor M. Lambers, David Christen, Duncan J. Webster, Gisela Kuhn, Ralph Müller*

Institute for Biomechanics, ETH Zurich, Zurich, Switzerland

Abstract

Bone is able to react to changing mechanical demands by adapting its internal microstructure through bone forming and resorbing cells. This process is called bone modeling and remodeling. It is evident that changes in mechanical demands at the organ level must be interpreted at the tissue level where bone (re)modeling takes place. Although assumed for a long time, the relationship between the locations of bone formation and resorption and the local mechanical environment is still under debate. The lack of suitable imaging modalities for measuring bone formation and resorption *in vivo* has made it difficult to assess the mechanoregulation of bone three-dimensionally by experiment. Using *in vivo* micro-computed tomography and high resolution finite element analysis in living mice, we show that bone formation most likely occurs at sites of high local mechanical strain ($p<0.0001$) and resorption at sites of low local mechanical strain ($p<0.0001$). Furthermore, the probability of bone resorption decreases exponentially with increasing mechanical stimulus ($R^2 = 0.99$) whereas the probability of bone formation follows an exponential growth function to a maximum value ($R^2 = 0.99$). Moreover, resorption is more strictly controlled than formation in loaded animals, and ovariectomy increases the amount of non-targeted resorption. Our experimental assessment of mechanoregulation at the tissue level does not show any evidence of a lazy zone and suggests that around 80% of all (re)modeling can be linked to the mechanical micro-environment. These findings disclose how mechanical stimuli at the tissue level contribute to the regulation of bone adaptation at the organ level.

Editor: Kornelius Kupczik, Friedrich-Schiller-University Jena, Germany

Funding: Funding from the European Union for the osteoporotic virtual physiological human project (VPHOP FP7-ICT2008-223865) is gratefully acknowledged. The funders had no role in study design, data collection and analysis, decision to publish, or preparation of the manuscript.

Competing Interests: The authors have declared that no competing interests exist.

* E-mail: ram@ethz.ch

Introduction

The shape, structure and material properties of living organs vary according to the function they fulfill in the organism [1]. One of the major tasks of the skeleton is load-bearing and it is known that external loads are able to change bone mass and architecture [2–5]. Functional bone adaptation to changes in the mechanical environment has been assumed for more than a century [6,7]. Since then, various mechanobiological experiments have been performed based on the concept of introducing controlled variations (either an increase or a decrease) in the mechanical loading and then measuring the corresponding bone response. Such studies have clearly indicated that bone mass and trabecular bone architecture are controlled by mechanical cues [3,8–14]. Moreover, trabecular bone, compared to cortical bone, has shown a higher response to changes in the loading environment since individual trabeculae have the freedom to reach an arrangement which optimizes load transfer [8,15,16].

The process of bone adaptation as a whole takes place locally by bone forming and resorbing cells. Bone adaptation to mechanical demands is also called bone modeling, i.e. uncoupled bone formation and resorption. Bone repair using spatially coupled bone resorption and formation has been referred to as bone remodeling. However, following the suggestion that the pathways governing both bone modeling and remodeling may be of the same origin [17], we will in the following refer to this process as bone (re)modeling.

Until recently, measurements of local bone formation were only possible with histology combined with fluorescent dyes injected into the animal. The dyes incorporate into bone matrix that is newly formed at the time of injection and the use of several labels injected at different time points allows measuring where and how much new bone is formed [18]. The disadvantage of histology is that the animal has to be sacrificed for the readout, and that the assessment can only be completed in two-dimensional slices. Furthermore, labels are lost due to bone turnover when the time between the injection and sacrifice of the animal is too long, hence allowing only relatively short observation periods. Trabecular bone formation rates as measured from histological slices have been linked to high loads, however with moderate R^2-values (0.13–0.42) [19]. A reason for the low correlation may be that comparisons were made on two-dimensional data, and that the data were from a cross-sectional study. A direct connection of local bone formation with high loads, or local bone resorption with low loads in the same animal, however, has been difficult to date due to the lack of suitable technologies to measure the locations of trabecular bone formation and resorption as well as to compute the local mechanical environment *in vivo* and in three dimensions.

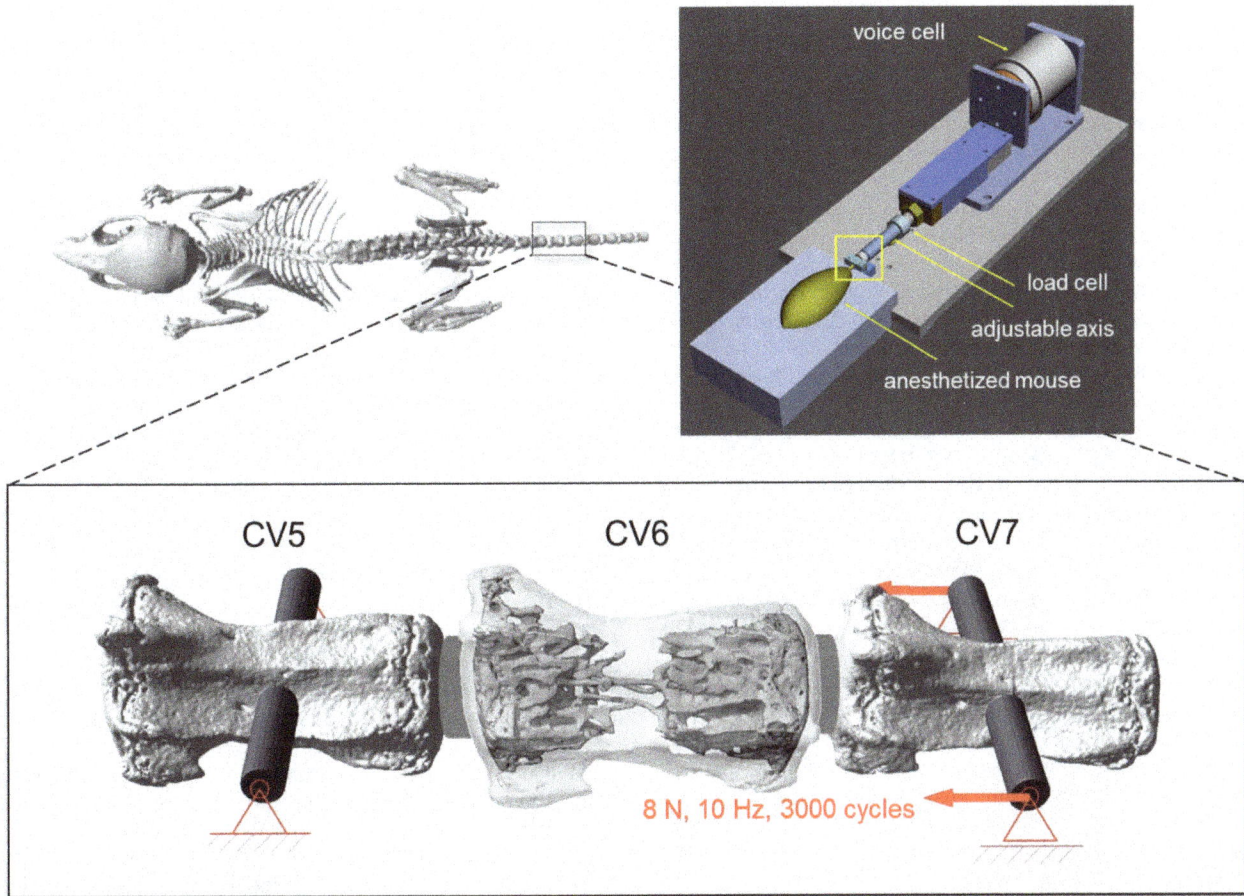

Figure 1. Experimental setup. The 6th caudal vertebra (CV6) is cyclically loaded (10 Hz) by a force of 8 N +1 N preload. Controlled application of the force is obtained through two pins inserted in the adjacent vertebrae (CV5 fixed, CV7 displaced) through a mechanical loading device [12].

This lack has given impetus to the field of computer simulations in bone research. In fact, the three-dimensional functioning of local mechanoregulation in trabecular bone at the tissue level has, so far, mostly been investigated with computational models by assuming different (re)modeling theories and comparing the resulting virtual trabecular architectures with experimental data or findings from the literature [17,20–23]. Such *in silico* models are certainly able to capture major aspects of the functioning of local mechanoregulation; however, current computational models are based on a number of (partly competing) assumptions. These include, for instance, whether resorption is caused by random micro-cracks or load-driven [21], whether formation depends linearly on the mechanical stimulus [21] or whether a step function (implying an activation barrier) should be used [22]. Another assumption presumes the existence of a so-called lazy zone, i.e. a range of strains in which only balanced bone formation and resorption occurs [2]. In an earlier publication, we developed a computer model for bone adaptation where the relationship between local changes in bone mass and the mechanical strain were described by two linear functions and a lazy zone [23]. Although the model predicted structural bone parameters reliably with less than 12.1% error, the simulated rates and sites of formation or resorption did not coincide well with the experiment.

In the present paper, to investigate further how formation and resorption can be linked to the mechanical environment,

we correlate sites of local bone formation, resorption and quiescence determined from time-lapsed *in vivo* micro-computed tomography (micro-CT), with the local strain distribution calculated by micro-finite element (micro-FE) models. The relation between the mechanical environment and bone formation/resorption was characterized in tail vertebrae of mice that were ovariectomized or subjected to mechanical loading. We hypothesized that regions of low local strains would, independent from the treatment, lead to site-specific bone resorption and regions of high local strains to site-specific bone formation.

Materials and Methods

Ethics Statement

During the animal experiments, all efforts were made to minimize suffering. All experiments were carried out under anesthesia and with the approval from the veterinary authority of the canton of Zurich, license number 171/2008 (Kantonales Veterinäramt Zürich, Zurich, Switzerland).

In vivo Experiments and Micro-computed Tomography

To induce a bone response under controlled loading conditions, a tail loading model [24] which permits the study of trabecular bone adaptation *in vivo* [16,25] was used. Specifically, in 15-week-old female C57Bl/6 (B6) mice (RCC, Füllinsdorf, Switzerland), the sixth caudal vertebra (CV6) was subjected to cyclic mechanical loading at 9 N (8 N +1 N preload, CML, n = 9) through stainless steel pins inserted in the adjacent vertebrae, 3 times/week for 4 weeks at 10 Hz and 3000 cycles (Figure 1). More details about the loading regime can be found in Lambers et al. [16]. A control group (CTR, n = 8) receiving the same pins was mounted into the loading device and was given the same amount of anesthesia for a period equivalent to the loading group; however, no force was applied to the vertebra. CV6 of all animals was scanned at the start of treatment and every week for the following 4 weeks with *in vivo* micro-CT (vivaCT 40, Scanco Medical, Brüttisellen, Switzerland) at an isotropic voxel resolution of 10.5 μm. After termination of the loading experiment, a subgroup of 5 loaded and 3 non-loaded vertebrae was dissected and scanned repetitively (5 times) with repositioning between the scans [26].

A systemic, catabolic bone response was induced by ovariectomy of 15-week old female B6 mice (OVX, n = 9). A control group underwent the same surgical procedure without removal of the ovaries, which is also called sham operation (SHM, n = 7). *In vivo* micro-CT scans of CV6 were performed on the day of operation and consecutively every two weeks over a twelve-week period [27]. The reason for the larger time intervals between two consecutive scans was that the ovariectomy experiment ran over a longer period than the loading experiment but the mice should not be exposed to more radiation than necessary.

The time-lapsed greyscale micro-CT measurements were registered using a rigid intensity-based, least-squared registration method, allowing arbitrary rotations and translations [28]. B-Splines were chosen as the interpolation technique where a registration error of less than 1.4% was found [26]. After registration, the three-dimensional (3D) volumes were Gaussian-filtered (sigma = 1.2, support = 1) and binarized at a global threshold of 22% of the maximum greyscale value [29]. Bone mineral phantoms were scanned weekly, to ensure that the threshold was at the same mineral density for each time point. Superimposition of the binarized *in vivo* micro-CT scans of the same vertebra taken at different time points allowed the identification of bone formation and resorption sites (Figure 2A) [25]. The assessment of bone formation and resorption from *in vivo* micro-CT was proposed in Schulte et al. [25] where a comparison with histomorphometry yielded correlation coefficients R of 0.68 and 0.78 for mineral apposition rate (MAR) and mineralizing surface (MS).

Finite Element Simulations

3D micro-FE models were generated by converting all voxels of the micro-CT image to 8 node hexahedral elements, with each model consisting of approximately 1.8 million elements (Figure 2B). A Young's modulus of 14.8 GPa and a Poisson's ratio of 0.3 [12] were assigned. To prevent the formation of unrealistically high strains situated on only a few nodes of the finite element mesh, loads were applied to CV6 through simulated intervertebral disks, having a circular cross sectional area of 260 μm and a height corresponding to maximum 10% of the full vertebral length. For numerical issues of the finite element solver, these disks were assigned the same Young's modulus and Poisson's ratio as bone. The top of the model was displaced by 1% of the total vertebral length, while the bottom was fixed. As our FE models were linear elastic, the resulting reaction force was rescaled to the value of the force applied in the experimental setup; consequently, all the FE outcomes were rescaled accordingly. Each model was solved with ParFE [30] running at the Swiss National Supercomputing Centre (CSCS, Lugano, Switzerland) with 128 CPUs in less than 60 seconds.

Strain energy density (SED), defined as the increase in energy associated with the tissue deformation per unit volume, was used as the mechanical signal. For simplicity, the time dependence of the loading was ignored because, as demonstrated by Huiskes et al. [31], (re)modeling simulations under static loading capture the main features of virtual bone evolution under dynamic loading.

The SED distribution in CV6 was calculated by applying simulated compressive loads (4 N for non-loaded [32] and 9 N for loaded mice). While applying a force of 9 N to the loaded mice is representative of the mechanical load applied in the experiment, the load magnitude for the control and ovariectomized mice requires more assumptions. Christen et al. [32] presented a technique to back-calculate prevalent loading forces (magnitudes and directions) by using the bone microstructure as input, and testing various loading scenarios with respect to the most uniform SED distribution. According to this study, the trabecular architecture of the mice which were not loaded in the loading device suggested a prevalent force in the z-direction (i.e. parallel to the long axis of the vertebra) of 4 N with negligible shear/bending forces for x- and y-directions (i.e. perpendicular to the long axis of the vertebra).

Analysis of Local Mechanoregulation

The local mechanical environment was derived from micro-FE simulations of the baseline scan (Figure 2B). Here, and in computer models of bone (re)modeling in general, SED is used as a mathematical term to describe the (re)modeling stimulus phenomenologically. To quantify the relationship between the local SED and cellular activity on the bone surface, the bone formation or resorption sites determined by rigid registration were projected onto the surface of the baseline scan (considering a 6-neighborhood topology), resulting in three masks, representing three different clusters of formed (F), quiescent (Q) or resorbed (R) bone. Figure 2A shows a three-colored image with formation sites in yellow, quiescent sites in grey and resorption sites in violet. The mean SED value was calculated in each of these three masks and for each mouse, and as absolute SED values differ per mouse and over time due to differences in BV/TV, normalization to the mean SED of the quiescent surface was performed. The amount of formed, quiescent or resorbed voxels was calculated for each group and at each time point to gain more insight into the effect of the different treatments. Furthermore, to establish a quantitative description of the mechanoregulatory system and following the same strategy adopted to describe previous (re)modeling theories [2,17,22,33], the relation between increasing mechanical stimuli and consequent (re)modeling events was investigated. Experimental strain-related (re)modeling rules for the behavior of osteoblasts and osteoclasts as a function of the local mechanical stimulus were obtained by analyzing and comparing the frequency distributions of SED in the three clusters of formed, resorbed and quiescent bone. Thus, for each value of SED (binned at 1% step size of the maximum SED), the relative percentage of voxels being formed, quiescent and resorbed can be interpreted as the probability for a given (re)modeling event to occur. Furthermore, all SED values in this analysis were normalized by the maximum value observed in each animal to allow for a comparison among individual animals and treatments. Additionally, before computing the (re)modeling rules, it was assumed that each (re)modeling event has the same occurrence probability (i.e., formation, resorption and quiescent regions were virtually rescaled to have the same amount of voxels) to rule out the dependence on the imbalance between bone

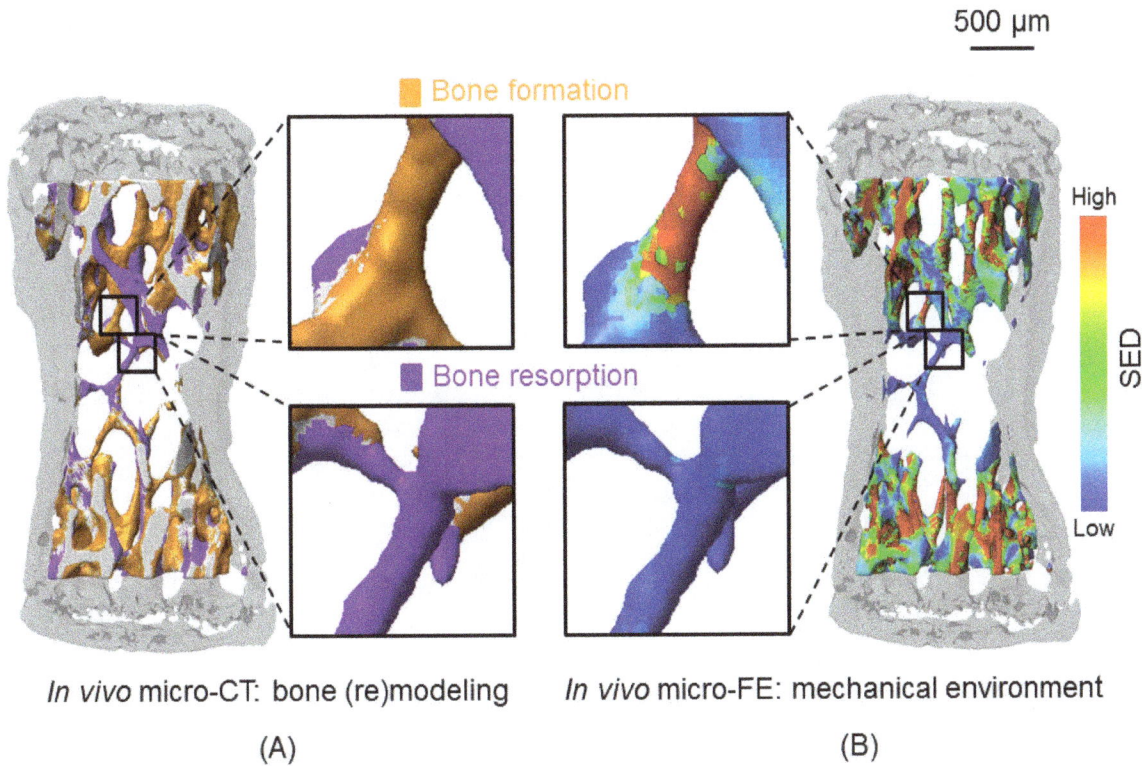

Figure 2. Comparison of local bone formation and resorption sites with the mechanical environment. (A) Three-dimensional trabecular bone formation and resorption sites measured with *in vivo* micro-CT over 4 weeks. The inset shows a magnified view of formation and resorption locations in individual trabeculae. (B) Corresponding SED computed with micro-FE in the basal scan. The same regions as in (A) are enlarged. A visual comparison reveals that high SED (red) matches with sites of bone formation (yellow), while low SED (blue) is found at locations of bone resorption (violet).

formation and bone resorption which may be due to bone growth, loading or OVX. Therefore, in mathematical terms, the (re)modeling rules are equivalent to the conditional probability of a (re)modeling event taking place within a given time interval. The (re)modeling probabilities were fitted by exponential functions using non-linear regression analysis.

As mechanoadaptation curves have typically been represented on a two-dimensional graph showing on the horizontal axis the mechanical stimulus and on the vertical axis the net bone response (i.e. net effect of (re)modeling), our data were furthermore converted into this format by subtracting the probability of resorption from the probability of formation at every binned SED value.

To ensure the effects seen were true and not due to measurement error, the reproducibility of formed/quiescent/resorbed surfaces was determined. To this purpose, the repeated *ex vivo* scans of the reproducibility study were superimposed onto the first scan repetition and the amount of erroneous formation/resorption voxels per bone volume or bone surface was calculated.

Statistics

A two-tailed paired (within animal) or unpaired (between groups) Student's t-test with Bonferroni-correction for multiple comparisons was performed after testing for equal variance of sample by the Kolmogorov-Smirnof test. Over time, two-tailed repeated measures ANOVA with Bonferroni-correction for multiple comparisons was used. All statistical tests were performed with R (R, Auckland, New Zealand, [34]). $p < 0.05$ was considered significant.

Results

Cyclic mechanical loading (CML, n = 9) of CV6 over four weeks of experiment led to a significant increase in the trabecular bone volume fraction (BV/TV) compared to CTR (n = 8) and over time (19.5%, repeated measures ANOVA, $p < 0.05$, Figure 3A, B). From week 0 to 1, mean SED at formation sites was 13.7±4.5% higher than mean SED at quiescent sites and, at the same time, mean SED at resorption sites was 22.8±3.4% lower than mean SED at quiescent sites (n = 9, Student's t-test with Bonferroni-correction, both $p < 0.0001$, Figure 3C). Table 1 contains the mean SED values in each cluster and all animal groups. In the remaining week intervals, a similar pattern was found, i.e. mean SED at formation sites was between 11.7%–16.1% higher ($p < 0.0001$) and at resorption sites between 23.8–26.4% lower ($p < 0.0001$) than mean SED at quiescent sites. The absolute values of SED in each cluster decreased slightly over time. This can be explained by the increasing BV/TV of the loaded animals. The percent difference to the mean SED of quiescence, however, is in the same range over all week intervals. These findings indicate that both osteoblastic and osteoclastic activities are controlled by local mechanical stimuli.

An analogous pattern with respect to mechanoregulation was observed in the animals of the CTR group (n = 8, repeated measures ANOVA with Bonferroni-correction, both $p < 0.05$, Figure 3D, Table 1). Again, the absolute values of mean SED reported in Table 1 differed as with respect to the loaded animals, since a loading force of 4 N instead of 9 N was assumed. The fact that the same pattern was found not only for loaded but also for

Table 1. Mean and standard deviation for absolute SED in formed (F), quiescent (Q) and resorbed (R) bone surfaces in all control and treatment groups.

Group	Week	F [kPa]	Q [kPa]	R [kPa]	|F–Q| in %	|R–Q| in %	p-value (F vs. Q)	p-value (R vs. Q)
CML	0–1	7.52±0.90	6.62±0.73	5.11±0.67	13.67	−22.83	1.5e-6	6.6e-5
	1–2	6.93±0.53	6.09±0.56	4.64±0.64	13.81	−23.76	5.4e-5	6.1e-7
	2–3	6.46±0.64	5.57±0.55	4.10±0.48	16.12	−26.41	2.6e-6	2.7e-7
	3–4	5.81±0.50	5.20±0.49	3.90±0.47	11.70	−25.00	4.1e-7	4.5e-7
CTR	0–1	1.60±0.30	1.40±0.27	1.17±0.24	13.87	−16.41	3.0e-5	0.00550
	1–2	1.55±0.31	1.37±0.26	1.11±0.22	12.64	−19.43	0.00090	0.00037
	2–3	1.47±0.30	1.31±0.27	0.99±0.20	12.00	−24.29	6.4e-5	0.00058
	3–4	1.41±0.30	1.27±0.25	0.97±0.19	10.96	−23.81	0.00161	0.00042
OVX	0–2	1.10±0.20	1.00±0.17	0.83±0.17	9.77	−17.24	0.00021	0.00307
	2–4	1.06±0.18	0.99±0.14	0.82±0.15	7.85	−16.95	0.010	6.1e-06
	4–6	1.22±0.24	1.13±0.18	0.99±0.17	8.36	−12.29	0.01979	0.00015
	6–8	1.69±0.37	1.40±0.27	1.29±0.20	20.48	−7.88	0.00023	0.02351
	8–10	2.00±0.36	1.64±0.26	1.35±0.19	22.15	−17.29	0.00015	0.00227
	10–12	1.96±0.28	1.67±0.24	1.38±0.16	17.46	−17.08	6.1e-05	0.0029
SHM	0–2	1.33±0.44	1.21±0.37	0.92±0.21	10.34	−23.57	0.013	0.014
	2–4	1.25±0.32	1.16±0.30	0.85±0.21	7.58	−27.35	0.00041	0.00057
	4–6	1.18±0.30	1.13±0.27	0.83±0.17	4.34	−27.13	n.s.	0.0012
	6–8	1.14±0.26	1.10±0.27	0.82±0.18	3.74	−26.07	0.01804	0.00095
	8–10	1.14±0.29	1.10±0.27	0.80±1.53	3.98	−27.32	n.s.	0.0016
	10–12	1.13±0.29	1.07±0.25	0.79±0.18	5.13	−26.76	0.04208	0.00032

Furthermore, the average difference from the mean SED in quiescent surfaces is given in %, as well as the p-values between F vs. Q and R vs. Q.

control mice, indicates that the same mechanism which allows bone to adapt to strong changes in the loading conditions also seems to control bone (re)modeling in daily activities.

Next, we investigated to what extent ovariectomy interferes with the local regulatory mechanism of bone (re)modeling. Following ovariectomy (OVX, n = 9), a significant loss of BV/TV caused by estrogen deficiency was measured after 12 weeks in the CV6 compared to the sham (SHM, n = 7) group and over time (31.7%; repeated measures ANOVA; p<0.0001; Figure 3E, F).

BV/TV in the SHM group, shown in Figure 3F, increased continuously and at week 4 was slightly higher than the percentage increase in BV/TV of the CTR group (Figure 3B). The absolute value of BV/TV did not differ significantly between CTR and SHM at 4 weeks. A reason for the small difference in BV/TV at week 4 may be that mice in the CTR group were anesthetized three times a week and scanned every week whereas the SHM operated animals were only anesthetized once during the operation and every two weeks afterwards for the measurements. BV/TV in SHM mice increased 16% after 12 weeks (Figure 3F). The increase in BV/TV in SHM mice is in line with the fact of continued growth in mice throughout their lifetime, with a stagnation in fast growth reported in the literature from 3 months (third lumbar vertebra of C57Bl/6 J mice) [35] to 20 months (seventh caudal vertebra in BALB/C mice) [36].

For OVX, mean SED values in the regions of bone formation (resorption) were significantly above (below) values at quiescent bone surfaces over all 2-week-time intervals (n = 9, repeated measures ANOVA with Bonferroni-correction, both p<0.05, Figure 3G). The same behavior characterized the SHM group (n = 7, repeated measures ANOVA with Bonferroni-correction;

p<0.01 and p<0.001, Figure 3H; Table 1). It is worth noticing that at week interval 4–6, the percent difference between quiescence and formation increased (Figure 3G) meaning that formation occurs in an even more targeted fashion at times of high bone loss. At the same time, the difference between resorption and quiescence became smaller, indicating that high bone loss occurred through less targeted bone resorption. When bone mass stabilized after 10 weeks, the differences in mean SED returned to their original levels (Fig. 3G, week 10–12 is similar to week 0–2). Comparing Figure 3G and 3H, the resorption in SHM mice occurred at lower SED values (percent difference at approximately −25%) than in ovariectomy (approximately −18%). On the other hand, formation in OVX seemed to be more targeted than in SHM (rising up to 21% in OVX, compared to less than 10% in SHM). Taken together, our results suggest that the local mechanical regulation mechanism is still active in the case of estrogen deficiency but less targeted during times of high bone loss.

Figure 4 shows the absolute probability for a (re)modeling event to occur at the bone surface, computed by counting the relative amounts of voxels in the formation, quiescent and resorption clusters. In both the loading and control group, more formed surfaces than resorbed surfaces were found. Compared to CTR, CML showed an increased bone mass by increasing the probability of bone formation and decreasing the probability of bone resorption (p<0.001, Figure 4A), indicating that a larger part of the surface was occupied by formation, and a smaller part by resorption sites. Figure 4A also points out that after four weeks of mechanical loading, bone adaptation has not been fully accomplished. However, it is known that the probability of formation and resorption will look similar between CML and CTR once the

Loading experiment

Ovariectomy experiment

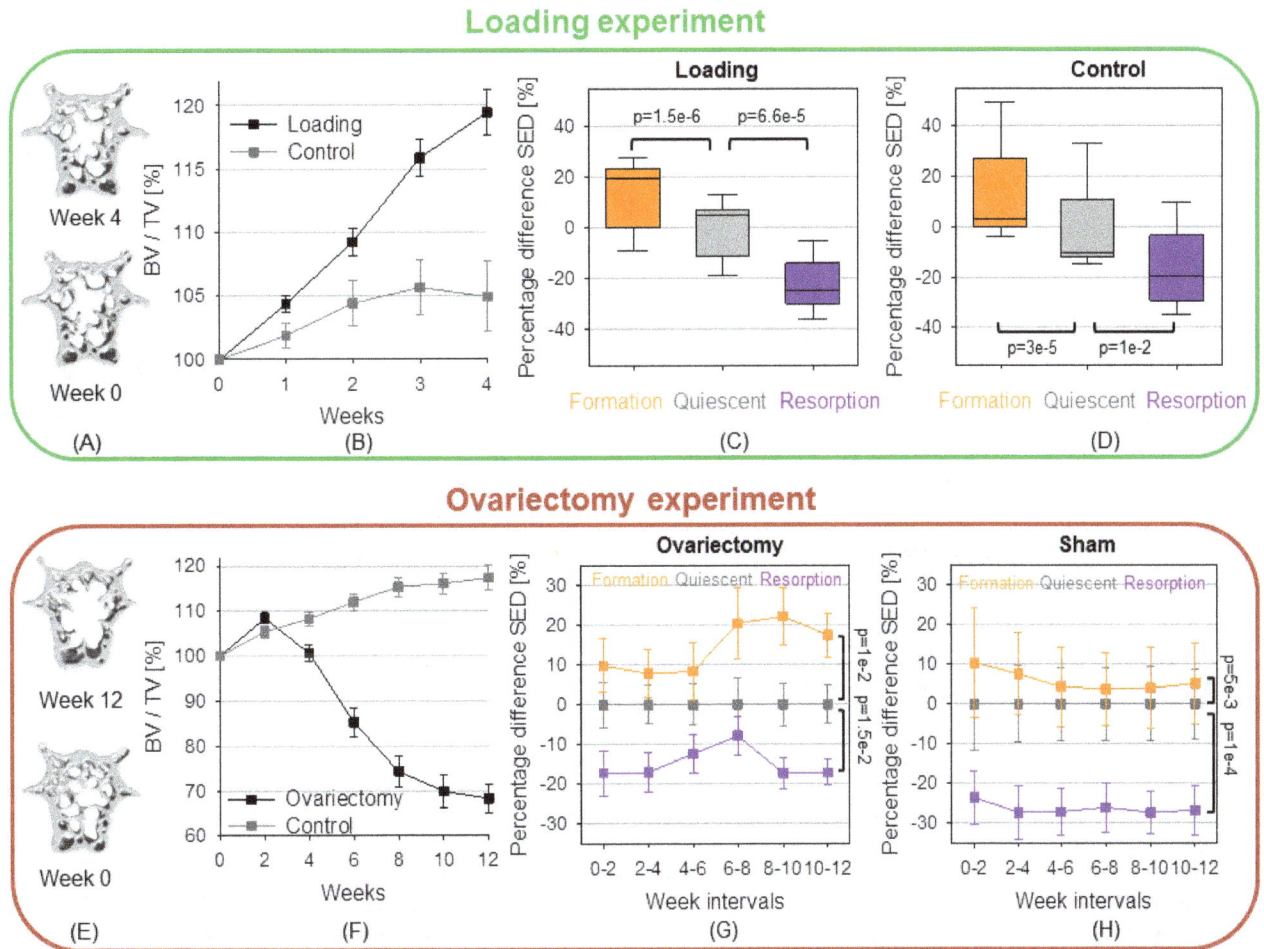

Figure 3. Evidence of local mechanical control for bone formation and resorption. (A) Sagittal sections through the diaphyseal CV6 at (bottom) the basal scan and (top) after 4 weeks of mechanical loading. (B) Percentage increase in trabecular BV/TV in the loading and control group. (C) Mean SED at formation, quiescent and resorption sites expressed as percent difference from the mean (over all animals) quiescent SED within the first time interval. Data are represented by boxplots, i.e. the inner box contains 50% of all data, the whisker bars denote the full range and the black line represents the median value (over all animals). (D) Mechanical regulation in the CTR group. (E) Sagittal sections through CV6 at (bottom) 0 and (top) 12 weeks after ovariectomy (OVX) (F) OVX causes a strong decrease in BV/TV compared to SHAM-operation (SHM). (G) Evidence of the mechanical regulation of the (re)modeling process in OVX. (H) Evidence of the mechanical regulation of the (re)modeling process in SHM. All data points in (F) to (H) are presented as mean ± standard error.

loaded bone is fully adapted to the new loading conditions, as indicated by recent data on formation and resorption surfaces [37]. In the ovariectomy group, the probability of bone resorption peaked at the third week interval whereas in the same time interval the probability of formation reached a minimum, indicating that more resorbed than formed bone surfaces can be found.

Figure 5A shows the conditional probability curves of bone formation, P_f, resorption, P_r, and quiescent, P_q, as a function of SED. The graph was created by averaging the individual probability curves of (re)modeling of the single animals over the first week interval. At low SED ($0 < SED/SED_{MAX} < 6\%$) it was more likely for bone to be resorbed, whereas in regions of higher SED ($SED/SED_{MAX} > 12\%$), the probability of bone formation became higher. Moreover, P_f and P_q were very close to each other for SED $\leq 12\%$; above this value, P_f constantly increased whereas P_q did not vary, suggesting that bone formation may be activated only above a given stimulus. The (re)modeling curves for the CTR group were similar but not as evident as the CML curve (Figure 5B). This indicates that a strong mechanical signal

enhanced bone cell response and made the mechanical control more "detectable". The (re)modeling probabilities in SHM and OVX animals, computed at the time point when OVX showed the highest bone loss (i.e. week 4–6 in Figure 3F), showed a similar pattern as the controls of the loading experiment, thus confirming the presence of mechanical control also after ovariectomy (Figure 5C and D).

The (re)modeling rules for bone formation and resorption were fitted by exponential functions (nonlinear regression analysis, $R^2 = 0.99$, $p < 0.001$, Figure 5) where the fitting parameters can be directly linked to the functioning of the mechanosensory system. The fitting functions and resulting parameters for formation and resorption in all groups can be found in Table 2.

The offset-parameter $y0$ indicates that bone resorption and formation show a certain probability to occur over the full range of mechanical stimuli which would be a fingerprint of non-targeted (re)modeling [38]. Regarding the loaded animals, the probability for bone formation and resorption independent of mechanical stimuli (i.e. offset parameter) was 25.6% and 21.0%, respectively.

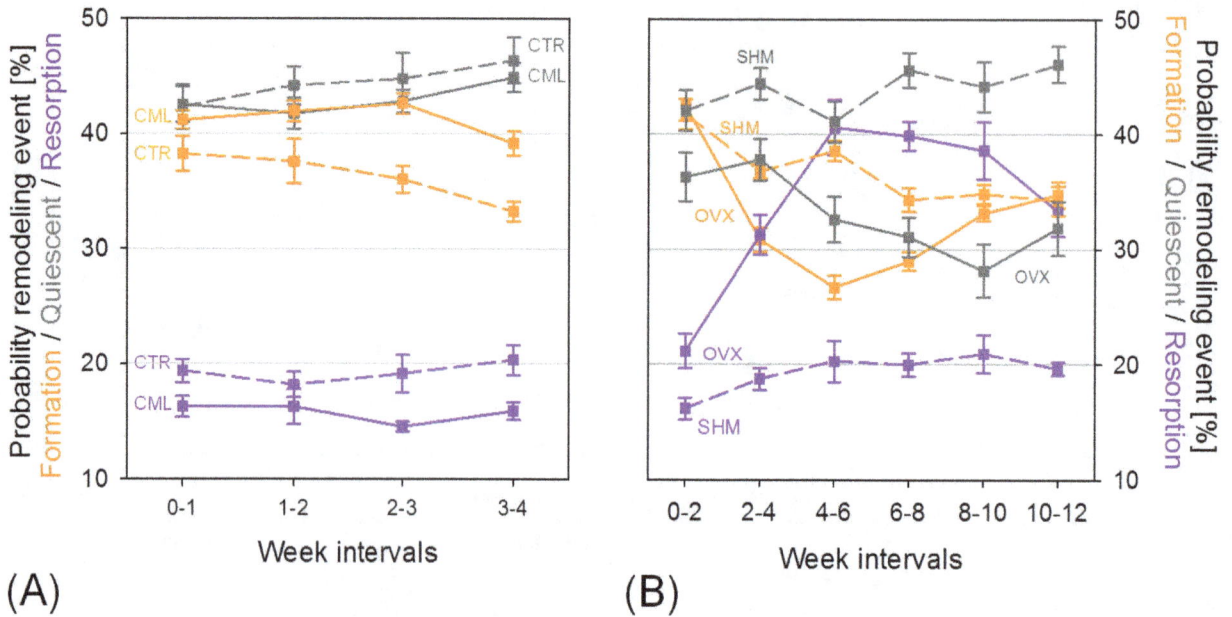

(A)   (B)

Figure 4. Absolute (re)modeling probabilities. (A) Mechanical loading experiment: The probability for bone formation on the bone surface is higher than for bone resorption in both the CML and the CTR group, with the effects in the CML group being more pronounced than in the control group. (B) Ovariectomy experiment: The probability of bone resorption on the bone surface of OVX increases in the first four weeks after ovariectomy, with decreasing probability for bone formation at the same time. After 4 weeks, the effects are reverted.

The amount of non-targeted (re)modeling in the estrogen depleted condition was 29.5% (formation) and 25.2% (resorption), compared to 29.1% non-targeted formation and 15.2% non-targeted resorption in the SHM group.

The parameter a in front of the exponent quantifies the mechanical sensitivity of the system. For OVX, a reduced mechanical sensitivity may be assumed as the extent of variation of P_r and P_f was approximately a factor 2 and a factor 4 smaller compared to the SHM animals. Lastly, the coefficient b inside the exponent can be interpreted as the amount of mechanical control in the (re)modeling process. In all animal groups, the coefficients for bone resorption were more than a factor 2 higher than those for bone formation, suggesting that the sites of bone resorption are more strictly mechanically controlled than the sites of bone formation. This means that small decreases in SED will have large effects on the resorption response while large increases in SED will only have moderate changes in the formation response. Moreover, OVX showed less mechanical control both for formation (-55%) and resorption (-39%) with respect to CML. Figure 6 shows the (re)modeling probabilities relative to all time intervals for CML and OVX. It can be seen that for mechanical loading, the (re)modeling curves lay very close together and were all located within the one standard deviation range of the first week interval (except for the last measurement, Figure 6A). In ovariectomy, the (re)modeling curves presented some variation over the different time intervals, most probably due to the acute phase of bone loss (Figure 6B).

Figure 7 shows the net bone response as a function of the local mechanical stimulus. The graphs agree with the current knowledge that net bone loss occurs at low mechanical stimulus and net bone gain in the zone of high mechanical stimulus. However, a disagreement with the current understanding of bone (re)modeling is that our data give no evidence of a lazy zone. This is true for ovariectomized, loaded and control animals. Moreover, for the loading group, the relation between SED/SED$_{MAX}$ and net bone

response is well described by an exponential growth function to a maximum value ($R^2 = 0.99$) whereas for CTR, SHM and OVX more linear fitting functions could also be used. Again, all exponential fitting function parameters can be found in Table 2. Furthermore, the point where CML crosses zero was shifted to the left in respect to the other groups (see inset of Figure 7) which means that only values of lower SED are leading to resorption in loaded mice. The zero-crossing-points for the CTR, SHM and OVX group were similar. For all groups, approximately the lower third of normalized SED values resulted in net bone loss, and the upper two-thirds in net bone gain, supporting the assumption that resorption is controlled more strongly than formation. The curves for CTR and SHM differ slightly in their shape. A possible reason could be that there was a time interval of two weeks for SHM, and a time interval of one week for CTR, where higher noise from erroneous voxels can be expected.

In respect to the experimental error, the amount of erroneously "formed" or "resorbed" voxels was determined from the repeated *ex vivo* measurements. The amount of erroneously formed bone volume (surface) was $5.3\% \pm 2.1\%$ ($17.8\% \pm 5.1\%$) and the erroneously resorbed bone volume (surface) was $4.9\% \pm 1.8\%$ ($16.5\% \pm 4.3\%$). It is assumed that this error mainly comes from the partial volumes effect and the registration error.

Discussion

The objective of the current study was to investigate the relationship between the local mechanical environment and the bone (re)modeling process at the tissue level. Our results demonstrate that the loads applied globally control local bone formation and resorption at the tissue level. On top of confirming the well-known assumption stating that "bone is formed where needed and resorbed where not needed" [6], we quantified the relation between SED and bone adaptation using novel evaluation methods.

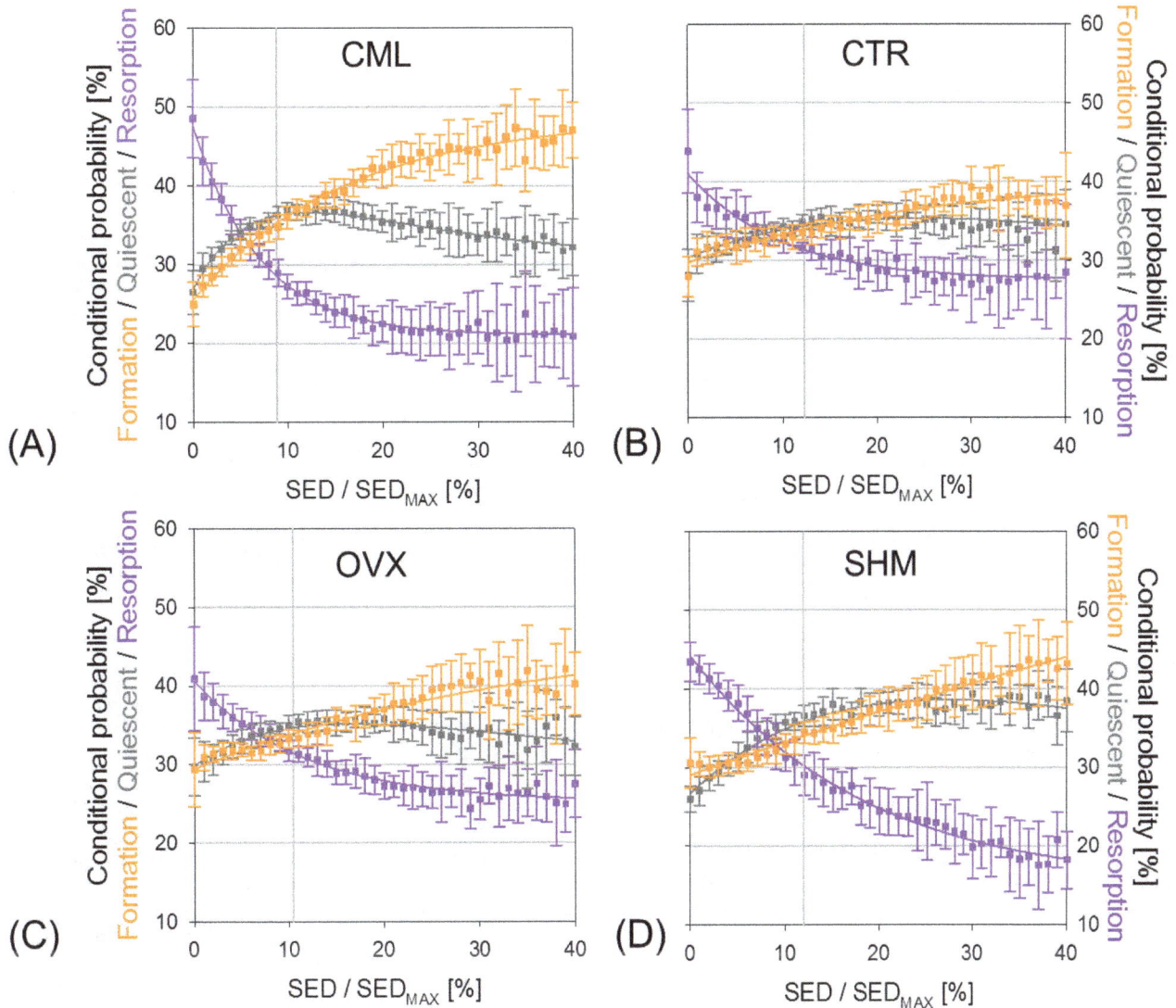

Figure 5. Conditional (re)modeling probabilities connecting the mechanical environment (SED) with the (re)modeling events. The mechanical regulation of bone (re)modeling is characterized by probability functions describing the so-called (re)modeling rules. When computing the (re)modeling rules, it is assumed that each (re)modeling event has the same probability of occurring to rule out the dependence on the time interval or the imbalance between bone formation and bone resorption (which may be due to bone growth, loading or OVX as shown in Figure 3). The normalized SED is truncated at 40% due to the very small number of voxels above this threshold (less than 1% of the total surface voxels). The plots show the data points (mean ± standard deviation) as well as the exponential fitting functions for bone formation, resorption and quiescence in all four experimental groups: (A) Mechanical loading group, (B) control group, (C) ovariectomy group, (D) sham-operated group.

The theory of bone gain in sites of high mechanical stimuli and loss in sites of low stimuli has existed for more than 100 years [6], and the theory of local control of this process for more than 20 years [2]. Several mechanobiological experiments have been designed to connect the loading at a single skeletal location, such as the ulna or the caudal vertebra, with the subsequent changes in bone mass, as well as in the cortical and trabecular architecture [11,12,24,39,40]. For instance, cortical bone responds to sustained loading mainly by increasing periosteal bone formation at the highly strained sites [41]. This, in turn, increases the resistance to bending and torsional loading [5]. Trabecular bone also shows an adaptive response to increased mechanical loading in terms of changes in trabecular bone volume and morphometry [1,5,11]. For instance, Sugiyama et al. [42] and Ellman et al. [43] suggested a linear relationship between mechanical signal and

bone mass changes by applying various loads (also unloading leading to bone loss) to different groups of mice. It is well accepted that unloading causes bone loss in both cortical and trabecular bone [5,41–43]. Nevertheless, experimental evidence that low mechanical stimuli control bone resorption at a single site is lacking, mostly because the characterization of the spatial locations of bone resorption by histomorphometry has proved more challenging in the past than for bone formation.

The novelties of our approach are i) formation and resorption can be analyzed in 3D in a longitudinal fashion, ii) they can be analyzed separately from each other and iii) they can be linked to the mechanical signal in a single animal and at a single site. This approach allows us to determine exponential formation vs. SED and resorption vs. SED relationships with very high confidence ($R^2 = 0.91–0.99$).

Table 2. Coefficients of the (re)modeling curves for formation, resorption and the net bone response.

Coefficient	CML (week 0–1)	CTR (week 0–1)	OVX (week 4–6)	SHM (week 4–6)
Formation: $F = y0 + a*(1 - \exp(-b*SED/SED_{MAX}))$				
y0	25.567	29.735	29.462	29.068
a	22.949	10.933	17.881	62.696
b	0.062	0.040	0.0275	0.0069
R^2	0.987	0.912	0.927	0.985
Resorption: $R = y0 + a*(\exp(-b*SED/SED_{MAX}))$				
y0	20.973	27.551	25.231	15.154
a	26.643	13.542	15.433	29.123
b	0.142	0.104	0.087	0.055
R^2	0.992	0.916	0.975	0.990
Net bone response: $N = y0 + a*(1 - \exp(-b*SED/SED_{MAX}))$				
y0	−21.458	−11.095	−10.796	−14.762
a	47.405	22.766	29.379	53.438
b	0.101	0.075	0.057	0.035
R^2	0.990	0.923	0.965	0.992

Fitting functions and coefficients of the (re)modeling curves for formation, resorption and the net bone response in all animal groups.

Our results agree with Sugiyama et al. [42] and Ellman et al. [43] that no lazy zone exists but in contrast to their suggestions, our findings favor an exponential relationship between (re)modeling and increasing SED. A reason for this disagreement might be that in previous studies it was not possible to investigate formation and resorption as two separate processes. The current data provide evidence about the local response of bone tissue to mechanical signals which is due to a coordinated action of several bone forming and resorbing cells. However, a downside of our technique may be that information on single cell behavior is still not accessible due to current limitations in image resolution.

Defining a relationship between local SED and the following bone response can also be helpful for computer simulations of bone (re)modeling. *In silico* modeling has, in the past decades, been considered a valuable tool for testing various assumptions on how the local mechanical environment can control bone formation and resorption [17,20–22]. Huiskes and colleagues [17,21], for example, proposed bone resorption to be independent of the mechanical signal and bone formation to take place linearly and only once the mechanical stimulus has exceeded a certain threshold. Adachi et al. [20,44] simulated the net effect of bone formation and resorption by an apparent movement of the trabecular surface, driven by local stress gradients, i.e. if the local stress value is higher/lower than its direct neighborhood. Weinkamer and coworkers [22,45] described bone formation and resorption as two separate stochastic processes and suggested the existence of an activation barrier for the mechanical stimulus above which bone formation is switched on. Though such computational studies all provided possible realistic scenarios for the regulation of trabecular bone (re)modeling in response to loading, they were not able to exclude any of the competing theories. Our experimental findings, in particular the (re)modeling rules, may be used as an input for future simulation models.

In our study, both bone forming and resorbing cell types seem to respond to the local mechanical stimulus. Here we used SED to describe the mechanical environment; however, it should be noted that SED is just one possible mathematical description of the deformation stated and must not be necessarily considered as "the" mechanical signal sensed by the cells. When converting the SED values into effective strain (which is a scalar value summarizing the strain tensor), the mean value for formation, quiescence and resorption in the loading case amounted to 1008, 945, 690 µstrain, respectively. With this, the SED values for bone formation reported here correspond to effective strains which are a bit lower than formation thresholds reported in the literature (between 1050 and 3074 µstrain) [46–48]. However it should be noted that it is difficult to directly compare these values with values in literature in absolute terms as different skeletal locations (e.g. tibia, femur or vertebra) are considered. Moreover, experimental measures of strain are conducted via strain gauges attached to the cortical shell (not the trabecular compartment) which typically account for only one component of the strain tensor [47,48].

Here, we showed that both the probabilities of bone formation and resorption can be described by exponential functions of SED, with small increases in SED evoking large decreases in the probability of bone resorption. For this reason, we conclude that bone resorption is more strictly controlled than bone formation. This is mechanically sound as it is more critical when bone is resorbed at the "wrong" place than when bone is formed at the "wrong" place.

Furthermore, our results revealed a considerable portion of bone (re)modeling (15.2%–29.7%) which was not related to SED (Table 2 and Figure 4). In principle, the experimental error may influence the amount of non-targeted (re)modeling; nevertheless, assuming that the wrong voxels are uniformly distributed in each SED interval and considering the normalization that each (re)modeling event has the same probability to occur, such error mainly affects the absolute probability of a (re)modeling event taking place and not the conditional probability describing the dependence with SED (i.e. (re)modeling rules). Also, we found that with estrogen depletion, the resorptive portion not related to SED increases considerably. This finding is in line with the increasing evidence that estrogen receptors are involved in the bone cell response to strain and that the removal of estrogen may influence the availability of estrogen receptors which in turn could reduce the "accuracy" of targeted bone resorption [49–51]. On the other

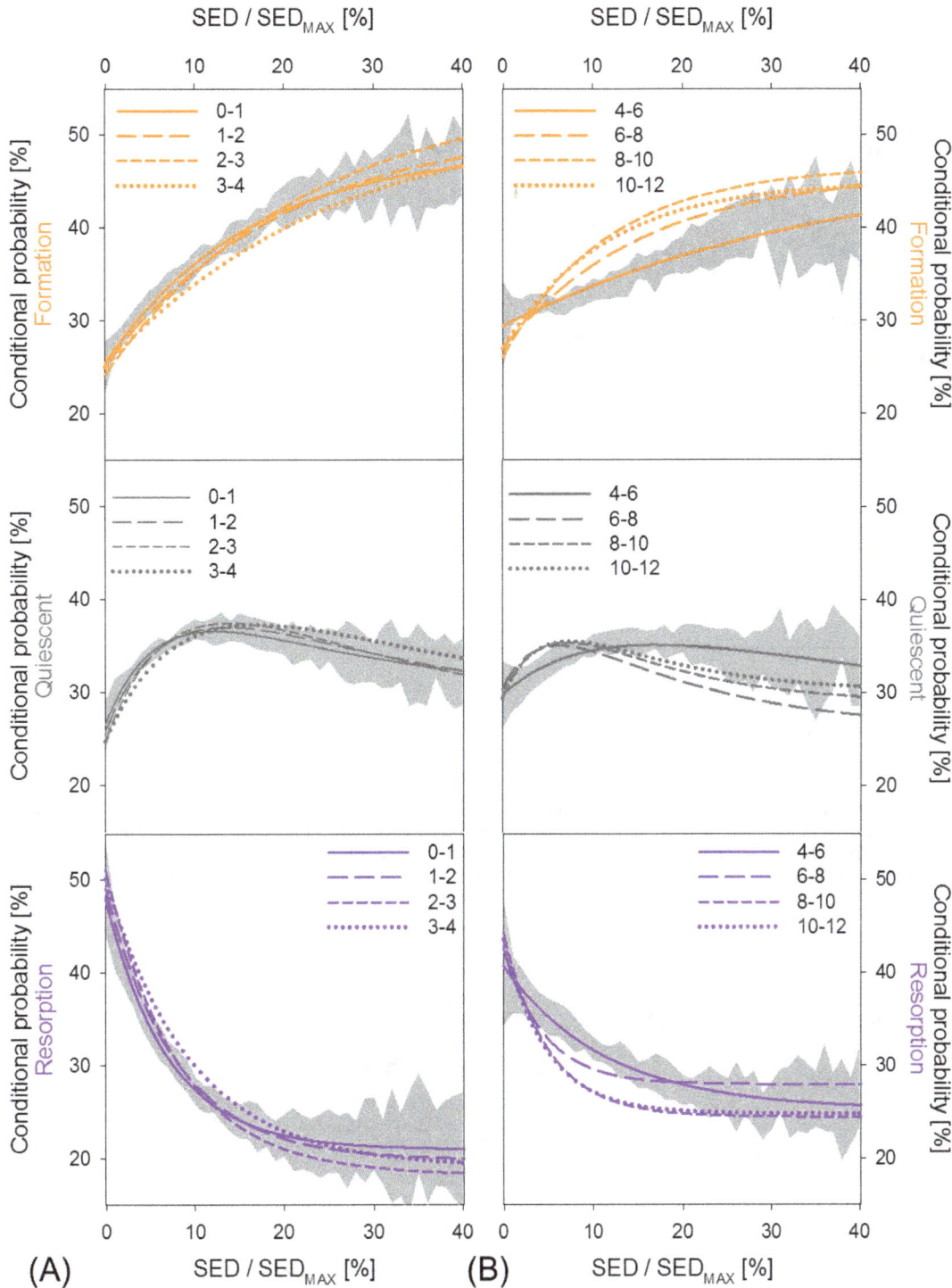

Figure 6. (Re)modeling probabilities of all week intervals in the CML and the OVX group. (A) For CML, the curves lie within or very close to the one standard deviation range of the first week interval, which is denoted as the grey area. (B) In the OVX group, the interval of highest bone loss (week 4–6) is characterized by an almost linear slope of bone formation. With time, this slope changes into an exponential curve. For OVX, also the curve of bone resorption becomes more exponential over time.

hand, this outcome also means that about 80% of all (re)modeling can be linked to mechanical demands.

In conclusion, we showed quantitatively how bone (re)modeling is regulated at the local tissue level, and how the regulation changes with different mechanical stimuli/estrogen deficiency. We

believe that our findings on the mechanoregulation of trabecular bone are of major importance for a better understanding of bone diseases and the development of potential pharmacological therapies. Furthermore, we anticipate that our results close a large gap between *in silico* and experimental research of bone

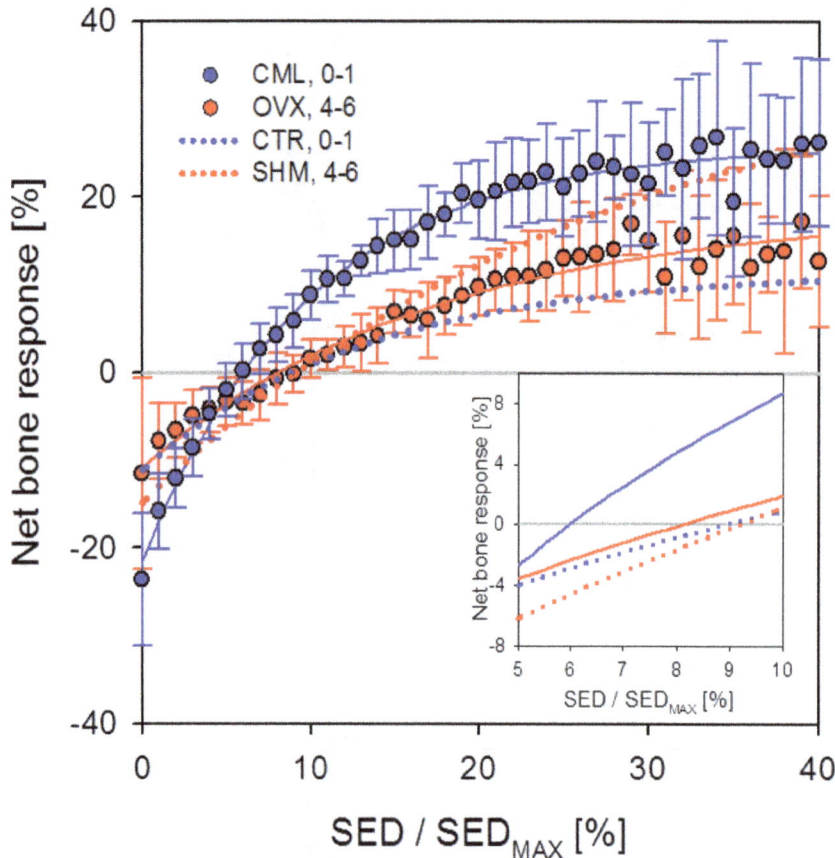

Figure 7. Experimental net bone response. Experimental net bone response versus normalized SED (i.e. SED/SED$_{max}$). The curves for the CML and CTR group are shown in blue and for OVX and SHM in red. The dotted lines denote the fitting functions of the control and sham groups. It can be noted that in the CML group, the exponential behavior is more pronounced than in CTR, SHM and OVX. Also, the zero crossing point of the loading group is shifted towards the left as indicated in the magnified view in the inset. Compared to OVX, this means that in a highly controlled loading environment only lower SED values are used for bone resorption. No indication of a "lazy zone" around the x-axis can be found.

(re)modeling. With this insight, the research of *in silico* bone (re)modeling can further advance and develop so that better long-term predictions of bone change and the outcomes of pharmacological intervention can be gained.

Acknowledgments

The authors would like to thank Dr. Andreas J. Wirth for his introduction to and support in micro-finite element analysis and Kathleen Koch for her assistance with the animal experiments. Furthermore, the authors acknowledge Dr. Richard Weinkamer for helpful discussions and Catherine Palmer for proofreading of the manuscript. Computational time was granted by the Swiss National Supercomputing Centre (CSCS, Lugano, Switzerland).

Author Contributions

Conceived and designed the experiments: GK RM. Performed the experiments: FML GK. Analyzed the data: FAS DR FML. Contributed reagents/materials/analysis tools: FAS DR FML GK DC DJW. Wrote the paper: FAS DR.

References

1. Fratzl P, Weinkamer R (2007) Nature's hierarchical materials. Progress in Materials Science 52: 1263–1334.
2. Frost HM (1987) The mechanostat: a proposed pathogenic mechanism of osteoporoses and the bone mass effects of mechanical and nonmechanical agents. Bone Miner 2: 73–85.
3. Turner CH, Akhter MP, Raab DM, Kimmel DB, Recker RR (1991) A noninvasive, in vivo model for studying strain adaptive bone modeling. Bone 12: 73–79.
4. Frost HM (2003) Bone's mechanostat: a 2003 update. Anat Rec A Discov Mol Cell Evol Biol 275: 1081–1101.
5. Robling AG, Castillo AB, Turner CH (2006) Biomechanical and molecular regulation of bone remodeling. Annu Rev Biomed Eng 8: 455–498.
6. Wolff J (1892) Das Gesetz der Transformation der Knochen (The Law of Bone Remodelling). Berlin, Germany: Springer-Verlag.
7. Roux W (1881) Der Kampf der Theile im Organismus. Leipzig: Wilhelm Engelman.
8. Lanyon LE (1973) Analysis of surface bone strain in the calcaneus of sheep during normal locomotion. Strain analysis of the calcaneus. J Biomech 6: 41–49.
9. Lanyon LE (1992) Control of Bone Architecture by Functional Load Bearing. Journal of Bone and Mineral Research 7: S369–S375.
10. Bikle DD, Sakata T, Halloran BP (2003) The impact of skeletal unloading on bone formation. Gravit Space Biol Bull 16: 45–54.
11. De Souza RL, Matsuura M, Eckstein F, Rawlinson SCF, Lanyon LE, et al. (2005) Non-invasive axial loading of mouse tibiae increases cortical bone formation and modifies trabecular organization: A new model to study cortical and cancellous compartments in a single loaded element. Bone 37: 810–818.
12. Webster DJ, Morley PL, van Lenthe GH, Müller R (2008) A novel in vivo mouse model for mechanically stimulated bone adaptation - a combined experimental and computational validation study. Comput Methods Biomech Biomed Engin 11: 435–441.

13. Webster D, Wasserman E, Ehrbar M, Weber F, Bab I, et al. (2010) Mechanical loading of mouse caudal vertebrae increases trabecular and cortical bone mass-dependence on dose and genotype. Biomech Model Mechanobiol 9: 737–747.

14. Basso N, Jia YH, Bellows CG, Heersche JNM (2005) The effect of reloading on bone volume, osteoblast number, and osteoprogenitor characteristics: Studies in hind limb unloaded rats. Bone 37: 370–378.

15. Fritton JC, Myers ER, Wright TM, van der Meulen MCH (2005) Loading induces site-specific increases in mineral content assessed by microcomputed tomography of the mouse tibia. Bone 36: 1030–1038.

16. Lambers FM, Schulte FA, Kuhn G, Webster DJ, Müller R (2011) Mouse tail vertebrae adapt to cyclic mechanical loading by increasing bone formation rate and decreasing bone resorption rate as shown by time-lapsed in vivo imaging of dynamic bone morphometry. Bone 49: 1340–1350.

17. Huiskes R, Ruimerman R, van Lenthe GH, Janssen JD (2000) Effects of mechanical forces on maintenance and adaptation of form in trabecular bone. Nature 405: 704–706.

18. Parfitt AM, Drezner MK, Glorieux FH, Kanis JA, Malluche H, et al. (1987) Bone histomorphometry: standardization of nomenclature, symbols, and units. Report of the ASBMR Histomorphometry Nomenclature Committee. J Bone Miner Res 2: 595–610.

19. Kim CH, Takai E, Zhou H, von Stechow D, Müller R, et al. (2003) Trabecular bone response to mechanical and parathyroid hormone stimulation: the role of mechanical microenvironment. J Bone Miner Res 18: 2116–2125.

20. Adachi T, Tsubota K, Tomita Y, Hollister SJ (2001) Trabecular surface remodeling simulation for cancellous bone using microstructural voxel finite element models. J Biomech Eng 123: 403–409.

21. Ruimerman R, Hilbers P, van Rietbergen B, Huiskes R (2005) A theoretical framework for strain-related trabecular bone maintenance and adaptation. J Biomech 38: 931–941.

22. Dunlop JW, Hartmann MA, Brechet YJ, Fratzl P, Weinkamer R (2009) New suggestions for the mechanical control of bone remodeling. Calcif Tissue Int 85: 45–54.

23. Schulte FA, Zwahlen A, Lambers FM, Kuhn G, Ruffoni D, et al. (2013) Strain-adaptive in silico modeling of bone adaptation - A computer simulation validated by in vivo micro-computed tomography data. Bone 52: 485–492.

24. Chambers TJ, Evans M, Gardner TN, Turner-Smith A, Chow JW (1993) Induction of bone formation in rat tail vertebrae by mechanical loading. Bone Miner 20: 167–178.

25. Schulte FA, Lambers FM, Kuhn G, Müller R (2011) In vivo micro-computed tomography allows direct three-dimensional quantification of both bone formation and bone resorption parameters using time-lapsed imaging. Bone 48: 433–442.

26. Schulte FA, Lambers FM, Mueller TL, Stauber M, Müller R (2012) Image interpolation allows accurate quantitative bone morphometry in registered micro-computed tomography scans. Comput Methods Biomech Biomed Engin.

27. Lambers FM, Kuhn G, Schulte FA, Koch K, Müller R (2011) Longitudinal assessment of in vivo bone dynamics in a mouse tail model of postmenopausal osteoporosis. Calcif Tissue Int 90: 108–119.

28. Thevenaz P, Ruttimann UE, Unser M (1998) A pyramid approach to subpixel registration based on intensity. IEEE Trans Image Process 7: 27–41.

29. Bouxsein ML, Boyd SK, Christiansen BA, Guldberg RE, Jepsen KJ, et al. (2010) Guidelines for assessment of bone microstructure in rodents using micro-computed tomography. J Bone Miner Res 25: 1468–1486.

30. Arbenz P, van Lenthe GH, Mennel U, Müller R, Sala M (2008) A scalable multi-level preconditioner for matrix-free μ-finite element analysis of human bone structures. International Journal for Numerical Methods in Engineering 73: 927–947.

31. Huiskes R (2000) If bone is the answer, then what is the question? J Anat 197 (Pt 2): 145–156.

32. Christen P, van Rietbergen B, Lambers FM, Müller R, Ito K (2012) Bone morphology allows estimation of loading history in a murine model of bone adaptation. Biomech Model Mechanobiol 11: 483–492.

33. Beaupre GS, Orr TE, Carter DR (1990) An approach for time-dependent bone modeling and remodeling–theoretical development. J Orthop Res 8: 651–661.

34. R Development Core Team (2007) R: A language and environment for statistical computing. Vienna, Austria: R Foundation for Statistical Computing.

35. Buie HR, Moore CP, Boyd SK (2008) Postpubertal architectural developmental patterns differ between the L3 vertebra and proximal tibia in three inbred strains of mice. J Bone Miner Res 23: 2048–2059.

36. Willinghamm MD, Brodt MD, Lee KL, Stephens AL, Ye J, et al. (2010) Age-related changes in bone structure and strength in female and male BALB/c mice. Calcif Tissue Int 86: 470–483.

37. Lambers FM, Koch K, Kuhn G, Weigt C, Schulte F, et al. (2012) Bone structure and strength adapt to long-term cyclic overloading in an in vivo mouse model. J Biomech 45: S89.

38. Parfitt AM (2002) Targeted and nontargeted bone remodeling: relationship to basic multicellular unit origination and progression. Bone 30: 5–7.

39. Gross TS, Srinivasan S, Liu CC, Clemens TL, Bain SD (2002) Noninvasive loading of the murine tibia: an in vivo model for the study of mechanotransduction. J Bone Miner Res 17: 493–501.

40. Rubin CT, Lanyon LE (1985) Regulation of bone mass by mechanical strain magnitude. Calcif Tissue Int 37: 411–417.

41. Gross TS, Edwards JL, McLeod KJ, Rubin CT (1997) Strain gradients correlate with sites of periosteal bone formation. J Bone Miner Res 12: 982–988.

42. Sugiyama T, Meakin LB, Browne WJ, Galea GL, Price JS, et al. (2012) Bones' adaptive response to mechanical loading is essentially linear between the low strains associated with disuse and the high strains associated with the lamellar/woven bone transition. J Bone Miner Res 27: 1784–1793.

43. Ellman R, Spatz J, Cloutier A, Palme R, Christiansen B, et al. (2012) Partial reductions in mechanical loading yield proportional changes in bone density, bone architecture, and muscle mass. J Bone Miner Res.

44. Tsubota K, Adachi T, Tomita Y (2002) Functional adaptation of cancellous bone in human proximal femur predicted by trabecular surface remodeling simulation toward uniform stress state. J Biomech 35: 1541–1551.

45. Weinkamer R, Hartmann MA, Brechet Y, Fratzl P (2004) Stochastic lattice model for bone remodeling and aging. Phys Rev Lett 93: 228102.

46. Turner CH, Forwood MR, Rho JY, Yoshikawa T (1994) Mechanical loading thresholds for lamellar and woven bone formation. J Bone Miner Res 9: 87–97.

47. Hsieh YF, Robling AG, Ambrosius WT, Burr DB, Turner CH (2001) Mechanical loading of diaphyseal bone in vivo: the strain threshold for an osteogenic response varies with location. J Bone Miner Res 16: 2291–2297.

48. Cullen DM, Smith RT, Akhter MP (2000) Time course for bone formation with long-term external mechanical loading. J Appl Physiol 88: 1943–1948.

49. Bonnelye E, Kung V, Laplace C, Galson DL, Aubin JE (2002) Estrogen receptor-related receptor alpha impinges on the estrogen axis in bone: potential function in osteoporosis. Endocrinology 143: 3658–3670.

50. Zaman G, Jessop HL, Muzylak M, De Souza RL, Pitsillides AA, et al. (2006) Osteocytes use estrogen receptor alpha to respond to strain but their ERalpha content is regulated by estrogen. J Bone Miner Res 21: 1297–1306.

51. Windahl SH, Saxon L, Borjesson AE, Lagerquist MK, Frenkel B, et al. (2013) Estrogen receptor-alpha is required for the osteogenic response to mechanical loading in a ligand-independent manner involving its activation function 1 but not 2. J Bone Miner Res 28: 291–301.

6

Sex Specific Association of Physical Activity on Proximal Femur BMD in 9 to 10 Year-Old Children

Graça Cardadeiro[1], Fátima Baptista[1]*, Rui Ornelas[2], Kathleen F. Janz[3], Luís B. Sardinha[1]

1 Exercise and Health Laboratory, Faculty of Human Movement, Technical University of Lisbon, Lisbon, Portugal, 2 Centre of Social Sciences, Department of Physical Education and Sport, University of Madeira, Funchal, Portugal, 3 Department of Health and Human Physiology, Department of Epidemiology, University of Iowa, Iowa City, Iowa, United States of America

Abstract

The results of physical activity (PA) intervention studies suggest that adaptation to mechanical loading at the femoral neck (FN) is weaker in girls than in boys. Less is known about gender differences associated with non-targeted PA levels at the FN or other clinically relevant regions of the proximal femur. Understanding sex-specific relationships between proximal femur sensitivity and mechanical loading during non-targeted PA is critical to planning appropriate public health interventions. We examined sex-specific associations between non-target PA and bone mineral density (BMD) of three sub-regions of the proximal femur in pre- and early-pubertal boys and girls. BMD at the FN, trochanter (TR) and intertrochanter (IT) regions, and lean mass of the whole body were assessed using dual-energy x-ray absorptiometry in 161 girls (age: 9.7 ± 0.3 yrs) and 164 boys (age: 9.7 ± 0.3 yrs). PA was measured using accelerometry. Multiple linear regression analyses (adjusted for body height, total lean mass and pubertal status) revealed that vigorous PA explained 3–5% of the variability in BMD at all three sub-regions in boys. In girls, vigorous PA explained 4% of the variability in IT BMD and 6% in TR BMD. PA did not contribute to the variance in FN BMD in girls. An additional 10 minutes per day of vigorous PA would be expected to result in a ~1% higher FN, TR, and IT BMD in boys ($p<0.05$) and a ~2% higher IT and TR BMD in girls. In conclusion, vigorous PA can be expected to contribute positively to bone health outcomes for boys and girls. However, the association of vigorous PA to sub-regions of the proximal femur varies by sex, such that girls associations are heterogeneous and the lowest at the FN, but stronger at the TR and the IT, when compared to boys.

Editor: Xing-Ming Shi, Georgia Health Sciences University, United States of America

Funding: Funded by Portuguese Science and Technology Foundation (http://www.fct.pt/index.phtml.en); grants SFRH/BD/38671/2007 and PTDC/DES/115607/2009. The funders had no role in study design, data collection and analysis, decision to publish, or preparation of the manuscript.

Competing Interests: The authors have declared that no competing interests exist.

* E-mail: fbaptista@fmh.utl.pt

Introduction

Mechanical loading by impact or muscle forces is a contributing factor to skeletal health throughout the life course; however, mechanical loading is particularly important during the transition period from childhood to adolescence. This may be due to the efficient response of bone to loading during middle childhood (elementary school years), since the magnitude of bone accrual associated with mechanical loading is reported to be greater when compared to early childhood and adulthood [1–6]. In addition, the evidence shows that the amount and intensity of PA levels are highest during middle childhood compared to other time points across childhood and adolescence [7]. The amount and intensity of PA during middle childhood is important since PA levels dramatically decrease during adolescence [7–9]. Consequently, a timely intervention in children's activity habits when bone appears to be most responsive to activity's effect and PA is more easily accepted could be an important public health strategy for optimal bone development.

Studies conducted with pre- and early-pubertal children have shown augmented bone mineral accrual in several skeletal sites in both girls [10–17] and boys [15–17] after high-impact mechanical loading (running and jumping) interventions. The effect of high-impact loading exercise on bone has been shown to be site specific,

with accrual occurring at weight bearing sites such as lumbar spine [11] and proximal femur [18–22]. The loads imposed during targeted exercise are likely to represent the "best case scenario" and might not generalize to the spontaneous PA choices of children.

Population-based observational studies provide important information on the relationship of bone accrual to the type and amount of PA children voluntarily choose to do. Twenty-five to forty minutes of vigorous PA per day has been suggested as a minimum daily dose for optimal bone growth [23–26] but the relationship between bone mineral accrual and PA during the growing years has not been thoroughly examined. Of the studies that exist, there is a lack of consensus on whether sex moderates the association between mechanical loading and bone accrual at the proximal femur [18,19,27,28]. However, it has been suggested that the proximal femur of boys' is more sensitive to mechanical loading than girls' [27]. This idea was not derived from work specifically powered to examine sex differences [23,27–31]. Furthermore, most studies reporting a lower responsiveness of PA in girls (when compared to boys) analysed only the femoral neck and not other sub-regions of the proximal femur (i.e. trochanter and intertrochanter).

A potential explanation for sex-specific bone response at the proximal femur is the lower intensity mechanical loading on the skeleton by muscle or impact forces in girls due to their lower lean body mass and less weight-bearing PA. However, given sex differences in body composition [32,33], in PA [7,8], maturation timings [34], morphology and gait kinematic parameters [35], it is plausible that bone response sensitivity to PA differs among the proximal femur sub-regions for both boys and girls. This possibility has not been examined and is clinically relevant because women suffer more fragility fractures in old age [36–38] and have a higher incidence of fractures at the femoral neck region compared to men who have higher incidence of trochanteric femoral fractures [39,40]. Therefore, the purpose of this cross-sectional study was to investigate the sex-specific association between non-targeted PA and bone mineral density (BMD) of sub-regions of the proximal femur in boys and girls.

Methods

Sample

This cross sectional cohort study included 325 pre and early pubertal subjects (Tanner stage 1 and 2), aged 9–10 years (164 boys and 161 girls) living in the island of Madeira (Portugal) and drawn from the European Youth Heart Study (EYHS). Selection procedures and methods are described in detail elsewhere [24]. None of the subjects were taking any medication affecting bone and none reported a history of bone fracture in lower limbs. The research protocol was in accordance with the Helsinki Declaration. Parents or legal guardians provided written informed consent and the study was approved by the Ethics Committee attached to the scientific board of the Faculty of Human Movement.

Physical Activity. PA was assessed with a uniaxial accelerometer (model WAM 6471, Manufacturing Technology Incorpo-

rated, Fort Walton Beach, FL), over two weekdays and two weekend days. The subjects were asked to wear the accelerometer all day except during water activities, in a representative week of their normal activity, and the procedure was repeated in all cases in which any abnormal event was reported. MAHUffe software (www.mrc-epid.cam.ac.uk) was used to analyze and process activity data. Outcome variables were time (minute/day) spent in light, moderate and vigorous intensity of PA. The intensity of PA was defined according the counts per minute (cpm) as follows: light intensity from 501 to 1999 cpm; moderate intensity from 2000 to 2999 cpm; and vigorous intensity over 2999 cpm. All of the activity data were averaged over the 4-day period and subjects who failed to provide a minimum of 3 days of ≥600 minutes of accelerometer data were excluded. PA procedures are detailed in previous report [41].

Clinic Measures

Standing height (to the nearest millimetre) was measured on a stadiometer (Secca 770, Hamburg, Germany) without shoes. Body weight (kilograms), total fat (kilograms), and lean mass without bone (kilograms) were determined from a total body scan by dual x-ray absorptiometry (DXA) (QDR-1500, high-speed performance mode, software 5.7) (Hologic, Waltham, MA; pencil beam, software 5.73). Sexual maturity was assessed using self-report and Tanner's 5-stage scale for breast development in girls and pubic hair in boys. Children were stratified as prepubertal (Tanner stage 1) or having started puberty (Tanner stage 2) [42].

BMD from three proximal femur sub-regions, i.e., femoral neck, trochanter and intertrochanter, were measured with DXA (QDR-1500, high-speed performance mode, software 4.76). Quality

Table 1. Characteristics of participants as mean±standard deviation.

	Girls (n = 161)	Boys (n = 164)	p*
Age, y	9.7±0.3	9.7±0.3	0.780
Tanner Stage (1/2), %	40/60	4/96	<0.001
Body Weight, kg	34.2±9.0	34.1±7.8	0.960[a,b]
Body Height, cm	137.2±0.1	137.0±0.1	0.813
Body Fat, kg	10.2±5.7	8.2±5.7	0.002[a]
Body Lean Mass, kg	23.1±3.2	25.1±2.9	<0.001
Calcium Intake, mg/d	1020±424	1048±407	0.553
Light PA, min/d	296±47	278±49	0.001
Moderate PA, min/d	142±47	169±55	<0.001
Vigorous PA, min/d	18±14.3	30±21	<0.001[a,b]
Moderate and Vigorous PA, min/d	159±56	198±70	<0.001
Total PA, min/d	456±77	476±90	0.030
PA Average Intensity, count/min/d	586±189	732±273	<0.001
Proximal Femur BMD, g/cm²	0.690±0.07	0.753±0.08	<0.001
Femoral Neck BMD, g/cm²	0.656±0.06	0.722±0.07	<0.001[b]
Trochanter BMD, g/cm²	0.544±0.06	0.591±0.07	<0.001[a,b]
Intertrochanter BMD, g/cm²	0.762±0.08	0.820±0.09	<0.001[a,b]
Femoral Neck BMD/Trochanter BMD	1.21±0.08	1.22±0.08	0.095
Femoral Neck BMD/Intertrochanter BMD	0.86±0.05	0.88±0.05	0.001[a]
Trochanter BMD/Intertrochanter BMD	0.72±0.05	0.72±0.05	0.678

*Student's t-test comparing boys to girls was performed when both variables have normal distribution with the same variance. In cases of no normality or no homogeneity of variances, Mann-Whitney nonparametric test was used. [a]Girl's variable without normal distribution; [b]Boy's variable without normal distribution. PA - physical activity BMD – bone mineral density.

Table 2. Standardized regression coefficients (β), level of significance (p) and coefficient of determination (R²) for proximal femur sub-region models, adjusted for sex, Tanner stage, body height and body lean mass, with data for boys and girls pooled together.

	Predictor Variables	β	p	R²
FN BMD	BLM	0.308	<0.001	0.204
	Sex	0.239	<0.001	0.100
	Body Height	−0.064	0.311	–
	Tanner Stage	0.110	0.038	0.009
	Vigorous PA	0.191	<0.001	0.033
	MVPA	−0.004	0.955	–
	Moderate PA	−0.003	0.955	–
TR BMD	BLM	0.372	<0.001	0.209
	Sex	0.170	0.001	0.024
	Body Height	−0.096	0.179	–
	Tanner Stage	0.077	0.156	–
	Vigorous PA	0.227	0.001	0.073
	MVPA	0.039	0.604	–
	Moderate PA	0.031	0.604	–
IT BMD	BLM	0.426	<0.001	0.242
	Sex	0.129	0.012	0.014
	Body Height	0.030	0.634	–
	Tanner Stage	0.082	0.030	–
	Vigorous PA	0.182	<0.001	0.046
	MVPA	0.019	0.798	–
	Moderate PA	0.015	0.798	–

BMD – bone mineral density; FN BMD - femoral neck BMD; TR BMD - trochanter BMD; IT BMD - intertrochanter BMD; PA - physical activity; MVPA – moderate-through-vigorous PA; BLM – body lean mass.

assurance tests were performed each morning. Precision errors were estimated from 2 measurements in 14 subjects [43]. The coefficients of variation of femoral neck, trochanter and inter-trochanter BMD ranged from 1.2% to 1.5%. From the DXA scans, BMD ratios among sub-regions were calculated as indicators of BMD homogeneity in the proximal femur as follows [40]: FNTR is the ratio between femoral neck BMD and trochanter BMD (FNBMD/TRBMD); FNIT is the ratio between femoral neck BMD and intertrochanter BMD (FNBMD/ITBMD); TRIT is the ratio between trochanter BMD and intertrochanter BMD (TRBMD/ITBMD).

Calcium intake were calculated from a semi-quantitative Food Frequency Questionnaire assessing regular intake of a wide set of typical Portuguese foods using the Food Processor SQL software (ESHA Research, Salem OR).

Statistical Analysis

Data were analysed using the SPSS statistical software package (Version 18.0 for Windows; SPSS, Chicago, IL, USA). Distribution properties of all variables were examined using the Kolmogorov-Smirnov test and appropriate measures of central tendency and variability were selected. Differences between groups (girls and boys) were analysed by Independent-samples T-tests in case of normality and equally of variance and Mann-Whitney

nonparametric test otherwise. The Chi-square test of homogeneity was used to compare Tanner stage distributions across sexes. Stepwise regressions were used to analyse associations between PA variables, (i.e., time spent at moderate intensity, vigorous intensity, and moderate-through-vigorous PA - MVPA) and BMD or BMD ratios of proximal femur sub-regions, adjusted for Tanner stage, body height, and total body lean mass. Data were initially analysed with boys and girls pooled together to test the significance of sex as predictor variable and then separately for each sex. All the assumptions for the linear regression analysis were verified (normality and linearity of the residuals, multicollinearity and homoscedasticity). The hypothetical effect of PA intensity on the BMD of the proximal femur sub-regions was estimated by regression analyses (enter approach, p<0.05) calculating the percentage of BMD change associated with an additional 10 minutes per day of PA at two different intensities (MVPA and vigorous PA) by multiplying the unstandardized regression coefficients by 10 and dividing by the correspondent BMD mean at each sub-region of proximal femur. Significance level was set at p<0.05.

Results

The characteristics of the children are presented in Table 1. There were no differences in age, body weight and height between boys and girls. However, lean mass was higher and fat mass was lower in boys, who were also more active than girls. Boys spent more time in moderate and vigorous PA than girls, whereas girls spent more time in light activities. The proportion of participants in early puberty (Tanner 2) was higher in boys than in girls (40% of the girls and 4% of the boys were in the Tanner stage 1). The BMD of the proximal femur and of its three sub-regions was higher in boys than in girls, but statistically significant sex differences in BMDs ratios were not found, with the exception of the FNIT, with boys revealing a higher ratio than girls.

Associations between PA and BMD of the proximal femur sub-regions were analysed using multiple regression models, first with boys and girls polled together (Table 2) and after separated (Table 3). Among PA variables (time spent at moderate, MVPA, and vigorous PA), vigorous PA was the one with the highest contribution to the R squared of each model (3–7%, p<0.001) (Table 2). None of the other two PA variables showed additional explanatory power once vigorous PA had entered the model. In the same table, body lean mass explained 20–24% of variance in all BMD models (p<0.001) while Tanner stage was responsible for ~1% variability of femoral neck BMD (p = 0.038). In all the three regression models ran with boys' and girls' data pooled together (Table 2), sex turned out to be a significant predictor variable, giving empirical ground for subsequent separated data treatment.

Three other similar models were run for the proximal femur BMD ratios with boys and girls together but none of them complied with the assumptions for regression analysis, having therefore been rejected. Conversely, the models of proximal femur BMD ratios were added to the initial three ones when data was considered separately for boys and girls to analyse associations between BMD and PA variables adjusted for Tanner stage, lean body mass and body height (Table 3). Among all the PA intensity variables examined, vigorous PA was the best predictor: it explained ~3–5% of the BMD variance (p<0.05) in boy's femoral neck, trochanter and intertrochanter. However, none of the variation of BMD ratios in boys was predicted by PA intensity variables. In girls, vigorous PA was also the best PA predictor variable explaining 6% of the trochanter BMD and 4% of the intertrochanter BMD variance; a 3% variation in the FNTR and

Table 3. Standardized regression coefficients (β), level of significance (p) and coefficient of determination (R^2) for proximal femur sub-region models, adjusted for Tanner stage, body height, and body lean mass, with data for boys and girls treated separately.

	Predictor Variables	Girls			Predictor Variables	Boys		
		β	p	R^2		β	p	R^2
FN BMD	BLM	0.483	<0.001	0.233	BLM	0.277	<0.001	0.071
	Body Height	−0.075	0.462	–	Body Height	−0.047	0.639	–
	Tanner Stage	0.100	0.194	–	Tanner Stage	0.100	0.179	–
	Vigorous PA	0.135	0.060	–	Vigorous PA	0.225	0.003	0.051
	MVPA	0.125	0.072	–	MVPA	−0.092	0.440	–
	Moderate PA	0.109	0.116	–	Moderate PA	−0.073	0.440	–
TR BMD	BLM	0.511	<0.001	0.306	BLM	0.238	0.002	0.052
	Body Height	−0.139	0.138	–	Body Height	−0.085	0.407	–
	Tanner Stage	0.062	0.236	–	Tanner Stage	−0.007	0.925	–
	Vigorous PA	0.241	<0.001	0.056	Vigorous PA	0.214	0.005	0.046
	MVPA	0.076	0.400	–	MVPA	−0.016	0.893	–
	Moderate PA	0.064	0.400	–	Moderate PA	−0.013	0.893	–
IT BMD	BLM	0.514	<0.001	0.305	BLM	0.324	<0.001	10.1
	Body Height	−0.017	0.861	–	Body Height	0.061	0.543	–
	Tanner Stage	0.134	0.060	–	Tanner Stage	−0.058	0.434	–
	Vigorous PA	0.213	0.001	0.044	Vigorous PA	0.159	0.033	0.025
	MVPA	0.102	0.263	–	MVPA	−0.063	0.499	–
	Moderate PA	0.086	0.263	–	Moderate PA	−0.081	0.499	–
FNTR BMD	BLM	−0.178	0.024	0.043	BLM	0.027	0.731	–
	Body Height	0.079	0.483	–	Body Height	0.061	0.429	–
	Tanner Stage	0.034	0.690	–	Tanner Stage	0.168	0.031	0.028
	Vigorous PA	−0.172	0.029	0.029	Vigorous PA	−0.033	0.673	–
	MVPA	−0.038	0.730	–	MVPA	−0.072	0.357	–
	Moderate PA	−0.032	0.730	–	Moderate PA	−0.078	0.314	–
					–	–	–	–
FNIT BMD	BLM	−0.250	0.001	0.068	BLM	−0.009	0.932	–
	Body Height	−0.119	0.288	–	Body Height	−0.208	0.006	0.043
	Tanner Stage	0.093	0.272	–	Tanner Stage	0.260	0.001	0.064
	Vigorous PA	−0.095	0.385	–	Vigorous PA	0.057	0.445	–
	MVPA	0.184	0.016	0.029	MVPA	0.041	0.595	–
	Moderate PA	0.313	0.385	–	Moderate PA	0.029	0.706	–
TRIT BMD	–	–	–	–	BLM	0.029	0.781	–
					Body Height	0.223	0.004	0.005
					Tanner Stage	0.076	0.325	–
					Vigorous PA	0.068	0.381	–
					MVPA	0.082	0.296	–
					Moderate PA	0.078	0.324	–

PA - physical activity; BMD – bone mineral density; FN BMD - femoral neck BMD; TR BMD - trochanter BMD; IT BMD - intertrochanter BMD; FNTR – BMD ratio of femoral neck for trochanter; FNIT – BMD ratio of femoral neck for intertrochanter; TRIT- BMD ratio of trochanter for intertrochanter; BLM – body lean mass.

FNIT was also associated with vigorous and MVPA, respectively. In girls, with exception of femoral neck BMD, PA (vigorous and MVPA) explained 3–6% of all BMD variances. Unlike boys, in girls there was a negative association between PA variables and FNTR and FNIT (p<0.05).

Table 3 also shows that body lean mass was a significant predictor variable in all girls and boy's models for the proximal femur's regional BMDs, except in the girls TRIT BMD model.

Table 4 presents regression models using the three PA intensity variables mostly widely used in the literature, moderate PA, vigorous PA and MVPA. In our analysis, there was a higher absolute effect (estimated by unstandardized regression coefficients) of one minute per day of vigorous PA on the BMD than one minute per day of MVPA or moderate PA (only in girls). The effect of PA was not homogeneous for all proximal femur sub-regions and was dissimilar between boys and girls. For example, in girls the hypothetical BMD increase associated with an additional

Table 4. Effects of 10 minutes per day of additional physical activity on femoral neck, trochanter, and intertrochanter BMD, adjusted for Tanner stage, body height, and body lean mass.

	β					
	Girls			Boys		
	FN BMD	TR BMD	IT BMD	FN BMD	TR BMD	IT BMD
Moderate PA	ns	0.00022	0.00031	ns	ns	ns
		(p = 0.008)	(p = 0.008)			
ModVig PA	0.00014	0.00022	0.00030	0.00015	0.00016	0.00013
	(p = 0.072)	(p = 0.002)	(p = 0.002)	(p = 0.048)	(p = 0.025)	(p = 0.186)
Vigorous PA	0.00059	0.00101	0.00127	0.00075	0.00056	0.00066
	(p = 0.056)	(p<0.001)	(p = 0.001)	(p = 0.003)	(p = 0.016)	(p = 0.033)
	Δ BMD (%) associated to Δ 10 min/day of physical activity					
	Girls			Boys		
	FN BMD	TR BMD	IT BMD	FN BMD	TR BMD	IT BMD
Moderate PA	ns	0.4	0.4			
ModVig PA	ns	0.4	0.4	0.2	0.3	ns
Vigorous PA	ns	1.9	1.7	1.0	1.1	0.8

PA – physical activity; BMD – bone mineral density; FN - femoral neck; TR - trochanter; IT – intertrochanter; ns – non-significant regression coefficient.

10 min/day of PA was comparable (~2%) for trochanter and intertrochanter regions (with no effect on femoral neck). The effect for boys was lower (~1%) but the response was similar among all three sub-regions of proximal femur.

Discussion

PA showed a positive contribution to the BMD variation of the three sub-regions of the proximal femur in boys but in girls PA did not help to explain femoral neck BMD variance. For the same duration of PA, the regression coefficients of more intense PA (vigorous PA) were always higher than those corresponding to a less intense PA (MVPA) in boys and girls. The extrapolation of our results suggest a ~2% higher BMD in the trochanter and intertrochanter regions in girls at the studied age range if an additional 10 minutes per day of vigorous PA is achieved. In boys, the corresponding gain is a ~1% higher femoral neck, trochanter and intertrochanteric BMDs.

The higher regression coefficients for PA of highest intensity – vigorous PA –compared to lower levels of intensity – MVPA or moderate PA – when regional BMDs of femoral neck, trochanter and intertrochanter are in stake underline the relevance of the PA intensity to bone mineral accrual during the studied pediatric years. The PA threshold under which the effects on bone mass could be modest has been proposed [24–26]. Given that boys are usually more active than girls [7–9], this could partially explain BMD differences between sexes at proximal femur sub-regions. However this difference seems not be homogeneous among sub-regions. Our results are consistent with studies that reported a positive response of girls' proximal femur BMD (or bone mineral content - BMC) to PA but also with studies that revealed a response of femoral neck BMD or BMC to PA only in boys. Particularly, our site specific response of girls' proximal femur in

the trochanteric region is in line with the Iowa Bone Development Study which reported 5% and 14% more BMC at the total body and trochanteric region in the most active pre- and early pubertal boys and girls, when compared to inactive peers [29]. Similar effects regarding skeletal regions were also found by Stear et al., who reported greater BMC accrual at the trochanter (4.8%) than in the whole body (0.8%) or lumbar spine (1.9%) in 144 adolescent girls enrolled in a 45-min exercise-to-music classes programme, three times per week, after 15.5-month [44]. Witzke et al. reached analogous findings at the trochanter BMC in adolescent girls using a plyometric jump training programme with no significant differences for the femoral neck, spine or whole body BMC [45]. McKay et al. who examined the effect of an 8-month school-based jumping programme in pre and early pubescent girls, found that the intervention group showed a significantly greater change in trochanter BMD than the control group [16,18]. Additionally, increments (4.3%) for femoral neck BMC of 8 to 12 years old boys (compared to controls) were reported after 2 years of a high-impact circuit intervention [19]. These observations contradict the idea that girls proximal femur is not responsive to PA, although, notably none of these studies reported a positive effect of PA on girl's femoral neck.

The positive associations that we found between PA and the BMD of the three proximal femur sub-regions in boys and only at the trochanter and at the intertrochanter region in girls is similar with the results of those studies that suggested no effect of PA in girls' proximal femur, whose analyses were focused in the femoral neck region [6,10,11]. The exception, seems to be the study conducted by Petit et al. [17] that showed significant gains in the BMD at the intertrochanter (1.7%) and at the femoral neck (2.6%) region in early pubertal girls (Tanner stages 2 and 3) when compared to controls after a 10-minute jumping programme, 3 times per week during 7 months.

The analysis of all sub-regions of proximal femur in both sexes was a distinctive aspect of our study that provided a more comprehensive examination of bone's response to PA. Compared to boys, girls showed inferior BMD in the different sub-regions of the proximal femur, which is not new. However, we observed a lower or a tendency to a lower BMD in the femoral neck relative to other sub-regions (FNIT, girls: 0.86 vs. boys: 0.88, p = 0.001; FNTR, girls: 1.21 vs. boys: 1.22, p = 0.095), i.e. the proximal femur sub-region where we did not find any positive association with MVPA or vigorous PA in girls. Our study showed that the pattern of proximal femur responsiveness to PA was more homogeneous in boys, when compared to girls. Conversely, in girls, there were negative relationships between PA and FNTR and FNIT, suggesting a heterogeneous responsiveness favouring the trochanteric and intertrochanteric sub-regions of the proximal femur. If our response pattern findings were generalizable, it is not surprising that researchers using the neck region to represent the entire proximal femur suggest that boys' femur is more responsive to mechanical loading than girls' at this age.

In addition to the well-known limitations of DXA technology in the assessment of bone, our study may have an additional limitation due to the self-report of children's maturity status. The sample selection was based on chronological age (9–10 yrs) and not to assure a representative maturational profile. At these ages, girls usually demonstrate a more advanced biological maturity than boys which did not happen in our study. However we conducted our analyses with and without adjustments for maturational status obtaining similar results (data not shown).

In conclusion, although a large proportion of bone mineralization is attributable to growth during late childhood, MVPA and especially vigorous PA can have an additional osteogenic effect in the proximal femur. The effect is not homogeneous throughout all bone regions in girls. Our work was not designed to detect why the femoral neck appears non responsive to PA in girls. Further research designed to simultaneously compare site-specific bone responses to PA in boys and girls at a wider age range and level of sexual maturity is needed. In addition, sample sizes should be large enough to allow investigators to test interactions among PA and hip biomechanical factors (as opposed to systemic factors as nutritional, hormonal or sun exposure factors) that can affect differently the BMD of specific regions of proximal femur. Our study shows a region-specific bone response to vigorous PA in pre and early pubertal girls and boys. More active girls have greater BMD in the trochanter and intertrochanter while more active boys have greater BMD in all sub-regions of the proximal femur.

Author Contributions

Conceived and designed the experiments: FB LBS GC. Performed the experiments: RO GC FB. Analyzed the data: GC RO LBS FB KFJ. Contributed reagents/materials/analysis tools: RO LBS FB GC. Wrote the paper: GC FB LBS KFJ.

References

1. Bass SL (2000) The prepubertal years: a uniquely opportune stage of growth when the skeleton is most responsive to exercise? Sports Med 30: 73–8.
2. Khan K, McKay H, Haapasalo H, Bennell K, Forwood M, et al. (2000) Does childhood and adolescence provide a unique opportunity for exercise to strengthen the skeleton? J Sci Med Sports 3: 150–64.
3. Hughes J, Novotny S, Wetzsteon R, Petit M (2007) Lessons learned from school based skeletal loading intervention trials: putting research into practice. Med Sport Sci 51: 137–58.
4. MacKelvie K, Khan K, McKay H (2002) Is there a critical period for bone response to weight-bearing exercise in children and adolescents? A systematic review. Br J Sports Med 36: 250–7.
5. Bass S, Saxon L, Daly R, Turner C, Robling A, et al. (2002) The effect of mechanical loading on the size and shape of bone in pre-, peri-, and postpubertal girls: a study in tennis players. J Bone Miner Res 17: 2274–80.
6. Romann M, Zahner L, Schindler C, Puder J, Kraenzlin M, et al. (2011) Effect of a general school-based physical activity intervention on bone mineral content and density: A cluster-randomized controlled trial. Bone 48: 792–7.
7. Troiano R, Berrigani D, Dodd K, Masse L, Tilert T, et al. (2008) Physical activity in the United States measured by accelerometer. Med Sci Sports Exerc 40: 181–8.
8. Baptista F, Santos DA, Silva AM, Mota J, Santos R, et al. (2012) Prevalence of the Portuguese population attaining sufficient physical activity. Med Sci in Sports Exerc 44: 466–73.
9. Nader PR, Bradley RH, Houts RM, McRitchie SI, OBrien M (2008) Moderate-to-vigorous physical activity from ages 9 to 15 years. JAMA 300: 295–305.
10. Morris F, Naughton G, Gibbs J, Carlson J, Wark J (1997) Prospective ten month exercise intervention in premenarchal girls: positive effects on bone and lean mass. J Bone Miner Res 12: 1453–62.
11. Heinonen A, Sievanen H, Kannus P, Oja P, Pasanen M, et al. (2000) High-impact exercise and bones of growing girls: A 9-month controlled trial. Osteoporos Int 11: 1010–7.
12. MacKelvie K, McKay H, Khan K, Crocker P (2001) A school-based loading intervention augments bone mineral accrual in early pubertal girls. J Pediatr 139: 501–8.
13. MacKelvie K, Khan K, Petit M, Janssen P, McKay H, et al. (2003) A school-based exercise intervention elicits substantial bone health benefits: a 2- year randomised controlled trial in girls. Pediatrics 112: 447–52.
14. Linden C, Ahlborg H, Besjakov J, Gardsell P, Karlsson M (2006) A school curriculum-based exercise program increases bone mineral accrual and bone size in prepubertal girls: two-year data from the Pediatric Osteoporosis Prevention (POP) study. J Bone Miner Res 21: 829–35.
15. Fuchs R, Bauer J, Snow C (2001) Jumping improves hip and lumbar spine bone mass in prepubescent children: A randomized controlled trial. J Bone Miner Res 16: 148–56.
16. McKay H, Petit M, Schutz R, Prior J, Barr S, et al. (2000) Augmented trochanteric BMD after modified physical education classes: a randomized school based exercise intervention study in prepubescent and early pubescent children. J Pediatr 136: 156–62.
17. Petit M, McKay H, MacKelvie K, Heinonen A, Khan K, et al. (2002) A randomised school-based jumping intervention confers site and maturity specific benefits on one structural properties in girls: a hip structural analysis study. J Bone Miner Res 17: 363–72.
18. McKay H, MacLean L, Petit M, MacKelvie-O'Brien K, Janssen P, et al. (2005) Bounce at the Bell': a novel program of short bouts of exercise improves proximal femur bone mass in early pubertal children. Br J Sports Med 39: 521–6.
19. MacKelvie K, Petit M, Khan K, Beck T, McKay H (2004) Bone mass and structure are enhanced following a 2-year randomised controlled trial of exercise in prepubertal boys. Bone 34: 755–64.
20. Courteix D, Jaffre C, Lespessailles E, Benhamou L (2005) Cumulative effects of calcium supplementation and physical activity on bone accretion in premenarchal children: a double-blind randomised placebo-controlled trial. Int J Sports Med 26: 332–8.
21. Van Langendonck L, Claessens A, Vlietinck R, Derom C, Beunen G (2003) Influence of weight-bearing exercises on bone acquisition in prepubertal monozygotic female twins: a randomized controlled prospective study. Calcif Tissue Int 72: 666–74.
22. Gunter K, Baxter-Jones A, Mirwald R, Almstedt H, Fuchs R, et al. (2008) Impact exercise increases BMC during growth: An 8-year longitudinal study. J Bone Miner Res 23: 986–93.
23. Baptista F, Barrigas C, Vieira F, Santa-Clara H, Mil-Homens P, et al. (2011) The role of lean body mass and physical activity in bone health in children. J Bone Miner Metab 30: 100–8.
24. Sardinha LB, Baptista F, Ekelund U (2008) Objectively measured physical activity and bone strength in 9 year old boys and girls. Pediatrics 122: 728–36.
25. Janz K, Burns T, Torner J, Levy S, Paulos R, et al. (2001) Physical activity and bone measures in young children: the Iowa Bone Development Study. Pediatrics 107: 1387–93.
26. Gracia-Marco L, Moreno L, Ortega F, León F, Sioen I, et al. (2011) Levels of physical activity that predict optimal bone mass in adolescents: the HELENA study Am J Prev Med 40: 599–607.
27. Kriemler S, Zahner L, Puder J, Braun-Fahrländer C, Schindler C, et al. (2008) Weight-bearing bones are more sensitive to physical exercise in boys than in girls during pre- and early puberty: a cross-sectional study. Osteoporos Int 19: 1749–58.
28. Sundberg M, Gardsell P, Johnell O, Karlsson M, Ornstein E, et al. (2001) Peripubertal moderate exercise increases bone mass in boys but not in girls: a population-based intervention study. Osteoporos Int 12: 230–8.
29. Janz K, Gilmore J, Burns T, Levy S, Torner J, et al. (2006) Physical activity augments bone mineral accrual in young children: the Iowa Bone Development study. J Pediatr 148: 793–9.

30. Jones G, Dwyer T (1998) Bone mass in prepubertal children: gender differences and the role of physical activity and sunlight exposure. J Clin Endocrinol Metab 83: 4274–9.

31. Cardadeiro G, Baptista F, Zymbal V, Rodrigues L, Sardinha L (2010) Ward's area location, Physical Activity and body composition in 8 and 9 years old boys and girls. J Bone Miner Res 25: 1–10.

32. Kelly T, Wilson K, Heymsfield S (2009) Dual Energy X-Ray Absorptiometry Body Composition. Reference Values from NHANES 4: 1–8.

33. Garnett S, Hogler W, Blades B, Baur L, Peat J, et al. (2004) Relation between hormones and body composition, including bone, in prepubertal children. Am J Clin Nutr 80: 966–72.

34. Chumanov ES, Wall-Scheffler C, Heiderscheit BC (2008) Gender differences in walking and running on level and inclined surfaces. Clin Biomech 23: 1260–8.

35. Cummings S, Melton L (2002) Epidemiology and outcomes of osteoporotic fractures. Lancet 359: 1761–7.

36. Guerra-Garcia M (2011) Incidence of hip fractures due to osteoporosis in relation to the prescription of drugs for their prevention and treatment in Galicia, Spain Atencion Primaria 43: 82–8.

37. El Maghraoui A, Koumba B, Jroundi I, Achemlal L, Ahmed B, et al. (2005) Epidemiology of hip fractures in 2002 in Rabat, Morocco. Osteoporos Int 16: 597–602.

38. Shao C (2009) A nationwide seven-year trend of hip fractures in the elderly population of Taiwan. Bone 44: 125–9.

39. Lin WP, Wen CJ, Jiang CC, Hou SM, Chen CY, et al. (2011) Risk factors for hip fracture sites and mortality in older adults. J Trauma 71: 191–7.

40. Sardinha LB, Ornelas R, Andersen LB, Froberg K, Anderssen S et al. (2008) Objectively measured time spent sedentary is associated with insulin resistance independent of overall and central body fat in 9-to 10-Year-Old Portuguese children. Diabetes Care 31: 569–75.

41. Tanner JM (1962) Growth at adolescence. Oxford, Blackwell Scientific Publications.

42. Bonnick SL, Lewis LA (2002) Bone Densitometry for technologists. Totowa, New Jersey, Humana Press. 169–181.

43. Stear S, Prentice A, Jones S, Cole T (2003) Effect of a calcium and exercise intervention on the bone mineral status of 16–18-y-old adolescent girls. Am J Clin Nutr 77: 985–92.

44. Witzke K, Snow C (2000) Effects of plyometric jump training on bone mass in adolescent girls. Med Sci Sports Exerc 32: 1051–7.

7

Regeneration of Limb Joints in the Axolotl (*Ambystoma mexicanum*)

Jangwoo Lee[1,2], David M. Gardiner[1,2]*

1 Department of Developmental and Cell Biology, University of California Irvine, Irvine, California, United States of America, **2** The Developmental Biology Center, University of California Irvine, Irvine, California, United States of America

Abstract

In spite of numerous investigations of regenerating salamander limbs, little attention has been paid to the details of how joints are reformed. An understanding of the process and mechanisms of joint regeneration in this model system for tetrapod limb regeneration would provide insights into developing novel therapies for inducing joint regeneration in humans. To this end, we have used the axolotl (Mexican Salamander) model of limb regeneration to describe the morphology and the expression patterns of marker genes during joint regeneration in response to limb amputation. These data are consistent with the hypothesis that the mechanisms of joint formation whether it be development or regeneration are conserved. We also have determined that defects in the epiphyseal region of both forelimbs and hind limbs in the axolotl are regenerated only when the defect is small. As is the case with defects in the diaphysis, there is a critical size above which the endogenous regenerative response is not sufficient to regenerate the joint. This non-regenerative response in an animal that has the ability to regenerate perfectly provides the opportunity to screen for the signaling pathways to induce regeneration of articular cartilage and joints.

Editor: Elizabeth G. Laird, University of Liverpool, United Kingdom

Funding: Research funded by a Defense Advanced Research Projects Agency (DARPA) subcontract from Tulane University (TUL 519-05/06),a US Army Multidisciplinary University Research Initiative (MURI) subcontract from Tulane University (TUL 589-09/10), and the National Science Foundation through its support of the Ambystoma Genetic Stock Center at the University of Kentucky, Lexington. The funders had no role in study design, data collection and analysis, decision to publish, or preparation of the manuscript.

Competing Interests: The authors have declared that no competing interests exist.

* E-mail: dmgardin@uci.edu

Introduction

Many different approaches utilizing a variety of model systems have attempted to regenerate joint structures. Most of these efforts have focused on engineering specific joint tissues, articular cartilage in particular, that can be used for grafting to repair damaged joints. These efforts have been limited by the reality that cartilage has a limited endogenous regenerative response and forms fibrocartilage (scar tissue) in the joint in response to injury (see [1]). We already know from studies of salamanders that tetrapod limb joints in fact can regenerate perfectly during regeneration of an amputated limb (see [2,3]). In addition, surgical defects to the articular cartilage of the axolotl (Mexican Salamander) knee joint made by resection of the medial femoral condyle to the level of the metaphysis regenerate intrinsically [4]. Thus the intrinsic regenerative response of the axolotl provides an opportunity to discover the mechanisms for inducing repair and regeneration of articular cartilage and joints.

Although development of limb joints has been studied extensively, very little is know about the regeneration of joints. Given the conservation of mechanism for development of tetrapod limbs, it is reasonable to assume that axolotl limb joint development is regulated by the same mechanisms as in more widely studied model systems such as the chick and mouse (see [5,6]). Given the conserved morphology of tetrapod limb joints, along with the observation that a regenerated joint is morphologically the same as the joint that develops in the larva, it also is

reasonable to assume that the mechanisms of joint development and regeneration are conserved. It is important to test the extent to which these assumptions are correct in order to justify utilizing the axolotl regeneration model system to provide insights for inducing repair and regeneration of joints in humans.

The global skeletal pattern of regenerating limbs has been analyzed repeatedly to draw conclusions about the mechanisms controlling pattern formation (see [7,8]); however, little has been published regarding the details of the anatomy of either uninjured or regenerating joints in salamander limbs. The basic anatomy of axolotl joints with apposed articular surfaces between adjacent long bones that are encapsulated by connective tissues is very similar to mammals [4,5,9]. The expression patterns of the relatively few marker genes for mature joints that have been analyzed in the axolotl also are comparable to those in mammalian synovial joints [4,9]. At the same time, some of the joints (e.g. knee) are different in the axolotl in that the synovial cavity is filled with fibro-cellular tissue rather than acellular synovial fluid as in the typical diarthrodial mammalian joint [4,5,9]. The possible function of these synovial cells is unknown, though when grafted into a skeletal defect in the diaphysis, they can participate in a regenerative response and appear to differentiate as both chondrocytes and synovial cells [9].

In spite of the ability to regenerate entire amputated limbs, including joints, there are injuries to the limb skeleton of axolotls that fail to regenerate. As in mammals, a skeletal defect that exceeds a critical size (CSD, critical size defect) is not regenerated in axolotls [10,11,12]. In both axolotls and mammals there is a localized chondrogenic response that results in callus formation, but this healing response is not adequate to regenerate the defect. In contrast to defects in the diaphyseal region, axolotls and mammals exhibit different responses to injuries to the articular cartilage and the epiphysis of the knee joint [4]. In mammals, injury to the epiphysis results in formation of fibrocartilage rather than regeneration of articular cartilage (see [1]). Similar injuries in the axolotl knee joint are repaired by regeneration of the defect [4]. One of the goals of the current study was to further characterize this intrinsic ability of axolotls (and presumably other salamanders) to regenerate surgical defects in the joint region.

In this paper, we describe the morphology and the expression patterns of marker genes during joint regeneration in response to limb amputation. These data are consistent with the hypothesis that the mechanisms of joint formation whether it be development or regeneration are conserved. We also have determined that defects in the epiphyseal region of both forelimbs and hind limbs in the axolotl are regenerated only when the defect is small. Thus, as is the case with defects in the diaphysis, there is a critical size above which the endogenous regenerative response is not sufficient to regenerate the joint. Since axolotl joints can regenerate perfectly in response to signaling associated with limb amputation, a non-regenerative CSD joint excision provides the opportunity to screen for the signaling pathways that control the regeneration of articular cartilage and joints.

Materials and Methods

Ethics Statement

This study was carried out in strict accordance with the recommendations in the Guide for the Care and Use of Laboratory Animals of the National Institutes of Health. The protocol was approved by the Institutional Animal Care and Use Committee of the University of California Irvine (Protocol # 2007–2705). All surgeries were performed under MS222 anesthesia, and all efforts were made to minimize suffering.

Animals and Surgical Procedures

Experiments were performed on axolotls (Ambystoma mexicanum) measuring 8–14 cm from snout to tail tip that were spawned at the University of California, Irvine or the Ambystoma Genetic Stock Center at the University of Kentucky. For all surgeries, animals were anesthetized in a 0.1% solution of MS222 (Ethyl 3-aminobenzoate methanesulfonate salt, Sigma), pH 7.0. Animals were kept anesthetized and covered with moist lab tissues for one hour post-surgery. Regeneration was induced by amputation through proximal (mid-humerus or femur) or distal (mid-radius/ulna or tibia/fibula) levels of the limb. For induction of metamorphosis, 10–12 cm axolotls were treated with thyroid hormone (L-Thyroxine, Sigma) according to methods described by Page and Voss [13].

To create a defect in the proximal epiphysis of the radius or tibia, we made three incisions in the skin overlying the elbow/knee joint area so as to create a skin flap that was still attached to the arm skin on the forth side of the square. We reflected the flap back to expose the underlying soft tissues, and then reflected the muscle fibers to expose the joint. We then used microforceps and iridectomy scissors to dissect the adherent connective tissues, muscle and tendon between the proximal radius/tibia and distal humerus/femur, and removed a 1-mm segment of the epiphysis from the radius/tibia. We then repositioned the soft tissues and the

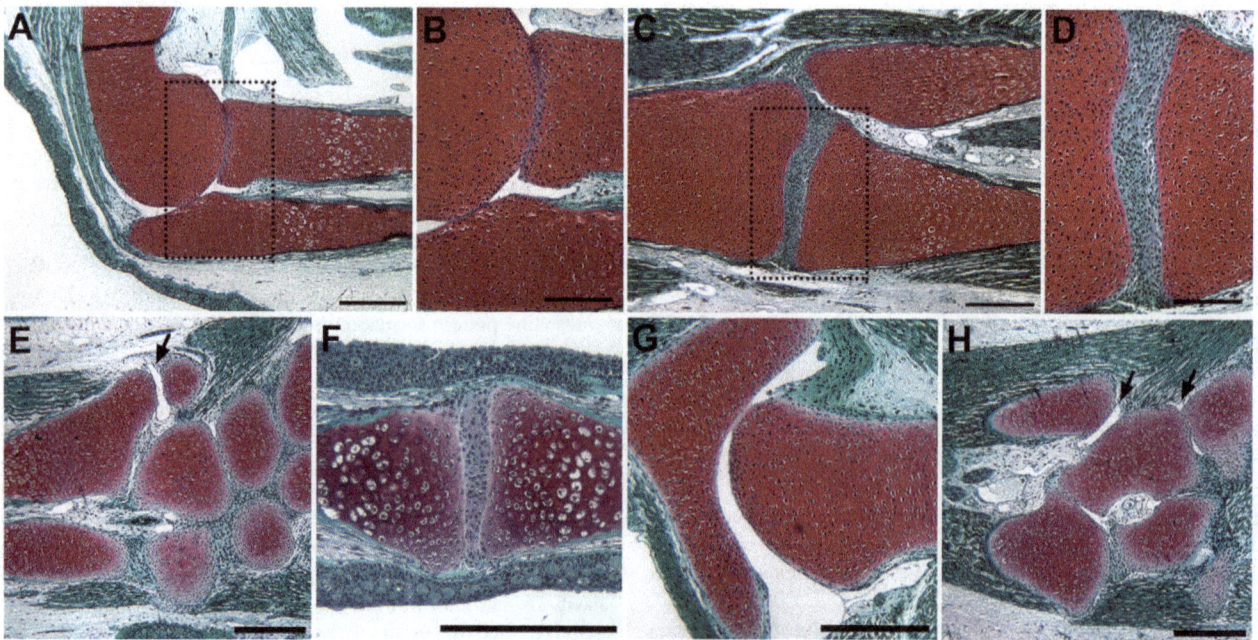

Figure 1. Joints in the forelimb and hind limb of the axolotl. The elbow joint was comparable to a mammalian synovial joint and lacked fibro-cellular tissue in the synovial cavity (A, B). Although the basic anatomy of the knee joint was similar to mammalian synovial joints, it differed in that the synovial space was filled with fibro-cellular tissues (C, D). The wrist (E) and ankle (H) joints exhibited a mixed phenotype in that the synovial space between some bones was fibro-cellular, but was acellular at other articulations (arrows in E and H). The interphalangeal joints of the fingers (E) and toes (not illustrated) were fibro-cellular; whereas, the hip (G) and shoulder (not illustrated) joints were acellular. Tissue sections were stained with Fast Green/Safranin O/Weigert's Iron Hematoxylin. Boxed areas in (A) and (C) are illustrated at higher magnification in (B) and (C) respectively. The conserved anatomy was confirmed by analysis of the joints in two different animals. Scale bars in (A, C, E, G and H) = 500 μm; (B, D and F) = 200 μm.

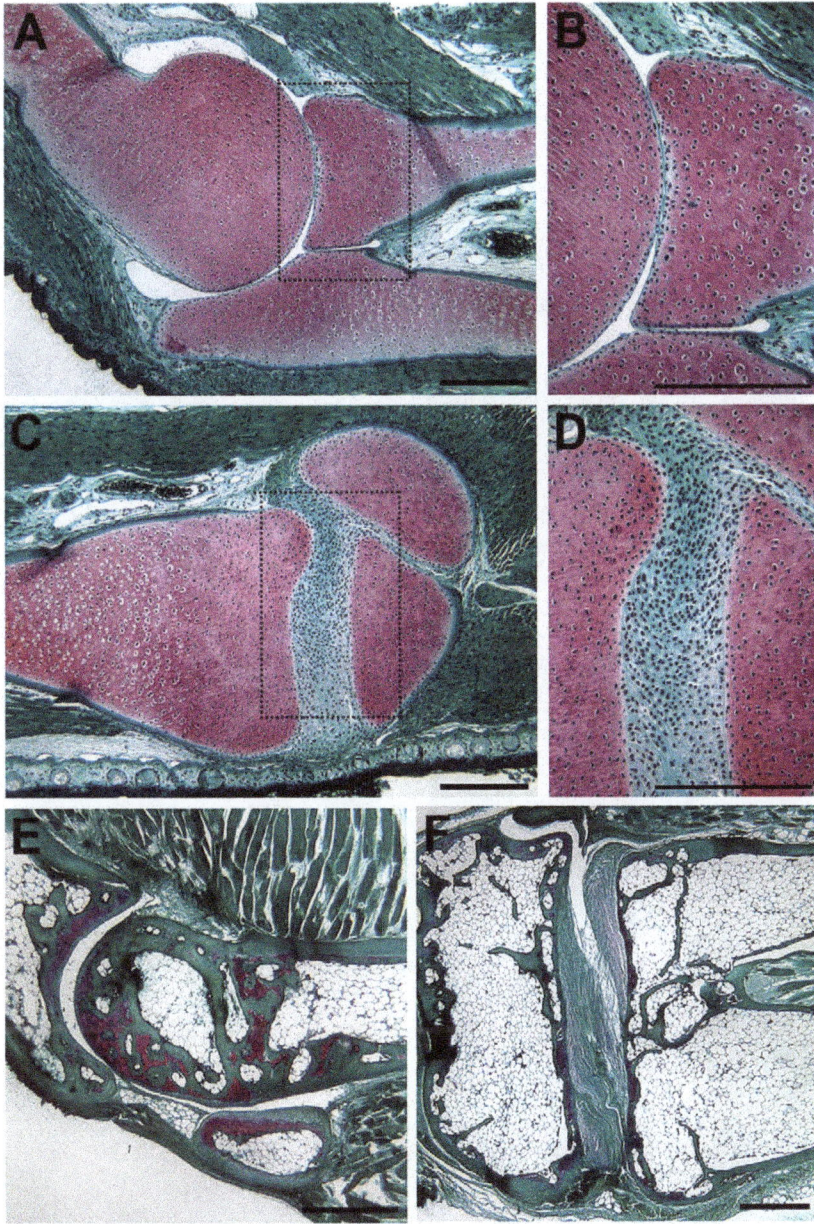

Figure 2. Joints in the forelimb and hind limb of the post-metamorphic axolotl and frog (*Xenopus tropicalis*). The elbow joint (A, B) and knee joint (C, D) of a post-metamorphic axolotl (treated with thyroid hormone to induce metamorphosis) were morphological the same as in neotenous adult axolotls (compare to Fig. 1 A–D). In adult *X. tropicalis*, the synovial cavity of the elbow joint was acellular (E); whereas, and in the knee the synovial cavity contained fibro-cellular tissue (F). This is the same pattern as observed in axolotl elbow and knee joints (compare to Fig. 1 A–D; Fig. 2 A–D). Tissue sections were stained with Fast Green/Safranin O/Weigert's Iron Hematoxylin. Boxed areas in (A) and (C) are illustrated at higher magnification in (B) and (C) respectively. The conserved anatomy was confirmed by analysis of the joints in two different post-metamorphic axolotls and two individual *X. tropicalis*. Scale bars = 500 µm.

skin flap, which healed into place without sutures by reepithelialization within 6–8 hr [14,15].

Histological Analysis

To visualize skeletal elements in whole-mount preparations, we fixed samples in 10% Z-Fix (Anatech) diluted with 40% Holtfreter's solution overnight. The skin was removed manually after fixation, the samples were refixed in 30% Z-Fix/70% ethanol for 2 hours, and then stained with 0.1% Alizarin Red S in 95% ethanol : 0.3% Alcian Blue 8GX in 70% ethanol : 70% ethanol

(1:1:8) for 1–2 days at room temperature. Stained tissue samples were treated with 2% KOH to digest the connective tissues to the point where the skeletal tissues could be visualized. The samples were cleared stepwise by soaking in 2% KOH/25% Glycerol, 2% KOH/50% Glycerol, and 2% KOH/75% Glycerol for overnight each, followed by storage in 100% Glycerol.

For staining sectioned samples, collected tissues were fixed in Z-FIX, and treated with Decalcifier I (Surgipath),to decalcify the bones of the limb tissues in order to facilitate subsequent sectioning. Tissues were then dehydrated with a ascending series

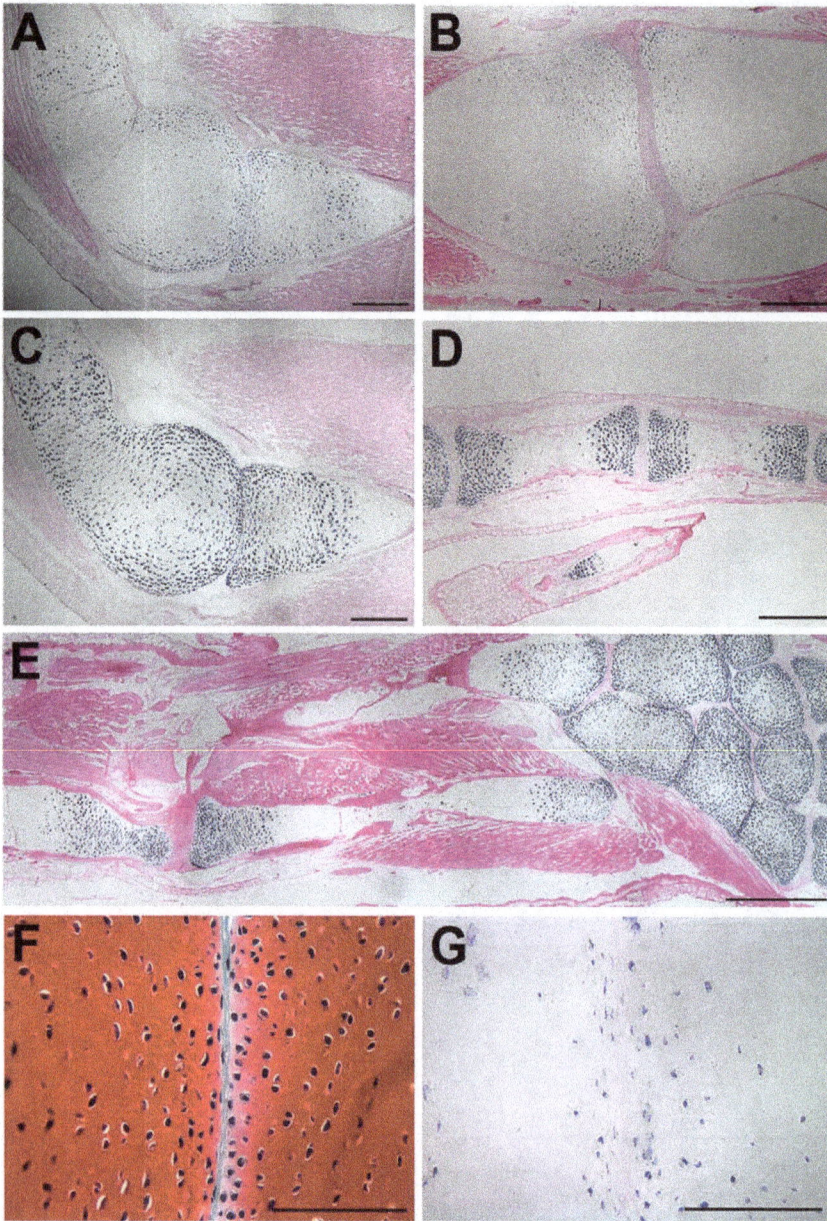

Figure 3. Joint marker gene expression in uninjured axolotl joints. Aggrecan was expressed in the articular cartilage of the epiphysis that was associated with the synovial cavity, as well as at the boundary between the epiphysis and diaphysis. This pattern was the same in both the elbow (A) and the knee (B). Type II-collagen (Col2a1) was expressed throughout the epiphyseal regions of all the joints of the elbow (C), fingers (D), and the hind limb (E), but was not localized to the articular cartilage. CD44 was expressed in cells scattered throughout the epiphyseal cartilage, but was mostly localized to the superficial layers of cells of the articular cartilage associated with the synovial cavity (F, G). Tissue sections for *in situ* hybridization were counterstained with Eosin Y (A–E, G). Tissue section in (F) was stained with Fast Green/Safranin O/Weigert's Iron Hematoxylin. The conserved patterns of expression were confirmed by analysis of tissues sections from seven different animals. Scale bars in (A–C) = 500 μm; (D, F and G) = 200 μm; (E) = 1 mm.

of ethanol (25%, 50%, 75% and 100%), cleared in xylene, embedded in paraplast, and sectioned at 6 μm thickness. For Alcian Blue staining, the sectioned samples were stained with a solution of 0.03% Alcian blue/0.1% HCl/70% ethanol for 30 min, followed by standard hematoxylin and eosin Y staining. For Fast Green/Safranin O staining, the sectioned samples were stained with 1% Fast Green and 0.1% Safranin O, followed by Weigert's Iron Hematoxylin Solution.

Cloning of the Coding Region of Axolotl CD44

We used sequence and homology data from the *Ambystoma* EST database for cloning and generating probes for target axolotl genes. Homologies to the human RefSeq database were annotated as "best hit" with a BLASTX threshold of $E = 10-7$ (www. ambystoma.org). To obtain full-length coding sequence for the axolotl ortholog of *CD44*, 5′ and 3′ mRNA sequences that contain the start and stop codons of axolotl *CD44* were retrieved from the *Ambystoma* EST Database (ID#: 5′- C100076, 3′-C261340).

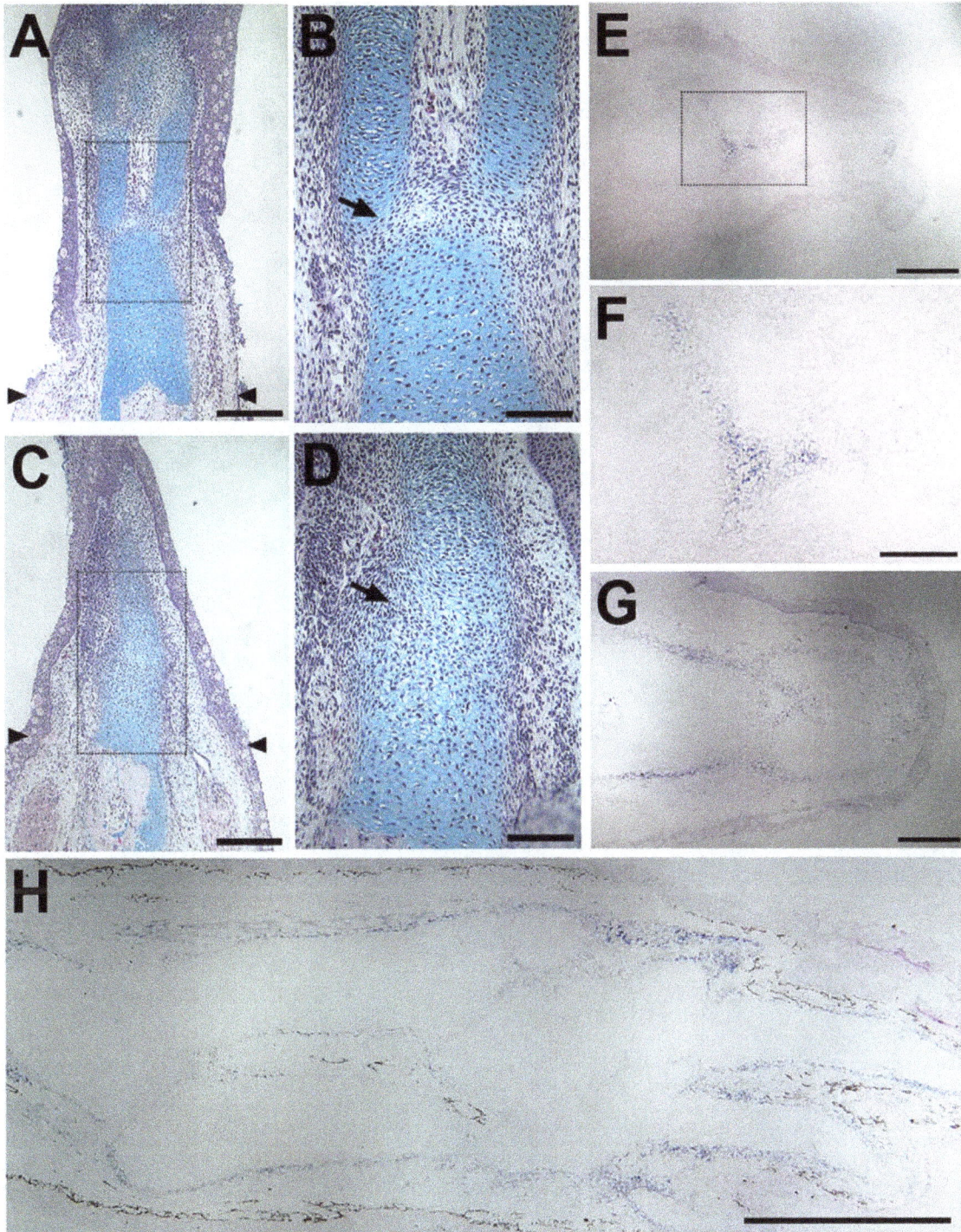

Figure 4. Morphology and patterns of gene expression during elbow and knee regeneration. After an amputated limb had formed a blastema (3 weeks post-amputation), cells in the central region began to condense and differentiate as chondrocytes, at which point they stained with Alcian Blue (Fig. 4A–D). Blastema cells at more distal regions of the blastema (toward the top of Figs. A and C) had not begun to form cartilage condensations. At the same stage of regeneration, joints began to appear at more proximal levels (toward the bottom of Figs. A and C) as the cartilage condensations segregated into discrete skeletal elements with an interzone region (arrows in B, D) that expressed the marker gene *Gdf5* (E, F). Tenascin-C was an early marker (3 weeks post-amputation) for both the perichondrium of the diaphyseal region and the regenerating articular cartilage associated with formation of the interzone (G). Expression of tenascin-C remained high in the perichondrium at later stages of regeneration (5 weeks post-amputation), but decreased in the regions where the joints were regenerated (H). Tissue sections in (A–D) were stained with Alcian Blue/hematoxylin/Eosin Y. Sections for *in situ* hybridization were counterstained with Eosin Y (E–H). Boxed areas in (A) and (C) are illustrated at higher magnification in (B) and (D) respectively. Arrowheads in (A) and (C) indicate the level at which the limbs were amputated. The conserved patterns of expression were confirmed by analysis of tissues sections from five different animals. Scale bars in (A, C, E, and G) = 500 μm; (B, D and F) = 200 μm; (H) = 1 mm.

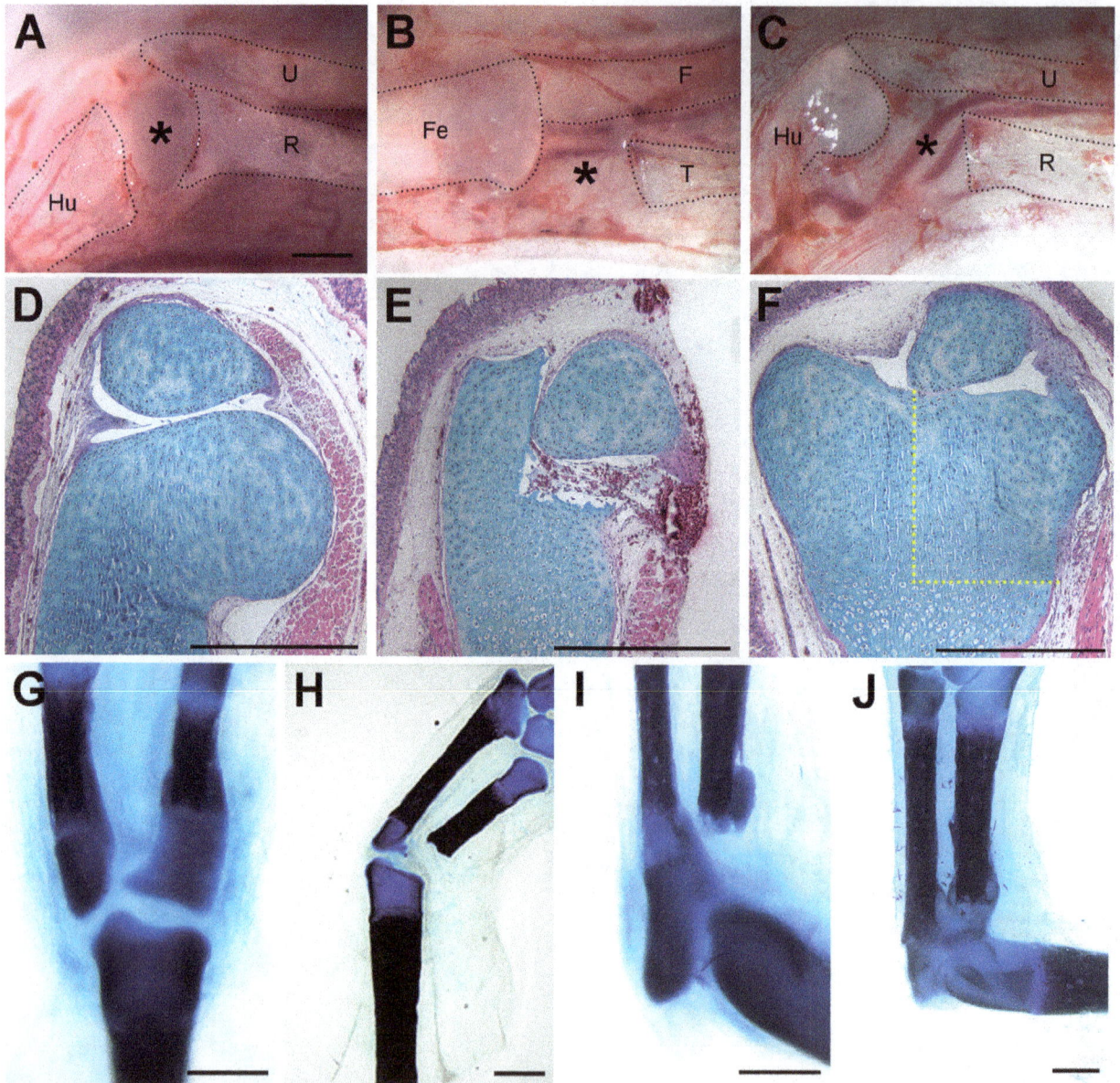

Figure 5. Regeneration of sub-critical size defects in axolotl limb joints. A defect in the distal region of the epiphysis of the humerus (A, D–E) was regenerated (F) within 8 weeks after surgery (macroscopic view of the surgery is illustrated in A). Similarly, a 1 mm defect in the proximal epiphysis of the tibia (surgery illustrated in B) regenerated (G), as did the same size defect in the proximal radius (J; surgery illustrated in C). The regenerative response to excision of 1 mm of the proximal radius was variable. In smaller animals (8–10 cm snout to tail tip), a 1 mm defect failed to regenerate (I); however, in larger animals (12–14 cm), a 1 mm defect was regenerated within 6 weeks post-amputation (J). Defects greater than 1 mm failed to regenerate in either the radius (not illustrated) or the tibia (H). Tissue sections were stained with Alcian blue/Hematoxylin/Eosin Y (D–F). Whole-mount limbs were stained with Alcian Blue/Alizarin Red (G–J). The black dotted lines in A–C demarcate the skeletal elements that remained after the surgical defect was created (Hu, humerus; U, ulna; R, radius; Fe, femur; F, fibula; T, tibia). The yellow dotted lines in (F) indicate the region of the defect that was regenerated (compare to E). Results were confirmed by experimental replication in three animals for distal humerus excisions, seven animals for proximal tibia excisions, and eight animals for proximal radius excisions. The magnification is the same in A–C. Scale bars = 1 mm.

Axolotl *CD44*-specific primers were designed based on the EST sequences to amplify the intervening sequence by RT-PCR. Total RNA from axolotl larvae was isolated using miRNeasy Mini Kit (Qiagen) following the manufacturer's recommended protocol. To synthesize first-strand cDNA from the isolated total RNA, Oligo(dT)12–18 primer and SuperScriptIII reverse transcriptase (Invitrogen) were used. Polymerase chain reaction (PCR) was performed using ExTaq DNA polymerase (Takara).with the axolotl *CD44* specific primers: CD44-F; 5′-AACTTCCAGC-

TAACTCTGCCTG-3′, CD44-R; 5′-CTTTAAGTTC-CAGTCCCAGTCC-3′. The full-length coding sequence of axolotl CD44 has been submitted to Genbank, accession # JX457476.

RNA *in situ* Hybridization (*Aggrecan, CD44, Col2a1, GDF-5, Tenascin-C*)

RNA *in situ* hybridization was performed on paraffin-sectioned axolotl limb tissues. Digoxigenin (DIG)-labeled antisense RNA probes for axolotl *Aggrecan* (*Ambystoma* EST database gene ID#: C065974; 900 bp), *CD44* (1082 bp), *Col2a1* (ID#: C081592; 944 bp), *GDF-5* (ID#: C030457 and ID#: C733258; 1203 bp), and *Tenascin* (ID#: C064822; 905 bp) were used to perform *in situ* hybridization. RT-PCR was performed to amplify the sequences for antisense RNA probes with gene specific PCR primers. The specific PCR primers for the individual genes were as follows:

Aggrecan forward: 5′-GATATGCGAAGAAGGATGGACC-3′
Aggrecan reverse: 5′-GTCTTCTTCGTTCTTCCCTTGG-3′
CD44 forward: 5′-AACTTCCAGCTAACTCTGCCTG-3′
CD44 reverse: 5′-CTTTAAGTTCCAGTCCCAGTCC-3′
Col2a1 forward: 5′-CACCTATGGATATTGGTGGAGC-3′
Col2a1 reverse: 5′-GTACATCATCCACTTGGCTACC-3′
GDF-5 forward: 5′-GTCAACGTGCACGCAGATTCTA-3′
GDF-5 reverse: 5′-ATTAGGTTGGGTTCCATCCCG-3′
Tenascin-C forward: 5′-TACTGGGCTCTACACCATCTAC-3′
Tenascin-C reverse: 5′-CCAAGAGGATGACAAGTCTGTG-3′

All template clones were subcloned into the pCRII vector (Invitrogen). To synthesize antisense RNA probes for each of the genes, the plasmid DNA templates were linearized by *BamHI* (for *Col2a1* and *GDF-5*) or *XhoI* (for *Aggrecan, CD44, Tenascin*). The RNA probes were synthesized with the DIG RNA Labeling Kit (Roche) according to the manufacturer's protocol using T7 (*Col2a1, GDF-5*) or Sp6 (*Aggrecan, CD44, Tenascin*) RNA polymerases. For tissue sample collection, tissues were collected and fixed in MEMFA (0.1M MOPS, pH7.4, 2 mM EGTA, 1 mM MgSO$_4$, 3.7% formaldehyde), skeletal tissues were decalcified in Decalcifier I, dehydrated with a ascending series of ethanol (25%, 50%, 75% and 100%), cleared in xylene, and embedded in paraplast. Paraffin sections were cut at 6 μm thickness. Sections were treated with 7.5 μg/ml of Proteinase K (Invitrogen) for 20 min at 37°C, refixed with 4% paraformaldehyde, and then hybridized with antisense RNA probes at 60°C overnight. After hybridization, the section were washed with the buffer #1 (Formamide: water: 20X SSC = 2:1:1), and buffer #2 (5:4:1), and blocked with 2% blocking reagent (Roche) in TBST for 30 min. The sections were then incubated with 1:2,000 diluted alkaline phosphatase (AP)-conjugated anti-DIG antibody (Roche) overnight at 4°C. The color staining reaction was performed using BM purple (Roche) as a substrate for AP.

Results

Morphology and Gene Expression of Uninjured Limb Joints of the Axolotl

As reported previously [4,9], the basic anatomy of axolotl joints with apposed articular surfaces between adjacent long bones that are encapsulated by connective tissues was very similar to mammals (Fig. 1). The axolotl elbow joint was most similar to the equivalent mammalian joint in that the synovial space between the apposed skeletal elements was fluid-filled and acellular (Fig. 1A, B). In contrast, the synovial space of the axolotl knee joint was filled by a dense fibro-cellular tissue (Fig. 1C, D), as reported previously [4]. Based on the initial observations of the knee joint [4], these synovial cells were suggested to be equivalent to interzone cells (a developmentally transient population of joint-forming cells in the limb bud) that persisted as a consequence of the neotenous mode of development of the axolotl. Although the axolotl progresses through metamorphosis to the point of developing appendages and becomes sexually mature (neoteny), it does not complete metamorphosis and become terrestrial as occurs in most other salamanders. Therefore it has been

hypothesized that the persistence of "interzone-like" cells in the knee joint is a consequence of this arrested development, and that these cells maintain chondrogenic potential and contribute to the repair of joint defects [4,9].

In order to explore the relationship between the presence and absence of synovial cell and the ability to regenerate joint structures, we examined all the joints in both the axolotl forelimb and hind limb to determine the variability in morphology, and the distribution of acellular and fibro-cellular synovial joints. As reported recently [9], we observed that both types of joint morphologies were present in both forelimbs and hind limbs (Fig. 1). In general, there was a trend with the proximal joints being acellular and more distal joints being fibro-cellular. Thus the shoulder and elbow joints were acellular (Fig. 1A, B; not illustrated for the shoulder joint), and the wrist (Fig. 1E) and interphalangeal joints (Fig. 1F) were fibro-cellular. Although many of the joints between the carpals in the wrist were fibro-cellular, some of the articulations between the distal zeugopod and proximal autopod were acellular (Fig. 1E, arrow), and thus the wrist joints exhibited an intermediate phenotype. In the hind limb, the hip joint was acellular (Fig. 1G); whereas, the knee (Fig. 1C, D) and interphalangeal joints (same morphology as in the forelimb, Fig. 1F) were fibro-cellular. As observed in the wrist region, the joints of the ankle were a mixed phenotype of both acellular and fibro-cellular (Fig. 1H, arrow) even though the knee joint was fibro-cellular. Thus the presence of mixed phenotype joints was associated with the boundary between the zeugopod and autopod, rather than being a transitional phenotype between acellular and fibro-cellular joints along the proximal-distal limb axis. Since all the joints in both the forelimb and hind limb regenerate when the limb is amputated, there does not appear to be a relationship between joint regeneration and the presence or absence of interzone-like cells.

In order to test whether the persistence of interzone-like cells is a consequence of neoteny in axolotls, and whether they are unique to axolotl limbs, we examined the morphology of limb joints in axolotls that had undergone metamorphosis, as well as the joints in a post-metamorphic frog (*Xenopus tropicalis*) (Fig. 2). Although axolotls typically do not complete metamorphosis, individuals will occasionally undergo metamorphosis spontaneously. Experimentally, metamorphosis can be induced by adding thyroid hormone to the aquarium water [13]. The morphology of both the elbow (Fig. 1A, B and Fig. 2A, B) and knee joints (Fig. 1C, D and Fig. 2C, D) were the same in both pre-metamorphic axolotls (Fig. 1) and post-metamorphic axolotls (Fig. 2). The same pattern of joint morphology observed in the axolotl was present in post-metamorphic frog limb joints (acellular elbow joint, Fig. 2E and fibro-cellular knee joint, Fig. 2F). Thus the persistence of interzone-like cells is not related to metamorphosis, and is not a novel feature of axolotl joints.

In addition to a conserved morphology, uninjured limb joints in the axolotl expressed a number of joint marker genes in spatial patterns that were comparable to what is observed in mammalian joints (see [5]). The proteoglycan, Aggrecan was expressed in the articular cartilage of the epiphysis that was associated with the synovial cavity, as well as at the boundary between the epiphysis and diaphysis (Fig. 3A, B). This pattern was the same in both the elbow (Fig. 3A) and the knee (Fig. 3B). Type II-collagen (Col2a1) expression was localized to the epiphyseal regions of all the joints of both the forelimb (Fig. 3C, D) and hind limb (Fig. 3E). Finally, CD44 was expressed in cells scattered throughout the epiphyseal cartilage, but was mostly localized to the superficial layers of cells of the articular cartilage associated with the synovial cavity (Fig. 3F, G). Previous studies of the fibro-cellular knee joint reported similar but not identical expression patterns for Aggrecan and Type II-collagen proteins [4,9]. Our *in situ* hybridization data provide validation of the specificity of the heterologous antibodies used for immunohistochemistry in those studies. In addition, Type I-collagen, GDF5 and BOC (Brother of CDO) were reported to be expressed in association with the fibro-cellular tissues of the uninjured knee joint [4,9].

Morphology and Gene Expression of Regenerating Limb Joints in the Axolotl

The events associated with the regeneration of joints of amputated limbs appeared to be the same as occur during limb development (see [5]). During the later stages of blastema growth, cells in the central region began to condense and differentiate as chondrocytes, at which point they stained with alcian blue (Fig. 4A–D). As the blastema continued to grow distally, joints began to form more proximally as chondrogenic cells began to form an interzone as evidenced by the expression of *Gdf5* (Fig. 4E, F). *Gdf5* expression was transient in joints that subsequently underwent cavitation to form an acellular synovial cavity (e.g. elbow). In contrast, *Gdf5* expression was reported to persist in joint that have fibro-cellular tissue within the synovial cavity (e.g. knee; [9]). Tenascin-C was an early marker for both the perichondrium of the diaphyseal region and the regenerating articular cartilage associated with formation of the interzone (Fig. 4G). Expression of tenascin-C remained high in the perichondrium at later stages of regeneration, but decreased in the regions where the joints were regenerated (Fig. 4H). In addition, the expression patterns of the joint marker genes described above for uninjured joints (Fig. 3) were reestablished by the end of regeneration.

The Regenerative Response to Injury was Dependent on the Extent of the Defect

We determined that the ability of the axolotl to regenerate surgical defects to articular cartilage of joints was dependent on the size of the defect, and to some extent the size and age of the animal. We confirmed the previous report that a defect in the femoral condyle was repaired endogenously (data not shown; [4]). This intrinsic regenerative response was not restricted to the knee joint that has fibro-cellular tissue, but also occurred in the elbow joint that has an acellular synovial cavity (Fig. 5A, D–F). A resection of the distal humerus comparable to what was reported previously for the distal femur [4], also regenerated endogenously Fig. 5F). Similarly, a 1 mm defect in the proximal epiphysis of the tibia regenerated (Fig. 5B, G), as did the same size defect in the proximal radius (Fig. 5C, J). Thus the bones that were apposed in the knee and elbow joints responded the same. In our initial experiments with smaller animals (8–10 cm snout to tail tip), a small (1 mm) defect in the proximal radius failed to regenerate (Fig. 5I; n = 10 animals); however, when we repeated this experiment with larger animals (12–14 cm), nearly all the 1 mm defects were regenerated (Fig. 5J; n = 5 of 6 animals). Thus the ability to regenerate a small defect in the articular cartilage of the epiphysis was not related to the presence or absence of interzone-like cells in the synovial cavity.

The intrinsic ability of the axolotl to regenerate joint defects was limited to relatively small defects. For both the forelimb and hind limb, defects that were greater than 1 mm failed to regenerate (e.g. Fig. 5H; n = 3 forelimb; n = 3 hind limb). Thus the response to joint injuries in the axolotl was comparable to what occurs when defects were created in the diaphyseal region in both the axolotl and in mammals [11,12,16]. Defects heal when they are smaller than a critical size (Critical Size Defect, CSD), but form a callus and fail to regenerate when greater than the CSD. In the axolotl, a diaphyseal defect that exceeds a CSD can be induced to regenerate in response to signaling from a deviated nerve and wound epithelium [12]. In spite of attempts to induce regeneration of proximal radial defects with a deviated nerve and wound epithelium (n = 4 animals), we did not observe evidence of an enhanced regenerative response relative to control defects (Fig. 5I).

Discussion

The anatomy of limb joints in the axolotl, particularly the acellular joints such as the elbow is conserved compared to mammals and other tetrapods. Similarly, the expression patterns of marker genes in both the uninjured and regenerating axolotl joints are comparable to what is observed in uninjured and developing tetrapod limbs. In addition, the sequence of events during joint regeneration leading to interzone formation and cavitation is the same as during limb development. Taken together, these findings suggest that a similar molecular mechanism is used during the development of many joints [17], and this same mechanism is used again during regeneration of limbs and joints [18].

The presence of fibro-cellular tissues in some of the axolotl joints has raised the question of the functional relationship of these interzone-like cells to regeneration [4,9]. At this point, it appears that interzone cells are neither necessary nor sufficient for joint regeneration. These cells are not necessary for joint regeneration in an amputated limb since all the different limb joints are regenerated perfectly whether or not they have interzone-like cells. These cells also are not sufficient for regeneration of a joint defect in the absence of limb amputation. Both the elbow and knee joints regenerate a sub-critical size defect, and neither can regenerate a defect greater than the CSD, even though interzone-like cells are present in the knee joint. Finally, both acellular and fibro-cellular joints are present in post-metamorphic *Xenopus* froglets that do not regenerate any joints when a limb is amputated [19,20].

Although interzone-like cells may not be directly involved in an endogenous regenerative response, they may have the potential to be a source of cells for the repair of skeletal tissues. When grafted into a non-regenerating defect in the diaphysis they participate in the formation of an ectopic joint-like structure (pseudarthrosis) and contribute both to the chondrocytes at the ends of the skeletal elements and the intervening fibro-cellular tissue [9]. A major challenge in the field of regeneration biology is understanding how specific progenitor cell types are recruited to participate in regeneration. It has been difficult to address this issue given the diversity of cell types involved and the lack of molecular markers for regeneration-competent cells. The homogeneity of axolotl interzone-like cells and the markers that we have validated provide the opportunity to test this population of cells for the ability to respond to regeneration-inducing signals [12,21] and function as a multipotent progenitor cell for the repair and regeneration of joints in response to both acute and chronic damage.

Acknowledgments

We wish to thank the members of the Bryant/Gardiner Lab for help with and encouragement of the research.

Author Contributions

Conceived and designed the experiments: JL DMG. Performed the experiments: JL. Analyzed the data: JL DMG. Contributed reagents/materials/analysis tools: JL. Wrote the paper: JL DMG.

References

1. Lorenz H, Richter W (2006) Osteoarthritis: cellular and molecular changes in degenerating cartilage. Progress in Histochemistry and Cytochemistry 40: 135–163.
2. Bryant SV, Endo T, Gardiner DM (2002) Vertebrate limb regeneration and the origin of limb stem cells. Int J Dev Biol 46: 887–896.
3. Wallace H (1981) Vertebrate Limb Regeneration. Chichester: John Wiley and Sons.
4. Cosden RS, Lattermann C, Romine S, Gao J, Voss SR, et al. (2011) Intrinsic repair of full-thickness articular cartilage defects in the axolotl salamander. Osteoarthritis and Cartilage/OARS, Osteoarthritis Research Society 19: 200–205.
5. Khan IM, Redman SN, Williams R, Dowthwaite GP, Oldfield SF, et al. (2007) The development of synovial joints. Current Topics in Developmental Biology 79: 1–36.
6. Koyama E, Shibukawa Y, Nagayama M, Sugito H, Young B, et al. (2008) A distinct cohort of progenitor cells participates in synovial joint and articular cartilage formation during mouse limb skeletogenesis. Dev Biol 316: 62–73.
7. Bryant SV, French V, Bryant PJ (1981) Distal regeneration and symmetry. Science 212: 993–1002.
8. French V, Bryant PJ, Bryant SV (1976) Pattern regulation in epimorphic fields. Science 193: 969–981.
9. Cosden-Decker RS, Bickett MM, Lattermann C, Macleod JN (2012) Structural and functional analysis of intra-articular interzone tissue in axolotl salamanders. Osteoarthritis and cartilage/OARS, Osteoarthritis Research Society.
10. Goss RJ (1969) Principles of Regeneration. New York: Academic Press.
11. Hutchison C, Pilote M, Roy S (2007) The axolotl limb: a model for bone development, regeneration and fracture healing. Bone 40: 45–56.
12. Satoh A, Cummings GM, Bryant SV, Gardiner DM (2010) Neurotrophic regulation of fibroblast dedifferentiation during limb skeletal regeneration in the axolotl (Ambystoma mexicanum). Dev Biol 337: 444–457.
13. Page RB, Voss SR (2009) Induction of metamorphosis in axolotls (Ambystoma mexicanum). Cold Spring Harbor Protocols 2009: pdb prot5268.
14. Carlson MRJ, Bryant SV, Gardiner DM (1998) Expression of *Msx-2* during development, regeneration, and wound healing in axolotl limbs. J Exp Zool 282: 715–723.
15. Satoh A, Graham GM, Bryant SV, Gardiner DM (2008) Neurotrophic regulation of epidermal dedifferentiation during wound healing and limb regeneration in the axolotl (Ambystoma mexicanum). Dev Biol 319: 321–335.
16. Schmitz JP, Hollinger JO (1986) The critical size defect as an experimental model for craniomandibulo-facial nonunions. Clin Orthop Relat Res: 299–308.
17. Crotwell PL, Mabee PM (2007) Gene expression patterns underlying proximal-distal skeletal segmentation in late-stage zebrafish, Danio rerio. Dev Dyn 236: 3111–3128.
18. Muneoka K, Bryant SV (1982) Evidence that patterning mechanisms in developing and regenerating limbs are the same. Nature 298: 369–371.
19. Muneoka K, Holler-Dinsmore G, Bryant SV (1986) Intrinsic control of regenerative loss in *Xenopus laevis* limbs. J Exp Zool 240: 47–54.
20. Satoh A, Suzuki M, Amano T, Tamura K, Ide H (2005) Joint development in Xenopus laevis and induction of segmentations in regenerating froglet limb (spike). Dev Dyn 233: 1444–1453.
21. Endo T, Bryant SV, Gardiner DM (2004) A stepwise model system for limb regeneration. Dev Biol 270: 135–145.

8

Functional Relationship between Skull Form and Feeding Mechanics in *Sphenodon*, and Implications for Diapsid Skull Development

Neil Curtis[1]*, Marc E. H. Jones[2], Junfen Shi[1], Paul O'Higgins[3], Susan E. Evans[2], Michael J. Fagan[1]

1 Medical and Biological Engineering Research Group, Department of Engineering, University of Hull, Hull, United Kingdom, 2 Research Department of Cell and Developmental Biology, University College London, London, United Kingdom, 3 Hull-York Medical School, University of York, York, United Kingdom

Abstract

The vertebrate skull evolved to protect the brain and sense organs, but with the appearance of jaws and associated forces there was a remarkable structural diversification. This suggests that the evolution of skull form may be linked to these forces, but an important area of debate is whether bone in the skull is minimised with respect to these forces, or whether skulls are mechanically "over-designed" and constrained by phylogeny and development. Mechanical analysis of diapsid reptile skulls could shed light on this longstanding debate. Compared to those of mammals, the skulls of many extant and extinct diapsids comprise an open framework of fenestrae (window-like openings) separated by bony struts (e.g., lizards, tuatara, dinosaurs and crocodiles), a cranial form thought to be strongly linked to feeding forces. We investigated this link by utilising the powerful engineering approach of multibody dynamics analysis to predict the physiological forces acting on the skull of the diapsid reptile *Sphenodon*. We then ran a series of structural finite element analyses to assess the correlation between bone strain and skull form. With comprehensive loading we found that the distribution of peak von Mises strains was particularly uniform throughout the skull, although specific regions were dominated by tensile strains while others were dominated by compressive strains. Our analyses suggest that the frame-like skulls of diapsid reptiles are probably optimally formed (mechanically ideal: sufficient strength with the minimal amount of bone) with respect to functional forces; they are efficient in terms of having minimal bone volume, minimal weight, and also minimal energy demands in maintenance.

Editor: Andrew A. Farke, Raymond M. Alf Museum of Paleontology, United States of America

Funding: Funding was provided by the Biotechnology and Biological Sciences Research Council (BBSRC-http://www.bbsrc.ac.uk) - grant numbers: BB/E007465/1, BB/E009204/1 and BB/E007813/1. The funders had no role in study design, data collection and analysis, decision to publish, or preparation of the manuscript.

Competing Interests: The authors have declared that no competing interests exist.

* E-mail: n.curtis@hull.ac.uk

Introduction

There is a longstanding debate as to whether bone in the skull is minimised in relation physiological loading [1,2], or whether skulls are 'over-designed' and constrained by phylogeny, development, and the need to accommodate functions in addition to normal loading [3–5]. The skull provides a structure for jaw and neck muscle attachment and should be rigid enough to withstand the forces these muscles apply, along with accompanying feeding and other forces [6–8]. Exactly how the skull responds to these forces in tandem with accommodating the brain and sense organs is not fully understood. Adaptation to loads consistent with Wolff's law [9] would result in minimisation of bony material with respect to functional loading, and following a long held theory [10] the term *bone functional adaptation* [11–13] is often used to describe the mechanism by which bone is modelled and remodelled. Briefly, it is proposed that bone strain is the stimulus for bone modelling/remodelling [14,15], and there is an *equilibrium window* of strain, above which bone is deposited and below which bone is removed [16–18]. The rules regulating bone adaptation and the exact levels at which bone is remodelled are however likely more complex, being dependent on more than just pure strain magnitudes. Strain rate, load history, bone age, disease, initial bone shape, bone

developmental history, hormonal environment, diet, and genetic factors have all been highlighted as potential factors that could impact bone form [15–25].

The skull of *Sphenodon*, a New Zealand reptile, is not dominated by a large vaulted braincase like mammals, but instead comprises an open arrangement of fenestrae (windows or openings) and bony rods or struts [26,27]. Without the constraint of a large brain and associated forces [28–31], the dominant loads applied to the frame-like skull of *Sphenodon* are most likely linked to feeding (i.e. muscle forces, bite forces, and jaw joint forces). This is probably also true for other diapsids that lack large brains, such as lizards, crocodiles, and theropod dinosaurs, which share comparable skull morphologies (Figure 1). Without the effect of neurocranial expansion, these frame-like skulls may be useful for investigating the correlation between skull form and bone strain under loading. Some insight into this relationship would provide new perspectives towards understanding skull form in other amniotes.

Finite element analysis (FEA) is a virtual technique that is used to predict how a structure will deform when forces and constraints are applied to it, and has been used previously to predict stress and strain distribution within skulls [4,27–29,31–33,35,36]. However, such studies tend to apply limited loading data and are used to investigate particular aspects of skull morphology or the impact of

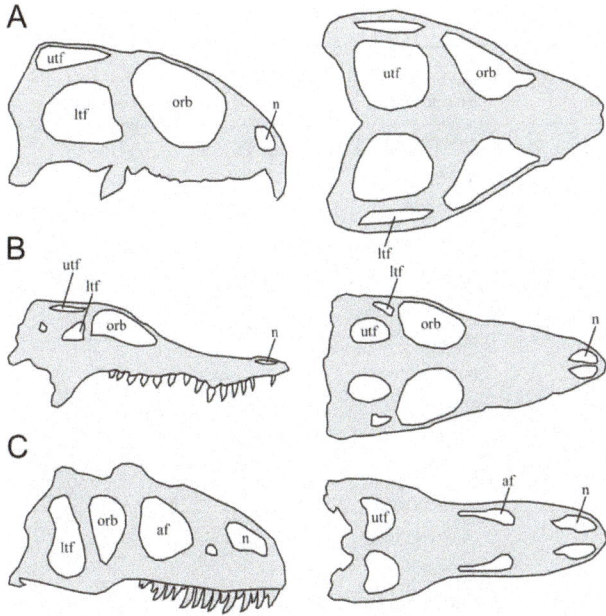

Figure 1. The diapsid skull form. Simplified schematic lateral and dorsal skull views of **A.** *Sphenodon* (redrawn [87]), **B.** *Crocodylus siamensis* (original drawing), **C.** *Allosaurus fragilis* (redrawn [92]). All skulls are scaled to the same length. af – antorbital fenestra; ltf – lower temporal fenestra; n – nasal opening; orb – orbital opening; utf – upper temporal fenestra.

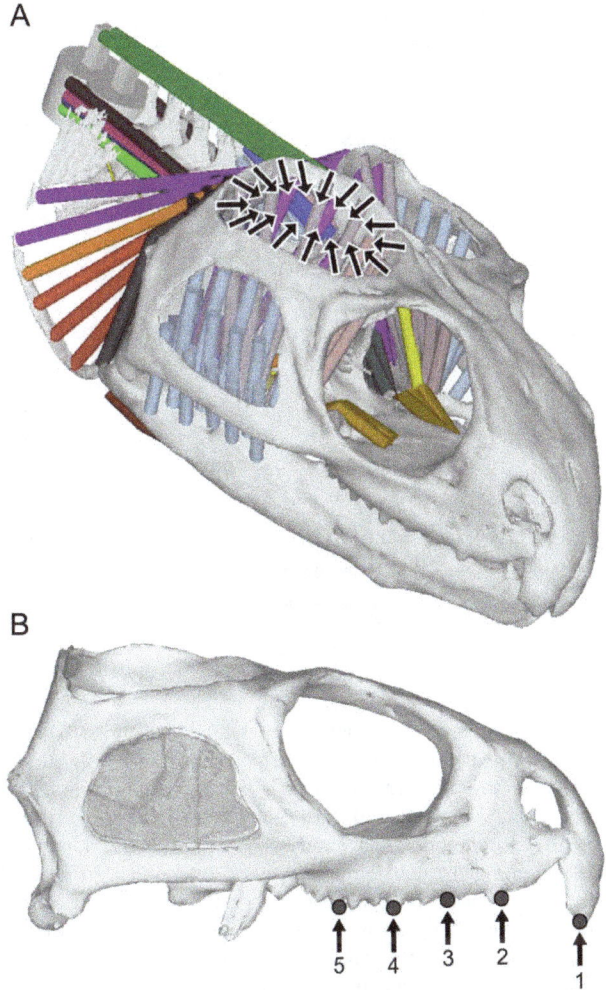

Figure 2. MDA model. A. Multibody computer model used to calculate the muscle, joint and biting forces for a series of biting simulations. Black arrows represent the location and direction of the fascial force vectors applied to the finite element model over one temporal opening. **B.** Bite locations. Bilateral (biting on both sides simultaneously) and unilateral biting (biting on one side only) at locations 2–5; bilateral biting only at location 1; ripping bites at location 2 only. Skull measures approximately 68 mm long from the tip of the premaxilla to the posterior end of the quadrate condyles.

single bites. To fully evaluate skull form it is important to take into account several different load cases, because skull form is most likely to be related to the range of physiological loads experienced by an animal rather than a single load case. We investigated the relationship between skull form and bone strain in *Sphenodon* by carrying out a series of static finite element analyses (FEAs), applying bite forces at several different bite positions. We combine the powerful computational techniques of multibody dynamics analysis (MDA) [32–34] and FEA, to first predict the forces acting on the skull of *Sphenodon*, and in turn analyse the strains within the skull under these forces. This enables us to evaluate the degree of correlation between skull form and three strain modes: tensile (also known as maximum and 1st principal), compressive (also know as minimum and 3rd principal) and von Mises (also known as equivalent and mean). Multibody dynamics analysis has recently been applied to study skull biomechanics [32–38], and was used here to predict muscle forces, joint forces, and bite forces in *Sphenodon* during fifteen separate biting simulations. These simulations covered a range of biting types and locations. They include four bilateral and eight unilateral bites at different tooth positions, a bite on the anterior-most chisel-like teeth, and two ripping bites that incorporate neck muscles (MDA model shown in Figure 2 and a summary of all biting simulations is given in Table 1). A corresponding set of fifteen separate FEAs was carried out to investigate the total mechanical performance of the skull under these predicted forces. Each separate FEA applied a peak static bite force and corresponding muscle and joint forces.

Results

MDA

Total bite and quadrate-articular joint forces (i.e. working and balancing sides combined) are similar whether the animal is biting

unilaterally or bilaterally. However, the bite force on each side of the skull during bilateral biting is half that of unilateral biting (i.e. the total bite force is shared over both sides of the skull). Also, forces located at the balancing side joint during unilateral biting are always in excess of those at working side joint (Table 2). Bite force at the most posterior bite location (location 5 – Figure 2B) is almost 80% greater than on the chisel-like teeth at the front of the skull (location 1), whereas during unilateral biting the balancing side joint force is approximately 50% greater than the working side joint force at the most posterior bite location (location 5). Total muscle forces applied during the MDA are presented in Table 3.

FEA

Bite location has a considerable effect on the way the skull deforms. During individual bites, strain gradients (or heterogeneous strain magnitudes) are apparent over the skull, with some regions subject to high strains and others subject to low strains

Table 1. The 15 load cases simulated during the MDA and applied in the FEA.

Load case	Type of bite	Side of skull	Bite location	Bite Location
1	unilateral	right	anterior	2
2	unilateral	right	middle	3
3	unilateral	right	posterior	4
4	unilateral	right	posterior-most	5
5	unilateral	left	anterior	2
6	unilateral	left	middle	3
7	unilateral	left	posterior	4
8	unilateral	left	posterior-most	5
9	bilateral	both	anterior	2
10	bilateral	both	middle	3
11	bilateral	both	posterior	4
12	bilateral	both	posterior-most	5
13	bilateral	both	chisel-like tooth	1
14	neck ripping bite (left)	both	anterior	2
15	neck ripping bite (right)	both	anterior	2

See Figure 2 for explanation of bite locations.

(example von Mises strain plots are presented in Figure 3). As the skull deforms it experiences both compressive and tensile strains (dominant strains over all bites at specific skull locations is presented in Figure 4), and during unilateral biting these strains tend to reach their peak magnitudes (Figure 5A). In addition to the peak strains generated during unilateral bites, high strain also occurs in the nasal bone when biting on the large anterior-most chisel-like teeth, a distinctive feature of *Sphenodon* ([39]; Figure 5B, bilateral location 1). Ripping bites in which the neck muscles are highly active also strain the posterior aspects of the skull and braincase more than non-ripping bites (Figure 5B, ripping location 2). Across all simulations unilateral bites account for approximately 79% of the peak strains generated across the skull, with the posterior-most unilateral bite accounting for 60% of peak strains. Biting on the anterior-most chisel-like teeth generates approximately 9% of the peak strains in the skull, while the ripping bites were attributable for 10%. Bilateral bites (excluding biting on the anterior-most teeth) accounted for less than 2% of peak strains

across the skull when all biting simulations were assessed. Strains vary over the skull at any one bite location (including those yielding the highest strains), with approximately 30% of the skull at low levels of strain below 200 microstrain, and 65% of the skull at strains of below 500 microstrain during separate bites (Figure 6).

When the individual peak element strains (i.e. the highest strain any one element ever experienced) are extracted from all fifteen individual biting analyses to generate a combined loading peak strain map, the obvious strain gradients (or heterogeneous strain magnitudes) noted during separate bites are considerably reduced (Figure 7). During combined loading 94.6%, 96.7%, and 98.0% of the skull experiences tensile, compressive, and von Mises strains of above 200 microstrain respectively when the peak element strains over all bites are considered (Figure 6). This compares to an average of approximately 70% during separate bites for all strain modes. Moreover, during combined loading 85.3%, 87.9%, and 91.1% of the skull in our model is at strains of between 400 and 2500 microstrain for tensile, compressive, and von Mises strain

Table 2. Bite forces and jaw joint forces predicted by the MDA.

Bite Type	Bite Location	Bite Force (N)	Working Joint Force (N)	Balancing Joint Force (N)
bilateral	1	121	540	-
bilateral	2	150	524	-
bilateral	3	165	510	-
bilateral	4	185	490	-
bilateral	5	214	462	-
unilateral	2	150	249	276
unilateral	3	166	232	276
unilateral	4	187	212	277
unilateral	5	216	183	278

Total forces are shown for bilateral bites, therefore the force on each side of the skull is approximately half that presented. Working refers to the force on the same side as the bite occurs, while balancing refers to the opposite side to which biting occurs. See Figure 2 for explanation of bite locations.

Table 3. Total muscle forces applied to each side of the skull during the MDA.

Muscle	Total Muscle Force (N)
Depressors (defined as 2 groups)	40
Adductors (defined as 14 groups)	448
Neck (defined as 11 groups)	158

The depressor muscles were represented by two muscle groups, the adductor muscles were represented by fourteen muscle groups, and the neck muscles were represented by eleven muscle groups. This arrangement of muscles accurately depicts the anatomy of Sphenodon. Muscle sections are visually presented in Figure 3A, while detailed descriptions of all muscle groups are published elsewhere [18,39].

respectively, implying that the majority of the skull is shaped (remodelled) to keep strains within a specific tolerance range (Figure 6). Mean tensile, compressive, and von Mises strain over the entire skull (average strain across all individual finite elements in the model) is 784 microstrain, 887 microstrain, and 1140 microstrain when peak strains over all load cases are assessed. This value is typically only 500 microstrain during separate bites.

Overall strain distributions over the skull remain largely unchanged with the addition of a fascial sheet over the upper temporal fenestra, but there were some striking reductions in localised peak strains, as highlighted in Figure 8. In particular, there is a reduction of peak strain on the lateral aspect of the postorbital bar where the jugal and postorbital meet, but the most obvious reductions in peak strains are on the posterior surface of the quadrate (encircled in Figure 8B), the temporal bar (squamosal and parietal, encircled in Figure 8B) and the posterior edges of the parietals where they meet in the midline (also encircled in Figure 8B). Localised peak strain areas around the perimeter of the upper fenestra were unaffected, with the exception of a small region on the posterior part of the postorbital.

Discussion

The results of our comprehensive analysis implies that the form of the diapsid skull of Sphenodon is strongly linked to feeding forces. We show that both tensile and compressive peak strains are relatively evenly distributed throughout the skull when several loading cases are analysed (Figure 7). Although tensile strains are dominant in some regions of the skull, compressive strains are dominant in others (Figure 4). However, when analysing von Mises strain, which takes into account all principal strains, the distribution of strain is even more uniform when compared to tensile and compressive strains alone (Figure 7).

Our analyses show that over 91% of the skull is at von Mises strains of between 400 and 2500 microstrain when peak biting forces were analysed (Figure 6). While von Mises strain does not show which principal strain mode is dominant, making it difficult to interpret the exact response of the structure (e.g. whether or not it might fracture under tensile forces), von Mises strain does appear to be a good indicator of bone adaptation. In vivo studies predominantly on long bones have shown that both tensile and compressive strains are frequently experienced by bones during normal use, with peak strain during forceful loading ranging from 900 to 5200 microstrain [40–52]. In our analyses we find both high compressive and tensile strains over the skull, comparable in magnitude to those recorded experimentally in other animals (Figure 7), where compressive strains are dominant in approximately 60% of the skull (Figure 4). Focusing specifically on skulls, Herring et al. [53,54] recorded strains of 2000–3000 microstrain when the masseter muscle was maximally contacted in a pig skull, peak values very similar to those predicted in our study.

Most literature on bone adaptation only refers to strain without inferring a particular mode, or even magnitude to this regard. What we do know is that bone adapts to mechanical loading, for example in experimental studies on adult rats, Robling et al. [55] showed bone to be deposited on both the tensile and compressive sides of artificially loaded forearms. Also under 'normal' loading situations, Haapassalo et al. [56] used peripheral quantitative computed tomography to show mean bilateral asymmetries (between the racket holding arm and non-racket holding arm) in second moments of area of the humeral midshaft in male tennis players. Although such studies show bone adaptation to functional loading, it is difficult to infer the exact strain magnitudes that initiate a particular bone remodelling effect. A figure published in Martin [57] does provide some suggestion into the approximate strain magnitudes that could cause bone adaptation. In this case, strains of below 50 microstrain are thought to represent disuse and thus bone resorption, whereas strains of between 1500 and 3000

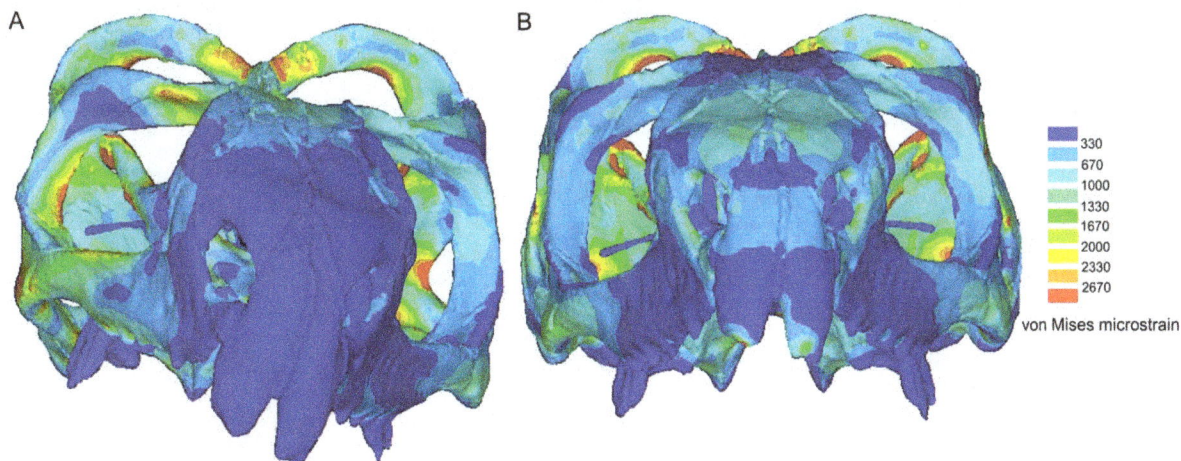

	von Mises microstrain
330	
670	
1000	
1330	
1670	
2000	
2330	
2670	

Figure 3. von Mises FEA plots during two single bites. Deformation and von Mises strain plots of the skull of Sphenodon during **A.** right unilateral biting and **B.** during bilateral biting on the anterior-most chisel-like teeth; (note the displacements are scaled by a factor of 50).

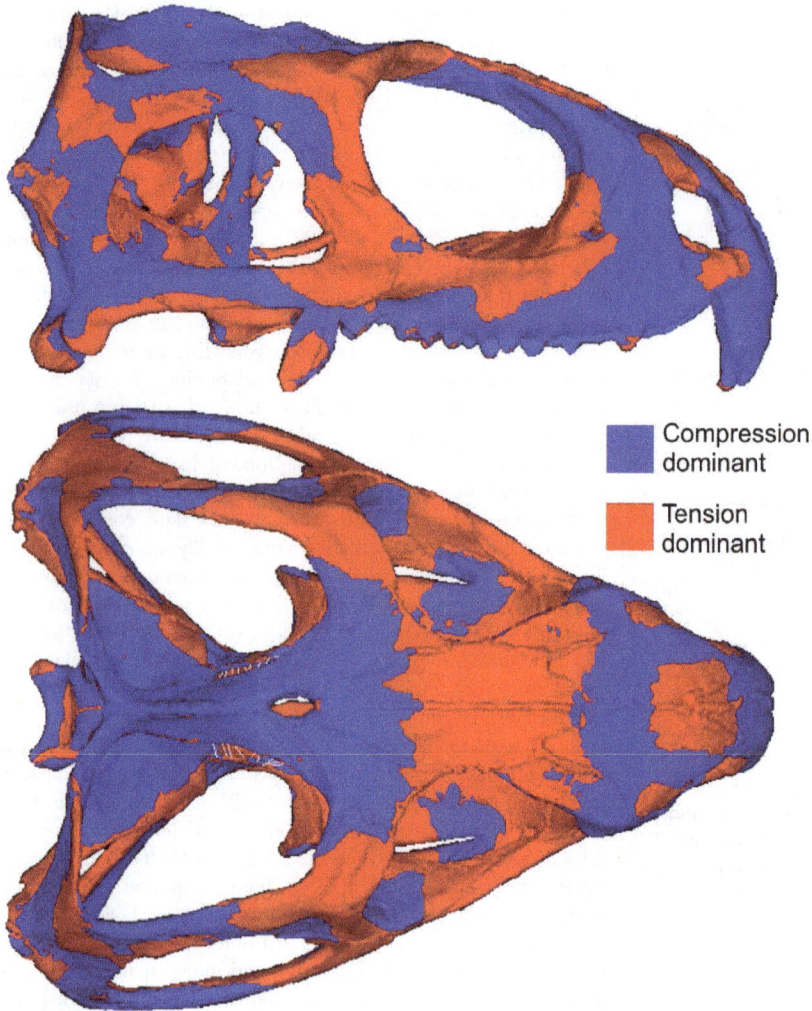

Figure 4. Plot of dominant strain regions. Cumulative map of peak dominant strains over all bites. Red represents regions of the skull where tensile strains are in excess of compressive strains (i.e. tensile strains are dominant), and blue represents regions where compressive strains are in excess of tensile strains (i.e. compressive strains are dominant).

cause some bone formation. Levels above 3000 microstrain are recognised as pathological overload and strains of between 50 and 1500 microstrain would generate equal bone resorption and formation rates (i.e. homeostasis). These values are only speculative and the strain mode or frequency is not specified, but our predicted von Mises strains in the skull of *Sphenodon* are comparable.

We simulated peak bite forces in our study (i.e. ~140 N at an anterior bite position [36,58]), and although bone needs to be able to withstand such forces without risk of failure, the majority of feeding forces will be significantly lower than these applied peak bite forces. For example, Aguirre et al. [59] showed that the approximate force needed to crush a beetle was 34 N, while Herrel et al. [60] recorded a value of 27 N to crush an egg. *Sphenodon* has a varied diet but it frequently includes beetles and occasionally sea bird eggs [61–63]. Thus, the force required to crush these foods is over four times lower than the peak bite force in *Sphenodon*. Scaling skull strains by a factor of four (i.e. in line with bite forces being four times lower) we show that over 91% of the skull is at strains of between 100 and 625 microstrain, well within the equilibrium window (i.e. equal bone resorption and deposition) as inferred by Martin [57].

The findings of this study imply that the skull of *Sphenodon* is adapted to feeding forces, with some regions adapted to tensile forces and others to compressive forces. Tendons and ligaments provide little resistance to compressive strains, and bone is necessary to provide compressive stability. We show that all regions of the skull experience compressive strain when all biting load cases are analysed, suggesting that it is mechanically necessary. However, while bone is necessary to resist compression, it must also be strong enough not to fail under tension. Therefore, once formed, bone must also adapt to tensile strains, and our results support this. Previous analyses, which include *in vivo* experimentation and FEAs suggest different functions for different regions of the skull based on stress and strain recordings/ predictions [5,64–69] (i.e. specific regions seem better suited to biting forces, bending strains, impact loads etc.). While our findings agree with this to some extent (e.g. a specific area of the skull may be linked to a specific bite point, or the forces generated at the jaw joints), they are not consistent with the conclusion that some regions of the skull are formed in relation to factors unrelated to functional strains (e.g. the idea that bone is formed to protect the brain and/or sensory organs from potential impact forces that have not yet occurred [5]). Previous studies did not take into

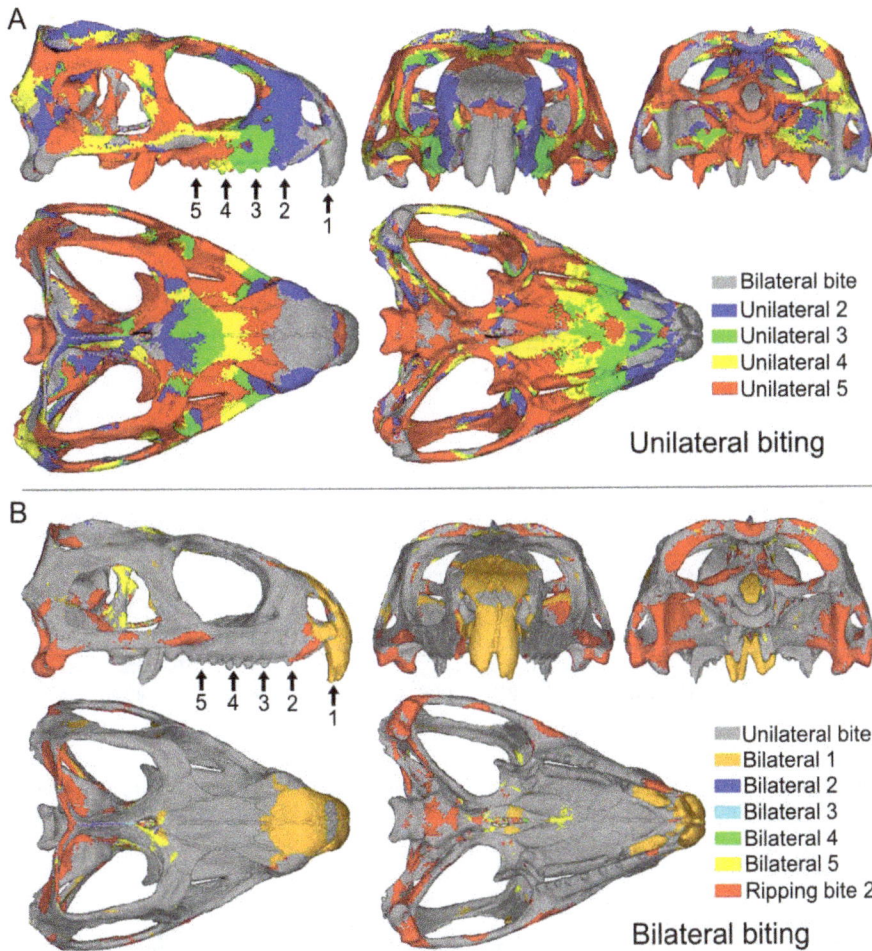

Figure 5. Models showing which bite location generated the highest strains in particular areas of the skull. Results based on von Mises strains. **A.** Unilateral bites and **B.** bilateral bites. (For example, in **A.** unilateral biting at location 2 was responsible for the highest strains in those areas coloured blue).

account the full range of possible and potential loadings, a point made by Mikic and Carter [70] "one difficulty that is encountered when using bone strain data in studies of functional adaptation is the reported data are often far from a complete record of strain over an experimental period". In relation to *in vivo* strain data, these authors further note that "reported results generally consist of a few average cyclic strain parameters that are extracted from a short period of recordings while an animal performs a very restricted task. Most investigators agree, however, that a much more complete record of strain history is required to relate bone biology and morphology to strain".

In our study of skull function we found that strains resulting from a single bite do provide a limited view of overall skull performance (Figure 3 and Figure 6). When we considered a more complete range of physiological loads we showed strains to be more uniform over the entire skull (Figure 7). This finding suggests that the skull is well adapted to a range of functional strains. Although some regions appear to be adapted to tensile strains and others to compressive strains, all regions of the skull seem to be equally important with respect to overall feeding forces. We have shown that unilateral bites, in particular the more posterior unilateral bites, generate the highest strains across the skull. This suggests that such bites are more important to the morphology of the skull of *Sphenodon* than the bilateral ones.

The extent to which general skull form is determined by selection or growth remains uncertain, but our findings show that the skull of *Sphenodon* is optimally suited (mechanically ideal - or at least very well suited) to deal with the full range of loadings applied here. The term 'optimally' refers to the minimum amount of material (i.e. bone) necessary to ensure sufficient skull strength. An optimally formed skull as defined here will be more efficient than a sub-optimal, e.g. heavier skull form, in ensuring minimal bone volume, minimal weight, and also minimal energy demands in maintenance. For clarity, we would predict a non-optimised skull to display one of two contrasting conditions. It would either appear weak in relation to the normal forces applied to it, and experience very high and potentially damaging stresses and strains during normal loading, or, conversely, it might appear overly robust, with very low stresses and strains during normal loading and with excess bone mass that is not mechanically necessary. Since our findings infer that the skull of *Sphenodon* is well formed to resist the everyday forces applied to it, it is not unreasonable to suggest this may also be true for other diapsids with a frame-like skull.

Within our analyses a few small regions of high and low strain are present even when all fifteen biting load cases were accounted for. However, although the muscle representation is detailed in our models, some additional soft tissue structures, such as fascia and ligaments, were not included. At first consideration these

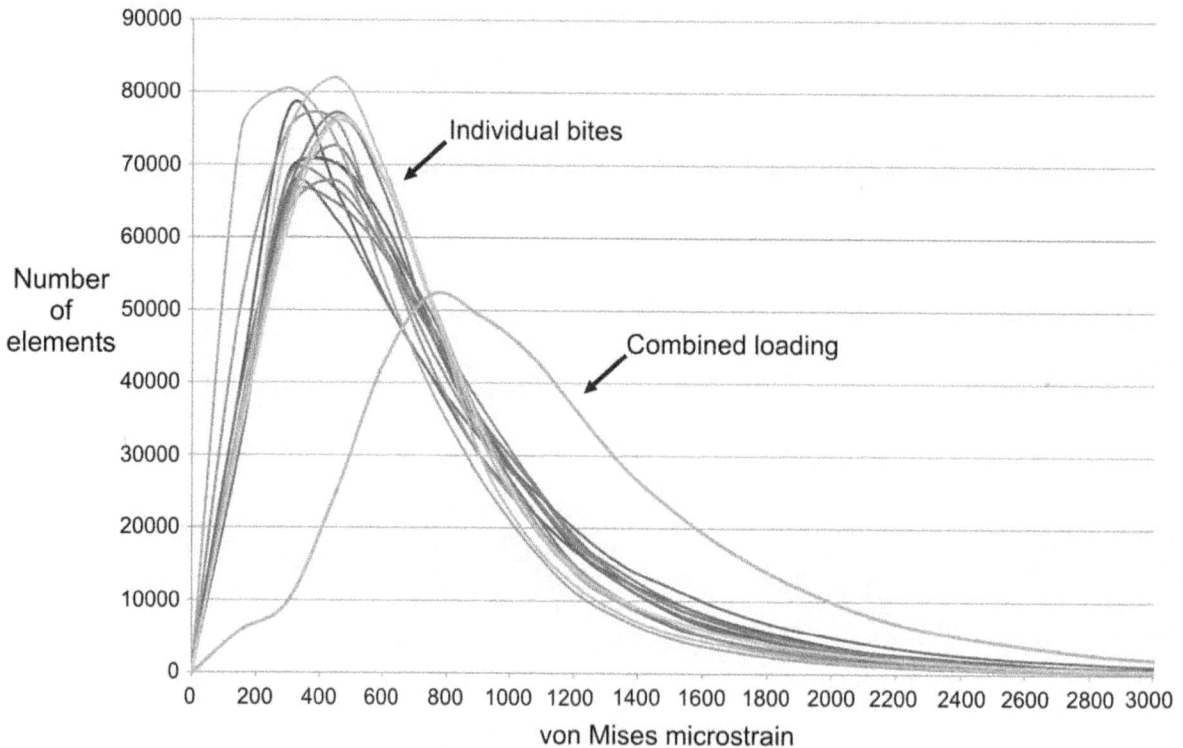

Figure 6. von Mises element strain distribution plots. Plot represents the number of elements within the finite element model that experience a specific strain magnitude. The plot shows the element strains from all fifteen biting simulations (labelled individual bites) and the combined loading model.

structures may appear unimportant, but a recent study investigating the influence of the temporal fascia in primates has revealed that it might play a major role in the function of the skull [71]. Our analyses indicate that the fascial sheet stretched over the upper temporal fenestra in *Sphenodon* may also be significant (Figure 8). This fascial sheet is apparently tensed by upward bulging of the jaw adductor muscles (notably pseudotemporalis superficialis and adductor mandibulae externus medialis) as *Sphenodon* bites down on food (personal observations at Chester Zoo, UK; Dallas Zoo, USA). In this case the fascia serves to reduce peak strains (Figure 8), creating a more uniform strain distribution throughout the skull. The finding that the muscles (including the neck muscles), other soft tissue structures (e.g. upper temporal fascia), bite location, and joint forces all influence the strains within the skull suggests that modifications to any of these anatomical structures has the potential to affect skull form. This may even be somewhat applicable to the formation of unusual skull features, such as crests in chameleons, ceratopsians, and theropod dinosaurs [72–75].

The skull of *Sphenodon*, and probably other non-avian diapsid reptiles without a vaulted braincase (both extant and extinct), is adapted (in the sense of bone adaptation, rather than evolutionary development) to resist a range of load cases, not just single biting loads. The lower temporal bar, secondarily acquired in *Sphenodon* [66,76–80] as well as in the common ancestor of archosaurs like crocodiles [66,80,81], is under compressive strain during all bites. This is consistent with previous suggestions that it provides a brace [66,79,82] that contributes to skull robusticity, and in large theropods such as *Tyrannosaurus rex* Osborn, 1905 and *Allosaurus fragilis* Marsh, 1877 this would be important as they would likely generate extremely large biting forces and experience heavy

cranial loading [4,83]. The corollary is that reptiles that lack a lower bar do not need a brace in this location. Early relatives of *Sphenodon* lack a lower temporal bar, the primitive condition for the group [76–79], but the dorsal position of the jaw joint in these small reptiles suggests that reaction forces would not have been directed along the lower temporal bar, had one existed [78,84].

To conclude, our analysis of the skull of *Sphenodon* indicates that the bone has adapted to tensile and compressive strains generated during normal feeding activities. The combined peak von Mises strain distribution over the skull is relatively uniform, showing that all regions are strong enough mechanically to withstand normal everyday forces, while no region is overly robust and 'over-designed'. Based purely on this finding, the skull form of *Sphenodon* can be considered optimal (mechanically ideal) in the sense that it comprises the minimal amount of bone material for the required skull strength. This optimal form is more efficient in terms of minimal bone volume, minimal weight, and minimal energy demands in maintenance over a sub-optimal, heavier skull form. While this study has not investigated potential forces associated with the brain, sense organs, and non-biting activities such as swallowing and tongue movements, its results are relevant to a broader understanding of skull form and not just to the skulls of diapsid reptiles. However, to test whether all skulls are optimally formed (sufficient strength with the minimal amount of material) with respect to bone strains (both tensile and compressive) would require the application of similar methods to other animal groups. Preliminary findings in macaques are encouraging in this regard (personal observations) but skulls with large vaulted braincases may be subject to additional quasi-static or high frequency low loads (e.g. associated with the brain) that could impact on skull form [28–30,85].

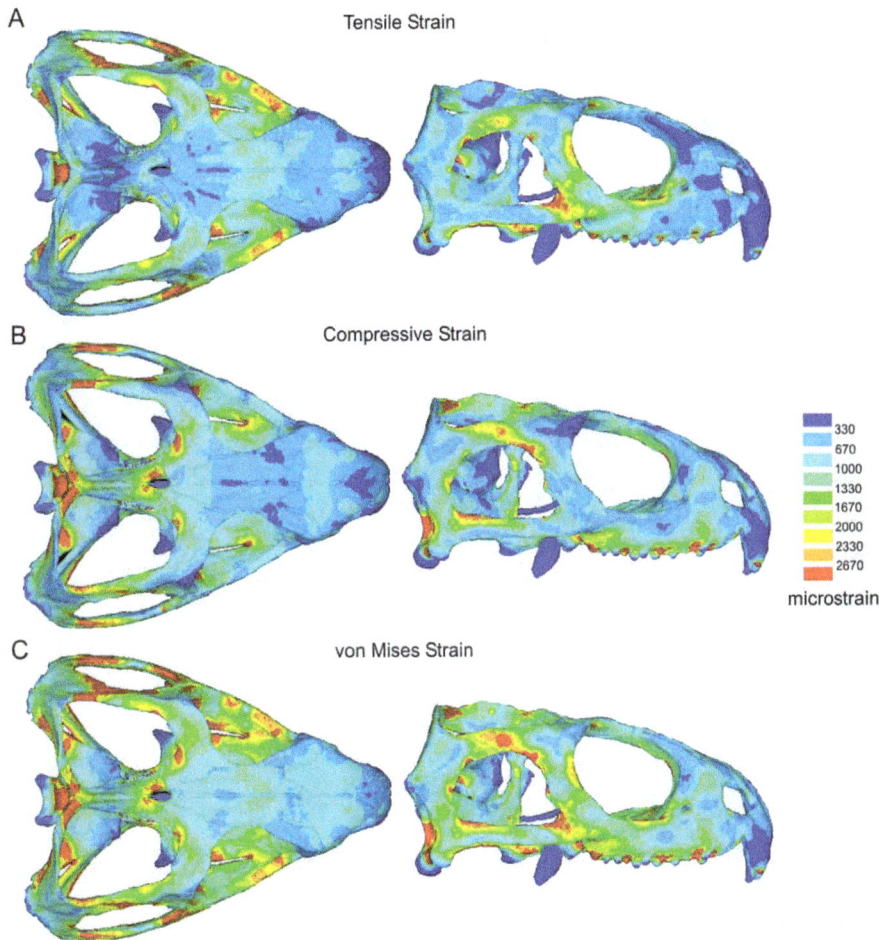

Figure 7. Combined loading tensile, compressive, and von Mises strain plots. Peak combined loading **A.** tensile, **B.** compressive, and **C.** von Mises strain plots.

Materials and Methods

MDA

Detailed descriptions of the MDA model development have been presented elsewhere [33,36,86]. Briefly, the skull and lower jaws (left and right parts) of a *Sphenodon* specimen (specimen LDUCZ x036; Grant Museum of Zoology, UCL, London, UK) were scanned in-house by micro-computed tomography (micro-CT), from which three-dimensional (3D) geometries were constructed using AMIRA image segmentation software (AMIRA 4.1, Mercury Computer Systems Inc., USA). Neck vertebral geometries were generated from additional micro-CT scans (specimen YPM 9194; Yale Peabody Museum of Natural History, New Haven, USA). These 3D geometries were imported into ADAMS multibody analysis software (version 2007 r1, MSC Software Corp., USA) in preparation for an MDA. Within ADAMS detailed muscle anatomy was incorporated onto the geometries, and accurate jaw joint and tooth contact surfaces were specified. Where the neck meets the skull a spherical joint was assigned that permitted the skull to rotate freely about all axes while constraining translational movements. The major adductor (jaw closing), depressor (jaw opening), and neck musculature were included, with each muscle group split into several sections and defined over the anatomical origin and insertions areas on the skull and lower jaws respectively [33,86,87] (Figure 2A). To permit

biting, a food bolus was modelled that could be located at any position along the jaw, and a specially developed motion technique, named dynamic geometric optimisation (DGO), was utilised to open the jaw and to simulate peak biting. This motion technique, along with the muscle forces and biting performance, has been described and validated elsewhere [33,36] (in reference to work carried out *in vivo* [58,88]).

The biting simulations covered a range of biting types and locations, including four bilateral and eight unilateral bites at different tooth positions, a bite on the anterior-most chisel-like teeth, and two ripping bites that incorporate neck muscles (MDA model shown in Figure 2 and a summary of the simulations is shown in Table 1). During the ripping bites the jaws closed on a fixed food bolus, upon which and neck muscles were activated to lift (or try to lift) the head up and to the left, and up and to the right. These two ripping simulations ensured full activation of the neck muscles. During each simulation peak bite force, quadrate-articular joint forces, and muscle forces were predicted.

FEA

The same 3D geometry constructed for the MDA skull was converted into a tetrahedral mesh consisting of 640,000 elements. The model was constructed from solid (ten node) higher order elements, which were specified with a Young's modulus of 17 GPa and a Poisson's ratio of 0.3 (consistent with direct

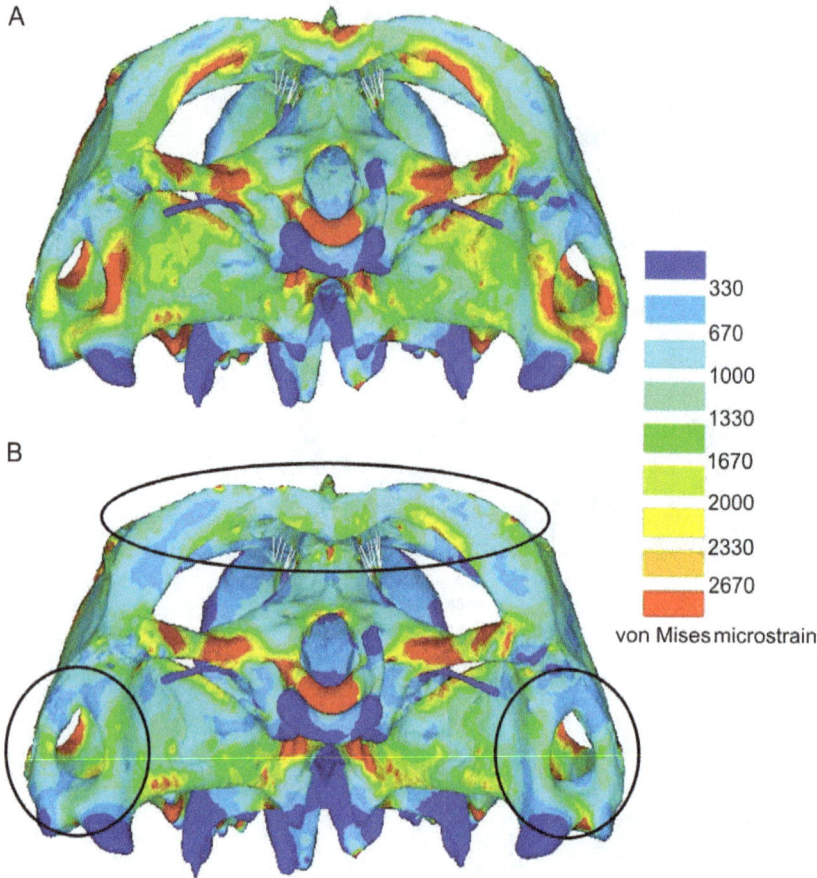

Figure 8. von Mises FEA plots with and without fascia forces. Posterior views of the skull showing von Mises strains predicted by the combined loading model. **A.** Without including fascial forces and **B.** including modelled fascial forces (see Figure 2A). Encircled regions highlight areas where strains have changed significantly due to the inclusion of the fascial forces.

measurements and within the ranges applied by others [1,89–91]. Using the MDA predicted forces, a series of fifteen FEAs were carried out. Although theoretically all forces within the system should be in equilibrium, due to the large number of individual forces even small variations from the exact MDA locations of these applied forces causes instability within the FEAs (i.e. there would be unconstrained full body motion of the model). To ensure a stable FE solution, fixed constraints were included at the joint and bite contacts as defined by the MDA (i.e. neck joint, jaw joints, and bite point). One node at the neck location was constrained in the medial-lateral and anterior-posterior directions (x and z axes), one node at each jaw joint and bite point was constrained in the vertical direction (y axis). These constraints were considered minimal, and restricted rigid body motion but not deformations of the skull. For example, the neck, bite, and joint contact locations could all deform with respect to each other, and both jaw joint contact locations could deform relative to each other. After the FE solutions were complete, tensile (also known as maximum and 1st principal), compressive (also known as minimum and 3rd principal), and von Mises (also known as equivalent and mean) element strains of all 640,000 elements in the model were stored in element tables. In addition, the peak strain recorded in any one particular element during the fifteen separate simulations was extracted and combined to map the peak strains across the skull. This is referred to as a combined loading model.

An additional investigation was carried out to understand the influence of other *non-bone* structures. To this end we simulated an upper temporal fascial sheet, which is likely tensioned by large superior bulging of the jaw adductor muscles during biting (personal observations from animals at Chester Zoo, UK; Dallas Zoo, USA). Here we applied a total force of 133 N around the perimeter of each upper temporal fenestra (7 N over 19 force vectors – see Figure 2A). This magnitude was based on an unrelated investigation [71], where the total fascial force was found to be approximately 85% of the muscle force applied by an associated muscle group(s). In this case the associated muscles were pseudotemporalis superficialis and adductor mandibulae externus medialis [36,87].

Acknowledgments

The authors thank Chester Zoo (UK) and Dallas Zoo (USA) for allowing us access to *Sphenodon* for filming and observation, and the Grant Museum of Zoology (UCL) for access to specimens.

Author Contributions

Conceived and designed the experiments: NC MEHJ JS PO SEE MJF. Performed the experiments: NC. Analyzed the data: NC MEHJ JS PO SEE MJF. Contributed reagents/materials/analysis tools: NC JS. Wrote the paper: NC. Contributed significantly to editing of the submitted manuscript: MEHJ JS PO SEE MJF. Created the computational models: NC MEHJ. Developed analysis techniques: NC JS.

References

1. Witzel U, Preuschoft H (2005) Finite-element model construction for the virtual synthesis of the skulls in vertebrates: case study of *Diplodocus*. The Anatomical Record 283: 391–401.
2. Witzel U (2011) Virtual synthesis of the skull in Neanderthals by FESS. In: Condemi S, Weniger G-C, eds. Continuity and discontinuity in the peopling of Europe: one hundred fifty years of Neanderthal study (vertebrate paleobiology and paleoanthropology). Verlag Berlin Heidelberg: Springer. pp 203–211.
3. Ross CF, Metzger KA (2004) Bone strain gradients and optimization in vertebrate skulls. Annals of Anatomy 186: 387–396.
4. Rayfield EJ, Norman DB, Horner CC, Horner JR, Smith PM, et al. (2001) Cranial design and function in a large theropod dinosaur. Nature 409: 1033–1037.
5. Hylander WL, Johnson KR (1997) *In vivo* strain patterns in the zygomatic arch of macaques and the significance of these patterns for functional interpretations of craniofacial form. American Journal of Physical Anthropology 102: 203–232.
6. Olson EC (1961) Jaw mechanisms in rhipidistians, amphibians, reptiles. American Zoologist 1: 205–215.
7. Moore WJ (1965) Masticatory function and skull growth. Journal of Zoology 146: 123–131.
8. Frazzetta TH (1968) Adaptive problems and possibilities in the temporal fenestration of tetrapod skulls. Journal of Morphology 125: 145–158.
9. Wolff J (1882) Das gesetz der transformation der Knochen: Hirchwild, Berlin. Translated as 'The Law of Bone Remodelling', Springer-Verlag Berlin 1986.
10. Roux W (1881) Der zuchtende Kampf der Teile, oder die "Teilauslee". Organismus (Theorie der "funktionellen Anpassung"). Leipzig: Wilhelm Engelmann.
11. Churches AE, Howlett CR (1982) Functional adaptation of bone in response to sinusoidally varying controlled compressive loading of the ovine metacarpus. Clinical Orthopaedics and Related Research 168: 265–280.
12. Cowin SC, Hart RT, Balser JR, Kohn DH (1985) Functional adaptation in long bones: establishing *in vivo* values for surface remodeling rate coefficients. Journal of Biomechanics 18: 665–684.
13. Lanyon LE, Rubin CT (1985) Functional adaptation in skeletal structures. In: Hildebrand M, Bramble DM, Liem KF, Wake DB, eds. Functional vertebrate morphology. Cambridge, MA: Belknap Press.
14. Lanyon LE (1982) Mechanical function and bone remodeling. In: Sumner-Smith G, ed. Bone in clinical orthopaedics. Philadelphia: Saunders.
15. Lanyon LE, Skerry T (2001) Postmenopausal osteoporosis as a failure of bone's adaptation to functional loading: a hypothesis. Journal of Bone and Mineral Research 16: 1937–1947.
16. Carter DR (1984) Mechanical loading histories and cortical bone remodeling. Calcified Tissue International 39: 19–24.
17. Frost HM (1987) Bone "mass" and the "mechanostat": a proposal. The Anatomical Record 219: 1–9.
18. Turner CH (1998) Three rules for bone adaptation to mechanical stimuli. Bone 23: 399–407.
19. Hsieh YF, Robling AG, Ambrosius WT, Burr DB, Turner CH (2001) Mechanical loading of diaphyseal bone *in vivo*: the strain threshold for an osteogenic response varies with location. Journal of Bone and Mineral Research 16: 2291–2297.
20. Lieberman DE, Devlin MJ, Pearson OM (2001) Articular area responses to mechanical loading: effects of exercise, age, and skeletal location. American Journal of Physical Anthropology 116: 266–277.
21. Currey JD Bones: structure and mechanics: Princeton University Press.
22. Lee K, Jessop H, Suswillo R, Zaman G, Lanyon LE (2003) Endocrinology: bone adaptation requires oestrogen receptor-alpha. Nature 424: 389.
23. Pearson OM, Lieberman DE (2004) The aging of Wolff's "law:" ontogeny and responses to mechanical loading in cortical bone. Yearbook of Physical Anthropology 47: 63–99.
24. Suuriniemi M, Mahonen A, Kovanen V, Alen M, Lyytikainen A, et al. (2004) Association between exercise and pubertal BMD is modulated by estrogen receptor alpha genotype. Journal of Bone and Mineral Research 19: 1758–1765.
25. Burr DB, Robling AG, Turner CH (2002) Effects of biomechanical stress on bones in animals. Bone 30: 781–786.
26. Preuschoft H, Witzel U (2002) Biomechanical investigations on the skulls of reptiles and mammals. Senckenbergiana Lethaea 82: 207–222.
27. Jones MEH, Curtis N, Fagan MJ, O'Higgins P, Evans SE (2011) Hard tissue anatomy of the cranial joints in *Sphenodon* (Rhynchocephalia): sutures, kinesis, and skull mechanics. Palaeontologia Electronica 14: 17A.
28. Moss ML (1954) Growth of the calvaria in the rat, the determination of osseous morphology. American Journal of Morphology 94: 333–361.
29. Moss ML, Young RW (1960) A functional approach to craniology. American Journal of Physical Anthroplogy 74: 305–307.
30. Heifetz MD, Weiss M (1981) Detection of skull expansion with increased intracranial pressure. Journal of Neurosurgery 55: 811–812.
31. Sun Z, Lee E, Herring SW (2004) Cranial sutures and bones: growth and fusion in relation to masticatory strain. The Anatomical Record 276A: 150–161.
32. Curtis N, Kupczik K, O'Higgins P, Moazen M, Fagan MJ (2008) Predicting skull loading: applying multibody dynamics analysis to a macaque skull. The Anatomical Record 291: 491–501.
33. Curtis N, Jones MEH, Evans SE, Shi J, O'Higgins P, et al. (2010) Predicting muscle activation patterns from motion and anatomy: modelling the skull of *Sphenodon* (Diapsida: Rhynchocephalia). Journal of the Royal Society Interface 7: 153–160.
34. Curtis N (2011) Craniofacial biomechanics: an overview of recent multibody modelling studies. Journal of Anatomy 218: 16–25.
35. Curtis N, Jones MEH, Evans SE, O'Higgins P, Fagan MJ (2010) Feedback control from the jaw joints during biting: an investigation of the reptile *Sphenodon* using multibody modelling. Journal of Biomechanics 43: 3132–3137.
36. Curtis N, Jones MEH, Lappin AK, O'Higgins P, Evans SE, et al. (2010) Comparison between *in vivo* and theoretical bite performance: using multi-body modelling to predict muscle and bite forces in a reptile skull. Journal of Biomechanics 43: 2804–2809.
37. Koolstra JH, van Eijden TMGJ (2004) Functional significance of the coupling between head and jaw movements. Journal of Biomechanics 37: 1387–1392.
38. Moazen M, Curtis N, Evans SE, O'Higgins P, Fagan MJ (2008) Rigid-body analysis of a lizard skull: modelling the skull of *Uromastyx hardwickii*. Journal of Biomechanics 41: 1274–1280.
39. Robinson PL, ed. Morphology and biology of the reptiles; how *Sphenodon* and *Uromastix* grow their teeth and use them: Academic Press, London. pp 43–64.
40. Nunemaker DM, Butterweck DM, Provost MT (1990) Fatigue fractures in thoroughbred racehorses: relationships with age, peak bone strain and training. Journal of Orthopaedic Research 8: 604–611.
41. O'Connor JA, Lanyon LE, MacFie H (1982) The influence of strain rate on adaptive remodelling. Journal of Biomechanics 15: 767–781.
42. Rubin CT, Lanyon LE (1982) Limb mechanics as a function of speed and gait: a study of functional strains in the radius and tibia of horse and dog. Journal of Experimental Biology 101: 187–211.
43. Biewener AA, Thomason JJ, Lanyon LE (1983) Mechanics of locomotion and jumping in the forelimb of the horse (*Equus*): *in vivo* stress developed in the radius and metacarpus. Journal of Zoology, London 201: 67–82.
44. Biewener AA, Thomason JJ, Lanyon LE (1988) Mechanics of locomotion and jumping in the horse (*Equus*): *in vivo* stress in the tibia an metatarsus. Journal of Zoology, London 214: 547–565.
45. Biewener AA, Taylor CR (1986) Bone strain: a determinant of gait and speed? Journal of Experimental Biology 123: 383–400.
46. Lanyon LE, Bourne S (1979) The influence of mechanical function on the development of remodeling of the tibia: an experimental study in sheep. Journal of Bone and Joint Surgery 61-A: 263–273.
47. Hylander WL (1979) Mandibular function in *Galago crassicaudatus* and *Macaca fascicularis*: an *in vivo* approach to stress analysis of the mandible. Journal of Morphology 159: 253–296.
48. Rubin CT, Lanyon LE (1984) Regulation of bone formation by applied dynamic loads. Journal of Bone and Joint Surgery 66-A: 397–402.
49. Swartz SM, Bennett MB, Carrier DR (1992) Wing bone stresses in free flying bats and the evolution of skeletal design for flight. Nature 359: 726–729.
50. Biewener AA, Dial KP (1992) *In vivo* strain in the pigeon humerus during flight. American Journal of Zoology 32: 155A.
51. Burr DB, Milgrom C, Fyrhie D, Forwood M, Nyska M, et al. (1996) *In vivo* measurement of human tibial strains during vigorous activity. Bone 18: 405–410.
52. Blob RW, Biewener AA (1999) *In vivo* locomotor strain in the hindlimb bones of *Alligator mississipiensis* and *Iguana iguana*: implications for the evolution of limb bone saftey factor and non-sprawling limb posture. Journal of Experimental Biology 202: 1023–1046.
53. Herring SW, Mucci RJ (2000) *In vivo* strain in cranial sutures: the zygomatic arch. Journal of Morphology 207: 225–239.
54. Herring SW, Pedersen SC, Huang X (2005) Ontogeny of bone strain: the zygomatic arch in pigs. Journal of Experimental Biology 208: 4509–4521.
55. Robling AG, Hinant FM, Burr DB, Turner CH (2002) Improved bone structure and strength after long-term mechanical loading is greatest if loading is separated into short bouts. Journal of Bone Mineral Research 17: 1545–1554.
56. Haapasalo H, Kontulainen S, Sievanen H, Kannus P, Jarvinen M, et al. (2000) Exercise-induced bone gain is due to enlargement in bone size without a change in volumetric bone density: a peripheral quantitative computed tomography study of the upper arms of male tennis players. Bone 27: 351–357.
57. Martin RB (2000) Toward a unifying theory of bone remodeling. Bone 26: 1–6.
58. Jones MEH, Lappin AK (2009) Bite-force performance of the last rhynchocephalian (Lepidosauria: *Sphenodon*). Journal of the Royal Society of New Zealand 39: 71–83.
59. Aguirre LF, Herrel A, van Damme R, Matthysen E (2003) The implications of food hardness for diet in bats. Functional Ecology 17: 201–212.
60. Herrel A, Wauters P, Aerts P, De Vree F (1997) The mechanics of ovophagy in the beaded lizard (*Heloderma horridum*). Journal of Herpetology 31: 189–393.
61. Ussher GT (1999) Tuatara (*Sphenodon punctatus*) feeding ecology in the presence of kiore (*Rattus exulans*). New Zealand Journal of Zoology 26: 117–125.
62. Walls GY (1978) Influence of the tuatara on fairy prion breeding on Stephens Island, Cook Strait. New Zealand Journal of Ecology 1: 91–98.
63. Walls GY (1981) Feeding ecology of the tuatara (*Sphenodon punctatus*) on Stephens Island, Cook Strait. New Zealand Journal of Ecology 4: 89–97.
64. Dumont ER, Davis JL, Grosse IR, Burrows AM (2011) Finite element analysis of performance in the skulls of marmosets and tamarins. Journal of Anatomy 218: 151–162.
65. McHenry CR, Wroe S, Clausen PD, Moreno K, Cunningham E (2007) Supermodeled sabercat, predatory behavior in *Smilodon fatalis* revealed by high-

resolution 3D computer simulation. Proceedings of the National Academy of Sciences 104: 16010–16015.

66. Moazen M, Curtis N, O'Higgins P, Evans SE, Fagan MJ (2009) Biomechanical assessment of evolutionary changes in the lepidosaurian skull. Proceedings of the National Academy of Sciences 106: 8273–8277.

67. Ross CF, Berthaume MA, Dechow PC, Iriarte-Diaz J, Porro LB, et al. (2011) *In vivo* bone strain and finite-element modeling of the craniofacial haft in catarrhine primates. Journal of Anatomy 218: 112–141.

68. Degrange FJ, Tambussi CP, Moreno K, Witmer LM, Wroe S (2010) Mechanical analysis of feeding behavior in the extinct "terror bird" *Andalgalornis steulleti* (Gruiformes: Phorusrhacidae). PLoS ONE 5: doi:10.1371/journal. pone.0011856.

69. Ravosa MJ, Noble VE, Hylander WL, Johnson KR, Kowalski EM (2000) Masticatory stress, orbital orientation and the evolution of the primate postorbital bar. Journal of Human Evolution 38: 667–693.

70. Mikic B, Carter DR (1995) Bone strain gauge data and theoretical models of functional adaptation. Journal of Biomechanics 28: 465–469.

71. Curtis N, Witzel U, Fitton L, O'Higgins P, Fagan MJ (in press) The mechanical significance of the temporal fasciae in *Macaca fascicularis*: an investigation using finite element analysis. The Anatomical Record.

72. Hammer WR, Hickerson WJ (1994) A crested theropod dinosaur from Antarctica. Science 264: 828–830.

73. Rieppel O, Crumly C (1997) Paedomorphosis and skull structure in Malagasy chamaeleons (Reptilia: Chamaeleoninae). Journal of Zoology 243: 351–380.

74. Bickel R, Losos JB (2002) Patterns of morphological variation and correlates of habitat use in chameleons. Biological Journal of the Linnean Society 76: 91–103.

75. Farlow JO, Dodson P (1975) The behavioral significance of frill and horn morphology in ceratopsian dinosaurs. Evolution 29: 353–361.

76. Evans SE (2008) The skull of lizards and tuatara. In: Gans C, Gaunt AS, Adler K, eds. Biology of the Reptilia. Ithaca New York: Society for the Study of Amphibians and Reptiles. pp 1–344.

77. Evans SE, Jones MEH, eds (2010) The origin, early history and diversification of lepidosauromorph reptiles. Verlag Berlin Heidelberg: Springer. pp 27–44.

78. Jones MEH (2008) Skull shape and feeding strategy in *Sphenodon* and other Rhynchocephalia (Diapsida: Lepidosauria). Journal of Morphology 269: 945–966.

79. Whiteside DI (1986) The head skeleton of the Rhaetian sphenodontid *Diphydontosaurus avonis* gen. et sp. nov., and the modernising of a living fossil.

Physiological Transactions of the Royal Society of London, Series B 312: 379–430.

80. Müller J (2003) Early loss and multiple return of the lower temporal arcade in diapsid reptiles. Naturwissenschaften 90: 473–476.

81. Nesbitt SJ (2011) The early evolution of archosaurs: relationships and the origin of major clades. Bulletin of the American Museum of Natural History 352: 1–292.

82. Rieppel O, Gronowski RW (1981) The loss of the lower temporal arcade in diapsid reptiles. Zoological Journal of the Linnean Society 72: 203–217.

83. Erickson GM, Van Kirk SD, Su J, Levenston ME, Caler WE, et al. (1996) Bite-force estimation for *Tyrannosaurus rex* from tooth-marked bones. Nature 382: 706–708.

84. Jones MEH, Curtis N, Evans SE, O'Higgins P, Fagan MJ (2010) Cranial joints in *Sphenodon* (Rhynchocephalia) and its fossil relatives with implications for lepidosaur skull mechanics. Journal of Vertebrate Paleontology, Program and Abstracts 2010: 113A.

85. Sun Z, Lee E, Herring SW (2004) Cranial sutures and bones: growth and fusion in relation to masticatory strain. The Anatomical Record 276A: 150–161.

86. Curtis N, Jones MEH, Evans SE, O'Higgins P, Fagan MJ (2009) Visualising muscle anatomy using three-dimensional computer models - an example using the head and neck muscles of *Sphenodon*. Palaeontologia Electronica 12.3.7T.

87. Jones MEH, Curtis N, O'Higgins P, Fagan MJ, Evans SE (2009) The head and neck muscles associated with feeding in *Sphenodon* (Reptilia: Lepidosauria: Rhynchocephalia). Palaeontologia Electronica 12.

88. Gorniak GC, Rosenberg HI, Gans C (1982) Mastication in the tuatara, *Sphenodon punctatus* (Reptilia: Rhynchocephalia): structure and activity of the motor system. Journal of Morphology 171: 321–353.

89. Strait DS, Wang Q, Dechow PC, Ross CF, Richmond BG, et al. (2005) Modeling elastic properties in finite element analysis: how much precision is needed to produce an accurate model? The Anatomical Record 283: 275–287.

90. Dumont ER, Grosse IR, Slater GJ (2009) Requirements for comparing the performance of finite element models of biological structures. Journal of Theoretical Biology 256: 96–103.

91. Wang Q, Wright BW, Smith A, Chalk J, Byron CD (2010) Mechanical impact of incisor loading on the primate midfacial skeleton and its relevance to human evolution. The Anatomical Record 293: 607–617.

92. Madsen JH (1976) *Allosaurus fragilis*: a revised osteology: Utah Geological Survey Bulletin 109: 1–163.

9

The Skeletal Phenotype of Chondroadherin Deficient Mice

Lovisa Hessle[1,⑨], Gunhild A. Stordalen[2*,⑨], Christina Wenglén[1], Christiane Petzold[3], Elizabeth K. Tanner[4,5], Sverre-Henning Brorson[2], Espen S. Baekkevold[2], Patrik Önnerfjord[1], Finn P. Reinholt[2], Dick Heinegård[1]

1 Sections of Molecular Skeletal Biology and Rheumatology, Department of Clinical Sciences Lund, Lund University, Lund, Sweden, 2 Department of Pathology, University of Oslo, and Oslo University Hospital, Rikshospitalet, Oslo, Norway, 3 Faculty of Odontology, University of Oslo, Oslo, Norway, 4 School of Engineering, University of Glasgow, Glasgow, United Kingdom, 5 Section of Orthopaedics, Department of Clinical Sciences Lund, Lund University, Lund, Sweden

Abstract

Chondroadherin, a leucine rich repeat extracellular matrix protein with functions in cell to matrix interactions, binds cells via their $\alpha2\beta1$ integrin as well as via cell surface proteoglycans, providing for different sets of signals to the cell. Additionally, the protein acts as an anchor to the matrix by binding tightly to collagens type I and II as well as type VI. We generated mice with inactivated chondroadherin gene to provide integrated studies of the role of the protein. The null mice presented distinct phenotypes with affected cartilage as well as bone. At 3–6 weeks of age the epiphyseal growth plate was widened most pronounced in the proliferative zone. The proteome of the femoral head articular cartilage at 4 months of age showed some distinct differences, with increased deposition of cartilage intermediate layer protein 1 and fibronectin in the chondroadherin deficient mice, more pronounced in the female. Other proteins show decreased levels in the deficient mice, particularly pronounced for matrilin-1, thrombospondin-1 and notably the members of the $\alpha1$-antitrypsin family of proteinase inhibitors as well as for a member of the bone morphogenetic protein growth factor family. Thus, cartilage homeostasis is distinctly altered. The bone phenotype was expressed in several ways. The number of bone sialoprotein mRNA expressing cells in the proximal tibial metaphysic was decreased and the osteoid surface was increased possibly indicating a change in mineral metabolism. Micro-CT revealed lower cortical thickness and increased structure model index, i.e. the amount of plates and rods composing the bone trabeculas. The structural changes were paralleled by loss of function, where the null mice showed lower femoral neck failure load and tibial strength during mechanical testing at 4 months of age. The skeletal phenotype points at a role for chondroadherin in both bone and cartilage homeostasis, however, without leading to altered longitudinal growth.

Editor: Nikos K. Karamanos, University of Patras, Greece

Funding: Grants were obtained from the European Union (OSTEOGENE, FP6-502491), the Swedish Research Council, Konung Gustaf V's 80-Årsfond, Sweden, and South-Eastern Regional Health Authority, Norway. The funders had no role in study design, data collection and analysis, decision to publish, or preparation of the manuscript.

Competing Interests: The authors have declared that no competing interests exist.

* E-mail: gunhild@stordalenfoundation.no

⑨ These authors contributed equally to this work.

Introduction

Bone and cartilage are both made up of relatively few cells embedded in an abundant extracellular matrix (ECM). In cartilage, collagen fibrils and the negatively charged proteoglycan aggrecan, forming large aggregates with hyaluronic acid, constitute the major structural assemblies of the matrix. These two components provide tissue with tensile strength and resistance against compressive forces, respectively. The members of the small leucine rich repeat proteins (SLRPs) regulate assembly and function of the ECM, particularly the collagen networks, and include decorin, biglycan, asporin, fibromodulin, lumican, keratocan, PRELP (proline arginine-rich end leucine-rich repeat protein), osteoadherin (OSAD) and chondroadherin (CHAD) [1]. Several SLRPs have roles in bridging between cells and matrix by providing for interactions with cell surface receptors such as syndecans (CHAD and PRELP) and integrins (CHAD and OSAD) at the same time as binding to structural matrix proteins, particularly fibril forming collagens exemplified in Camper et al.,

1997, Haglund et al., 2011, and Haglund et al., 2013. The important roles of the SLRP molecules in matrix organization are illustrated by the abnormalities in mice with inactivated SLRP genes showing signs of dysregulation of collagen fibril formation [2–5]. CHAD is a 38 kD protein, first isolated from bovine cartilage [6]. It contains 11 leucine rich repeats (LRRs) and is classified as a SLRP based on its primary structure [1]. CHAD is highly expressed in cartilaginous tissues and is primarily located close to the cells. Lower levels of expression are found in bone, tendon [6–8] and eye [9]. In bovine bone, CHAD is implicated in direct interaction with calcium phosphate mineral [10]. CHAD mediates adhesion of isolated chondrocytes via two mechanisms: one is binding via the $\alpha2\beta1$ integrin [11] an interaction that can mediate signalling between chondrocytes and their extracellular matrix [12]; the other interaction is between the C-terminal chondroadherin sequence and cell surface proteoglycans such as syndecans that can act as receptors (Haglund et al., 2013). Bone CHAD promotes attachment of osteoblastic cells (Mizuno et al.,

1996) and binds with high affinity to collagen types I and II [13]. Also, CHAD interacts tightly with both the N- and C-terminal globular domains of collagen type VI [14]. As CHAD can interact with structural extracellular matrix (ECM) molecules as well as with cells in the tissue, the protein may provide a mechanism for regulating cell activities in relation to ECM structure, and thus, play a role in both cartilage and bone homeostasis. CHAD has an unusually restricted tissue distribution: In rat femoral heads, CHAD is localized mainly in the territorial matrix at different stages of articular cartilage development, and CHAD mRNA is particularly prominent in the late proliferative cells in the epiphyseal growth plate at young age [15]. We now report the generation of a mouse with the CHAD gene inactivated (CHAD $-/-$) and have performed detailed studies of its phenotype with an emphasis on bone and cartilage homeostasis to reveal functions of CHAD *in vivo*. We found that CHAD plays roles in the cartilage development and maturation of the growth plate at young age and in the molecular composition of articular cartilage in adults as well as in bone homeostasis and function.

Results

2.1. Characterization of CHAD$-/-$ mice

CHAD null mice showed normal embryological development and appeared healthy after birth. Macroscopically no phenotypic abnormalities were visible and the mice appeared healthy up to more than one year of age. Both female and male mice were fertile and the CHAD$-/-$ breeding pairs did not differ in litter sizes compared to WT.

2.2. Demonstration of gene inactivation and loss of CHAD

In initial experiments a procedure for identification of CHAD$-/-$ and WT mice was established by the use of PCR of tail samples prepared by routine procedures (Svensson et al., 1999) with primers selected to give different products when CHAD was present or not. These products were distinguished by Agarose gel electrophoresis (fig. 1a). WT mice demonstrated one band of 650 bp while the null mice showed an expected band of 320 bp and the heterozygote showing both bands. The data clearly demonstrate disruption of the CHAD gene in the null animals. This was further substantiated by Western blotting confirming the absence of CHAD in extracts from the null mice as compared to wild-type cartilage, which showed robust expression of CHAD. Liver tissue, which does not normally express CHAD, was used as an additional negative control to confirm that the antibody was not recognizing non-specific bands (fig. 1b).

2.3. General morphology

Heart, lung including bronchial cartilage, kidney, liver and spleen showed no histopathological changes by systematic investigation at the light microscopic level of paraffin sections. Since chondroadherin is expressed in the eye, a more detailed study by semi thin epon sections was undertaken but showed no differences (data not shown).

2.4. Tissue screening by DXA scanning

BMD, lean and fat content were measured in mice 6 weeks, and 3, 5 and 8 months of age. Results showed only very small differences in 6 week-old males lacking CHAD compared to controls. In this group the BMD/mg body weight was slightly lower (CHAD$-/-$ = 1.79±0.1 mm-2, WT = 1.88±0.4 mm-2, p = 0.03), so was the fat content measured in the whole mouse

(CHAD$-/-$ = 11.04%±0.26, WT = 12.23%±0.40, p = 0.03). Apart from these differences, the DXA data did not reveal any abnormalities in CHAD-null mice (data not shown).

2.5. Cartilage

2.5.1. The epiphyseal growth plate. Overall the CHAD$-/-$ mice presented a 35% increase in mean height of the femoral epiphyseal growth plate at 3 weeks of age (p = 0.02, table 1), despite normal length of the femur. When the relative height of each zone was calculated, the resting zone was increased by 30% and the proliferating zone by 45% in CHAD$-/-$ mice aged 3 weeks (p = 0.04 and p = 0.007, respectively) (fig. 2). At 6 weeks of age the proliferating zone was increased by 20% compared to WT mice (p = 0.04).

2.5.2. Expression and localization of proteins in growth plate cartilage. We analysed the expression and organization of a number of proteins in the growth plate to discern differences in the tissue organization upon CHAD inactivation. In both null and wild type animals the distribution of mRNA showed the expected pattern of expression: Cartilage oligomeric protein (COMP) mRNA was primarily detected in the proliferative chondrocytes of the epiphyseal growth plate, with no significant difference between the groups. Osteopontin (OPN) and BSP were both detected in hypertrophic chondrocytes, while there was no detectable difference. Immunostaining for COMP was almost exclusively localized to the epiphyseal growth plate, in addition to some cartilaginous remnants in trabecular areas. The staining was most intense in the territorial matrix along the columns of proliferative chondrocytes, although pericellular staining was observed to various degrees in all zones, and weaker staining was observed in the interterritorial matrix (fig. S1). Histological scoring of COMP staining did not demonstrate any differences between CHAD$-/-$ and WT mice (data not shown).

2.5.3. Changes in the proteome of articular cartilage. Initially there were 226 proteins identified by MASCOT 2.1 from 5457 peptide matches above homology or identity threshold. After filtering the data removing obvious false positives (13 proteins) and protein hits with only one peptide hit, it was possible to measure and calculate the relative ratios of 178 proteins detected in all extracts (table S1). The proteins identified include the major components of the extracellular matrix in cartilage such as collagens, aggrecan and members of the SLRPs family. The comparisons were made without normalization although the average ratios for all proteins were 0.8–0.9 suggesting that the total protein content of the samples from the knockout animals may be slightly lower than the corresponding wild type samples. Overall most of the proteins were present at the same level in the null compared with the WT mice. However, some differences were noted (table 2). The null CHAD mouse vs. its wild type counterpart showed increased levels of fibronectin (ratio 1.85 and 1.61 vs wild type). Both secreted parts of the gene product of CILP 1 (CILP 1–1 and 1–2) were particularly elevated in female null mice (ratio of 1.84 and 1.63, respectively). Markedly decreased levels in the null mice were noted for alpha-1-antitrypsin 1 family members (ratio around 0.3) and apolipoprotein E. Expectedly, CHAD in the null mice were at background noise levels. In support no CHAD was detected in LC-MS analyses of the individual samples of the CHAD null mouse. Corresponding western blots showed no reactivity at all verifying that the sample preparation is free from cross-contamination. Most other proteins showed similar levels in null and wild type mice.

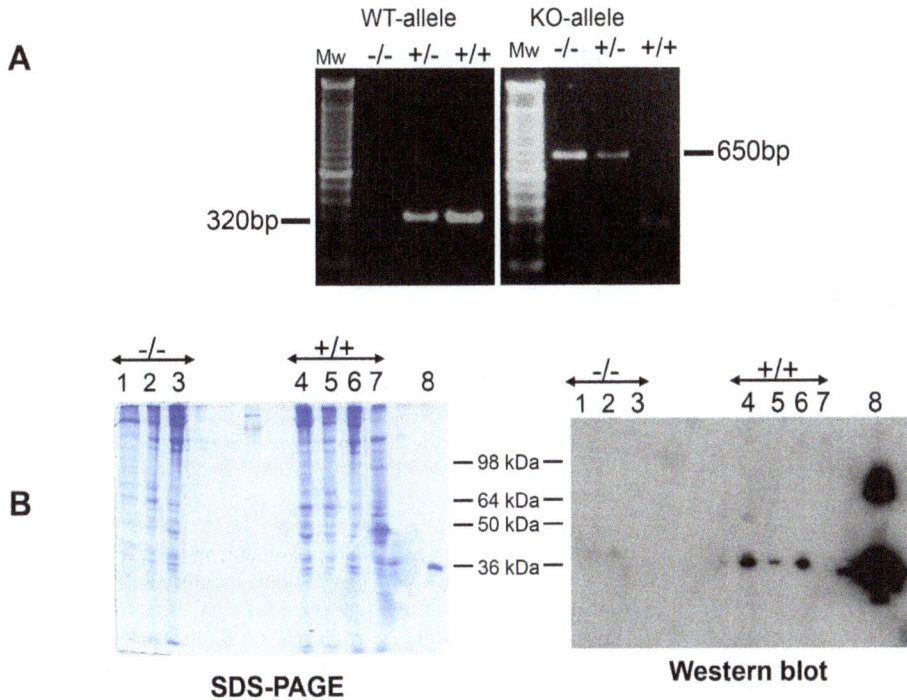

Figure 1. Analyses of message and proteins in CHAD−/− and WT mice. A: PCR and agarose gel electrophoresis of mouse tail samples. There is a faint, barely visible, reactivity at the position of the wild type allele (320 bp), but this is not observed in the CHAD−/− mice. It probably represents some weak reactivity of the wild type allele by the primers for the deleted allele B: Protein stained gel and Western blot of cartilages and liver as a control for non-specific reactions. Different cartilages were extracted with 4 M guanidine hydrochloride, proteins precipitated with ethanol and electrophoresed on 4–16% SDS-PAGE. Left picture represents a Coomassie stained gel and the right picture represents Western blots with the anti-CHAD antibody. The lanes represent extracts of 1. Trachea (−/−); 2. Nasal cartilage (−/−); 3. Knee cartilage (−/−); 4. Trachea (+/+); 5. Nasal cartilage (+/+); 6. Knee cartilage (+/+); 7. Liver (+/+); 8. Recombinant CHAD; −/− represents CHAD−/− and +/+ wild type mice.

2.6. Bone

2.6.1. Micro-CT. Screening of mice aged 5 days, 3 week and 4 months showed that the length of the femora (i.e. the distance between the distal growth plate and the gluteal tuberosity) increased with time but was not significantly different between CHAD−/− and WT mice (table S2).

The 4 months old mice were analysed further. A number of parameters expectedly showed significantly higher values for the male wild type mice (table 3). Particularly noticeable differences were the higher trabecular thickness and structure model index for the null animals (p<0.05). There were also some noticeable differences in the form of lower polar moment of inertia and cortical diameter at midshaft only apparent for the male null mouse (p<0.05), where values were more similar to those of the females. The value for trabecular spacing in the female null mice was lower than the wild type and more similar to those of the male, although values did not reach significance.

2.6.2. Collagen fibres in bone. Qualitative electron microscopic analysis of 6 weeks old CHAD−/− mice showed no abnormalities in the structure, tissue organization and thickness of collagen fibrils in calvarial bone compared to WT mice (data not shown).

2.6.3. Protein expression in bone. Tartrate resistant acid phosphatase (TRAP) mRNA was primarily detected in metaphyseal osteoclasts; in addition, some resting and hypertrophic chondrocytes in the growth plate also expressed TRAP. Cathepsin K (CTK) mRNA was solely present in multinucleated cells in the metaphyseal region. OPN mRNA was detected primarily in osteoblasts lining the metaphyseal trabecular surfaces. As for OPN, BSP mRNA was highly expressed in osteoblasts lining trabeculae.

The relative number of BSP mRNA expressing cells was significantly lower in the metaphysis of CHAD−/− mice compared with WT (p = 0.01) (fig. 3). Thus, the mean score in the metaphysis for null mice was 0.4 (SD 0.9) as compared to WT 1.8 (0.4); n = 5 in both groups. No differences were found for relative number of COMP, TRAP, CTK or OPN mRNA expressing cells (data not shown).

2.6.4. Protein localization in bone. The most intense accumulation of gold particles for both BSP (fig. S2) and OPN was observed at electron dense extracellular areas representing osteoid-bone interface/mineralization fronts, and to a lesser extent, diffusely spread in mineralized bone. BSP exhibited a characteristic pattern with labelling confined to discrete sites in bone matrix corresponding to areas of early mineral deposition. Semi-quantitative analysis revealed a trend towards increased signal intensity in most compartments for both proteins in CHAD−/− mice (table 4). However, when each compartment was compared between CHAD−/− and WT mice, only BSP labeling in osteoid was found to be significantly increased (table 4).

2.6.5. Mechanical properties of bone. All mechanical properties increase significantly from 6 weeks to 4 months of age (fig. 4). Femoral neck failure load (fig. 4a) was significantly lower in the 4 month old CHAD−/− mice compared to the same age and gender wild type (p<0.01). The difference between CHAD−/− and wild type female at 4 months was small and not significant although again the wild type may show somewhat higher bone strength. Males showed higher strength than

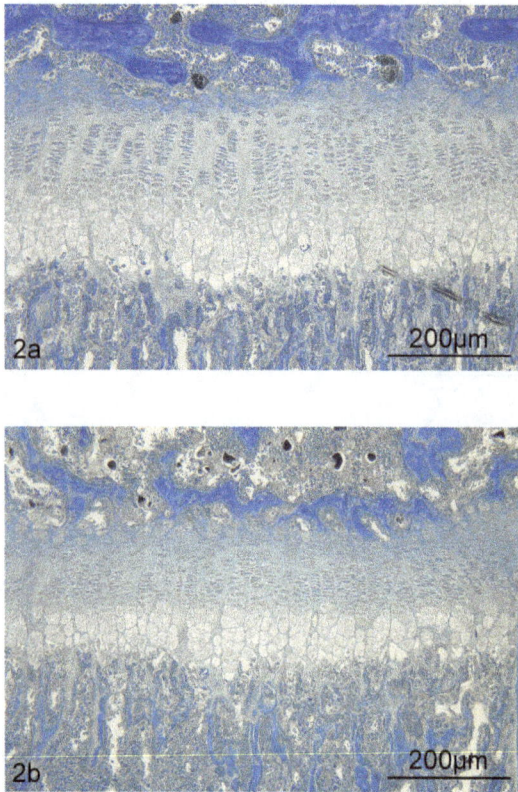

Figure 2. Micrographs of the epiphyseal plate in CHAD −/− and WT mice. Light micrographs of CHAD−/− (2a) and WT (2b) mice at 3 weeks of age. A small but significant increase in the height of the growth plate mostly confined to the proliferative zone was confirmed by histomorphometry (table 3). Epon-embedded tissue, toluidine blue staining (x 20).

Table 1. Histomorphometric analyses of the epiphyseal growth plate in 3 and 6 weeks old CHAD−/− mice and age-matched WT mice.

		3 weeks old		6 weeks old	
	Zone	CHAD−/−	WT	CHAD−/−	WT
Epiphyseal height	–	288±54 *	214±14	269±31	238±44
Height of zone	Resting	39±8 *	30±4	32±3	37±9
	Proliferating	155±27 **	107±6	151±16 *	126±23
	Hypertrophic	94±23	77±8	85±15	74±14
Osteoid surface	Metaphysis	–	–	21±6	17±5

Values are mean ± SD. For epiphyseal values (μm), n=6/4 and n=6/8 for CHAD−/− and WT mice after 3 and 6 weeks, respectively. For osteoid values (% of trabecular surface), n=7/6. *p<0.05, ** p<0.01.

joint or spine disease. An unexpected finding was the consistently low levels of several variants of alpha-1-antitrypsin. This might affect the susceptibility against proteolytic activity and thereby also overall tissue stability. Interestingly, it is rather obvious that none of the other SLRP proteins are differently expressed following the removal of CHAD.

3.1. Disturbances in the epiphyseal growth plate but normal collagen organization

The CHAD −/− mice presented a widened epiphyseal growth plate. This was most pronounced in the proliferative zone at 3 and 6 weeks of age which fits well with data showing that CHAD is synthesized mainly by late proliferative chondrocytes [7]. The balance between proliferation and differentiation of chondrocytes is an important regulatory step controlled by multiple signalling molecules, including the Indian hedgehog (Ihh)/parathyroid hormone related peptide (PTHrP) feedback loop [16]. Interestingly, Ihh, which is upstream in the signalling pathway of PTHrP, shows a similar distribution of expression to that of CHAD [17]. Ihh controls the transition from proliferating to hypertrophic chondrocytes [16], thereby regulating the height of the proliferative zone. Thus, based on the very distinct localization around a portion of the proliferative chondrocytes, the apparent absence of cell spreading and growth of chondrocytes on CHAD coated surfaces [18], as well as the observed widening of the proliferative zone in null mice, it could be speculated that CHAD may influence the Ihh/PTHrP feedback loop and/or participate in the control of chondrocyte development by promoting their differentiation into the hypertrophic stage. However, the lack of detectable differences in bone length between null and WT mice suggest that CHAD influences chondrocyte maturation only to a modest degree or that other processes compensate by modulating subsequent events. It has been documented that CHAD interacts with collagen and the protein is abundant in the territorial matrix, suggesting a role in early assembly and function of fibrillar collagen [13]. However, CHAD-null mice showed normal collagen organization and fibril diameter in the bone, indicating either that other molecules than CHAD play more prominent roles in the process or that CHAD differs from other SLRPs not only with respect to localization, but also regarding its function vis-à-vis collagen [4].

females. In contrast at 8 month there were no observable differences between the mice whether wild type-mutant or female-male were compared. Interestingly the strength of the 4 months wild type male mice appeared higher than that of 8 months animals. The tibial strength (fig. 4b) showed similar trends to the femoral neck strength. The null mice showed significantly lower strength than the wild type at 4 months both for males and females (p<0.001). At 8 months differences could not be discerned.

Discussion

CHAD deficient mice did not show gross anatomical defects, grew to normal size, were fertile, and had a normal life span up to 2 years of age, which is in line with other studies with SLRPs-null mice [2–5]. However, the CHAD deficient mice presented a distinct skeletal phenotype, demonstrating a role for CHAD in cartilage and bone turnover. With the exception of increased levels of CILP-1 and fibronectin in the female CHAD deficient mouse, alterations in the identifiable extracellular matrix proteins proper in articular cartilage were small. The altered levels of both CILP proteins in particularly the female mice, albeit still only by some 50%, are interesting but at this time the functional implications are not known. It can be noted that CILP is a protein up regulated in osteoarthritis (Bernardo et al., 2011), and there is a polymorphism that correlates to a higher incidence of lumbar disc disease (Seko et al., 2005). The mice at the ages studied showed no signs of either

Table 2. Changes in the proteome of cartilage from mice with the CHAD gene inactivated.

Acc. No.	Protein name	Fem KO vs Wt	Male KO vs Wt
P11276	Fibronectin	1,85	1,61
Q66K08	Cartilage intermediate layer protein 1	1,75	1,18
P31725	Protein S100-A9	1,02	0,21
Q9Z1F6	Chondromodulin-1	0,97	0,49
P29699	Alpha-2-HS-glycoprotein	0,82	0,46
P62259	14-3-3 protein epsilon	0,68	0,33
P63101	14-3-3 protein zeta/delta	0,66	0,32
P22599	Alpha-1-antitrypsin 1-2	0,39	0,27
Q00898	Alpha-1-antitrypsin 1-5	0,38	0,28
Q00897	Alpha-1-antitrypsin 1-4	0,38	0,27
Q00896	Alpha-1-antitrypsin 1-3	0,38	0,25
P07758	Alpha-1-antitrypsin 1-1	0,37	0,25
P08226	Apolipoprotein E	0,36	0,35

Acc. No.	Protein domains	Fem KO vs Wt	Male KO vs Wt
P28481	Collagen alpha-1(II) chain	1,23	1,40
P28481	Collagen C-propeptide	0,90	0,81
P28481	Collagen N-propeptide	1,00	1,37
Q66K08	Cartilage intermediate layer protein 1-2	1,84	1,27
Q66K08	Cartilage intermediate layer protein 1-1	1,63	1,06
D3Z7H8	Cartilage intermediate layer protein 2-2	1,46	1,08
SLRP proteins			
Q99MQ4	Asporin	0,90	1,11
P28653	Biglycan	1,16	1,11
P28654	Decorin	1,00	1,05
P70186	Epiphycan	0,84	0,90
P50608	Fibromodulin	1,01	1,05
P51885	Lumican	0,98	0,92
Q62000	Mimecan	0,80	0,89
O35103	Osteomodulin	0,77	1,02
Q9JK53	Prolargin	1,10	1,17

3.2. Altered mechanical properties, cortical/trabecular bone parameters and loss of sex-specific differences

Significant differences in trabecular/cortical parameters were apparent at the age of 4 months, where the null mice presented significantly higher BMD, lower cortical thickness, increased trabecular thickness, and increased structure model index (SMI) by micro-CT. SMI is a measure of the ratio of "plate-like" to "rod-like" trabecula within a trabecular bone specimen and higher density cancellous bone generally shows more "plate-like" trabecula. Interestingly, osteoporotic trabecular bone transits from plate-like to rod-like, increasing the SMI [19]. Thus, increased SMI of the trabecular bone in the null mice is consistent with impaired mechanical properties. In line with the micro-CT data indicating disturbed formation and/or remodelling of bone, the mechanical testing showed reduced mechanical strength of both femoral neck cancellous bone and tibial cortical bone at 4 months. This was most pronounced for the male mice, possibly reflecting different rates of bone turnover between male and female. The older 8 months mice showed no such discernable difference, perhaps indicating a lower bone metabolism at this age.

Noteworthy, this group of mice was based on the C57BL/6 strain which has been shown to have larger cortical cross-section areas but to be less responsive to increased mechanical loading than other used strains, e.g. the C3H/He and DBA/2 (Robling et al., 2002). Our data show that CHAD influences both cortical and trabecular bone formation and/or remodelling. The male null mice showed an appearance of the studied variables more similar to the parameters observed for the wild type female mice. Our data suggest that CHAD is important in the sex-specific development of the skeleton. Such loss of sex-specific differences has previously also been reported in OPN deficient mice [20].

3.3. Decreased number of cells expressing BSP mRNA

Non collagenous proteins of the SIBLING (small integrin-binding ligand, N-linked glycoprotein) family (Fisher et al., 2001), which includes OPN and BSP, are believed to play key biological roles in the development, turnover and mineralization of bone (reviewed in [21] and [22]). Both BSP and OPN are secreted by osteoblasts and have been shown to modulate osteoblast differentiation and mineralization *in vitro*. BSP for the most part promotes

Table 3. Micro-CT cortical/trabecular bone parameters in 4 month old mice.

Parameter	Male KO	Male WT	Female KO	Female WT	Average KO	Average WT
Bone length (mm)	7.04±0.36	6.82±0.52	6.75±0.28	7.27±0.35	6.89 ±0.34	7.10±0.45
Ct.Th (mm)	0.18±0.01	0.17±0.02[c]	0.18±0.01[b]	0.19±0.01	0.18±0.01	0.18±0.02
vBMD (g/cm^{-3})	1.57±0.08	1.51±0.04[c]	1.55±0.08	1.59±0.06	1.56±0.08	1.56±0.07
Ct.BV (mm^3)	99.61±0.58	99.66±0.17 [c]	99.62±0.44	99.82±0.07	99.61±0.50	99.76±0.14
Ct.P (%)	0.39±0.58	0.34±0.17 [c]	0.38±0.44	0.18±0.07	0.39±0.50	0.24±0.14
Tb.BV (%)	7.43±1.37	7.43±1.44 [d]	5.13±2.27	4.35±0.69	6.28±2.13	5.51±1.85
Tb.P (%)	92.57±1.37	92.57±1.44 [d]	94.87±2.28	95.65±0.69	93.72±2.13	94.49±1.85
Tb.Th (mm)	0.046±0.003	0.042±0.005	0.043±0.004	0.041±0.002	0.045±0.003[a]	0.041±0.003
Tb.Sp (mm)	0.27±0.01	0.27±0.01[c]	0.29±0.02	0.33±0.03	0.28±0.02	0.31±0.04
DA	2.14±0.20	2.16±0.13	2.24±0.35	2.56±0.54	2.19±0.27	2.41±0.46
SMI	2.43±0.11b	2.23±0.01[c]	2.53±0.36	2.42±0.07	2.48±0.25[a]	2.34±0.11
MMI (mm^4)	0.33±0.02[a]	0.72±0.15[d]	0.34±0.06	0.35±0.04	0.33±0.04	0.49±0.21
D (mm)	0.97±0.01[a]	1.15±0.06[d]	0.99±0.04	0.99±0.02	0.98±0.04	1.05±0.09

Bone length, cortical/trabecular thickness (Ct.Th/Tb.Th), volumetric BMD (vBMD), cortical/trabecular porosity (Ct.P/Tb.P), trabecular separation (Tb.Sp), degree of anisotropy (DA), structure model index (SMI), polar moment of inertia (MMI) at midshaft, and equivalent circle diameter (D) at midshaft of femur in CHAD−/− (KO) and WT mice. [a]: $p \leq 0.05$ KO vs. WT, [b]: $p \leq 0.05$ KO male/female vs. WT male/female, [c]: $p \leq 0.05$/[d]: $p \leq 0.01$ males vs. females in same group. Taken together the data clearly demonstrate an altered bone homeostasis in the mice with the chondroadherin gene inactivated.

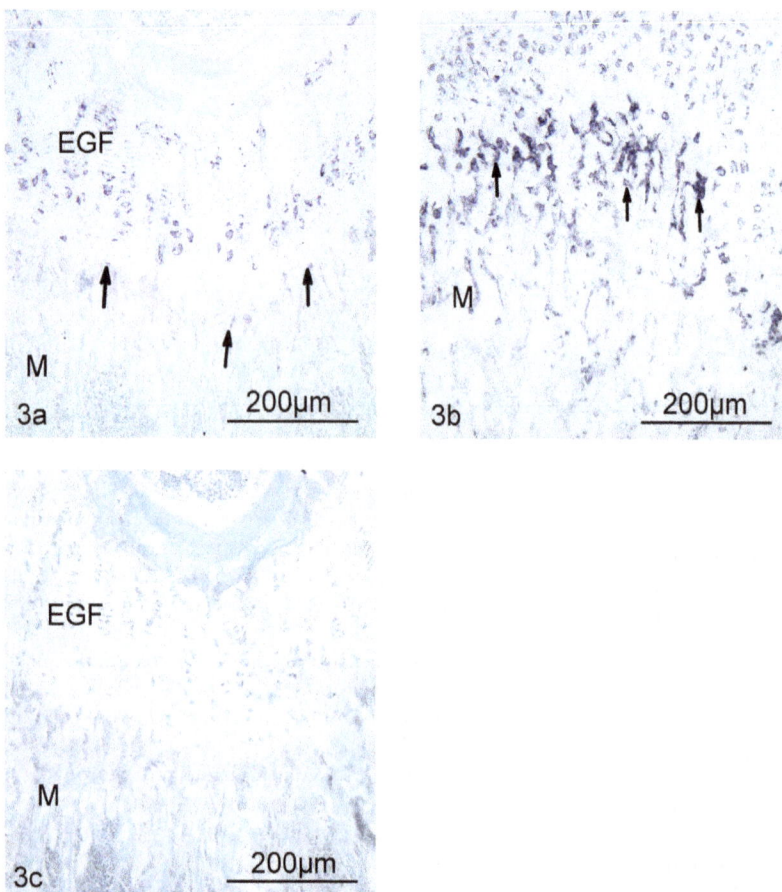

Figure 3. BSP mRNA expressing cells in the distal femur of CHAD−/− and WT mice at 6 weeks of age. *In situ* hybridization showed intense signal in multiple osteoblastic cells (arrows) in the metaphysis (M) of WT mice (3b), while CHAD−/− mice (3a) showed very sparse signal in cells in the corresponding area (p = 0.01). Also chondrocytes in the epiphyseal growth plate (EGF) showed signal, although no quantitative difference in number of cells were detected between the groups at this site (x 20). Negative control with sense probe was without signal (3c).

Table 4. Ultrastructural distribution of OPN and BSP in bone of CHAD−/− and WT mice.

Compartment	OPN		BSP	
	CHAD−/− (n = 6)	WT (n = 6)	CHAD−/− (n = 6)	WT (n = 6)
OB nucleus	1.53±0.46	1.32±0.41	0.94±0.19	1.08±0.19
OB cytoplasm	1.69±0.38	1.70±0.36	0.82±0.15	0.72±0.18
Osteoid	4.54±2.50	3.70±2.06	1.20±0.21*	0.93±0.14
Mineralization front	150.18±51.42	124.21±56.63	16.51±5.60	15.35±6.45
Mineralized bone	15.96±12.40	9.30±6.77	3.83±0.88	2.98±0.68
Capillary lumen	0.80±0.22	0.57±0.20	0.46±0.07	0.50±0.15

Semi quantitative analysis of the ultrastructural protein distribution in CHAD−/− and WT mice. Values are mean ± SD gold particles/um^2 per animal. OB = osteoblast.
* $p < 0.05$ when compartments are directly compared.

the process [23–25]. Interestingly, a considerable decrease in the number of BSP mRNA expressing cells was noted in the CHAD null mice. This decrease together with the slightly increased osteoid surface observed in the femoral metaphysis of these mice may imply impaired mineralization. On the background of altered cortical/trabecular parameters and decreased number of BSP mRNA positive cells in the distal femur metaphysis of CHAD-null mice, we extended the study of BSP and OPN and investigated their protein distribution in bone at the ultrastructural level. Despite a tendency towards increased signal intensity in osteoid, over osteoid-bone interfaces/mineralization fronts as well as in mineralized bone for both proteins in CHAD-null mice, there was no overall significant difference in the protein distribution pattern.

However, this observation is not necessarily contrary to the *in situ* hybridization data, as protein distribution in the tissue depends not only on synthesis but also on secretion and degradation in the ECM. Thus, although there are fewer cells expressing BSP mRNA in CHAD-null mice, the protein synthesis of those expressing the gene appears normal, and the number of cells expressing OPN is normal. Thus, taken together, the CHAD null mouse appears to have an altered and lower bone turnover.

Conclusions

The present study has provided the first evidence that the absence of CHAD leads to a distinct skeletal phenotype characterized by widening of the epiphyseal growth plate with possible impaired of hypertrophic differentiation of chondrocytes, reduced number of BSP expressing cells, disturbed molecular composition of articular cartilage and structural and functional alterations in trabecular and cortical bone tissue with alterations in bone turnover.

Materials and Methods

This study was carried out in strict accordance with the institutional guidelines for animal research at Lund University, Sweden. The protocols were approved the Committee on the Ethics of Animals at Lund University, Sweden (Permit Numbers: M31-09 and M177-11).

5.1. Generation of CHAD−/− mice

A mouse genomic cosmid library was screened using a 887-bp CHAD rat cDNA fragment as described [26]. A 31 kbp genomic DNA fragment was isolated and partly sequenced. Out of this a 3000 bp fragment including the ATG of the CHAD gene was

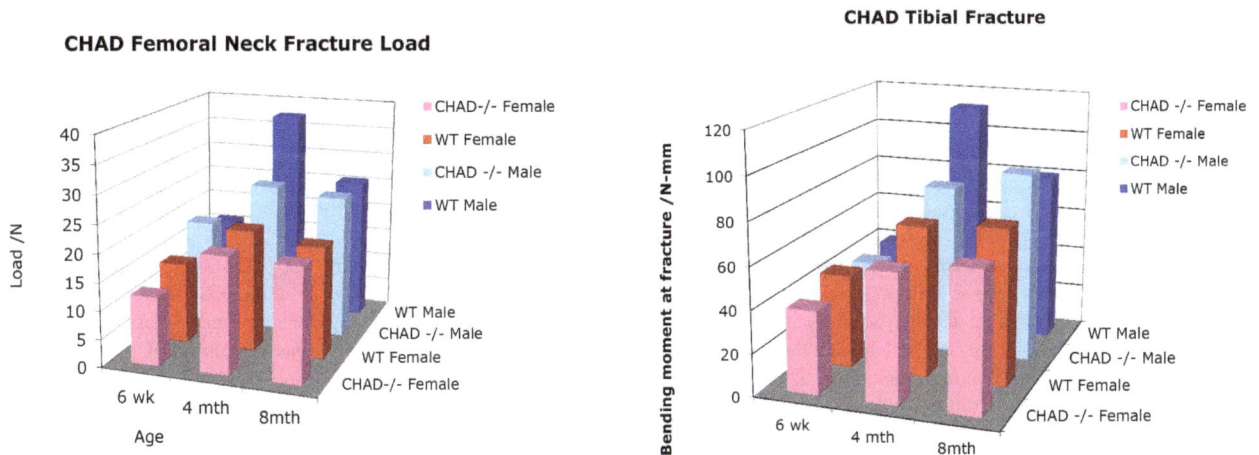

Figure 4. **Mechanical properties of bone.** 4a: Femoral neck failure load (Newtons) for 6 weeks, 4 months and 8 months old wild type (wt) and CHAD−/− male and female mice. The difference between CHAD−/− and wt in 4 months old male mice is significant (p<0.01). 4b: Tibial shaft failure load (Newtons) for 6 weeks, 4 months and 8 months old wild type (wt) and CHAD−/− male and female mice. The difference between CHAD−/− and wt in 4 months old mice is significant for both genders (p<0.001).

inserted to the pWH9 vector (kindly provided by Dr. R. Fässler) that carries a phosphoglycerate kinase-neomycin resistance cassette (pGKNeo). The 3000 bp fragment was inserted 5′ of the neocassette. In the 3′end of the cassette a 7000 bp CHAD fragment was inserted.

5.1.1. Generation of recombinant ES cell lines and chimeric mice. Approximately 20×10^6 semi-confluent R1 embryonic stem (ES) cells (kindly provided by professor Reinhard Fässler) (Nagy et al., 1993) were electroporated with 80 μg linearized targeting vector. The ES cells were cultured on feeder cells in DMEM supplemented with fetal calf serum and leukemia inhibitor factor (for reference see[27]). After 24 hours of culture, selection for positive clones was initiated by the addition of 500 μg/ml G418. Positive clones were picked, expanded and DNA was purified and subsequently analyzed by Southern Blot analysis for confirmation of the correct targeting events. Targeted ES-cells were injected into mouse blastocysts according to standard procedures. Chimeric males were mated with C57BL/ 6 females and males with germ line transmission were further bred with 129/sv females to establish a strain of CHAD-null mice. Before analyses, the mice were backcrossed for 10 generations into the C57BL/6 background.

5.1.2. Genotyping of CHAD−/− mice. Genomic DNA from tail tendon was purified, digested with EcoR1 and separated on an agarose gel using standard procedures. A 1000 bp XbaI-HindIII fragment was used as a probe in the hybridization. This probe detects a 16 kbp fragment in the wild type (WT) mouse and a 13 kbp fragment in the targeted mouse.

PCR was used to detect homologous recombination of the mouse CHAD gene. A 5′primer 5′CAG TCT GGT CTT TCT TGC CA was used together with a 3′primer 5′ATG TCG TTG TGG GAC AGG TA. This detects a 320 bp fragment in the WT mouse. An additional primer corresponding to the sequence 5′CGC CTT CTT GAC GAG TTC TT in the neo-cassette was used to detect a fragment of 650 bp corresponding to homologous recombination in the knock-out.

5.2. Skeletal X-ray analysis

Bone mineral density (BMD), fat and lean content were examined with dual-energy X-ray absorption (DXA) using the Lunar PIXImus Densitometer (GE Medical Systems). Measurements were performed on null and WT mice at the age of 6 weeks, and 3, 5, and 8 months of age (males and females separately). At each time-point at least 6 CHAD−/− and 6 WT animals were measured. The measurements were performed on anaesthetized living animals.

5.3. Micro computed tomography

Micro computed tomography (micro-CT) was performed as two experiments: first, femora of mice sacrificed at 5 days, 3 weeks and 4 months of age, respectively, were included as a general screening. In the second experiment, in depth analyses were performed in both sexes of 4 months old mice (4 male and 4 female CHAD−/−, and 5 male and 3 female WT mice). All specimens were scanned by the use of high-resolution micro-CT (SkyScan 1172; SkyScan, Kontich, Belgium). Dissected whole femora were affixed to the scanning stage and projection images were obtained at a resolution of 8.03 μm and reconstructed by use of manufacturer-provided software (NRecon, SkyScan). After calibration of the standard unit of X-ray CT density (Hounsfield unit, HU), the conversion from HU to volumetric bone mineral density (vBMD) was done. Reconstructed images were analyzed by use of manufacturer-supplied software. Three sections as shown in fig. S3 consisting of 63 slices or 0.5006 mm (5 days old mice) or

126 slices or 1.012 mm (3 weeks and 4 months old mice) were analyzed per bone for the following parameters: cortical thickness, cortical porosity, cortical bone volume, trabecular thickness, trabecular separation, trabecular bone volume, trabecular porosity, as well as degree of anisotropy (DA) (from mean intercept length analysis as an index of degree of preferred orientation of the structure [28]) and structure model index (SMI) (the amount of plates and rod composing the structure [19]). A threshold of 45, 66, and 86 to 255 was applied for 5 days, 3 weeks and 4 months old mice, respectively (fig. S4). Cortical vBMD was obtained after applying a threshold of 1–255 to sections of cortical bone.

5.4. Macroscopic and light microscopic analyses

At sacrifice the mice were subjected to macroscopic work up aiming at detection of malformation. Moreover, samples from heart muscle, kidney, spleen, liver and lung were fixed in formalin, paraffin-embedded, sectioned and stained with haematoxylin & eosin (H&E) according to a routine protocol. Intact eyes were fixed in 2% glutar aldehyde and embedded in an epoxy resin (Epon 812, Agar Scientific ltd., Stansted, Essex, UK) and equatorial semi thin sections were stained with toluidine blue according to a routine protocol. Three to five coded sections per organ and animal were subjected to conventional light microscopy by an experienced surgical pathologist and evaluated for structural tissue changes. Six animals (CHAD−/− and WT) were investigated at each age, i.e. 3 and 6 weeks as well as 3 and 8 months.

5.4.1. Bone histomorphometry. Femora from 3 and 6 weeks old animals were fixed in glutar aldehyde, decalcified in 7% EDTA for 15 days and embedded in an epoxy resin as above. Semi thin longitudinal sections of distal femoral metaphyses were cut and stained with toluidine blue, and histomorphometric analysis was performed on digital images (resolution 2576×1932 pixels) using image analysis software (AnalySIS pro, Digital Soft Imaging System, Münster, Germany). Mean height of the epiphyseal growth plate was calculated for each section using the mean of 10 randomly placed lines for measurement. The relative zonal distribution of the resting, proliferative and hypertrophic zone was estimated by point counting.

Femora for measurement of osteoid were fixed in 4% buffered formalin, embedded in a methyl methacrylate resin (K-plast, DiaTec Systems, Germany) without prior decalcification, sectioned and stained with Masson-Goldner's trichrome. Relative osteoid surface (% of trabecular surface) was estimated by point counting. For each animal, a minimum of 3 non-overlapping visual fields of vision were analyzed.

5.4.2. In situ hybridization. Five to six 6 weeks old animals from each group were subjected to in situ hybridization with riboprobes for OPN, CTK, BSP, TRAP and COMP. Gene sequences for TRAP, CTK, COMP, OPN and BSP were amplified by conventional PCR using cDNA from mouse osteoblasts (a generous gift from dr. Rune Jemtland, Oslo University Hospital, Norway) or IMAGE clones using the oligonucleotide primers listed in table S3. All sequences were subsequently cloned with a Dual Promoter TA Cloning Kit (Invitrogen) and sequenced. Digoxigenin (DIG)- conjugated complementary RNA cRNA) probes were synthesized with a DIG-labelling kit (Roche Diagnostics AS, Oslo, Norway) using T7 or Sp6 RNA polymerase to yield probes in the sense or antisense orientation. Hybridization of longitudinal sections of formalin-fixed femora embedded in paraffin was performed by modification of a previously described protocol [29]. Briefly, dewaxed and proteinase K-digested sections of paraffin-embedded samples were post-fixed in paraform aldehyde. Following prehybridization in formamide/$2 \times$ SSC, the sections were hybridized with 5 ng probe in 50% formamide/2× SSC/7.5% dextran sulphate. High stringency washing was performed, and unbound probe was removed by RNase-treatment. Hybridized probe was detected using an alkaline phosphatase (AP)-conjugated sheep anti-DIG antibody followed by the AP-substrate nitrobluetetrazolium chloride (NTB)/5-bromo-4- chloro-3-indolyl-phosphate (BCIP) (Roche Diagnostics GmbH, Mannheim, Germany). Coded sections of the epiphyseal growth plate and the metaphysis of the distal femur were micrographed and analyzed focusing on the resting zone, the proliferative zone, the hypertrophic zone, and the metaphysis. The following scoring system was used to semi-quantify mRNA positive cells; 0 = no positive cells, 1 = low concentration of positive cells, 2 = high concentration positive cells.

5.4.3. Immunohistochemistry. Ten 6 week old animals from each group were included in the analysis. Immunohisto-chemistry of COMP was performed using the peroxide technique with diaminobenzidine (DAB) as the chromogen according to a routine protocol (Hect et al., 2004). Longitudinal sections of formalin-fixed femoral bone sections embedded in paraffin were used. Following permeabilisation by digestion with chondroitinase ABC (Seikagaku Corporation, Tokyo, Japan) in tris/acetate buffer, the sections were incubated with rabbit polyclonal antiserum raised to rat COMP [30]. Bound antibodies were visualized using the Dako EnVision+ System (EnVision+ System, HRP K4010, DAKO, USA). The sections were counterstained with haematox-ylin and subsequently with a mixture of eosin and phloxine B. COMP-staining was confined to the articular cartilage and the epiphyseal growth plate, and the latter was subjected to semi – quantitative analysis. Thus, the growth plate was divided into the following zones; I (resting and proliferative zones) and II (hypertrophic zone). In zone I, scores (0 = no staining, 1 = weak staining and 2 = intense staining) for territorial matrix and interterritorial matrix were analyzed, while in zone II, pericellular, interterritorial and intracellular staining were graded.

5.5. Transmission electron microscopy

Tibias from 6 weeks old mice were immediately dissected free and fixed by immersion in a solution of 2% paraform aldehyde and 0.5% glutar aldehyde (GA). Subsequently, the tissue was embedded at low temperature in a freeze substitution device according to our established protocol [31].

5.5.1. Qualitative ultrastructural collagen analysis. Coded ultrathin sections from GA-fixed, epon-em-bedded samples of calvarial bone of 6 weeks old CHAD −/− and WT mice were subjected to electron microscopy of collagen fibrils. The fibrils were evaluated semi-quantitatively for thickness and spatial orientation. Sections from two blocks of each of 6 animals (3 CHAD−/− and 3 WT) were investigated and categorized as normal or pathological.

5.5.2. Immunogold labelling and semi-quantitative analysis. Immunogold labelling with antibodies against BSP and OPN was performed as previously described [32]. Micro-graphs were obtained by systematic random sampling of cells/surrounding matrix and analyzed using the semiautomatic interactive image analyzer software AnalySIS® pro (Soft Imaging System, Münster, Germany). In consensus with previous reports of the ultrastructural distribution of BSP [33,34] and OPN [31,35], regions of interests were confined to 1) osteoblast nucleus, 2) osteoblast cytoplasm, 3) osteoid, 4) osteoid-bone interface/mineralization fronts and 5) mineralized bone. Six animals from each group were included in the analysis, and 2 tissue blocks were sampled per animal. The results for OPN and BSP are based on the analysis of 60 osteoblasts and their surrounding microenvi-ronment in each group.

5.6. Protein contents of femoral head cartilage by proteomics

5.6.1. Dissection, pulverization and protein extraction of cartilage. Samples were obtained from G3 mice 4 months old. Full thickness femoral head cartilage was dissected from 8 female CHAD −/−, 7 female wild type, 7 male CHAD −/−, and 7 male wild type. The tissue from each group was pooled separately and homogeneous powder was made in liquid nitrogen. The samples were extracted with guanidine hydrochloride with added proteinase inhibitors (Larsson et al., 1991) and extracts were collected after centrifugation (IEC Micromax) at 13200 rpm for 30 min.

5.6.2. Preparation of proteins in extracts for quantitative proteomics and analyses. Procedures for quantitative proteo-mics were the same as those described for the analyses of a set of human cartilage tissues. Essentially, proteins in extracts were reduced and alkylated, followed by trypsin digestion (Onnerfjord et al., 2012). An isobaric 4-plex ITRAQ^{TM} was then used to enable simultaneous analysis of a mix of all the four samples. Trypsin digests of the four pools of cartilage were separately labelled using standard protocols according to the manufacturer. The trypsin digested labelled extracts were combined and chromatographed on a SCX cation exchange column. The 29 fractions collected were separately applied to and analyzed using a reversed phase C18 nano-LC column online with a QTOF mass spectrometer as described (Onnerfjord et al., 2012).

5.6.3. Database searching. The mass spectrometric raw data was processed using Protein Lynx 2.1 with internal calibration. The processed files were searched with taxonomy mus musculus using MASCOT 2.1. The ratios of individual peptides between female CHAD −/− vs. wild type, male CHAD −/− vs. wild type, female CHAD−/− vs. male CHAD −/−, and female wild type vs. male wild type mice were calculated by MASCOT.

5.6.4. Data analysis. iTRAQ quantification parameters: significant threshold p<0.05; weighted average ratios; minimum number of peptides of 2, minimum precursor charges of 2; at least homology of 0.05. The searched list was manually inspected for errors and a limited number of obviously incorrectly identified proteins were removed e.g. non-collagenous proteins identified with hydroxylation on proline residues.

5.7. Analysis of tissue protein pattern by SDS-polyacrylamide gel electrophoresis and presence of CHAD by Western blot

Cartilage from the femoral head, trachea and nose were dissected clean, cut into small pieces and extracted with guanidine-HCl (GuHCl) containing a proteinase inhibitor cocktail according to standard procedures [6]. For control of antibody specificity in the Western blot a liver sample was treated in the same way. Samples of extracts corresponding to 1 mg of wet weight tissue were precipitated with ethanol and electrophoresed on SDS-polyacrylamide 4–16% gradient gels followed by Western blotting as described [6,36]. The antibody used to stain the blot was raised in rabbits against bovine CHAD [6].

5.8. Testing of mechanical properties of bone by fracturing

Mice were sacrificed at 6 weeks, 4 or 8 months of age and frozen at −20°C. At the time of testing the mice were thawed, the femora and tibiae were dissected out and kept wet being wrapped in saline soaked tissue at 4°C overnight prior to testing. Two different mechanical tests were performed aimed at measuring femoral neck for the properties of cancellous bone and tibia for cortical bone. The strength of the cancellous bone of the femoral neck was measured using a miniaturised version of the mechanical test previously developed to test total hip replacements (Thompson et al., 2004). The bone was gripped in a cylindrical holder, which was then held at $\mathring{9}$ to the vertical, orientated so that the bone was vertical in the sagittal plane and in valgus in the frontal plane, thus similar to the position of the femur *in vivo*. Using an Instron® 8511.20 biaxial load frame with an MTS® TestStar II controller, displacement was applied to the femoral head, using a flat ended indenter, at 0.1 mm s^{-1} until fracture occurred. Care was taken to ensure that the load was applied to the top of the femoral head, such that the loading indenter was not touching the greater trochanter. Fracture lead to a drop in the applied load. After removal from the mechanical test machine the specimen was checked visually to ensure that the fracture had occurred through the femoral neck, the data was rejected ff the fracture had occurred outside the femoral neck. The load at fracture of the femoral neck was registered. Tibias were used for cortical bone testing after dissection and removal of the fibula just proximal to its insertion into the tibia. The tibia was then placed on two supports of an 8 mm span three point loading rig. The bone was positioned so the supports were under the curve in the proximal tibia and at the distal tibia so that when the load was applied at the mid point between the supports it was through the point of insertion of the fibula. Loading was applied at 0.1 mm s^{-1}. In preliminary tests it was found that this position was stable, reproducible and that the tibia did not rotate during the test. The stiffness was measured over the linear portion of the loading curve and the load at failure was recorded.

5.9. Statistics

Morphological and micro-CT results are given as mean ± standard deviation (SD) and differences between CHAD−/− and WT animals were tested using a two-tailed independent Student's t-test. A multivariate analysis of variance (MANOVA) was used to compare immunogold data. For the latter, interest was focused on whether the overall distribution pattern for each of the two proteins differed between the groups. Thus, for a protein, only differences in overall comparison between the groups using MANOVA, and not difference in tests between subjects, were considered. A p-value of <0.05 was considered significant for all analyses.

Supporting Information

Figure S1 Immunostaining for COMP in the epiphyseal growth plate at 6 weeks of age. The epiphyseal growth plate (EGF) showed intense staining for COMP in both interterritorial and territorial matrix, although the staining did not differ by histologic scoring between CHAD−/− (3a) and wild type mice

(3b). Non-immune control was negative (3c) (×10). Counterstained with H&E and phloxine B.

Figure S2 Ultrastructural protein distribution of BSP in bone at 6 weeks of age. Sections incubated with anti-BSP showed distinct accumulation of gold particles over cement lines/mineralization fronts (arrows) in areas of mineralized bone but quantitative comparison revealed no differences in distribution pattern between the CHAD−/− (4a), wild type mice (4b) and non-immune control (4c) (TEM, ×49,000).

Figure S3 Positions of the 3 sections analyzed in the femur of 4 months old mice by micro-CT.

Figure S4 Threshold levels in comparison to the original grey scale scan for the 3 different age groups in the screening micro-CT experiment.

Table S1 The proteins identified by MASCOT 2.1 from 5457 peptide matches above homology or identity threshold, and ratios in CHAD−/− (KO) versus wild type (WT) mice. Proteins identified by only one peptide were excluded. Proteins that changed ≥50% are underlined.

Table S2 Micro-CT cortical/trabecular bone parameters at different ages. Bone length (femur), cortical/trabecular thickness (Ct.Th/Tb.Th), cortical volumetric BMD (Ct. vBMD), cortical/trabecular bone volume (Ct.BV/Tb.BV) cortical/trabecular porosity (Ct.P/Tb.P), trabecular separation (Tb.Sp), degree of anisotropy (DA) and structure model index (SMI) in proximal (P), middle (M) or distal (D) femur. * $p<0.05$, $^{**}p<0.01$ between CHAD−/− and wild type (WT) mice in the age-group.

Table S3 Oligonucleotide primer sequence for DIG-labeled cRNA probes.

Acknowledgments

Aileen Murdoch-Larsen, Linda T. Dorg, Linda I. Solfjell, Areej Khabut and Kristin Holmgren are acknowledged for skilled technical assistance, Ahnders Franzén for advice and backcrossing of mice and Reinhard Fässler for valuable advice with the generation of the CHAD −/− mouse strain.

Author Contributions

Conceived and designed the experiments: FPR DH. Performed the experiments: LH GAS CW EKT FPR DH. Analyzed the data: LH GAS CW CP EKT SHB ESB PÖ. Contributed reagents/materials/analysis tools: SHB ESB. Wrote the paper: GAS LH FPR DH.

References

1. Neame PJ, Sommarin Y, Boynton RE, Heinegard D (1994) The structure of a 38-kDa leucine-rich protein (chondroadherin) isolated from bovine cartilage. J Biol Chem 269: 21547–21554.
2. Danielson KG, Baribault H, Holmes DF, Graham H, Kadler KE, et al. (1997) Targeted disruption of decorin leads to abnormal collagen fibril morphology and skin fragility. J Cell Biol 136: 729–743.
3. Xu T, Bianco P, Fisher LW, Longenecker G, Smith E, et al. (1998) Targeted disruption of the biglycan gene leads to an osteoporosis-like phenotype in mice. Nat Genet 20: 78–82.
4. Svensson L, Aszodi A, Reinholt FP, Fassler R, Heinegard D, et al. (1999) Fibromodulin-null mice have abnormal collagen fibrils, tissue organization, and altered lumican deposition in tendon. Journal of Biological Chemistry 274: 9636–9647.
5. Chakravarti S, Magnuson T, Lass JH, Jepsen KJ, LaMantia C, et al. (1998) Lumican regulates collagen fibril assembly: skin fragility and corneal opacity in the absence of lumican. J Cell Biol 141: 1277–1286.
6. Larsson T, Sommarin Y, Paulsson M, Antonsson P, Hedbom E, et al. (1991) Cartilage matrix proteins. A basic 36-kDa protein with a restricted distribution to cartilage and bone. J Biol Chem 266: 20428–20433.

7. Shen Z, Gantcheva S, Mansson B, Heinegard D, Sommarin Y (1998) Chondroadherin expression changes in skeletal development. Biochem J 330 (Pt 1): 549–557.

8. Mizuno M, Fujisawa R, Kuboki Y (1996) Bone chondroadherin promotes attachment of osteoblastic cells to solid-state substrates and shows affinity to collagen. Calcif Tissue Int 59: 163–167.

9. Johnson JM, Young TL, Rada JA (2006) Small leucine rich repeat proteoglycans (SLRPs) in the human sclera: identification of abundant levels of PRELP. Mol Vis 12: 1057–1066.

10. Zhou HY (2007) Proteomic analysis of hydroxyapatite interaction proteins in bone. Ann N Y Acad Sci 1116: 323–326.

11. Camper L, Heinegard D, Lundgren-Akerlund E (1997) Integrin alpha2beta1 is a receptor for the cartilage matrix protein chondroadherin. J Cell Biol 138: 1159–1167.

12. Haglund L, Tillgren V, Addis L, Wenglen C, Recklies A, et al. (2010) Identification and characterization of the integrin {alpha}2{beta}1 binding motif in Chondroadherin mediating cell attachment. Journal of Biological Chemistry.

13. Mansson B, Wenglen C, Morgelin M, Saxne T, Heinegard D (2001) Association of chondroadherin with collagen type II. J Biol Chem 276: 32883–32888.

14. Wiberg C, Heinegard D, Wenglen C, Timpl R, Morgelin M (2002) Biglycan organizes collagen VI into hexagonal-like networks resembling tissue structures. Journal of Biological Chemistry 277: 49120–49126.

15. Sommarin Y, Wendel M, Shen Z, Hellman U, Heinegard D (1998) Osteoadherin, a cell-binding keratan sulfate proteoglycan in bone, belongs to the family of leucine-rich repeat proteins of the extracellular matrix. J Biol Chem 273: 16723–16729.

16. Lanske B, Karaplis AC, Lee K, Luz A, Vortkamp A, et al. (1996) PTH/PTHrP receptor in early development and Indian hedgehog-regulated bone growth. Science 273: 663–666.

17. Vortkamp A, Lee K, Lanske B, Segre GV, Kronenberg HM, et al. (1996) Regulation of rate of cartilage differentiation by Indian hedgehog and PTH-related protein. Science 273: 613–622.

18. Sommarin Y, Larsson T, Heinegard D (1989) Chondrocyte-matrix interactions. Attachment to proteins isolated from cartilage. Exp Cell Res 184: 181–192.

19. Hildebrand T, Ruegsegger P (1997) Quantification of Bone Microarchitecture with the Structure Model Index. Comput Methods Biomech Biomed Engin 1: 15–23.

20. Franzen A, Hultenby K, Reinholt FP, Onnerfjord P, Heinegard D (2008) Altered osteoclast development and function in osteopontin deficient mice. J Orthop Res 26: 721–728.

21. Ganss B, Kim RH, Sodek J (1999) Bone sialoprotein. Crit Rev Oral Biol Med 10: 79–98.

22. Sodek J, Ganss B, McKee MD (2000) Osteopontin. Crit Rev Oral Biol Med 11: 279–303.

23. Gordon JA, Tye CE, Sampaio AV, Underhill TM, Hunter GK, et al. (2007) Bone sialoprotein expression enhances osteoblast differentiation and matrix mineralization in vitro. Bone 41: 462–473.

24. Hunter GK, Goldberg HA (1994) Modulation of crystal formation by bone phosphoproteins: role of glutamic acid-rich sequences in the nucleation of hydroxyapatite by bone sialoprotein. Biochem J 302 (Pt 1): 175–179.

25. Kojima H, Uede T, Uemura T (2004) In vitro and in vivo effects of the overexpression of osteopontin on osteoblast differentiation using a recombinant adenoviral vector. J Biochem 136: 377–386.

26. Landgren C, Beier DR, Fassler R, Heinegard D, Sommarin Y (1998) The mouse chondroadherin gene: characterization and chromosomal localization. Genomics 47: 84–91.

27. Fassler R, Meyer M (1995) Consequences of lack of beta 1 integrin gene expression in mice. Genes Dev 9: 1896–1908.

28. Whitehouse WJ (1974) The quantitative morphology of anisotropic trabecular bone. J Microsc 101: 153–168.

29. Baekkevold ES, Yamanaka T, Palframan RT, Carlsen HS, Reinholt FP, et al. (2001) The CCR7 ligand elc (CCL19) is transcytosed in high endothelial venules and mediates T cell recruitment. J Exp Med 193: 1105–1112.

30. Hedbom E, Antonsson P, Hjerpe A, Aeschlimann D, Paulsson M, et al. (1992) Cartilage matrix proteins. An acidic oligomeric protein (COMP) detected only in cartilage. J Biol Chem 267: 6132–6136.

31. Hultenby K, Reinholt FP, Oldberg A, Heinegard D (1991) Ultrastructural immunolocalization of osteopontin in metaphyseal and cortical bone. Matrix 11: 206–213.

32. Brorson SH, Roos N, Skjorten F (1994) Antibody penetration into LR-White sections. Micron 25: 453–460.

33. Bianco P, Riminucci M, Silvestrini G, Bonucci E, Termine JD, et al. (1993) Localization of bone sialoprotein (BSP) to Golgi and post-Golgi secretory structures in osteoblasts and to discrete sites in early bone matrix. J Histochem Cytochem 41: 193–203.

34. Hultenby K, Reinholt FP, Norgard M, Oldberg A, Wendel M, et al. (1994) Distribution and synthesis of bone sialoprotein in metaphyseal bone of young rats show a distinctly different pattern from that of osteopontin. Eur J Cell Biol 63: 230–239.

35. Reinholt FP, Hultenby K, Oldberg A, Heinegard D (1990) Osteopontin – a possible anchor of osteoclasts to bone. Proc Natl Acad Sci U S A 87: 4473–4475.

36. Heinegard D LP, Saxne T (1999) Noncollagenous proteins; glycoproteins and related proteins. In: Seibel MJ RS, Bilezikian JP, editor. Dynamics of bone and cartilage metabolism. New York: Academic Press. pp. 59–69.

10

Evaluation of Biological Properties of Electron Beam Melted Ti6Al4V Implant with Biomimetic Coating In Vitro and In Vivo

Xiang Li[1]◐, Ya-Fei Feng[2]◐, Cheng-Tao Wang[1], Guo-Chen Li[3], Wei Lei[2], Zhi-Yong Zhang[2,4]*, Lin Wang[2]*

1 School of Mechanical Engineering, Shanghai Jiao Tong University, State Key Laboratory of Mechanical System and Vibration, Shanghai, China, **2** Department of Orthopaedics, Xijing Hospital, The Fourth Military Medical University, Xi'an, China, **3** Department of Orthopaedics, Tangdu Hospital, The Fourth Military Medical University, Xi'an China, **4** Department of Plastic and Reconstructive Surgery, Shanghai 9th People's Hospital, Shanghai Key Laboratory of Tissue Engineering, School of Medicine, Shanghai Jiao Tong University, Shanghai, China

Abstract

Background: High strength porous titanium implants are widely used for the reconstruction of craniofacial defects because of their similar mechanical properties to those of bone. The recent introduction of electron beam melting (EBM) technique allows a direct digitally enabled fabrication of patient specific porous titanium implants, whereas both their in vitro and in vivo biological performance need further investigation.

Methods: In the present study, we fabricated porous Ti6Al4V implants with controlled porous structure by EBM process, analyzed their mechanical properties, and conducted the surface modification with biomimetic approach. The bioactivities of EBM porous titanium in vitro and in vivo were evaluated between implants with and without biomimetic apatite coating.

Results: The physical property of the porous implants, containing the compressive strength being 163 - 286 MPa and the Young's modulus being 14.5–38.5 GPa, is similar to cortical bone. The in vitro culture of osteoblasts on the porous Ti6Al4V implants has shown a favorable circumstance for cell attachment and proliferation as well as cell morphology and spreading, which were comparable with the implants coating with bone-like apatite. In vivo, histological analysis has obtained a rapid ingrowth of bone tissue from calvarial margins toward the center of bone defect in 12 weeks. We observed similar increasing rate of bone ingrowth and percentage of bone formation within coated and uncoated implants, all of which achieved a successful bridging of the defect in 12 weeks after the implantation.

Conclusions: This study demonstrated that the EBM porous Ti6Al4V implant not only reduced the stress-shielding but also exerted appropriate osteoconductive properties, as well as the apatite coated group. The results opened up the possibility of using purely porous titanium alloy scaffolds to reconstruct specific bone defects in the maxillofacial and orthopedic fields.

Editor: Jie Zheng, University of Akron, United States of America

Funding: The authors would like to thank Contract Grant Sponsor: XIJING ZHUTUI Foundation (XJZT10M07), the Project supported by National Key Technology R&D Program (2012BAI18B07), the Ph.D. Programs Foundation of the Ministry of Education of China (Grant No. 20100073120051) and the Research Project of State Key Laboratory of Mechanical System and Vibration (Grant No. MSV201114) for their support to this work. The funders played no role in study design, data collection and analysis, decision to publish, or preparation of the manuscript.

Competing Interests: The authors have declared that no competing interests exist.

* E-mail: wanglinxj@fmmu.edu.cn (LW); mr.zhiyong@gmail.com (ZYZ)

◐ These authors contribute equally to this work.

Introduction

Titanium and titanium alloys have been widely used in orthopedic and dental implants due to low density, excellent mechanical properties, favorable biocompatibility, and good corrosion resistance. However, clinical practices and studies have shown that the mechanical mismatch between metallic implant and natural bone may lead to stress-shielding, and thus cause bone resorption and eventually the failure of metallic implant fixation [1]. Porous metallic structure can be utilized to overcome this drawback, which not only reduced the mechanical mismatch but also achieved stable long-term fixation by promoting full bone ingrowth [2,3]. Many techniques have been investigated to

produce porous metallic structure, including powder sintering approach, space holder method, combustion synthesis, plasma spraying, and polymeric sponge replication [3–6]. However, these conventional techniques have very limited control of the internal pore architecture and the external shape of the porous titanium implants, which hinder further application of porous titanium.

Rapid prototyping (RP), generally known as solid freeform fabrication (SFF), is a type of technologies that can automatically construct physical models from computer-aided design (CAD) data. The applications of state-of-the-art RP techniques for fabricating polymeric tissue engineering (TE) scaffolds were reviewed by Leong and Hutmacher [7,8], with detailed illustration to show the superiority of RP techniques over the conventional fabrication methods. Direct fabrication of metallic components for biomedical application with RP approaches, such as three-dimensional fiber deposition [9], laser-engineered net shaping [10], direct laser forming [11] and so on, has been shown to become a feasible and promising manufacturing technology in producing porous Ti6Al4V scaffolds with interconnected porous networks. Recently, selective electron beam melting (EBM) approach as a metal rapid prototyping process has been studied for fabricating patient specific porous orthopedic implants [12]. The EBM porous titanium implant not only avoided the stress-shielding effects in vivo for similar mechanical properties with native bone [13], but also matched irregular defect at specific site such as skull, maxillofacial and bone joint region [Fig. 1]. Heinl et al. has reported that cellular Ti6Al4V structures with inter-connected macro porosity fabricated by EBM might have favorable long-term stability and were suitable for orthopedic applications [14].

In addition to the internal structure and external shape, the implant surface physicochemical properties are also critical for bone-implant integration. As solid titanium and alloys are generally encapsulated by fibrous tissue after implantation for their bioinert nature [15,16], many surface modification methods, including plasma spraying [17], sol-gel [18], electrophoretic deposition [19], sputter deposition [20] and micro-arc oxidation [21], were applied to improve the bone-implant integration of this kind of implants. The biomimetic approach, which via soaking implants in simulated body fluids (SBF) at a physiological temperature and pH, has been shown to be a superior option due to its low operation temperature and homogenously bone-like carbonated apatite deposition [22–24]. Barrere et al. [25] reported that the biomimetic coating enhanced the implant-bone integration and were beneficial for the long-term fixation of metal prostheses in the load-bearing applications. However, few studies have analyzed the biological performance of EBM porous titanium implants with biomimetic coating.

The aim of this study was to fabricate Ti6Al4V implants with well controlled porous structure using EBM process, and to evaluate the bioactivity of this porous implant with or without biomimetic apatite coating in vivo and in vitro. The implant

physical properties were examined by scanning electron microscopy, micro computer tomography, and digital microscope. The implant surfaces were modified by biomimetic approach and characterized by energy-dispersive X-ray analysis, X-ray diffraction and Fourier transform infrared spectroscopy. In vitro, experiments were performed to assess cell attachment, viability, cell proliferation and differentiation. In vivo, samples were implanted into 12 mm bone defects in calvaria of rabbits, and bone ingrowth and bone implant bonding was evaluated by histology and histomorphometry.

Materials and Methods

Preparation of Porous Ti6Al4V Implant

Porous Ti6Al4V implant was fabricated using electron beam melting process as previously described [12]. Briefly, the external shape and internal porous structures were designed with commercial CAD software (Unigraphics NX, EDS). The CAD data of the structures was converted into STL data which then imported into Materialise's Magics software, and converted into input file for EBM. The samples were produced on an Arcam's EBM machine (EBM S12, Arcam AB, Sweden).

Characterization of Porous Ti6Al4V Implant

The porous structures of the EBM produced samples were observed using a scanning electron microscope (SEM, JSM-6460, JEOL, Japan) equipped with an energy dispersive X-ray analyzer (EDX, Oxford, UK). The structural analysis using micro computer tomography (μCT, Locus SP, GE, USA) was conducted prior to the chemical treatments to determine the mean pore size and the porosity of the implants. The resolution was 20 μm. The three-dimensional (3D) surface topography of the samples was examined by optical profiler (OP, Wyko NT9300, Veeco, USA). Five samples were measured from each of the two different surfaces (top&bottom and lateral) to obtain an average roughness value R_a. The porosity of EBM produced samples was evaluated from the weight and the apparent volume of the specimen ($n = 5$). Compression tests were conducted to determine the mechanical properties of the EBM Ti6Al4V implants ($n = 5$) using a material

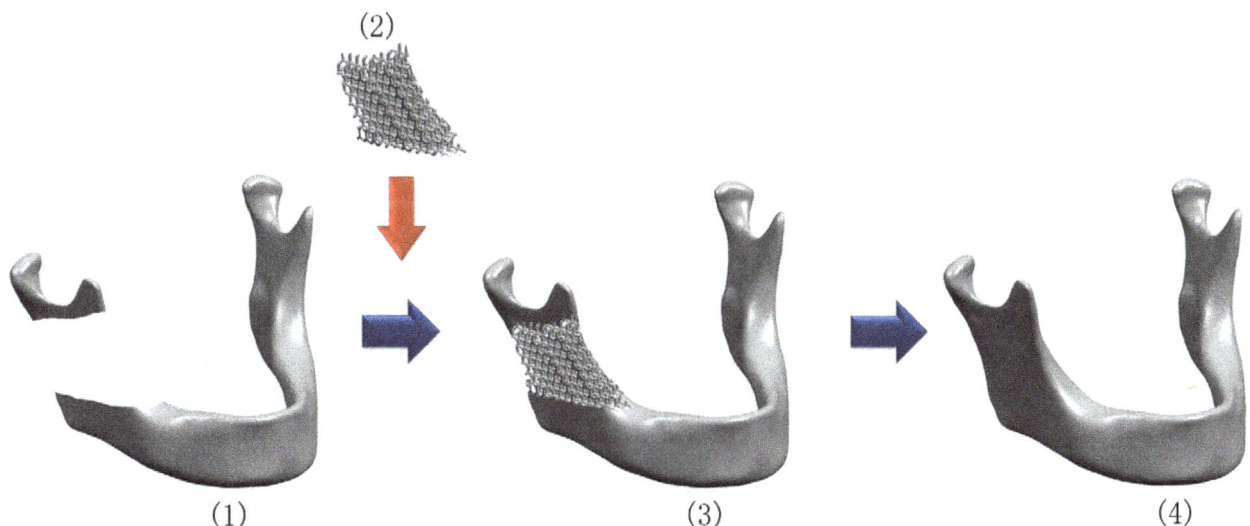

Figure 1. The flow diagram showed the design of electron beam melting (EBM) porous Ti6Al4V implant with CAD for repair of mandibular bone defect: (**1**) acquisition of the CT data of the patients; (**2**) design with CAD and fabrication of custom EBM porous titanium implant; (**3**) implantation of the patient specific porous implant; (**4**) reconstruction of the bone defect.

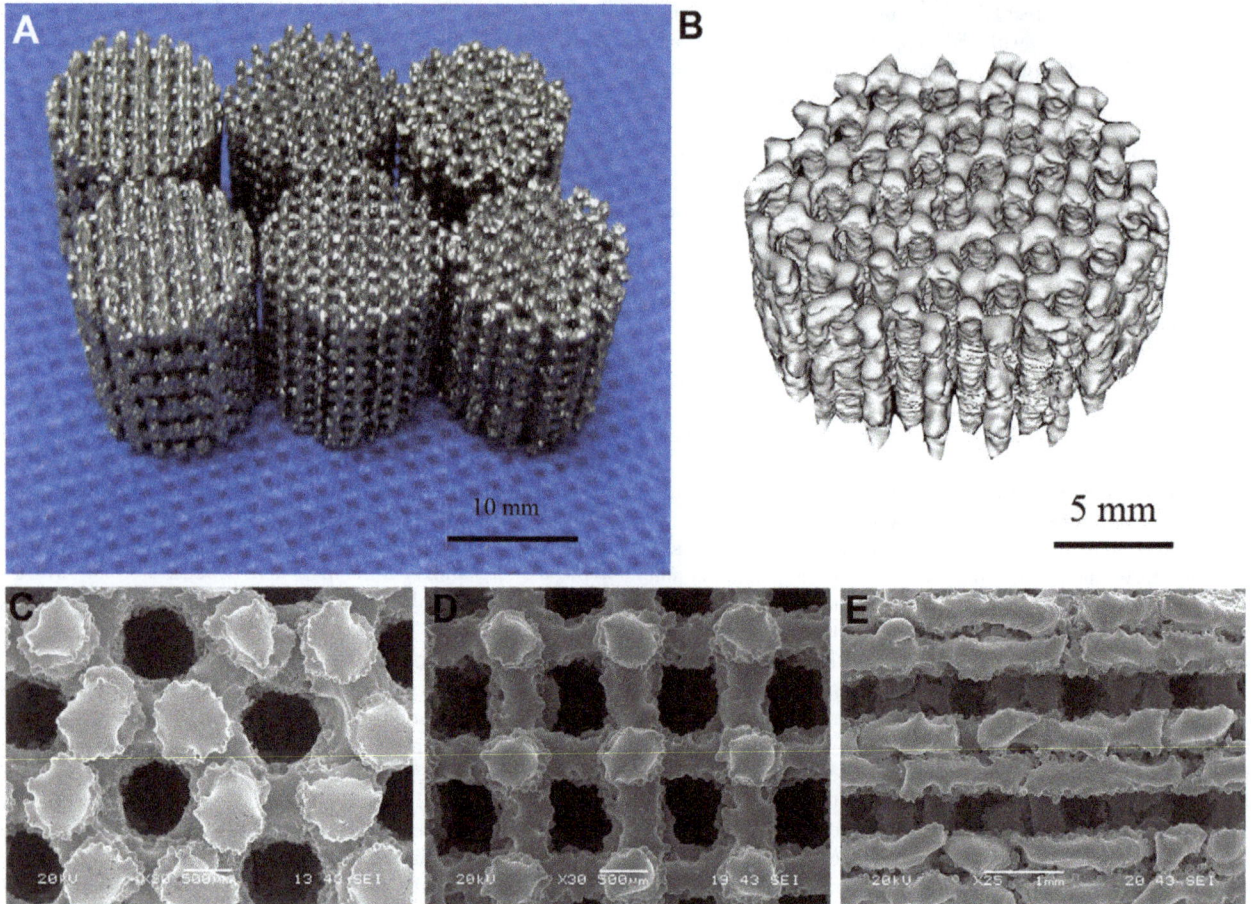

Figure 2. Characterization of porous Ti6Al4V samples. (A) Porous Ti6Al4V implants fabricated by electron beam melting process. **(B)** Reconstructed 3D micro-CT image of the porous implant with honey-like structure. SEM images of porous Ti6Al4V samples with **(C)** honeycomb-like structure, **(D)** orthogonal structure, **(E)** layer structure.

testing system (MTS 810) with a 10 kN load cell at room temperature. The cross-head speed was set at 0.5 mm/min. The elastic modulus E was calculated from the slope of the compressive stress–strain curve in the linear elastic region. The compressive yield strength $\sigma_{0.2}$ was determined from the stress–strain curve according to the 0.2% offset method, while the maximum strength σ_{max} was calculated by dividing the highest load by the initial cross-sectional area.

Biomimetic Coating

Samples for surface modification were machined to cylinder-shape with diameter in 11.6 mm and height in 5 mm. Before chemical treatment, samples were washed with isopropanol and distilled water in an ultrasonic cleaner, and then dried in air. Alkali treatment was performed by soaking these samples in 10 M NaOH aqueous solution at 60°C for 24h, then washed gently with distilled water. The alkali-treated samples were heated to 600°C (at a rate of 5°C min^{-1}) in an electric furnace, kept at 600°C for 1h, and then allowed to cool to room temperature in the furnace [23]. The alkali-heat treated samples were immersed in SBF solution under static conditions at 37°C for 14 days. The solution was changed every 2 days to maintain the solution concentrations. The SBF solution was prepared according to Kokubo et al's method [26]. Briefly, reagent grade chemicals of NaCl, NaHCO$_3$, KCl, K$_2$HOP$_3$·3H$_2$O, MgCl$_2$·6H$_2$O, CaCl$_2$, and Na$_2$SO$_4$ were

dissolved into distilled water, and buffered at pH 7.40 with hydrochloric acid and tris-hydroxymethyl aminomethane at 36.5±1.5°C.

Characterization of Implant Surface with Biomimetic Coating

After immersion in SBF for 14 days, the specimens were rinsed in distilled water and dried in vacuum desiccator. Surface morphologies of the samples after chemical treatment and soaking in SBF were examined by scanning electron microscope. The thin film X-ray diffraction (TF-XRD) measurement was performed on a Rigaku X-ray diffractometer (Cu Kα radiation, 40 kV, 30 mA). The data was collected in the 2θ ranges of 20–60° with a step size of 0.02°. FTIR spectra were measured in transmission using the KBr technique in the range from 4000 to 400 cm^{-1} at a resolution of 4 cm^{-1}. Approximately 1 mg of the coatings formed on the chemically treated titanium sample after soaking in SBF solutions for 14 days was removed from the substrate, then mixed with 500 mg of dry KBr powder and ground using an agate mortar and pestle. The resulting mixture was pressed into transparent pellets with diameter of 13 mm and applying force of 10^5 N [27]. Both the EBM Ti6Al4V implants (TiI) and biomimetic apatite coated implants (TiC) were used for in vitro and in vivo analysis in the present study.

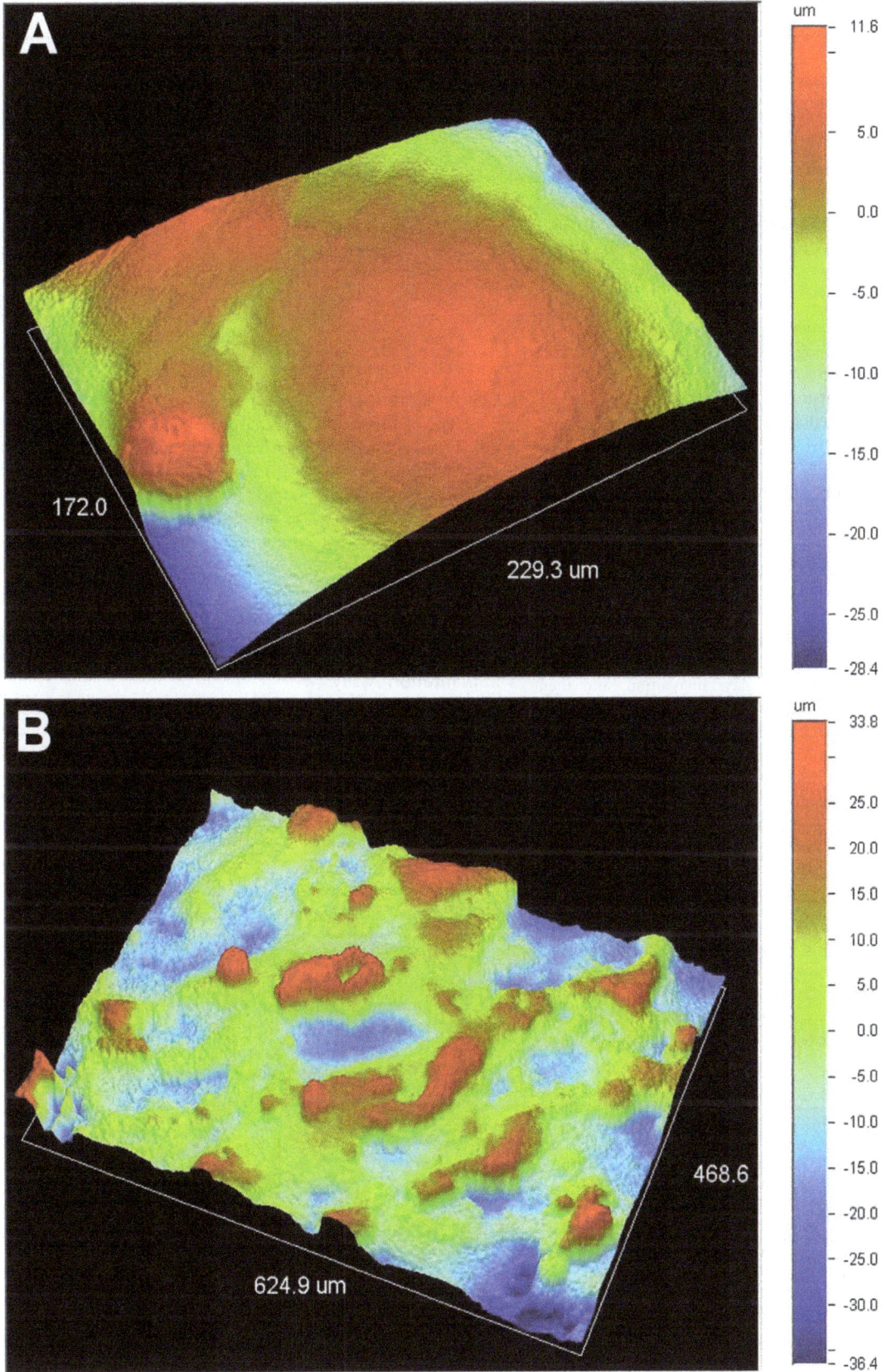

Figure 3. Digital topographic images of the sample surface. The images exhibited a rough anisotropic surface with the roughness R_a values of the (**A**) top and (**B**) lateral surfaces being in the range of 5–10 and 15–21 μm.

Table 1. Geometric characteristics and mechanical properties of porous titanium samples.

	Honey-like structure	Orthogonal structure	Layering structure	Cortical bone
Porosity (%)	51.4 ± 2.4	61.2 ± 3.7	55.3 ± 2.6	
Pore size (μm)	~600	~600	~500	
Strut size (μm)	~650	~500	~900	
E (GPa)	25.9 ± 2.8	14.5 ± 3.1	38.5 ± 6.7	10–30
$\sigma_{0.2}$ (MPa)	185.4 ± 15.9	138.2 ± 8.6	194.4 ± 12.1	~138
σ_{max} (MPa)	286.6 ± 35.5	163.8 ± 11.3	235.6 ± 18.3	~193

E, the elastic modulus.
$\sigma_{0.2}$, the compressive yield strength.
σ_{max}, the compressive maximum strength.

Cell Seeding and Co-culture with Implant

All samples were sterilized in ethylene oxide gas before cell seeding. Osteoblastic cells were isolated from new born New Zealand rabbit's calvaria [28]. Cells were cultured in DMEM supplemented with 10% BCS and 1% penicillin/streptomycin and incubated at 37°C in a 5% CO_2, 100% relative humidity incubator. The third passage cells were used in the present study. Osteoblasts were seeded on the scaffolds at a density of 6×10^5 cells/ml.

Cell Attachment and Morphology

Samples were removed from culture and washed with phosphate buffered saline after 14 days incubation, and then fixed in 2% v/v glutaraldehyde in 0.1 M sodium cacodylate buffer at 4°C overnight for attachment and morphological observation. The samples were washed in the buffer, dehydrated through an ethanol series, critical-point dried, lastly sputtered with gold, and were analyzed using a scanning electron microscope (SEM) operating at 15 kV and a semiautomatic interactive image analyzer.

Histological Examination in vitro

Samples for histological study were fixed in 10% formalin for 7 days, dehydrated through an ethanol series, cleared with toluene, and embedded in methylmethacrylate. After polymerization, thin sections were prepared with a modified sawing microtome technique. The sections were stained with Haematoxylin and Eosin (H&E) and examined with a standard microscope (Leica, Wetzlar, Germany).

Cell Proliferation

Cell proliferation was assessed using a methylthiazol tetrazolium (MTT) assay. Briefly, 800 μl serum free medium and 80 μl MTT solution were added to each sample. After 4 h of incubation at 37°C in a fully humidified atmosphere with 5% CO_2 in air, MTT was taken up by active cells and reduced in the mitochondria to insoluble purple formazan granules. Subsequently, the medium was discarded, and the precipitated formazan was then dissolved in dimethyl sulfoxide (DMSO). After 10 min of slow shaking, the absorbance was read at 570 nm using a Bio-Rad 500 spectrophotometric microplate reader.

Implantation Assay

All animal experiments were performed in accordance with the National Institutes of Health Guidelines for the Use of Laboratory Animals. The animals used in the current study were obtained from protocols approved by the Fourth Military Medical University Committee on Animal Care (Permit Number: 12090). Thirty male New Zealand white rabbits, weighing 2.5–3.0 kg, were used to examine the implant-bone integration and bone ingrowth of the EBM Ti6Al4V porous implants of orthogonal structure with or without biomimetic apatite coating in experimental cranial implantation. All animals were anaesthetized by intramuscular injection of Ketamine. Two 12 mm full-thickness calvaria defects were drilled. The implants were placed alternately in the right or left defect. The breath and heart rate of rabbits were carefully monitored to ensure that they were under anesthetic and painless before recovering from anesthesia. Each rabbit was administered 400,000 U of penicillin intraoperatively and on the first postoperative day to prevent infection. The animals were

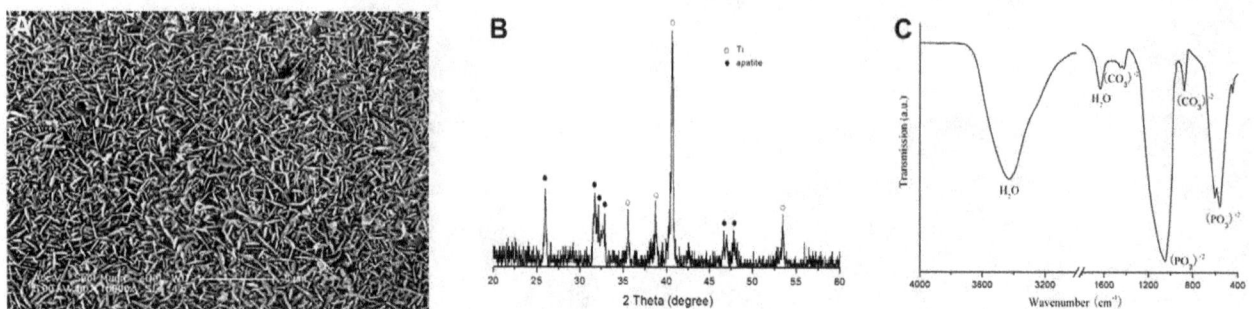

Figure 4. Characterization of porous titanium with biomimetic coating. (A) SEM image of the precipitate on sample surfaces soaked in SBF for 14 days. (**B**). TF-XRD pattern of the sample surface after alkali-heat treatment and subsequently immersion in SBF for 14 days. (**C**) FTIR spectra of chemically pretreated Ti6Al4V samples soaked in SBF for 14 days.

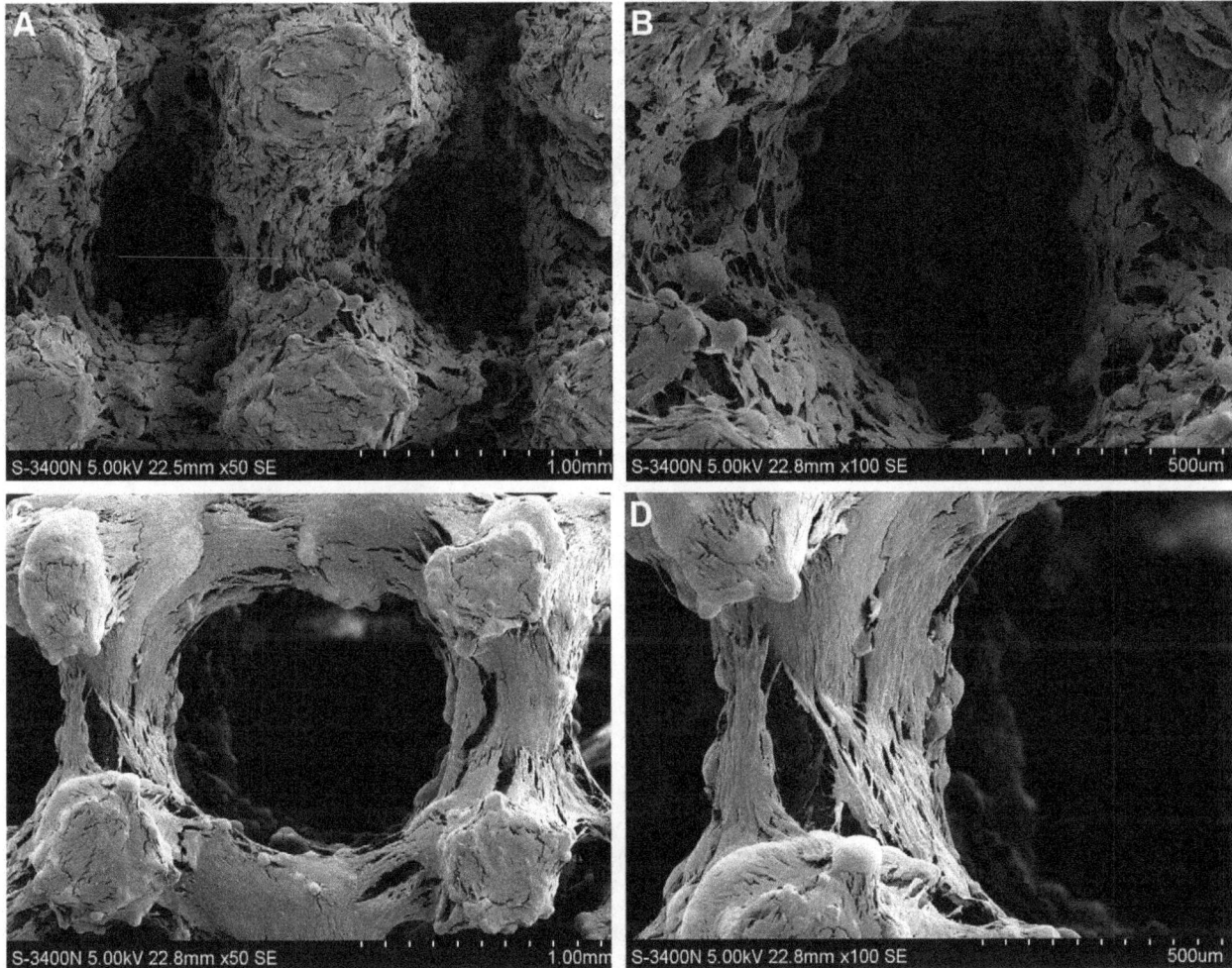

Figure 5. SEM morphologies of cells on porous titanium samples after 14 days of culture. A great number of osteoblasts attached to the **(A–B)** pure porous titanium scaffolds and **(C–D)** porous titanium scaffolds with biomimetic coating, and presented an elongated morphology with cytoplasmic extensions on scaffolds. There were no obvious differences in cell adhesion and morphology between the uncoated and coated samples.

anaesthetized and sacrificed by intra-cardiac overdose of sodium pentobarbital at 4, 8 and 12 weeks after implantation.

Fluorochrome Labeling

Sequential fluorochrome markers were administered after implantation. Tetracycline (30 mg/kg, Sigma, USA) was administered 2 weeks and Calcein Green (8 mg/kg, Sigma, USA) was administered 3 days before the animals were sacrificed using intramuscular injection. After the animals were sacrificed at 12 weeks post-surgery, the implants were retrieved for fluorescence analysis.

Histological and Histomorphometric Analysis

All retrieved specimens were fixed in 10% formalin for 7 days. After fixation, the samples were dehydrated in a graded series of ethanol and embedded in methylmethacrylate. After polymerization, thin sections were prepared with a modified sawing microtome technique and stained with 1.2% trinitrophenol and 1% acid fuchsin (Von-Gieson staining). The qualitative analysis of bone formation and fluorochrome markers were performed using a light/fluorescence microscope (Leica LA Microsystems, Bensheim, Germany). Image analysis was performed using the Image-

Pro Plus software (version 6.0, Media Cybernetics, USA). The amount of newly formed bone inside the defect was evaluated and expressed as percentage of bone area in total available pore space [(bone area/(total implant area − total scaffold area)×100%]. In addition, the bone mineralization apposition rate (MAR, vertical spacing between two fluorochrome markers/injection interval) was analyzed by from the images of fluorochrome labeling, which generally indicated the new bone growth rate.

Statistical Analysis

All the values in the experiment were expressed as the mean ± standard deviation, and were compared using one-way analysis of variance (ANOVA). Differences at $p<0.05$ were considered statistically significant.

Results

Structural Characterization

The Ti6Al4V samples produced by EBM process were shown in Fig. 2A. The pore architectures of samples were examined using micro-CT and SEM. The reconstructed 3D images from micro-CT measurements and micrographs from SEM examination were

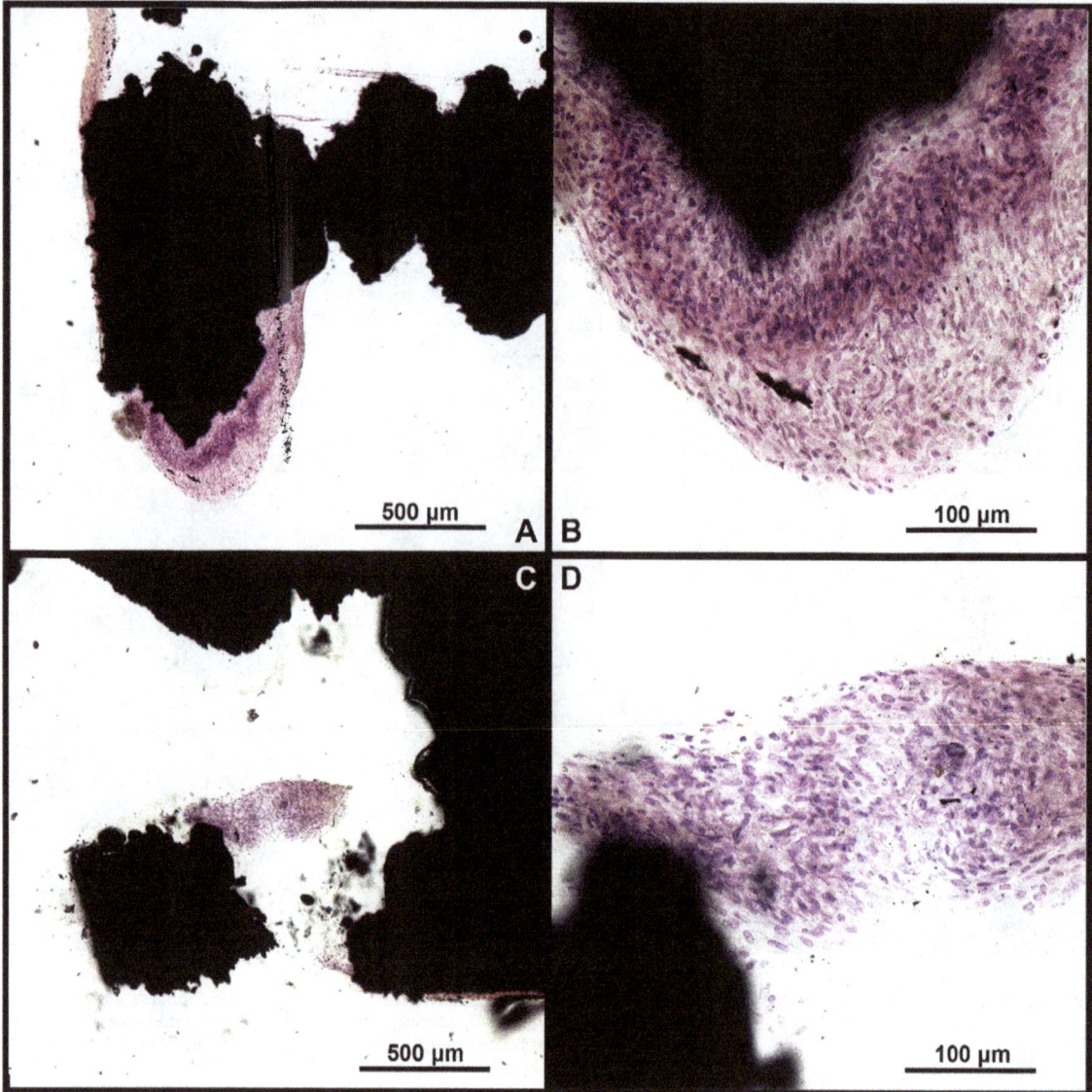

Figure 6. H&E stained sections of sample after 14 days of in vitro culture. Large amount of extracellular matrix deposited among the cells, and some cells migrated into the inner pore of both (**A–B**) uncoated and (**C–D**) coated samples.

shown in Fig. 2B and Fig. 2C–E, respectively. All samples exhibited fully interconnected porous networks. This is one of the most important requirements for tissue ingrowth. Some molten particles could be found in the pores. The three-dimensional topographies of the implants surfaces, illustrated in Fig. 3, exhibited a rough anisotropic surface. The roughness R_a values of the top and lateral surfaces were in the range of 5–10 and 15–21 μm. The sample lateral surfaces were rougher compared to the top surfaces due to adherent molten powder particles. For porous implants with different internal structure, the geometric characteristics and mechanical properties of the samples were shown in Table 1. The orthogonal structure implants, which exerted better mechanical properties than two other kinds of implants, were used for the in vitro and vivo analysis.

Surface Modification

In Figure 4A, the SEM photographs showed a homogeneous surface layer formed on the surface of implants. EDX analysis

revealed calcium, magnesium, sodium and phosphorus to be presented in the surface. The detected Ca/P molar ratio was 1.65. The TF-XRD pattern of the sample surfaces after alkali-heat treatment and subsequent immersion in SBF for 14 days was shown in Fig. 4B. The TF-XRD pattern exhibited broad diffraction lines. The position of these diffraction lines indicated that the deposits are formed of poor crystalline hydroxyapatite. The phase composition of the layer was analyzed in detail using FTIR (Fig. 4C). The spectra showed featureless phosphate and carbonate bands. Intense and broad bands were assigned to oxygen-hydrogen groups, and stretching and bonding were observed at 3435 and 1642 cm^1 respectively. A broad band at about 1420–1450 cm^{-1} from the carbonate groups was incorporated in the apatite structure. These carbonate groups gave rise to the band at 875 cm^{-1}. A broad band in the range of 960 to 1200 cm^{-1} from v_3 P-O antisymmetric stretching vibration, indicated the deviation of phosphate ions from the ideal tetrahedral structure. The bending modes of the O–P–O bonds

Cell proliferation

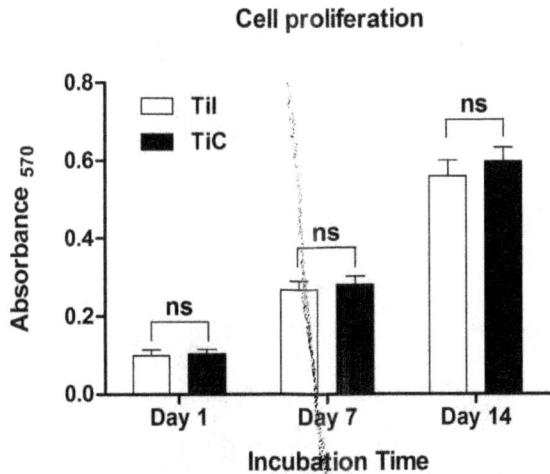

Figure 7. Cell viability of the sample over 14 days of in vitro culture. The cells on the porous titanium exerted a high and sustained proliferation rate within 14 days, and no significant difference was observed between the uncoated and coated samples at each timepoint (P>0.05). TiI, pure porous titanium implant; TiC, porous titanium implant with biomimetic coating.

in the phosphates were found in the spectral range from 560 to 602 cm^{-1}. HPO$_4{}^{2-}$ groups were detected at 1108 cm^{-1} (v_3 mode). Therefore, the precipitate formed in SBF was characterized in a poorly crystallized or amorphous carbonated Ca-P phase.

Cell Attachment and Morphology

After 14 days of culture, it was found that cells adhered to the modified surface, spread out and proliferated into the multilayer structures. The osteoblasts achieved confluence and cover the Ti surface completely with large amount of extracellular matrix (ECM) deposited among the cells, as demonstrated on the SEM images (Fig. 5) and H&E stained section (Fig. 6) of the cell/implant constructs. In addition, some cells have migrated into the inner of porous implants, but they did not span across the total implant. The cross-section histological staining showed a decrease of cell number from top to bottom of porous implants. Moreover, there was no obviously difference in morphology and dense of cells between porous implants with and without biomimetic coating.

Cell Proliferation

MTT assay was used to study the cellular viability and proliferation of the cell culture in the biomimetically modified samples. It showed the maintenance of high cellular viability and a rapid proliferation of cells within 14 days (Fig. 7). The cell viability on EBM porous implant was similar to that on the biomimetic apatite-coated implants at each timepoint (all P>0.05).

Fluorochrome Labeling

Bone remodeling was validated by interrupted fluorescent bone labeling and new bone deposition. The labeling was detected in the regenerated bone in the pores of EBM porous implants, revealing that the increase rate of bone deposition progressed into the pores of the scaffolds (Fig. 8A–D). The fluorescent labeling indicated similar bone ingrowth from the bone-implants interface to the center of the defect in pure porous implants compared with biomimetic apatite-coated porous implants, as well as the fluorescent labeling analysis showed that there were no differences in MAR between the coated (2.22±0.25 μm/d) and uncoated

implants (2.18±0.27 μm/d, P>0.05) (Fig. 8E), indicating similar rate of bone ingrowth within the porous titanium with and without biomimetic apatite coating.

Histological and Histomorphometric Analysis

To evaluate the tissue response to the implanted scaffolds and the defect healing progress, we performed histological analysis on the tissue/implants interface and the internal area of the implants. For all Ti6Al4V implants, no foreign body or inflammatory reaction was observed. Bone formation was found to start from the host bone bed towards the implant in all implants. Four weeks after surgery, histological observations revealed that a small amount of new immature bone tissue grew from the rim of bone defect and began to integrate with the periphery of porous implants, with the direct deposition of newly formed bone onto the surface of implants (Fig. 9A–B, bone tissue in red color while Ti6Al4V implant in black color). After 8 weeks of implantation, histological evaluation indicated progressive growth of more newly formed bone from the calvarial margins toward the center of the bone defect. The newly formed bone integrated well with implant surface (Fig. 9C–D). At 12 weeks, the newly formed bone has successfully bridged the calvarial defect, and the majority of the pores at the bottom part of the implants were filled with bone tissue, which were directly bridged with the calvarial defects (indicated in red color Fig. 9E–F). In addition, the process of bone remodeling was spotted in all the implant sites. Quantitative histomorphometric analysis demonstrated a rapid and continuous new bone formation in the defect region in both coated and uncoated implants at each timepoint during the whole experiment (all P>0.05) (Fig. 10).

Discussion

The EBM process is a new additive manufacturing technique with great capability to fabricate dense and porous Ti6Al4V implants with better control on the both internal structure and external shape [29,30,31,32]. In addition, the EBM production process is much faster comparing with conventional manufacturing methods using laser beam. In EBM process, the high power electron beam is used instead of conventional laser beam. With regards to energy utilization efficiency (EUE) during metal sintering, high power electron beam can be as high as 95 percent, which is 5 to 10 times higher than the EUE of the laser beam; whereas laser beam may lose up to 95% of its energy, due to the light reflection by the metal powder during the sintering. This leads to 3 to 5 times faster processing speed of EBM than other laser beam based metal additive fabrication methods. Furthermore, thanks to the high EUE of electron beam, EBM process provides a better and more reliable fabrication process compared to selective laser sintering (SLS), by fully melting the metal particles to manufacture high quality of implants, which is completely void-free. The whole EBM process occurs in an ultra high vacuum environment, eliminating any imperfections caused by oxidation.

The Young's modulus of solid Ti6Al4V implants produced by Arcam EBM is up to 120 GPa [32]. In comparison with the solid ones, the Young's modulus of porous Ti6Al4V implants in current study is similar to cortical bone, which could minimize stress shielding. It is obvious that the porous structures help to reduce the mechanical mismatches between metallic implants and host bone tissue. The ideal porous implants should provide a good mechanical environment for initial function and appropriate remodeling of regenerating tissue while concurrently providing sufficient porosity for cell migration and tissue ingrowth. Whereas, these

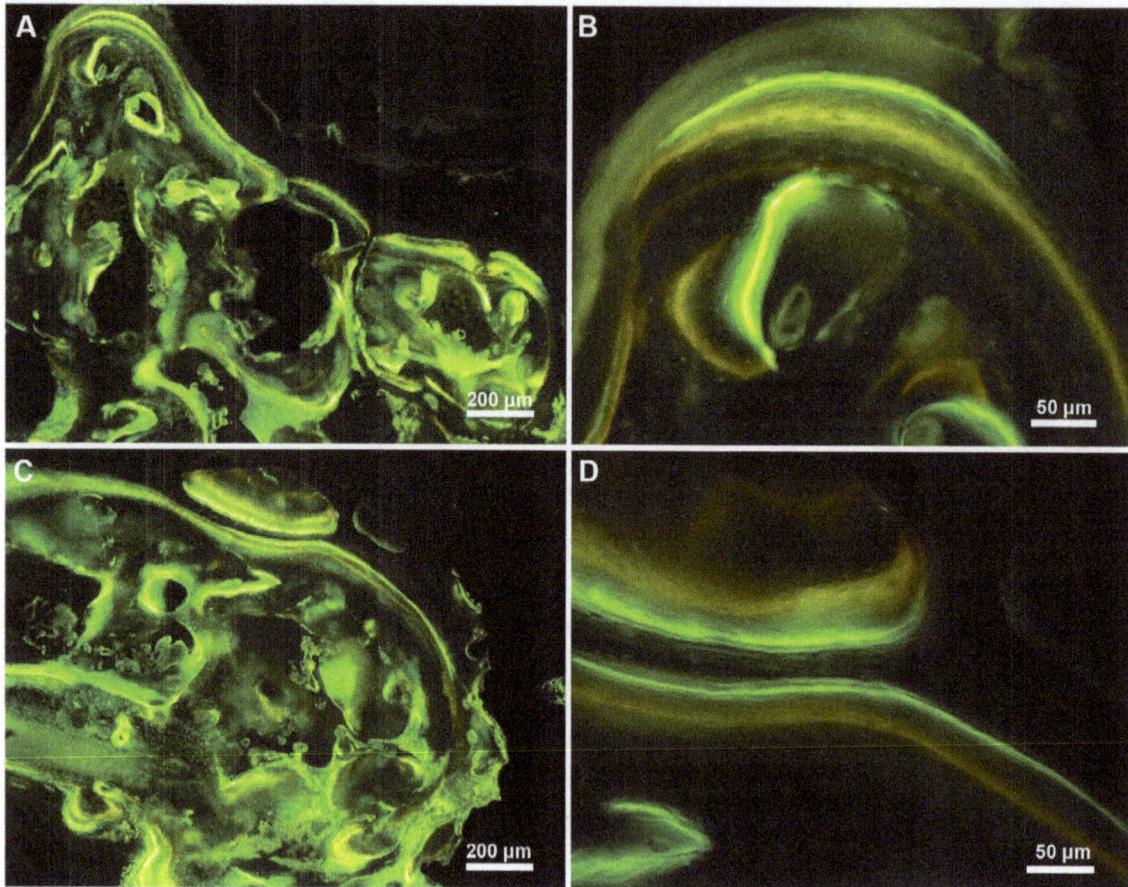

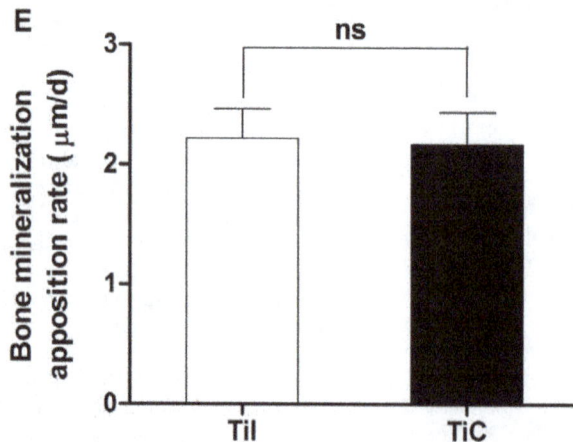

Figure 8. Fluorochrome labeling of bone regeneration at 12 weeks post-surgery. The fluorescent labeling indicated that abundant new bone growth into the porous titanium and continuous process of bone remodeling in both (**A–B**) uncoated and (**C–D**) coated samples. (**E**) The rate of bone mineralization apposition was similar in these two kinds of porous titanium at 12 weeks post-surgery ($P>0.05$). TiI, pure porous titanium implant; TiC, porous titanium implant with biomimetic coating.

requirements may lead to the conflicting design goals. Hollister et al. [33,34] developed a general design optimization scheme for the pore architecture to match desired elastic properties and porosity simultaneously, by introducing the homogenization-based topology optimization algorithm. Furthermore, they demonstrated the prototypes of the designed structures which can be fabricated using additive manufacturing technique. Various porous structures cause different mechanical properties. Generally speaking, with increase of the porosity, the mechanical properties such as stiffness and compressive strength decreased [14]. It was found that not only porosity has an influence on the mechanical properties of porous samples, but also the pore size and orientation, and strut size of samples affect the mechanical properties [12,14]. In current study, we pre-designed the porosity, pore size, and strut size of

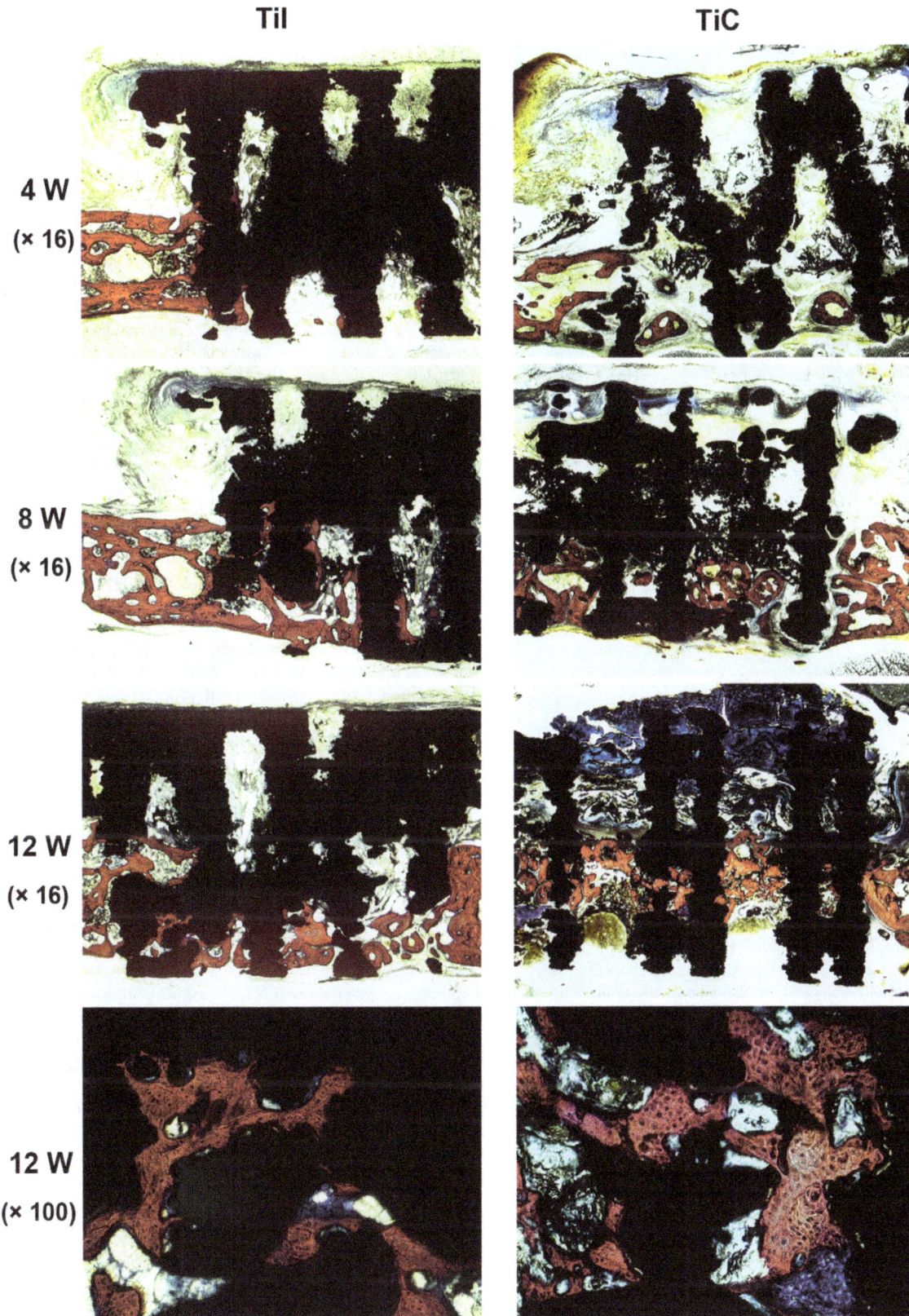

Figure 9. Histological staining for osteogenesis within porous titanium after implantation for 4, 8 and 12 weeks. The images showed that the rapid increase of bone ingrowth into the pores of titanium throughout the experiment and close contact between bone tissue and EBM implants at 12 weeks post-surgery, while there were no obvious differences between the uncoated and coated samples at each timepoint. TiI, pure porous titanium implant; TiC, porous titanium implant with biomimetic coating.

Bone Histomorphology

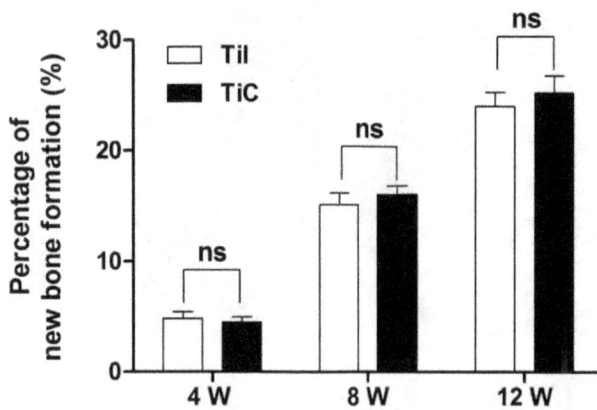

Figure 10. Histomorphometric analysis of new bone formation. The data showed a high percentage of newly formed bone in the defect region in both coated and uncoated implants during the whole experiment ($P>0.05$). TiI, pure porous titanium implant; TiC, porous titanium implant with biomimetic coating.

samples to adjust the mechanical properties to match that of cortical bone. The EBM-produced porous Ti6Al4V implants showed a favorable mechanical property (compressive strength range from 163–286 MPa, Young's modulus range from 14.5–38.5 GPa) and suitable porous structure (fully interconnected porous networks, porosity range from 51–61%, the pore size range from 500 - 600 μm) for the load bearing application in treating large segmental bone defects.

The biomimetic technique allowed the homogeneous deposition of a carbonated apatite layer onto the porous metallic implants [14,25]. This coating has a composition and crystallinity index similar to that of bone mineral [23]. Jalota et al. investigated the proliferation of osteoblast on heat and bone-like apatite coated surfaces of titanium foams and found that bone-like apatite coated surfaces exhibited the highest protein production and cell attachment [35]. It was revealed that the direct bone contact was significantly higher for bone-like apatite coated dense Ti6Al4V than the corresponding uncoated implants [25].

As one part of our study design, the biomimetic approach was used to deposit bone-like apatite on porous samples surfaces to promote the attachment, migration and proliferation of osteoblasts on the samples. However, no significant differences were observed in osteoblast function and morphology on the porous scaffolds between groups with or without biomimetic coatings. Ponader et al. reported that topographical surface modifications of electron beam melted Ti-6Al-4V titanium markedly influenced the function of human fetal osteoblasts [36]. The interactions between cells and microtopography of scaffold have been extensively studied, suggesting that microtopographies can promote bone-to-implant contact via such mechanisms as mechanical interlocking [37] and enhancement of osteoblast functions [38]. Cells can well communicate and in consequence proliferate as long as there is enough space to attach on the relatively smooth surface [39], while the peaks on the rough surfaces of implants may offer more favorable biological environment for the adhesion and differentiation of osteoblasts [40,41]. The EBM titanium in the present study consist of Ti6Al4V powders (particle size: 45–100 μm) [12]

which provide mild undulant surfaces without high peaks and scarped flanks, making it suitable matrix to trigger the attachment and proliferation of osteoblasts.

The in vivo experiment revealed that bone had started to grow from the host bone bed to the porous implants. The penetration depth of new bone regeneration increased rapidly with the implantation time, and eventually bridged the defect at week 12, demonstrating excellent osteoconductivity of the Ti6Al4V implants fabricated by EMB process. The amount of formed bone of EBM Ti-6Al-4V titanium in our study was much higher than that of porous titanium manufactured by other methods [29], and was comparable to that of EBM porous titanium reported by other study [31]. Takemoto et al. reported that the bioactive porous titanium achieved bone ingrowth to a depth of 3 mm within 4 weeks of implantation and continued to increase throughout the 16 weeks of implantation, whereas bone ingrowth into the nontreated implants tended to decrease between 4 and 8 weeks [42], whereas the pure EBM porous titanium used in our study exerted remarkable and continuous increase of bone formation throughout the whole experiment. Although several studies suggested that it might be necessary to improve the biological performance of titanium implant by being treated chemically and thermally, and/or coated with bioactive materials [43,44,45], it was similar for the new bone ingrowth within the implants with and without apatite coating. It was revealed that the properties that were considered as the important requirements for bone ingrowth contained favorable topographical surface for cell adhesion and proliferation, interconnected porous structures in all dimensions, appropriate mechanical strength corresponding to those of native bone [14,31,36]. In this study, it looks like the adequate architecture and mechanical properties of EBM porous scaffolds may contribute mainly to the optimized ingrowth of surrounding bone tissue in addition to the surface modification.

Conclusion

EBM technique can be used to fabricate Ti6Al4V implants with well controlled porous structures and favorable mechanical properties, which is close to native bone tissue. This Ti6Al4V implants presented good cell attachment and proliferation properties in vitro, as well as satisfied bone ingrowth and direct bonding between host bone tissue and the implants. Biomimetic approach can be utilized to coat the porous Ti6Al4V implants with a homogenous layer of bone-like apatite, whereas this biomimetic coating did not further enhance the bioactivities of the titanium scaffolds. Therefore, porous Ti6Al4V implant fabricated by EBM technique possesses great potential for the clinical applications, which can not only reduce the mechanical mismatch but also achieve stable long-term fixation by promoting bone ingrowth. This study opens up the possibility of using high strength porous scaffolds with appropriate osteoconductive and osteogenic properties to reconstruct bone defects on specific sites in the maxillofacial and orthopedic fields.

Acknowledgments

We thank Professor Rong Lv for providing helpful advice of the histology.

Author Contributions

Conceived and designed the experiments: LW XL YFF ZYZ. Performed the experiments: XL YFF GCL LW. Analyzed the data: YFF XL. Contributed reagents/materials/analysis tools: CTW GCL WL. Wrote the paper: XL YFF ZYZ LW. Revised the manuscript: XL LW WL.

References

1. Kroger H, Venesmaa P, Jurvelin J, Miettinen H, Suomalainen O, et al. (1998) Bone density at the proximal femur after total hip arthroplasty. Clin Orthop Relat Res: 66–74.
2. Fujibayashi S, Takemoto M, Neo M, Matsushita T, Kokubo T, et al. (2011) A novel synthetic material for spinal fusion: a prospective clinical trial of porous bioactive titanium metal for lumbar interbody fusion. Eur Spine J 20: 1486–1495.
3. Ryan G, Pandit A, Apatsidis DP (2006) Fabrication methods of porous metals for use in orthopaedic applications. Biomaterials 27: 2651–2670.
4. Kienapfel H, Sprey C, Wilke A, Griss P (1999) Implant fixation by bone ingrowth. J Arthroplasty 14: 355–368.
5. Li J, Habibovic P, Yuan H, van den Doel M, Wilson CE, et al. (2007) Biological performance in goats of a porous titanium alloy-biphasic calcium phosphate composite. Biomaterials 28: 4209–4218.
6. Pilliar RM (1987) Porous-surfaced metallic implants for orthopedic applications. J Biomed Mater Res 21: 1–33.
7. Hutmacher DW, Sittinger M, Risbud MV (2004) Scaffold-based tissue engineering: rationale for computer-aided design and solid free-form fabrication systems. Trends Biotechnol 22: 354–362.
8. Leong KF, Cheah CM, Chua CK (2003) Solid freeform fabrication of three-dimensional scaffolds for engineering replacement tissues and organs. Biomaterials 24: 2363–2378.
9. Li JP, de Wijn JR, Van Blitterswijk CA, de Groot K (2006) Porous Ti6Al4V scaffold directly fabricating by rapid prototyping: preparation and in vitro experiment. Biomaterials 27: 1223–1235.
10. Krishna BV, Bose S, Bandyopadhyay A (2007) Low stiffness porous Ti structures for load-bearing implants. Acta biomaterialia 3: 997–1006.
11. Hollander DA, von Walter M, Wirtz T, Sellei R, Schmidt-Rohlfing B, et al. (2006) Structural, mechanical and in vitro characterization of individually structured Ti-6Al-4V produced by direct laser forming. Biomaterials 27: 955–963.
12. Li X, Wang C, Zhang W, Li Y (2009) Fabrication and characterization of porous Ti6Al4V parts for biomedical applications using electron beam melting process. Materials Letters 63: 403–405.
13. Hahn H, Palich W (1970) Preliminary evaluation of porous metal surfaced titanium for orthopedic implants. J Biomed Mater Res 4: 571–577.
14. Parthasarathy J, Starly B, Raman S, Christensen A (2010) Mechanical evaluation of porous titanium (Ti6Al4V) structures with electron beam melting (EBM). J Mech Behav Biomed Mater 3: 249–259.
15. Heinl P, Muller L, Korner C, Singer RF, Muller FA (2008) Cellular Ti-6Al-4V structures with interconnected macro porosity for bone implants fabricated by selective electron beam melting. Acta biomaterialia 4: 1536–1544.
16. Goransson A, Jansson E, Tengvall P, Wennerberg A (2003) Bone formation after 4 weeks around blood-plasma-modified titanium implants with varying surface topographies: an in vivo study. Biomaterials 24: 197–205.
17. Nishiguchi S, Nakamura T, Kobayashi M, Kim HM, Miyaji F, et al. (1999) The effect of heat treatment on bone-bonding ability of alkali-treated titanium. Biomaterials 20: 491–500.
18. O'Hare P, Meenan BJ, Burke GA, Byrne G, Dowling D, et al. (2010) Biological responses to hydroxyapatite surfaces deposited via a co-incident microblasting technique. Biomaterials 31: 515–522.
19. Ben-Nissan B, Milev A, Vago R (2004) Morphology of sol-gel derived nano-coated coralline hydroxyapatite. Biomaterials 25: 4971–4975.
20. Ma J, Wang C, Peng KW (2003) Electrophoretic deposition of porous hydroxyapatite scaffold. Biomaterials 24: 3505–3510.
21. Sugita Y, Ishizaki K, Iwasa F, Ueno T, Minamikawa H, et al. (2011) Effects of pico-to-nanometer-thin TiO2 coating on the biological properties of micro-roughened titanium. Biomaterials 32: 8374–8384.
22. Li LH, Kong YM, Kim HW, Kim YW, Kim HE, et al. (2004) Improved biological performance of Ti implants due to surface modification by micro-arc oxidation. Biomaterials 25: 2867–2875.
23. Aparecida AH, Fook MV, Guastaldi AC (2009) Biomimetic apatite formation on Ultra-High Molecular Weight Polyethylene (UHMWPE) using modified biomimetic solution. J Mater Sci Mater Med 20: 1215–1222.
24. Jonasova L, Muller FA, Helebrant A, Strnad J, Greil P (2004) Biomimetic apatite formation on chemically treated titanium. Biomaterials 25: 1187–1194.
25. Kim HM, Miyaji F, Kokubo T, Nakamura T (1997) Bonding strength of bonelike apatite layer to Ti metal substrate. J Biomed Mater Res 38: 121–127.
26. Barrere F, van der Valk CM, Meijer G, Dalmeijer RA, de Groot K, et al. (2003) Osteointegration of biomimetic apatite coating applied onto dense and porous metal implants in femurs of goats. J Biomed Mater Res B Appl Biomater 67: 655–665.
27. Kokubo T, Takadama H (2006) How useful is SBF in predicting in vivo bone bioactivity? Biomaterials 27: 2907–2915.
28. Nebe JB, Muller L, Luthen F, Ewald A, Bergemann C, et al. (2008) Osteoblast response to biomimetically altered titanium surfaces. Acta biomaterialia 4: 1985–1995.
29. Zhao L, Wei Y, Li J, Han Y, Ye R, et al. (2010) Initial osteoblast functions on Ti-5Zr-3Sn-5Mo-15Nb titanium alloy surfaces modified by microarc oxidation. Journal of biomedical materials research Part A 92: 432–440.
30. Li JP, Habibovic P, van den Doel M, Wilson CE, de Wijn JR, et al. (2007) Bone ingrowth in porous titanium implants produced by 3D fiber deposition. Biomaterials 28: 2810–2820.
31. Lopez-Heredia MA, Goyenvalle E, Aguado E, Pilet P, Leroux C, et al. (2008) Bone growth in rapid prototyped porous titanium implants. Journal of biomedical materials research Part A 85: 664–673.
32. Ponader S, von Wilmowsky C, Widenmayer M, Lutz R, Heinl P, et al. (2010) In vivo performance of selective electron beam-melted Ti-6Al-4V structures. Journal of biomedical materials research Part A 92: 56–62.
33. Thomsen P, Malmstrom J, Emanuelsson L, Rene M, Snis A (2009) Electron beam-melted, free-form-fabricated titanium alloy implants: Material surface characterization and early bone response in rabbits. J Biomed Mater Res B Appl Biomater 90: 35–44.
34. Hollister SJ, Maddox RD, Taboas JM (2002) Optimal design and fabrication of scaffolds to mimic tissue properties and satisfy biological constraints. Biomaterials 23: 4095–4103.
35. Jalota S, Bhaduri SB, Tas A C (2007) Osteoblast proliferation on neat and apatite-like calcium phosphate-coated titanium foam scaffolds. Materials Science and Engineering C 27: 432–40.
36. Boonrungsiman S, Gentleman E, Carzaniga R, Evans ND, McComb DW, et al. (2012) The role of intracellular calcium phosphate in osteoblast-mediated bone apatite formation. Proc Natl Acad Sci U S A 109: 14170–14175.
37. Ponader S, Vairaktaris E, Heinl P, Wilmowsky CV, Rottmair A, et al. (2008) Effects of topographical surface modifications of electron beam melted Ti-6Al-4V titanium on human fetal osteoblasts. Journal of biomedical materials research Part A 84: 1111–1119.
38. Hansson S, Norton M (1999) The relation between surface roughness and interfacial shear strength for bone-anchored implants. A mathematical model. J Biomech 32: 829–836.
39. Zhao L, Mei S, Chu PK, Zhang Y, Wu Z (2010) The influence of hierarchical hybrid micro/nano-textured titanium surface with titania nanotubes on osteoblast functions. Biomaterials 31: 5072–5082.
40. Sader MS, Balduino A, Soares Gde A, Borojevic R (2005) Effect of three distinct treatments of titanium surface on osteoblast attachment, proliferation, and differentiation. Clin Oral Implants Res 16: 667–675.
41. Anselme K, Bigerelle M, Noel B, Dufresne E, Judas D, et al. (2000) Qualitative and quantitative study of human osteoblast adhesion on materials with various surface roughnesses. J Biomed Mater Res 49: 155–166.
42. Anselme K, Linez P, Bigerelle M, Le Maguer D, Le Maguer A, et al. (2000) The relative influence of the topography and chemistry of TiAl6V4 surfaces on osteoblastic cell behaviour. Biomaterials 21: 1567–1577.
43. Takemoto M, Fujibayashi S, Neo M, Suzuki J, Kokubo T, et al. (2005) Mechanical properties and osteoconductivity of porous bioactive titanium. Biomaterials 26: 6014–6023.
44. Goyenvalle E, Aguado E, Nguyen JM, Passuti N, Le Guehennec L, et al. (2006) Osteointegration of femoral stem prostheses with a bilayered calcium phosphate coating. Biomaterials 27: 1119–1128.
45. Lopez-Heredia MA, Sohier J, Gaillard C, Quillard S, Dorget M, et al. (2008) Rapid prototyped porous titanium coated with calcium phosphate as a scaffold for bone tissue engineering. Biomaterials 29: 2608–2615.

11

Dissociation of Bone Resorption and Bone Formation in Adult Mice with a Non-Functional V-ATPase in Osteoclasts Leads to Increased Bone Strength

Kim Henriksen[1]*, Carmen Flores[2], Jesper S. Thomsen[3], Anne-Marie Brüel[3], Christian S. Thudium[1], Anita V. Neutzsky-Wulff[1], Geerling E. J. Langenbach[4], Natalie Sims[5], Maria Askmyr[2], Thomas J. Martin[5], Vincent Everts[6], Morten A. Karsdal[1], Johan Richter[2]

1 Nordic Bioscience A/S, Herlev, Denmark, 2 Molecular Medicine and Gene Therapy, Lund University, Lund, Sweden, 3 Institute of Anatomy, University of Aarhus, Aarhus, Denmark, 4 Department of Functional Anatomy, Academic Centre of Dentistry Amsterdam (ACTA), University of Amsterdam and VU University Amsterdam, Research Institute MOVE, Amsterdam, The Netherlands, 5 St. Vincent's Institute for Medical Research, Melbourne, Australia, 6 Department of Oral Cell Biology, Academic Centre of Dentistry Amsterdam (ACTA), University of Amsterdam and VU University Amsterdam Research Institute MOVE, Amsterdam, The Netherlands

Abstract

Osteopetrosis caused by defective acid secretion by the osteoclast, is characterized by defective bone resorption, increased osteoclast numbers, while bone formation is normal or increased. In contrast the bones are of poor quality, despite this uncoupling of formation from resorption. To shed light on the effect of uncoupling in adult mice with respect to bone strength, we transplanted irradiated three-month old normal mice with hematopoietic stem cells from control or *oc/oc* mice, which have defective acid secretion, and followed them for 12 to 28 weeks. Engraftment levels were assessed by flow cytometry of peripheral blood. Serum samples were collected every six weeks for measurement of bone turnover markers. At termination bones were collected for μCT and mechanical testing. An engraftment level of 98% was obtained. From week 6 until termination bone resorption was significantly reduced, while the osteoclast number was increased when comparing *oc/oc* to controls. Bone formation was elevated at week 6, normalized at week 12, and reduced onwards. μCT and mechanical analyses of femurs and vertebrae showed increased bone volume and bone strength of cortical and trabecular bone. In conclusion, these data show that attenuation of acid secretion in adult mice leads to uncoupling and improves bone strength.

Editor: Irina Agoulnik, Florida International University, United States of America

Funding: CST received funding from Nordforsk, AVNW received funding from the Danish Research Foundation, CF is supported by a PhD fellowship from European Calcified Tissue Society. JR was supported by grants from The Swedish Childhood Cancer Foundation, a Clinical Research Award from Lund University Hospital, Magnus Bergvalls Foundation, the Georg Danielsson Foundation and The Foundations of Lund University Hospital. The Lund Stem Cell Center is supported by a Center of Excellence grant in life sciences from the Swedish Foundation for Strategic Research. The funders had no role in study design, data collection and analysis, decision to publish, or preparation of the manuscript.

Competing Interests: KH, CST, AVNW and MAK are employees of Nordic Bioscience A/S, MAK owns stock in Nordic Bioscience A/S. All other authors have no conflicts of interest. All authors have been involved in study design, data analysis and writing of the manuscript.

* E-mail: kh@nordicbioscience.com

Introduction

Bone remodeling is a continuous process that maintains calcium homeostasis, removes old bone and mediates microfracture repair, thereby ensuring bone quality [1]. Bone resorption is performed by osteoclasts, after which the osteoblasts form new bone matrix, leading to restoration of the removed bone [2]. These two processes are normally tightly balanced, a process referred to as coupling [3, 4]. Recent studies have indicated that the coupling of bone formation to bone resorption is more complex than originally thought [5, 6], and likely includes secretion of bone anabolic factors by the osteoclasts, independent of bone resorptive activity [2, 7].

Osteoclasts derive from hematopoietic stem cells which, in the presence of the osteoblast-derived molecules RANKL and M-CSF, develop into mature multinucleated bone resorbing osteoclasts [8, 9]. The osteoclasts resorb bone by secretion of hydrochloric acid and proteases which, in combination, dissolve the calcified bone matrix [8, 9]. Acidification of the resorption compartment is achieved by active proton transport mediated by the osteoclast specific V-ATPase, while chloride is secreted by the chloride-proton antiporter ClC-7 [10–14].

Loss of function mutations or gene knockouts in humans and mice of these two molecules lead to different types of osteopetrosis indicating their importance for dissolution of the inorganic bone matrix [10, 15, 16]. These forms of osteopetrosis are characterized by normal or even increased indices of bone formation despite the presence of high numbers of non-resorbing osteoclasts [17-20], indicating that bone resorption and bone formation are no longer coupled. Despite the high bone mass, a feature of osteopetrosis is poor bone quality, which has been speculated to be due to the extreme suppression of bone resorption [21, 22], the failure to resorb calcified cartilage [9], and to hyper-activity of the osteoblasts [23].

A recent study of ClC-7 deficient mice indicated uncoupling of bone formation from bone resorption [24]. However, further characterization failed to confirm these findings due to the severe developmental phenotype, where calcified cartilage completely occluded the marrow cavity of all long bones [25]. This illustrates the difficulty of investigating bone phenotypes in these very young mice.

The *oc/oc* mice exhibit very severe osteopetrosis due to a mutation in the a3 subunit of the V-ATPase, and these mice die of anemia 3–4 weeks after birth [26]. Recent studies in these mice have shown that the osteopetrotic phenotype can be rescued by neonatal transplantation of normal or gene-corrected hematopoietic stem cells into irradiated mice, in accordance with the hematopoietic nature of the defect [27–30].

In order to investigate the effect of osteopetrosis on bone quality in adult mice and also shed light on the uncoupling observed in some forms of osteopetrosis, we induced osteopetrosis in normal 3-month old mice by transplanting them with fetal liver derived hematopoietic stem cells from *oc/oc* mice or their corresponding control littermates, and then followed them for three or six months and characterized their bone and osteoclast phenotypes in detail.

Materials and Methods

Mice

Breeding pairs of (C57BL/6J _ C3HheB/FeJ) F1 oc/+ mice (CD45.2) were obtained from the Jackson Laboratory (Bar Harbor, ME) and maintained in the conventional animal facility at the Biomedical Centre, University of Lund.

All experiments were performed according to protocols approved by the local animal ethics committee in both Denmark (Rådet for Dyreforsøg (The Animal Experiments expectorate)) registration number 2007/561-1303 and Sweden (Malmö/Lunds Djurförsöksetiske Nämnd (The ethics committee for animal studies in Malmö/Lund) registration number M 128-09.

Genotyping of mice

Mice were genotyped on the day of birth using DNA extracted from the tip of the tail as described previously [27].

Harvest and isolation of fetal liver hematopoietic cells

On embryonic day 14.5, pregnant mice were killed by CO_2 poisoning, and embryos were removed. Fetal livers (FLs) were dissected out and put into PBS (Invitrogen) supplemented with 2% FCS (Invitrogen). Single-cell suspensions were prepared by drawing liver cells through a 23-gauge needle followed by filtering through a 50 µm cell strainer. Individual FLs were genotyped by lysing a cell sample and running the PCR described above. Cells from both wild type (+/+) and *oc/+* embryos were used as controls and henceforth designated as such, as oc/+ mice are phenotypically indistinguishable from +/+ littermates.

Transplantation and follow-up

Three-month-old mice (C57BL/6J _ C3HheB/FeJ)(CD45.1) were irradiated with 950 cGy administered from a 137Cs source. Four hours later mice received an intravenous transplant of 2×10^6 freshly thawed FL cells in 300 µL PBS. To avoid infection following transplantation the animals were treated for 14 days with Baytril in their drinking water. After transplantation the two groups of mice were followed for 3 months. Intraperitonal injections of calcein (20mg/kg) were given 10 and 3 days prior to sacrifice.

For the 12 week experiment a total of 10 mice were transplanted, 5 controls and 5 *oc/oc*, and for the 28 week experiment a total of 11 mice were transplanted, 5 controls and 6 *oc/oc*. Of all the mice 1

control died of the 12 week and 2 controls died of the 28 week experiment, excluding these from the analyses. The deaths did not appear to be related to the transplantation procedure.

At termination the bones for µCT and mechanical testing in the 12 week experiment were stored in Lilly's fluid until analysis after which they were transferred to 0.9% NaCl and 0.1% NaN_3, while the bones from the 28 week experiment we stored in 0.9% NaCl and 0.1% NaN_3 at all time points. A published study clearly showed that fixation does not impact measurements of bone strength (F_{max}) in mice [31], and thus all samples were treated equally in the mechanical test (see later).

Engraftment and lineage distribution analysis of peripheral blood

Peripheral blood (PB) was collected in heparin (LEO Pharma, Thornhill, ON) after tail clipping of mice, and mixed with equal volumes of PBS containing 2% FCS. Following centrifugation, the supernatant was poured off, erythrocytes were lysed with NH_4Cl, and the cells were washed twice with PBS containing 2% FCS. Subsequently, cells were incubated on ice for 20 to 30 minutes with APC-conjugated antibodies directed against B220, CD3, Gr-1, and Mac-1 multilineage analysis) (Becton Dickinson). The cells were suspended in 300 µL PBS containing 2% FCS followed by addition of 1 µg/mL 7-amino-actinomycin D (7-AAD, for detection of nonviable cells; Sigma, St Louis, MO) before analysis using a fluorescence-activated cell sorting (FACS) Calibur Instrument (Becton Dickinson).

Serum collection

All sera were collected by retro-orbital bleeding after overnight fasting of the mice 6, 12, 18, 23 and 28 weeks after transplantation.

Bone Resorption by Mature Osteoclasts

Isolated spleen cells from either genotype were differentiated into mature osteoclasts by 4 days of culture in αMEM + M-CSF (25 ng/mL), trypsinization, and reseeding at 900,000 cells/six-well plate, followed by 7 days of culture in αMEM containing RANKL (100 ng/ml) and M-CSF (25 ng/ml) with media exchanged every day as described by Neutzsky-Wulff et al. [39]. Mature osteoclasts from either transplantation group were lifted using trypsin and cell scraping and reseeded on cortical bone slices (see reference [39]), at 50,000 cells/bone slice. Culture supernatants were collected and stored at -20°C until further analysis.

Bovine cortical bone slices

Bovine cortical bone from cows of more than 3 years of age was cut into thin slices (0.5 cm diameter) as described by Neutzsky-Wulff et al. [39] and stored in 70% ethanol until use. Prior to seeding of cells, bone slices were washed thoroughly in the appropriate medium.

Measurement of TRAP Activity in Cell Culture Supernatants

TRAP activity in cell culture medium was measured as described previously [32]. Briefly, samples were incubated with TRAP reaction buffer, containing p-nitrophenyl phosphate and sodium tartrate, for 1 hour at 37°C in the dark. The reaction was stopped with 0.3 M NaOH. Absorbance was measured in an ELISA reader at 405 nm with 650 nm as reference.

Biochemical Markers of Bone Turnover in serum

TRAP5b activity in serum was measured by the Mouse-TRAP assay (SD-TR103, IDS) according to the manufacturer's protocol.

Serum samples from individual mice were diluted in PBS to obtain readings within the range of the kit.

Alkaline phosphatase (ALP) was measured by mixing serum samples or controls with substrate solution (0.95 ml AMP buffer [50 ml Milli Q water, 6.25 ml 2-amino-2-methyl-1-propanol 95% {A65182, Sigma}, pH adjusted to 10.0, volume adjusted to 62.5 ml by addition of Milli Q water], 9.5 ml Milli Q water, 40 mg PNPP [P5994, Sigma], 190 μL 1M $MgCl_2$) and incubating for 20 minutes in the dark. The reaction was stopped by addition of 0.5 M NaOH. Colorimetric changes were measured at 405 nm with 650 nm as reference using an ELISA reader.

C-terminal type I collagen fragments (CTX-I) were measured using the RatLaps ELISA (1RTL4000; IDS Nordic A/S, Herlev, Denmark), according to the manufacturer's protocol.

Serum P1NP was measured using an ELISA (IDS Nordic A/S, Cat#AC-33F1) according to the manufacturer's instructions.

Micro-computed tomography (micro-CT) imaging

Three-dimensional reconstructions of trabecular and cortical bone of the lumbar vertebrae and femurs were generated with a high-resolution micro-CT system (μCT 40; Scanco Medical AG, Brüttisellen, Switzerland). The bones were mounted in a cylindrical specimen holder to be captured in a single scan. They were secured with synthetic foam and were completely submerged in physiological saltwater containing 0.1% NaN_3. Scans with an isotropic resolution of 10 μm were made using a 55-kV peak-voltage X-ray beam. Each scan projection (300 ms) was performed four times and averaged to optimize the signal-to-noise ratio, thereby facilitating segmentation. The computed linear attenuation coefficient of the X-ray beam in each volume element (voxel) was stored in an attenuation map and represented by a gray value in the reconstruction. Specific volumes of interest (VOIs) were selected. The complete vertebral trabecular bone was selected for analysis. To analyze the femur trabecular bone, a region of 5% of the bone length distal of the metaphysis was evaluated. Cortical bone analysis was performed in the region between 45 and 55% along the length of the femurs. To discriminate between bone and background, the reconstructions were segmented using an appropriate fixed threshold. For cortical and trabecular bone this threshold was the grey value comparable to respectively 500 and 350 mg hydroxyapatite/cm^3. Multiple cortical and trabecular bone parameters were determined using morphometric software supplied by the manufacturer [for trabecular bone: bone volume fraction (BV/TV), trabecular thickness (Tb.Th), and degree of mineralization of the bone (DMB); for cortical bone: Cortical bone volume (Ct.BV), cortical thickness (Ct.Th), degree of mineralization of the bone (DMB), endocortical diameter (Ec.Dm), endocortical marrow volume (Ec.M.Vol), and periosteal diameter (P.Dm)].

Bone Strength Measurements

Femoral Diaphysis. The femora were carefully cleaned from muscles and soft connective tissue. The length of the left femora was measured using an electronic caliper and the mid-point of the femora was marked with a permanent marker pen. The femora were placed in a testing jig for three-points bending with their posterior surface resting on two lower supports located 6.6 mm apart, with their midpoint centered between the two lower supports. The testing jig was then placed in an Instron materials testing machine (model 5566, High Wycombe, UK) and load was applied at a constant deformation rate of 2 mm/min with a rod at the upper anterior midpoint of the femur. During compression testing load-deformation data were recorded using Merlin (version 3.21, Instron, High Wycombe, UK), stored on an attached PC for

later analysis. After testing, the fracture line was examined to ensure the fracture occurred perpendicular to the longitudinal axis of the bone. Maximum load (F_{max}, N) was determined from the load-deformation data using in-house developed software.

Femoral Neck. The proximal femur (the proximal half obtained after the three-point bending test) was mounted in a custom-made device for standardized fixation [33]. The fixation device holding the specimen was then placed into the material testing machine, and a vertical load exerted by a cylinder was applied to the top of the femoral head. The cylinder was directed parallel to the axis of the femoral diaphysis and moved at a constant rate of 2 mm/min until fracture of the femoral neck. During biomechanical testing, load-deformation values were obtained and stored on the PC for later analysis. Maximum load (F_{max}, N) was determined from the load-displacement data using in-house developed software.

Vertebral Body. The fourth lumbar vertebral body was dissected free from L3 and L5 and the posterior processes were carefully removed under a dissecting microscope using a fine electric saw and a small clipper.

The cartilaginous endplates were removed with a small scalpel in a fashion that left parallel planes at the cranial and caudal ends without removing excess bone, resulting in a bone specimen height of approximately 2.8 mm. The vertebral bone specimens were placed in the materials testing machine between two parallel plates and compression tested at a constant velocity of 2 mm/min until failure. During biomechanical testing, load-deformation values were obtained and stored on the PC for later analysis. Maximum load (F_{max}, N) was determined from the load-deformation data using in-house developed software.

Histomorphometry and staining of plastic embedded specimens

For specimens destined for plastic embedding, the hind legs were fixed in 3.7% formaldehyde in PBS and stored in 70% ethanol. Tibias were embedded in methylmethacrylate in a fully calcified state as previously described [34]. Sections of 5 μm thickness were cut, and stained with each of the following solutions: Toluidine blue, Safranin O/fast green, Goldner's trichrome, Xylenol Orange (counterstain for calcein labeled specimens) and TRACP stain. Histomorphometry was carried out according to standard procedures [35] in the proximal tibia using the Osteomeasure system (OsteoMetrics Inc.). Standard histomorphometric measurements were performed on toluidine blue stained sections in a region 1.1 mm long commencing 370 μm from the end of the hypertrophic zone of the growth plate. Calculations of mineral apposition rate (MAR) were based only on measurements of doubled labeled surfaces (dLS), which were measured in the same region.

Assessment of bone structure by histology

Humeri were decalcified in 15% EDTA and embedded in paraffin. Cutting was done on a HM360 microtome (Micron) at a 5 μm thickness. The sections were stained with hematoxylin and observed through an Olympus BX60 microscope using a 20x/0,40 objective polarized through filters U-ANT and U-POT. Images were obtained with a DP71 digital camera (Olympus) using the Cell^A software (Olympus).

Statistics

All statistical calculations were performed by Student's two-tailed unpaired *t*-test, assuming normal distribution and equal variance, with a significance level of P<0.05 (NS: not significant;

*:p<0.05, **:p<0.01, ***:p<0.001). Error bars indicate standard error of the mean (SEM).

Results

Experimental setup and engraftment analysis

Figure 1 shows the experimental setups. No signs of hepato-splenomegaly were observed in any of the experiments (data not shown).

At week 6 the ratio of CD45.2 (donor) cells to CD45.1 (host) cells in peripheral blood was approximately 95% in both groups (Figure 2A), and at 12, 18, and 28 weeks an engraftment level of approximately 98% was obtained in both groups, confirming successful transplantation. Since *oc/oc* mice have altered cellular composition of the hematopoietic compartment [36], an analysis of the major hematopoietic lineage cells was conducted. This showed no changes in the levels of B220+, CD3+, Mac1$_{High}$/Gr1$_{low}$, and Mac1$_{low}$/Gr1$_{high}$ cell populations between the two groups (Figure S1).

At termination splenocytes and bone marrow cells were isolated and cultured on cortical bone slices for 10 days to investigate osteoclastogenesis and function. As seen in figure 2B-C bone resorption measured by calcium release and CTX-I is significantly reduced in spleen-derived osteoclasts from mice transplanted with *oc/oc cells* when compared to osteoclasts derived from control animals. Furthermore, measurements of the osteoclast marker TRACP activity in the supernatants showed no changes in osteoclast numbers, as seen *in vitro* for both ClC-7 and *Atp6i* deficient mice (Figure 2D)[24;37]. Similar data were obtained with bone marrow derived osteoclasts (data not shown).

Assessment of bone volume

In alignment with attenuation of bone resorption in the *oc/oc* group the bone volume fraction (BV/TV) of the trabecular compartment of vertebrae was increased by 80%, and the trabecular thickness (Tb.Th.) by 50% at the 12 week time point, while no change in the mineralisation degree (DMB) was observed,

when comparing to controls (Figure 3A). In the 28 week experiment, the increases in BV/TV and Tb.Th. in the *oc/oc* group were of the same magnitude as in the 12-week experiment. With respect to DMB a 5% increase was seen in vertebrae of the *oc/oc* compared to the control group after 28 weeks of transplantation. These data were supported by bone histomorphometry on vertebrae showing increased bone volume, as BV/TV, Tb.Th and Tb.N all were increased, while Tb.Sp. was decreased in the *oc/oc* group compared to the control group at the 12-week time point (Figure 3B).

In the femoral cortex, an increase in bone volume (BV) of 12% and cortical thickness of 15% was observed when comparing *oc/oc* to control at 12 weeks, while after 28 weeks the increases were 25 and 30%, respectively (Figure 3C). DMB of the femoral cortex showed a trend towards an increase, but this was not significant. Finally, at the 28-week time point both endocortical diameter and marrow volume were significantly reduced in the *oc/oc* group compared to control, while no changes were seen at the 12-week time point. No changes in periosteal parameters were observed at any of the time points.

Biochemical markers of bone turnover

Serum samples were collected throughout both experiments to investigate bone turnover markers. To combine the experiments, and to focus on between-group differences, rather than aspects of age, the levels of all markers were normalized to 100% at all time points in the control groups.

As seen in Figure 4A the level of the bone resorption marker CTX-I is significantly lower in the *oc/oc* group compared to the control group at all time points, except week 28, where overall CTX-I levels are low due to the advanced age of the mice (baseline CTX-I 50.1±7.9 ng/mL, week 28 CTX-I 24.5±3.5 ng/mL). The marker of osteoclast number TRACP 5b [38;39], was highly elevated in the *oc/oc* group from week 12 and throughout, compared with the control group, indicating increased osteoclast numbers *in vivo* (Figure 4B), and the ratio between CTX-I and TRACP 5b, which is used as a index for resorption per osteoclast

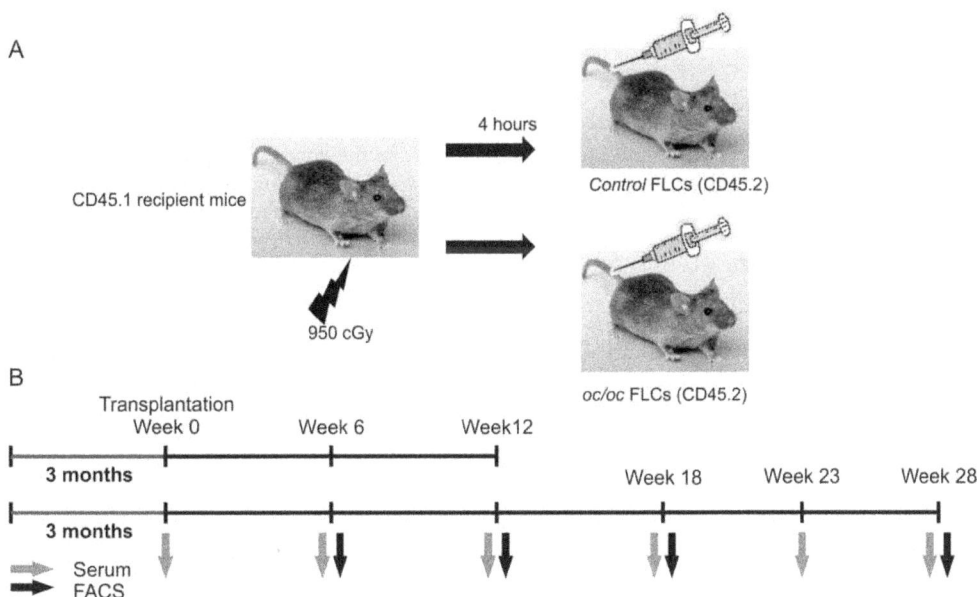

Figure 1. Schematic illustration of the experimental design. A) Illustration of the irradiation and transplantation setup. B) Overview of the timeline and sample collection times from the 12 week and 28 week experiments.

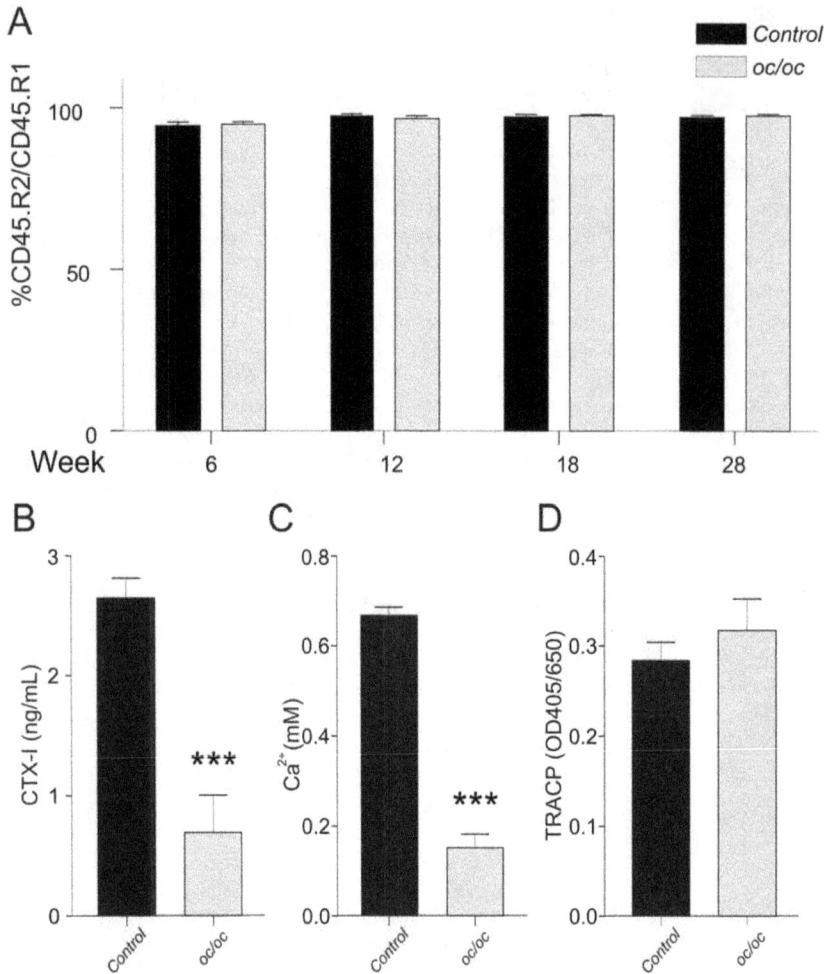

Figure 2. Engraftment analysis and *in vitro* bone resorption. A) Flow cytometry analysis of peripheral blood samples stained with an antibody against CD45.2 to quantify the level of engraftment. Flow cytometry was conducted in samples from all mice (see Methods section) and at the time points indicated. B-D) Splenocytes were isolated and cultured on bovine cortical bone in the presence of RANKL and M-CSF. At day 10 bone resorption was measured by CTX-I (B) and calcium release (C) release and osteoclast numbers measured by TRACP activity in the supernatants (D). Osteoclast cultures are representative of two individual experiments with 6 replicates of each condition.

[38], is markedly lower in the oc/oc group than the control, further confirming that activity per osteoclast is strongly reduced (Figure 4C). Interestingly, the bone formation markers PINP and ALP showed increased levels in the *oc/oc* group compared to the control group at week 6, while the levels returned to normal at week 12, and at the later time points were lower in the *oc/oc* group than the control group (Figure 4D&E). Finally, CTX-II levels, which are indicative of cartilage degradation, were similar in both groups (data not shown).

Histomorphometric analysis

Assessment of osteoclast and osteoblast numbers did not show any differences between the two groups (Figure 5A–D) at the 12-week time point. Furthermore, no differences in the dynamic parameters of bone formation, BRF/BS, MAR, MS/BS, and in osteoid volume (OV/BV) between groups were observed (Figure 5E-H) in the 12-week experiment.

Bone strength parameters

As osteopetrosis is associated with poor bone quality and fractures, we investigated the consequences of induction of

osteopetrosis in aged animals using mechanical testing. As for earlier data the values in the control group at both time points were normalized to 100% for comparative purposes. The 3-point bending test of the femoral mid-diaphysis showed a 33% increase in F_{max} when comparing *oc/oc* to control at the 12-week time point, while at the 28-week time point the difference was 55% (Figure 6A). At the femoral neck a significant increase of 60% in the *oc/oc* compared to control was seen at the 28-week time point, while at the 12-week time point a trend towards increased strength was seen (Figure 6B). In the vertebrae, no significant differences were observed, although the trends followed the other mechanical tests (Figure 6C).

Assessment of bone structure

To further understand the effects of transplantation with the *oc/oc* cells, bone structure was analyzed using polarized light microscopy. In figure 7 it is clearly shown that cortical bone is organized in well-structured lamellae indicating that transplantation has no detrimental effect on bone structure. Similar findings were obtained in trabecular bone (data not shown).

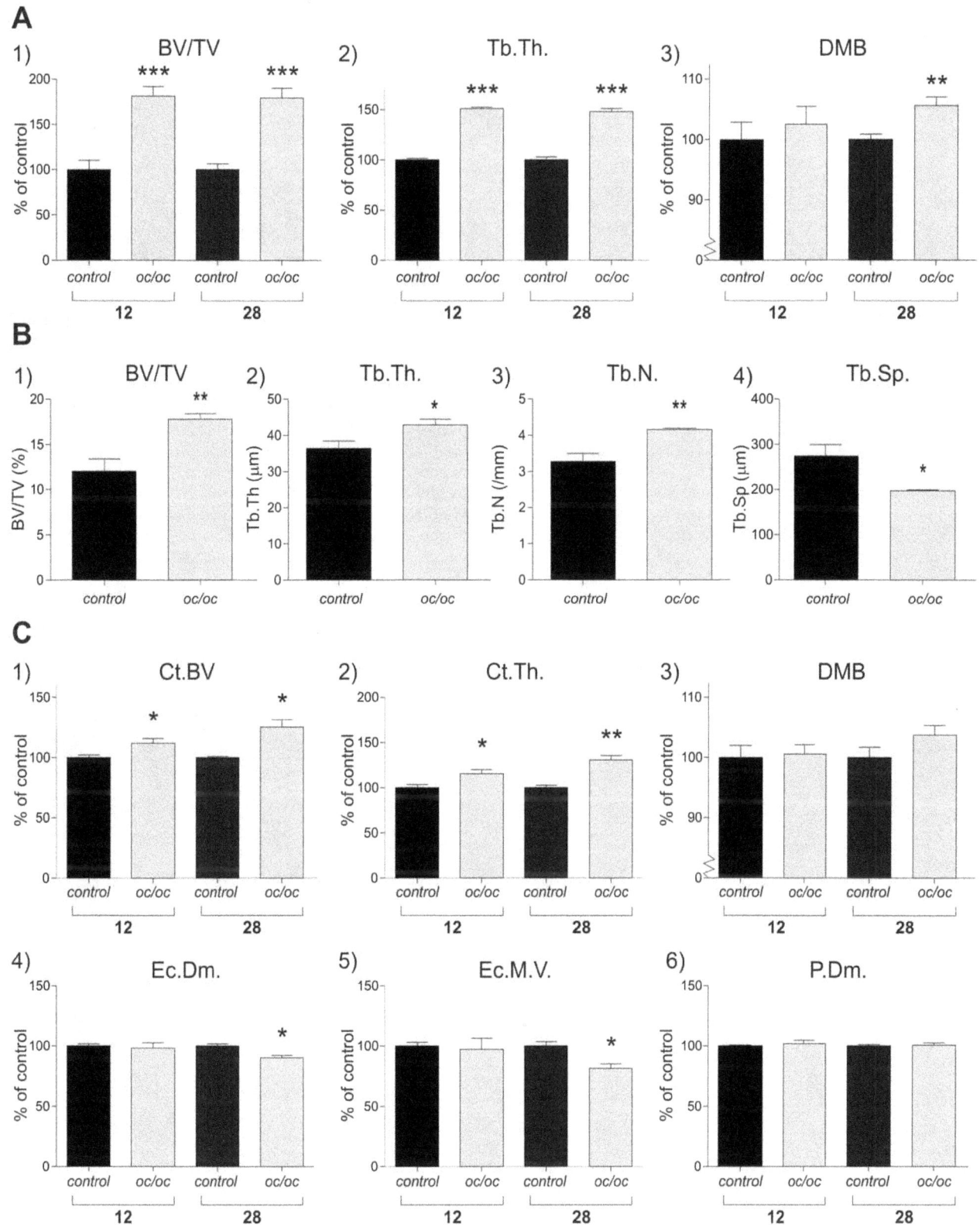

Figure 3. Assessment of bone volume. A) µCT analysis of the vertebrae from both the 12 and the 28-week experiment. For comparison the control group was normalized to 100%. 1) Bone volume/Total Volume (BV/TV) in % of control, 2) Trabecular Thickness (Tb.Th.) in % of control, and 3) Degree of Mineralization of the Bone (DMB) in % of control. B) Bone histomorphometry on vertebrae from the 12-week experiment. 1) Bone volume/Total Volume (BV/TV), 2) Trabecular Thickness (Tb.Th.), 3) Trabecular Number (Tb.N.), and 4) Trabecular Spacing (Tb.Sp.) C) µCT analysis of the femur diaphysis from both the 12 and the 28-week experiment. For comparison the control group was normalized to 100%. 1) Cortical Bone Volume (Ct.BV) in % of control, 2) Cortical Thickness (Ct.Th.) in % of control, 3) Cortical Degree of Mineralization of Bone (DMB) in % of control, 4) Endocortical

Diameter (Ec.Dm.) in % of control, 5) Endocortical Marrow Volume (Ec.M.V.) in % of control, 6) Periosteal Diameter (P.Dm.) in % of control. μCT was conducted on all bones from mice having completed the study (see methods section).

Discussion

The hematopoietic nature of osteopetrosis was established in the mid 1970s by transplantations of spleen cells from either healthy donor mice to osteopetrotic mice, or *vice versa* [40, 41, 41, 42]. Transfer of *oc/oc* splenocytes into healthy young mice led to increased bone weight [30], however other bone parameters were not examined.

Here we present novel data on the establishment of osteopetrosis in skeletally mature mice, in order to isolate the effect of non-resorbing osteoclasts on mature bone from the influence of non-resorbing osteoclasts on skeletal development and the resorption of mineralized cartilage in young mice.

Using fetal liver cells as a source of hematopoietic cells [27] an engraftment level in excess of 95% was obtained at 12 weeks, and the levels were around 98% 28 weeks after transplantation, confirming transplantation efficiency. No signs of hepatospleno-megaly were observed in any of the experiments, and no alterations in the cells of the hematopoietic lineages were observed, in contrast to haemopoietic defects observed in mice with lifelong osteopetrosis [10, 26, 36, 37]. This, not surprisingly, indicates that the haemopoietic phenotype of *oc/oc* mice is a developmental phenotype, in which the anemia effect is compounded by the complete lack of bone marrow cavities in mice with osteopetrosis due to defective acid secretion [10, 26, 36, 37]. These findings are further supported by studies in RANKL and RANK deficient mice, which have a less severe bone phenotype than *oc/oc*, *Atp6i*

and ClC-7 deficient mice, and accordingly have only mild changes in the hematopoietic system and show no sign of anemia [43, 44]; however, to fully understand these differences more detailed analyses are needed.

To validate that the osteoclasts were non-resorbing, osteoclas-togenesis and bone resorption were evaluated using spleen and bone marrow-derived osteoclasts from mice transplanted with either *oc/oc* or control hematopoietic cells. These data confirmed functional deficiency of the *oc/oc* osteoclasts, while showing no changes in osteoclastogenesis, as expected from a previous study of osteoclasts lacking the a3 subunit of the proton pump [37], as well as studies of osteoclasts with defective acid secretion [10, 11, 24, 37, 45]. These data also fit well with earlier findings showing that the increased numbers of osteoclasts in the acid secretion deficient mice are caused by increased survival of the osteoclasts, but not by changes in osteoclastogenesis [11, 46, 47].

In both human and murine osteopetrosis forms caused by defective acid secretion by the osteoclasts, bone quality is low and fractures are frequent [48-50]; however the explanation for this has never been clear, and the possibilities include over-suppression of bone turnover, accelerated osteoblast function, the presence of woven, and therefore immature, bone, and finally failure to resorb calcified cartilage [9, 21-23].

Our mechanical testing data of both trabecular and cortical bone indicate that induction of osteopetrosis in adult animals leads to increased bone strength. Since we found almost no remaining calcified cartilage, as well as no changes in cartilage

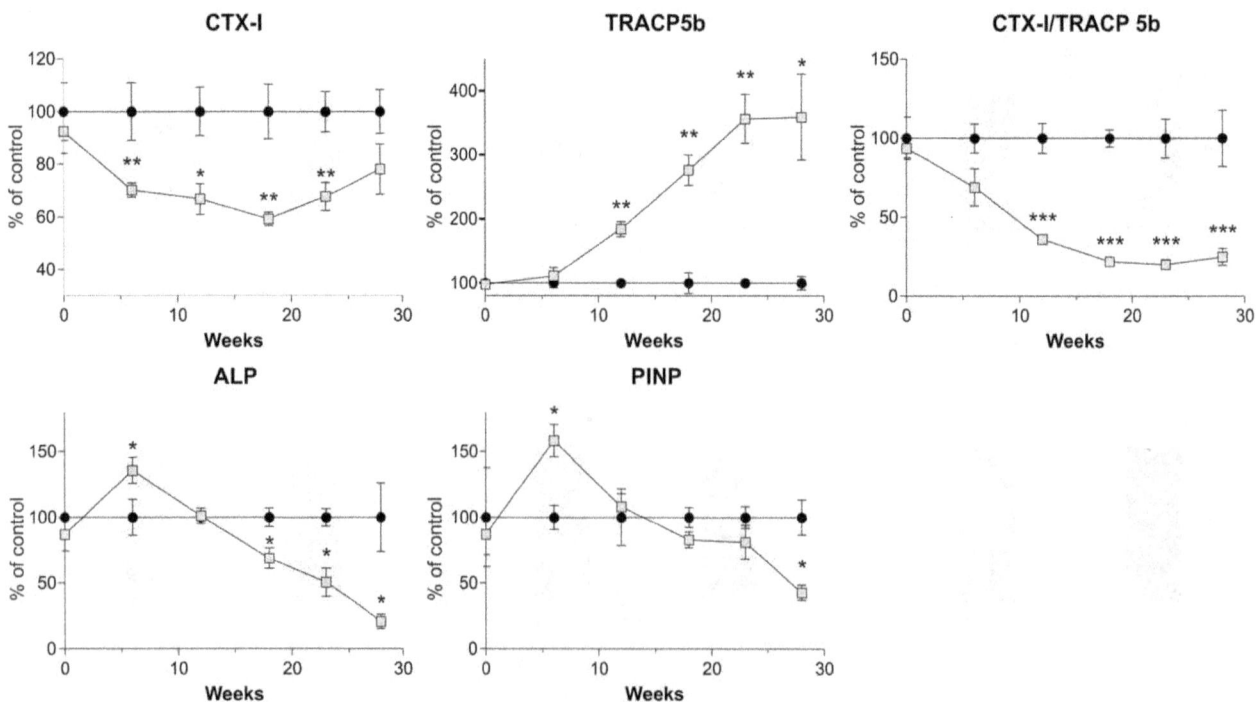

Figure 4. Biochemical markers of bone turnover. Serum samples were collected in both experiments and CTX-I (A), TRACP 5b (B), CTX/TRACP 5b (C), ALP (D), P1NP (E) were measured at baseline and at week 6, 10, 18, 22 and 28, post transplantation. The oc/oc data (gray squares) are plotted as percent of control (black circles) at all time points, and when samples from both experiments were present they were pooled after normalization. The biomarker measurements were conducted in samples from all mice, and for the samples collected during the first 12 weeks on pooled data from both experiments as described in the Methods section.

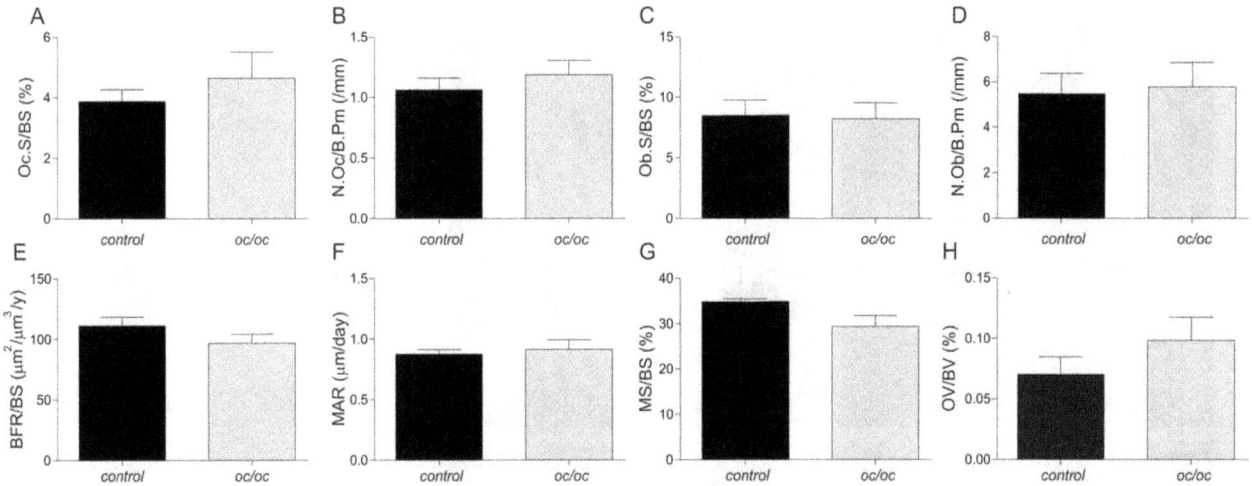

Figure 5. Bone histomorphometry. At termination of the 12-week experiment vertebrae were collected for bone histomorphometry. No significant differences were observed in osteoclast surface per unit bone surface (Oc.S/BS), number of osteoclasts per unit bone perimeter (N.Oc.Pm), osteoblast surface per unit bone surface (Ob.S/BS), number of osteoblasts per unit bone perimeter (N.Ob.Pm), bone formation rate (BFR/BV), mineral appositional rate (MAR), mineralizing surface (MS/BS) or osteoid volume (OV/BV). Bone histomorphometry was conducted on all specimens from the 12-week experiment (see Methods section).

degradation markers, these data suggest that it is the remaining calcified cartilage in the bones of young osteopetrotic mice that is the basis of the poor bone strength [51]. However, the gained bone was notably devoid of woven bone, a phenomenon observed in classical osteopetrosis, and thus the increase in lamellar bone volume is likely to also contribute the increased bone strength observed in the adult osteopetrotic mice. The tests performed do not take into account whether the bones from the transplanted osteopetrotic mice are more brittle at the tissue level; however, as the degree of mineralization only increases modestly and more slowly than breaking strength, this does not appear to be the cause. Furthermore, the normal bone structure observed in the oc/oc groups also supports the notion that the gained bone is normal at all levels. Importantly, these experiments do not take into account whether the poor bone quality observed in young oc/oc mice is due to expression of the a3 subunit of the V-ATPase in non-hematopoietic cells, i.e. gastric the parietal cells which are involved in calcium

homeostasis [52]; however, as the fragility of osteopetrotic bone is common to multiple types of osteopetrosis this does not appear to be likely. Increased bone strength has been observed in cortical, but not vertebral bone, of cathepsin K deficient mice [53], and in cortical bone of $Ae2_{a,b}$ deficient mice [54]. However, these mice also have thickened cortices, as opposed to acid secretion deficient mice, which have very little if any normal cortex [25]. Furthermore, the $Ae2_{a,b}$ and cathepsin K deficient mice also show less dramatic accumulation of calcified cartilage in the bone marrow cavities [54;55]. CT analysis of the bones showed increased bone volume in both trabecular and cortical compartments. Interestingly, the increase in bone volume in the vertebrae appeared to plateau after only three months, while the increase in femoral bone volume was continuous. Furthermore, the increase in cortical bone volume appeared to be mainly caused by a reduction in endocortical resorption, as endocortical diameter was reduced, but periosteal parameters were not changed.

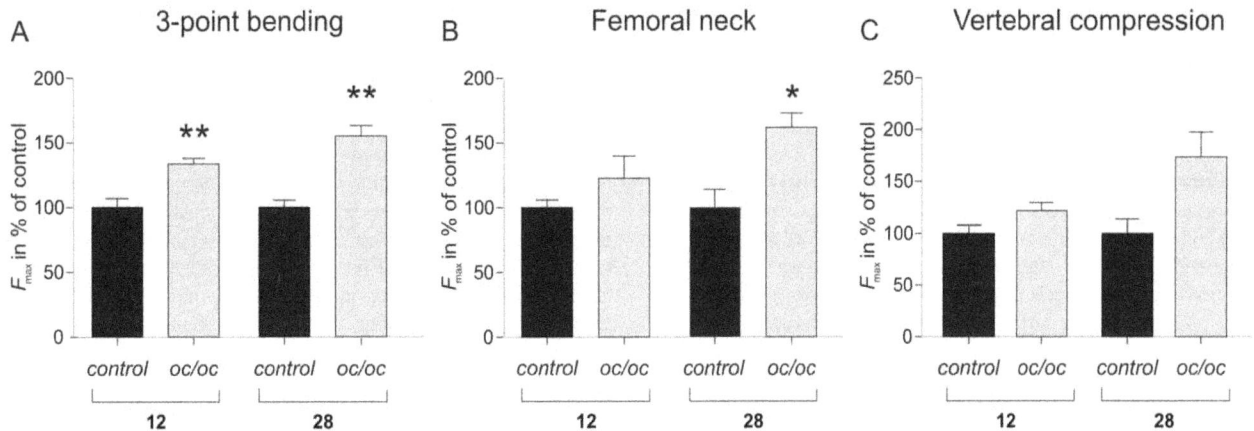

Figure 6. Bone strength analysis. Maximal force achieved at failure (F_{max}) as determined by 3-point bending test of the femoral cortex (A) or femoral neck (B). In C F_{max} was determined by vertebral compression. Bone strength testing was conducted on all bone specimens collected as described in the methods section.

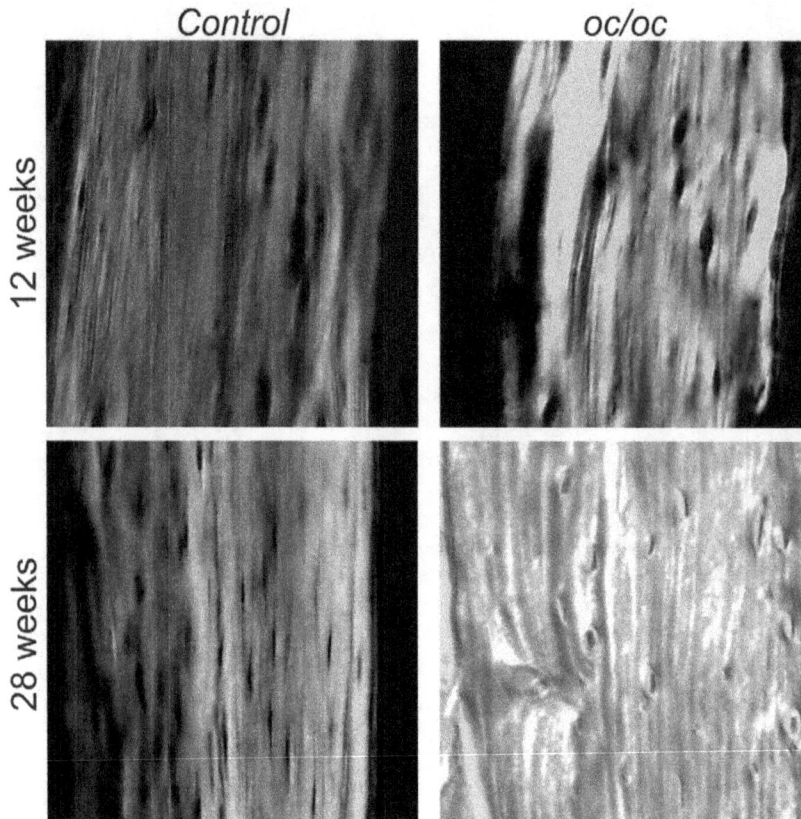

Figure 7. Analysis of bone structure. Bone structure was assessed using polarized light microscopy as described in the methods section.

The increase in bone volume is explained by the changes observed in biochemical markers. Bone resorption CTX-I was significantly reduced, which is as expected from the *in vitro* data, and this reduction in bone resorption most likely explains most of the increase in bone volume and bone strength. This differs from data presented in osteopetrosis models where the defect is present during bone development [24, 56]. However, a study conducted in ClC-7 deficient mice, which have a phenotype closely matching that of the *oc/oc* mice, indicated that the high resorption marker levels originate from resorption of non-mineralized matrices, which have not been removed correctly during endochondral ossification. The reasoning being that CTX-I release occurred completely independent of acid secretion by the osteoclast, and thus independent of resorption of calcified bone [24, 25].

As expected from previous studies, osteoclasts numbers increase with defective acid secretion [17, 18, 20, 46, 47, 57, 58]. In confirmation of a large reduction in resorptive capacity per osteoclast, the CTX-I/TRACP5b ratio was suppressed strongly [38]. The bone formation markers PINP and ALP were both increased by 6 weeks after transplantation, by 12 weeks they had returned to control levels, and at the later stage both these markers were decreased. The effect of this transient increase in bone formation on bone volume and strength is not clear, but the lower level of bone formation after 12 weeks may explain why the vertebral bone volume plateaus from that time, despite the ongoing reduction in resorption.

Taken together, the biochemical markers show that in early stages of induced osteopetrosis, bone formation is uncoupled from bone resorption, corresponding well to previous data from osteoclast-rich forms of osteopetrosis caused by defective acid

secretion [17, 19, 20]. In contrast, in osteoclast-poor forms of osteopetrosis bone formation is low from the starting point [59, 60], and in bisphosphonate or OPG-treated animals bone formation levels decrease rapidly after onset of treatment [61].

With respect to histomorphometry, we could neither confirm an increase in osteoclast numbers, nor a change in bone formation at week 12; and we speculate that it may require more time to see these differences by histomorphometry, as the early effects are mainly driven by the reduction in resorption, while the increased osteoclast survival is not seen until week 12 and at this time point the effect on the osteoclast marker TRACP 5b is not very dramatic. Furthermore, the biomarkers reflect the whole skeleton, whereas histomorphometry reflects only the vertebrae, and thus the markers will accumulate systemic changes. These biomarkers have, on the other hand, been shown to clearly reflect larger changes observed by histomorphometrical analysis [38, 61, 62].

Although bone formation decreases at later stages, these data indicate that when acid secretion by the osteoclast is attenuated a period of anabolic activity occurs. However, the duration and extent of this activity will need further investigation as osteoclast-rich osteopetrosis patients appear to have normal or increased levels of bone formation, even though bone resorption per osteoclast is significantly reduced [17–20].

The mechanisms controlling the coupling of bone formation to bone resorption have long been under debate, and several recent lines of evidence have indicated that the osteoclasts themselves, rather than their activity, are essential for the control of bone formation [2, 4, 17–20, 59, 60, 63–67]. In addition to the acid secretion deficient mice and patients, studies in cathepsin K deficient mice, and cathepsin K inhibitors in monkeys, have shown increased bone formation, despite reduced bone resorption, although the effects appear to be bone type dependent [53, 68, 69]. One study showed that inhibition of cathepsin K in osteoclasts *in vitro* led to augmented release of anabolic factors from the resorption compartment, while inhibition of acid secretion by bafilomycin prevented the release of anabolic factors [70]. All these data strongly indicate that the osteoclasts possess the ability to induce an anabolic response in osteoblasts. In addition, evidence has been provided that osteoclast-derived ephrinB2 might promote bone formation by acting upon receptor EphB4 in the osteoblast lineage, by a contact-dependent mechanism [71]. However, whether these are the factors involved in the uncoupling seen in these mice, and to what extent the coupling molecules originate from either bone resorption or directly from the osteoclasts, remain to be studied.

In conclusion, we here show an increase in bone volume and bone strength when osteopetrosis due to impaired acid seretion from osteoclasts is induced in adult mice. This suggests that the low bone quality seen in osteopetrosis in young animals most likely is due to the developmental nature of the phenotype. Furthermore, these data support that an ''uncoupling'' between bone resorption and bone formation can be obtained when attenuating acid secretion by the osteoclasts. Finally, the substantial increase in bone volume and bone strength observed in otherwise healthy mice with attenuated osteoclast acidification warrant further investigation of the osteoclastic V-ATPase as a therapeutic target for osteoporosis.

Author Contributions

Conceived and designed the experiments: KH CF MAK JR. Performed the experiments: CF JST AMB CST AVNW GEJL NS MA. Analyzed the data: KH JST AMB CF GEJL VE NS TJM MAK JR. Wrote the paper: KH JR. Read, commented and approved the final version of the manuscript: KM CF JST AMB CST AVNW GEJL NS MA TJM VE MAK JR.

References

1. Seeman E, Delmas PD (2006) Bone quality--the material and structural basis of bone strength and fragility. N Engl J Med 354: 2250–2261.
2. Karsdal MA, Martin TJ, Bollerslev J, Christiansen C, Henriksen K (2007) Are nonresorbing osteoclasts sources of bone anabolic activity? J Bone Miner Res 22: 487–494.
3. Martin TJ, Sims NA (2005) Osteoclast-derived activity in the coupling of bone formation to resorption. Trends Mol Med 11: 76–81.
4. Henriksen K, Neutzsky-Wulff AV, Bonewald LF, Karsdal MA (2009) Local communication on and within bone controls bone remodeling. Bone 44: 1026–1033.
5. Takahashi H, Epker B, Frost HM (1964) Resorption precedes formative activity. Surg Forum 15: 437–438.
6. Howard GA, Bottemiller BL, Turner RT, Rader JI, Baylink DJ (1981) Parathyroid hormone stimulates bone formation and resorption in organ culture: evidence for a coupling mechanism. Proc Natl Acad Sci U S A 78: 3204–3208.
7. Henriksen K, Bollerslev J, Everts V, Karsdal MA (2011) Osteoclast Activity and Subtypes as a Function of Physiology and Pathology--Implications for Future Treatments of Osteoporosis. Endocr Rev 32: 31–63.
8. Teitelbaum SL (2007) Osteoclasts: what do they do and how do they do it? Am J Pathol 170: 427–435.
9. Segovia-Silvestre T, Neutzsky-Wulff AV, Sorensen MG, Christiansen C, Bollerslev J, et al. (2008) Advances in osteoclast biology resulting from the study of osteopetrotic mutations. Hum Genet 124: 561–77.
10. Kornak U, Kasper D, Bosl MR, Kaiser E, Schweizer M, et al. (2001) Loss of the ClC-7 chloride channel leads to osteopetrosis in mice and man. Cell 104: 205–215.
11. Henriksen K, Gram J, Schaller S, Dahl BH, Dziegiel MH, et al. (2004) Characterization of osteoclasts from patients harboring a G215R mutation in ClC-7 causing autosomal dominant osteopetrosis type II. Am J Pathol 164: 1537–1545.
12. Graves AR, Curran PK, Smith CL, Mindell JA (2008) The Cl(-)/H(+) antiporter ClC-7 is the primary chloride permeation pathway in lysosomes. Nature 453: 788–92.
13. Yuan FL, Li X, Lu WG, Li CW, Li JP, et al. (2010) The vacuolar ATPase in bone cells: a potential therapeutic target in osteoporosis. Mol Biol Rep 37: 3561–3566.
14. Xu J, Cheng T, Feng HT, Pavlos NJ, Zheng MH (2007) Structure and function of V-ATPases in osteoclasts: potential therapeutic targets for the treatment of osteolysis. Histol Histopathol 22: 443–454.
15. Kornak U, Schulz A, Friedrich W, Uhlhaas S, Kremens B, et al. (2000) Mutations in the a3 subunit of the vacuolar H(+)-ATPase cause infantile malignant osteopetrosis. Hum Mol Genet 9: 2059–2063.
16. Frattini A, Orchard PJ, Sobacchi C, Giliani S, Abinun M, et al. (2000) Defects in TCIRG1 subunit of the vacuolar proton pump are responsible for a subset of human autosomal recessive osteopetrosis. Nat Genet 25: 343–346.
17. Del Fattore A, Peruzzi B, Rucci N, Recchia I, Cappariello A (2006) Clinical, genetic, and cellular analysis of 49 osteopetrotic patients: implications for diagnosis and treatment. J Med Genet 43: 315–325.
18. Alatalo SL, Ivaska KK, Waguespack SG, Econs MJ, Vaananen HK, et al. (2004) Osteoclast-derived serum tartrate-resistant acid phosphatase 5b in Albers-Schonberg disease (type II autosomal dominant osteopetrosis). Clin Chem 50: 883–890.
19. Bollerslev J, Steiniche T, Melsen F, Mosekilde L (1989) Structural and histomorphometric studies of iliac crest trabecular and cortical bone in autosomal dominant osteopetrosis: a study of two radiological types. Bone 10: 19–24.
20. Bollerslev J, Marks SC, Jr., Pockwinse S, Kassem M, Brixen K (1993) Ultrastructural investigations of bone resorptive cells in two types of autosomal dominant osteopetrosis. Bone 14: 865–869.
21. Chavassieux P, Seeman E, Delmas PD (2007) Insights into material and structural basis of bone fragility from diseases associated with fractures: how determinants of the biomechanical properties of bone are compromised by disease. Endocr Rev 28: 151–164.
22. Leeming DJ, Henriksen K, Byrjalsen I, Qvist P, Madsen SH, et al. (2009) Is bone quality associated with collagen age? Osteoporos Int 20: 1461–70.
23. Del FA, Cappariello A, Teti A (2008) Genetics, pathogenesis and complications of osteopetrosis. Bone 42: 19–29.
24. Neutzsky-Wulff AV, Karsdal MA, Henriksen K (2008) Characterization of the bone phenotype in ClC-7-deficient mice. Calcif Tissue Int 83: 425–437.
25. Neutzsky-Wulff AV, Sims NA, Supanchart C, Kornak U, Felsenberg D (2010) Severe developmental bone phenotype in ClC-7 deficient mice. Dev Biol 344: 1001–1010.
26. Scimeca JC, Franchi A, Trojani C, Parrinello H, Grosgeorge J, et al. (2000) The gene encoding the mouse homologue of the human osteoclast-specific 116-kDa V-ATPase subunit bears a deletion in osteosclerotic (oc/oc) mutants. Bone 26: 207–213.
27. Johansson MK, de Vries TJ, Schoenmaker T, Ehinger M, Brun AC, et al. (2007) Hematopoietic stem cell targeted neonatal gene therapy reverses lethally progressive osteopetrosis in oc/oc mice. Blood 109: 5178–85.
28. Askmyr M, Holmberg J, Flores C, Ehinger M, Hjalt T, et al. (2009) Low-dose busulphan conditioning and neonatal stem cell transplantation preserves vision and restores hematopoiesis in severe murine osteopetrosis. Exp Hematol 37: 302–308.
29. Tondelli B, Blair HC, Guerrini M, Patrene KD, Cassani B, et al. (2009) Fetal liver cells transplanted in utero rescue the osteopetrotic phenotype in the oc/oc mouse. Am J Pathol 174: 727–735.
30. Wiktor-Jedrzejczak W, Szczylik C, Ratajczak MZ, Ahmed A (1986) Congenital murine osteopetrosis inherited with osteosclerotic (oc) gene: hematological characterization. Exp Hematol 14: 819–826.
31. Nazarian A, Hermannsson BJ, Muller J, Zurakowski D, Snyder BD (2009) Effects of tissue preservation on murine bone mechanical properties. J Biomech 42: 82–86.
32. Sorensen MG, Henriksen K, Neutzsky-Wulff AV, Dziegiel MH, Karsdal MA (2007) Diphyllin, a Novel and Naturally Potent V-ATPase Inhibitor, Abrogates Acidification of the Osteoclastic Resorption Lacunae and Bone Resorption. J Bone Miner Res 22: 1640–1648.
33. Mosekilde L, Thomsen JS, Orhii PB, McCarter RJ, Mejia W, et al. (1999) Additive effect of voluntary exercise and growth hormone treatment on bone strength assessed at four different skeletal sites in an aged rat model. Bone 24: 71–80.
34. Sims NA, Clement-Lacroix P, Da PF, Bouali Y, Binart N, et al. (2000) Bone homeostasis in growth hormone receptor-null mice is restored by IGF-I but independent of Stat5. J Clin Invest 106: 1095–1103.

35. Parfitt AM, Drezner MK, Glorieux FH, Kanis JA, Malluche H, et al. (1987) Bone histomorphometry: standardization of nomenclature, symbols, and units. Report of the ASBMR Histomorphometry Nomenclature Committee. J Bone Miner Res 2: 595–610.

36. Blin-Wakkach C, Wakkach A, Sexton PM, Rochet N, Carle GF (2004) Hematological defects in the oc/oc mouse, a model of infantile malignant osteopetrosis. Leukemia 18: 1505–1511.

37. Li YP, Chen W, Liang Y, Li E, Stashenko P (1999) Atp6i-deficient mice exhibit severe osteopetrosis due to loss of osteoclast-mediated extracellular acidification. Nat Genet 23: 447–451.

38. Rissanen JP, Suominen MI, Peng Z, Halleen JM (2008) Secreted Tartrate-Resistant Acid Phosphatase 5b is a Marker of Osteoclast Number in Human Osteoclast Cultures and the Rat Ovariectomy Model. Calcif Tissue Int 82: 108–115.

39. Henriksen K, Tanko LB, Qvist P, Delmas PD, Christiansen C, et al. (2007) Assessment of osteoclast number and function: application in the development of new and improved treatment modalities for bone diseases. Osteoporos Int 18: 681–685.

40. Walker DG (1975) Spleen cells transmit osteopetrosis in mice. Science 190: 785–787.

41. Walker DG (1975) Bone resorption restored in osteopetrotic mice by transplants of normal bone marrow and spleen cells. Science 190: 784–785.

42. Walker DG (1975) Control of bone resorption by hematopoietic tissue. The induction and reversal of congenital osteopetrosis in mice through use of bone marrow and splenic transplants. J Exp Med 142: 651–663.

43. Kong YY, Yoshida H, Sarosi I, Tan HL, Timms E, et al. (1999) OPGL is a key regulator of osteoclastogenesis, lymphocyte development and lymph-node organogenesis. Nature 397: 315–323.

44. Li J, Sarosi I, Yan XQ, Morony S, Capparelli C, et al. (2000) RANK is the intrinsic hematopoietic cell surface receptor that controls osteoclastogenesis and regulation of bone mass and calcium metabolism. Proc Natl Acad Sci U S A 97: 1566–1571.

45. Taranta A, Migliaccio S, Recchia I, Caniglia M, Luciani M, et al. (2003) Genotype-phenotype relationship in human ATP6i-dependent autosomal recessive osteopetrosis. Am J Pathol 162: 57–68.

46. Nielsen RH, Karsdal MA, Sorensen MG, Dziegiel MH, Henriksen K (2007) Dissolution of the inorganic phase of bone leading to release of calcium regulates osteoclast survival. Biochem Biophys Res Commun 360: 834–839.

47. Karsdal MA, Henriksen K, Sorensen MG, Gram J, Schaller S, et al. (2005) Acidification of the osteoclastic resorption compartment provides insight into the coupling of bone formation to bone resorption. Am J Pathol 166: 467–476.

48. Tolar J, Teitelbaum SL, Orchard PJ (2004) Osteopetrosis. N Engl J Med 351: 2839–2849.

49. Waguespack SG, Hui SL, Dimeglio LA, Econs MJ (2007) Autosomal dominant osteopetrosis: clinical severity and natural history of 94 subjects with a chloride channel 7 gene mutation. J Clin Endocrinol Metab 92: 771–778.

50. Bollerslev J (1989) Autosomal dominant osteopetrosis: bone metabolism and epidemiological, clinical, and hormonal aspects. Endocr Rev 10: 45–67.

51. Neutzsky-Wulff AV, Karsdal MA, Henriksen K (2008) Characterization of the bone phenotype in ClC-7-deficient mice. Calcif Tissue Int 83: 425–437.

52. Schinke T, Schilling AF, Baranowsky A, Seitz S, Marshall RP, et al. (2009) Impaired gastric acidification negatively affects calcium homeostasis and bone mass. Nat Med 2009, 15: 674–681.

53. Pennypacker B, Shea M, Liu Q, Masarachia P, Saftig P, et al. (2009) Bone density, strength, and formation in adult cathepsin K (-/-) mice. Bone 44: 199–207.

54. Jansen ID, Mardones P, Lecanda F, de Vries TJ, Recalde S, et al. (2009) Ae2a,b-Deficient mice exhibit osteopetrosis of long bones but not of calvaria. FASEB J 23: 3470–81.

55. Gowen M, Lazner F, Dodds R, Kapadia R, Feild J, et al. (1999) Cathepsin K knockout mice develop osteopetrosis due to a deficit in matrix degradation but not demineralization. J Bone Miner Res 14: 1654–1663.

56. Kiviranta R, Morko J, Alatalo SL, NicAmhlaoibh R, Risteli J, et al. (2005) Impaired bone resorption in cathepsin K-deficient mice is partially compensated for by enhanced osteoclastogenesis and increased expression of other proteases via an increased RANKL/OPG ratio. Bone 36: 159–172.

57. Gram J, Antonsen S, Horder M, Bollerslev J (1991) Elevated serum levels of creatine kinase BB in autosomal dominant osteopetrosis type II. Calcif Tissue Int 48: 438–439.

58. Waguespack SG, Hui SL, White KE, Buckwalter KA, Econs MJ (2002) Measurement of tartrate-resistant acid phosphatase and the brain isoenzyme of creatine kinase accurately diagnoses type II autosomal dominant osteopetrosis but does not identify gene carriers. J Clin Endocrinol Metab 87: 2212–2217.

59. Demiralp B, Chen HL, Koh AJ, Keller ET, McCauley LK (2002) Anabolic actions of parathyroid hormone during bone growth are dependent on c-fos. Endocrinology 143: 4038–4047.

60. Koh AJ, Demiralp B, Neiva KG, Hooten J, Nohutcu RM, et al. (2005) Cells of the osteoclast lineage as mediators of the anabolic actions of parathyroid hormone in bone. Endocrinology 146: 4584–4596.

61. Samadfam R, Xia Q, Goltzman D (2007) Pretreatment with anticatabolic agents blunts but does not eliminate the skeletal anabolic response to parathyroid hormone in oophorectomized mice. Endocrinology 148: 2778–2787.

62. Schaller S, Henriksen K, Sveigaard C, Heegaard AM, Helix N, et al. (2004) The chloride channel inhibitor n53736 prevents bone resorption in ovariectomized rats without changing bone formation. J Bone Miner Res 19: 1144–1153.

63. Sobacchi C, Frattini A, Guerrini MM, Abinun M, Pangrazio A, et al. (2007) Osteoclast-poor human osteopetrosis due to mutations in the gene encoding RANKL. Nat Genet 39: 960–962.

64. Guerrini MM, Sobacchi C, Cassani B, Abinun M, Kilic SS, et al. (2008) Human osteoclast-poor osteopetrosis with hypogammaglobulinemia due to TNFRSF11A (RANK) mutations. Am J Hum Genet 83: 64–76.

65. Dai XM, Zong XH, Akhter MP, Stanley ER (2004) Osteoclast deficiency results in disorganized matrix, reduced mineralization, and abnormal osteoblast behavior in developing bone. J Bone Miner Res 19: 1441–1451.

66. Pederson L, Ruan M, Westendorf JJ, Khosla S, Oursler MJ (2008) Regulation of bone formation by osteoclasts involves Wnt/BMP signaling and the chemokine sphingosine-1-phosphate. Proc Natl Acad Sci U S A 105: 20764–20769.

67. Karsdal MA, Neutzsky-Wulff AV, Dziegiel MH, Christiansen C, Henriksen K (2008) Osteoclasts secrete non-bone derived signals that induce bone formation. Biochem Biophys Res Commun 366: 483–488.

68. Pennypacker B, Wesolowski G, Heo J, Duong LT, et al. Effects of Odanacatib on Central Femur Cortical Bone in Estrogen-Deficient Adult Rhesus Monkeys. J Bone Miner Res 24 Suppl 1, Avstrat 1171.

69. Scott K, Cusick T, Duong LT, Pennypacker B, Kimmel D (2009) Effects of Odanacatib on Bone Turnover and Osteoclast Morphology in the Lumbar Vertebra of Ovariectomized Adult Rhesus Monkeys. J Bone Miner Res 24 Suppl 1, Abstract SU0227.

70. Fuller K, Lawrence KM, Ross JL, Grabowska UB, Shiroo M, et al. (2008) Cathepsin K inhibitors prevent matrix-derived growth factor degradation by human osteoclasts. Bone 42: 200–211.

71. Zhao C, Irie N, Takada Y, Shimoda K, Miyamoto T, et al. (2006) Bidirectional ephrinB2-EphB4 signaling controls bone homeostasis. Cell Metab 4: 111–121.

12

Low Intensity, High Frequency Vibration Training to Improve Musculoskeletal Function in a Mouse Model of Duchenne Muscular Dystrophy

Susan A. Novotny[1]*, **Tara L. Mader**[1], **Angela G. Greising**[1], **Angela S. Lin**[2], **Robert E. Guldberg**[2], **Gordon L. Warren**[3], **Dawn A. Lowe**[1]

1 Program in Physical Therapy and Rehabilitation Sciences, University of Minnesota, Minneapolis, Minnesota, United States of America, 2 Institute for Bioengineering and Bioscience, Georgia Institute of Technology, Atlanta, Georgia, United States of America, 3 Department of Physical Therapy, Georgia State University, Atlanta, Georgia, United States of America

Abstract

The objective of the study was to determine if low intensity, high frequency vibration training impacted the musculoskeletal system in a mouse model of Duchenne muscular dystrophy, relative to healthy mice. Three-week old wildtype (n = 26) and *mdx* mice (n = 22) were randomized to non-vibrated or vibrated (45 Hz and 0.6 *g*, 15 min/d, 5 d/wk) groups. *In vivo* and *ex vivo* contractile function of the anterior crural and extensor digitorum longus muscles, respectively, were assessed following 8 wks of vibration. *Mdx* mice were injected 5 and 1 days prior to sacrifice with Calcein and Xylenol, respectively. Muscles were prepared for histological and triglyceride analyses and subcutaneous and visceral fat pads were excised and weighed. Tibial bones were dissected and analyzed by micro-computed tomography for trabecular morphometry at the metaphysis, and cortical geometry and density at the mid-diaphysis. Three-point bending tests were used to assess cortical bone mechanical properties and a subset of tibiae was processed for dynamic histomorphometry. Vibration training for 8 wks did not alter trabecular morphometry, dynamic histomorphometry, cortical geometry, or mechanical properties (P≥ 0.34). Vibration did not alter any measure of muscle contractile function (P≥0.12); however the preservation of muscle function and morphology in *mdx* mice indicates vibration is not deleterious to muscle lacking dystrophin. Vibrated mice had smaller subcutaneous fat pads (P = 0.03) and higher intramuscular triglyceride concentrations (P = 0.03). These data suggest that vibration training at 45 Hz and 0.6 *g* did not significantly impact the tibial bone and the surrounding musculature, but may influence fat distribution in mice.

Editor: Diego Fraidenraich, Rutgers University -New Jersey Medical School, United States of America

Funding: The research has been supported by grants from the Muscular Dystrophy Association grant 114071 (DAL, http://mda.org/), National Institute of Health (http://nih.gov/) grants T32-AR07612 (SAN), P30-AR0507220 (University of Minnesota Muscular Dystrophy Center), and K02-AG036827 (DAL), and the Patrick and Kathy Lewis Fund (SAN, http://www.cehd.umn.edu/gradsehd/grants/default.html). The funders had no role in study design, data collection and analysis, decision to publish, or preparation of the manuscript.

Competing Interests: The authors have declared that no competing interests exist.

* Email: golne003@umn.edu

Introduction

Duchenne muscular dystrophy (DMD) is an X-chromosome-linked disease characterized by progressive muscle weakness [1,2,3]. Bone strength, or mechanical properties, are compromised in these patients as evident by the occurrence of fragility fractures upon falling from standing or sitting height [4,5,6,7]. Compromised bone strength in DMD is multi-factorial, likely including effects of failure to accumulate peak bone strength during growth as well as declines in bone health secondary to the muscle disease. Furthermore, patients are recommended to avoid moderate- to high-intensity physical activity to prevent possible muscle damage and acceleration of the disease [8,9,10]. The absence of exercise, however, may result in the bone failing to increase in width, thus impacting bone strength. Preliminary data suggest that bone size is reduced in various skeletal sites in boys with DMD [11,12], and those data are supported by reports that that these patients have low bone mass across their lifespan [4,13]. Paralleling suboptimal attainment of bone strength, continual declines in muscle function

associated with disease progression (i.e., reduced magnitude and frequency of muscle-induced mechanical loads) likely initiates disuse-mediated bone remodeling. This is supported by evidence that the discrepancies in bone mass between boys with DMD and their age-matched peers are accentuated with age, especially following the loss of ambulation where skeletal regions such as the hip and calcaneus experience dramatic bone loss [4,13]. Therefore, effective bone-sparing interventions are warranted to thwart declines in bone health of boys with DMD in effort to preserve bone strength and prevent fractures.

Major determinants of bone health and interventions to preserve bone are related to mechanical loading [14]. Low-intensity loads (~5–10 µε) applied thousands of times per day is hypothesized to be just as effective as high-intensity loads (≥ 1500 µε) applied a few times per day[15,16]. Thus in the case of DMD, where high-intensity loads may be injurious to the inherently fragile muscle, utilizing low-intensity loads more often may be a reasonable approach to maintain bone health. Low intensity (i.e., ≤1.0 *g* of acceleration), high frequency vibration applies such stimulus to bone and has been shown to initiate an anabolic bone response [17], slow bone loss [18] [19], and improve bone mechanical properties [20]. Specifically, vibration has prevented bone loss associated with bed rest [21], as well as improved skeletal health in disabled children [22]. This suggests that vibration can have an osteogenic effect even in the presence of reduced mechanical loading (i.e., magnitude or spectrum of loads applied to the bone) or in the presence of disease. The benefits of vibration on skeletal muscle, however, remains ambiguous [21,22,23,24,25,26,27,28,29], and reports of contraindications raise concern [30,31]. Consequently, vibration may be efficacious for bone health in patients with a muscle disease such as DMD; however it is important to confirm its simultaneous safety in skeletal muscle.

The *mdx* mouse is a widely used model of DMD, and like patients has alterations in bone health [32,33,34,35,36] and is relatively physically inactive over a 24-hr period particularly during active hours [37]. However, the mouse model is widely recognized to have a mild phenotype compared to boys with DMD, for instance *mdx* mice are non-distinguishable from wildtype mice in their ability to bear weight or locomote. *Mdx* mice, therefore, provide an appropriate model to determine the efficacy of low intensity, high frequency vibration to improve musculoskeletal function because while this function is compromised due to the disease, mice are fully capable of weight bearing during vibration bouts. The extent of bone's response to vibration in mice is influenced by various factors including transmissibility of the vibration stimulus, the parameters of vibration used (i.e., acceleration and frequency), as well as genetic background of the mice [38,39,40]. These factors likely contributed to the lack of vibration-induced alterations in trabecular [41,42,43] and cortical bone [14,44,45]; highlighting that parameters of vibration are not universally effective across all mice. Therefore, 'optimization' specific to the model of interest may be necessary to maximize musculoskeletal benefits. Recently, we compared six different pairs of vibration parameters and identified 45 Hz at 0.6 g to best initiate increased expression of osteogenic genes in male *mdx* mice aged 5–7 weeks at the mRNA level [46]. It remains to be determined if those acute increases in gene expression would translate to improved bone structure and function with prolonged vibration training in dystrophic mice.

The objective of the present study, therefore, was to determine the extent to which low intensity, high frequency vibration training impacted the musculoskeletal system in mice modeling DMD, relative to healthy mice. Specifically, we sought to determine if trabecular morphometry, cortical geometry, and mechanical properties are better in tibia of vibrated than non-vibrated mice. Previous studies in mice showed that at least 3–6 weeks of vibration training is necessary to evoke structural adaptations within bone [39,44,47,48,49]. Consequently, we hypothesized that 8 weeks of vibration would improve the tibial bone of *mdx* mice. Specifically, three-point bending tests were utilized at the mid-diaphysis of the tibia to assess changes in mechanical properties, and micro-computed tomography (μCT) was performed to elucidate the possible underlying mechanical determinants of altered strength (i.e., geometry, mechanical properties and intrinsic material properties). Dynamic histomorphometry was also used as a direct measure of osteoblast activity in tibiae from *mdx* mice. In addition, we hypothesized that vibration training would not be injurious to dystrophic muscle as indicated by assessments of anterior crural muscle strength, contractility of extensor digitorum longus (EDL) muscle, muscle morphology, and plasma creatine kinase activity.

Methods

Animals and Experimental Design

Male wildtype (C57Bl/10) and *mdx* mice were obtained from our SPF-maintained breeding colony at the University of Minnesota. Mice were housed in standard cages, 3–4 mice per cage, on a 12:12-h light-dark cycle at 20–23°C and were provided food and water ad libitum. Mice were randomly assigned to either a non-vibrated group (wildtype non-vibrated n = 12, *mdx* non-vibrated n = 11) or vibration group (wildtype vibrated n = 14, *mdx* vibrated n = 11). Mice allocated to the vibration groups were exposed to 15-min bouts of vibration 5 d/wk for 8 wk (range 55–58 d) starting when mice were 3 wk of age. The vibration stimulus consisted of a 45-Hz stimulus with 0.6 g of acceleration (where 1 g is equivalent to the acceleration due to gravity) based on our preliminary work in *mdx* mice [46]. This vibration stimulus was well tolerated by *mdx* mice as previously reported [46] as well as for wildtype mice [50]. Specifically, in this study behaviors, ambulation patterns, and activities were indistinguishable between genotypes during (see Video S1 and S2) and immediately after bouts of vibration. The height of the vibration cage was set to 5 cm, to limit rearing and ensure mice consistently bore weight on their hindlimbs during the entire bout of vibration. This was verified during each vibration bout as mice were continually monitored by an investigator. The combination of these factors gives us confidence that an equivalent vibration stimulus was transmitted to the bone of *mdx* and wildtype mice.

Relatively young mice were selected for this study in order to determine the impact of prolonged vibration training while the disease pathology in this mouse model is apparent (i.e., 3–12 wk of age in *mdx* mice). *Mdx* mice, unlike boys with DMD, do not have progressive muscle pathology past the age of about 12 weeks, thus limiting the ages in which the mouse model mimics the disease. Mice in the non-vibrated group were placed on the vibration platform for the same duration of time, but with the machine turned off.

Mdx mice were injected subcutaneously with 15 mg/kg body mass (BM) Calcein (Sigma, St. Louis, MO) 5 days prior to sacrifice, and 1 day prior to sacrifice with 90 mg/kg BM Xylenol orange (Sigma) to quantify dynamic trabecular bone histomorphometry, as adapted from [51]. At 11 wk of age, mice were sacrificed by first anesthetizing with a cocktail of: fentanyl citrate (0.2 mg/kg body mass (BM)), droperidol (10 mg/kg BM) and diazepam (5 mg/kg BM). Plasma was collected via retro-orbital bleed and flash frozen in liquid nitrogen to assess creatine kinase activity. Functional capacity of the left anterior crural muscles (i.e., tibialis anterior (TA), extensor digitorum longus (EDL), and extensor hallucis longus muscles) were then assessed *in vivo* by quantifying maximal isometric torque and susceptibility to contraction-induced injury. The anterior crurals were selected because we previously showed vibration training to improve contractility of this muscle group [50]. Immediately following *in vivo* analyses, mice were injected with supplemental anesthesia intraperitoneal (i.e., 75 mg/kg BM sodium pentobarbital for wildtype mice and 37.5 mg/kg BM for *mdx* mice). The EDL muscle from the right hindlimb was excised and used to assess *ex vivo* force-producing capacity. This muscle was chosen because in *mdx* mice it is sensitive to disease progression, eccentric contraction-induced injury, and can adapt in response to intervention [52,53]. Prior to exsanguination, TA, soleus, and gastrocnemius muscles were also excised and weighed. These muscles were selected due to their proximity to the vibration platform and hence their potential ability to be affected by vibration training.

The subcutaneous and visceral fat pads were also excised and weighed, as consistent reductions in fat pad masses have been reported following long-term vibration training [47,50,54]. The TA, EDL, gastrocnemius, and soleus muscles were dissected and snap frozen in liquid nitrogen or mounted in Tissue-Tek OCT (Sakura, Torrance, CA). Tibial bones were removed and stored in either phosphate-buffered saline at $-20°C$ until the time of mechanical testing or in 70% alcohol at $4°C$ until the time of dynamic histomophometric processing. The tibial bone was selected, rather than the femur, due to its proximity to the vibration plate. That is, the range of transmissibility of vibration stimulus is reduced with increasing distance from the platform [55,56], and consequently, bone's response to vibration may be more robust in the tibia compared to the femur.

This study was carried out in strict accordance with the recommendations in the Guide for the Care and Use of Laboratory Animals of the National Institutes of Health and all procedures were approved by the Institutional Animal Care and Use Committee at the University of Minnesota (Permit Number: 1109A04549). Anesthetic regimes used were recommended and approved by veterinarian staff. For each of the musculoskeletal assessments, one investigator performed that specific assessment on all mice and all investigators were blinded to the genotype and training group of each mouse when performing the assessments.

In Vivo Assessments of Anterior Crural Muscle Functional Capacity

Mice underwent *in vivo* contractile testing of the anterior crural muscles of the left hindlimb. Outcome measures of interest included peak isometric dorsiflexor torque production [57] and peak eccentric and isometric torque loss following contraction-induced injury [53,58]. Muscle injury was induced as previously described [59], by performing 100 electrically-stimulated eccentric contractions evoked using 250 Hz at a constant optimal voltage, with an angular excursion of 38 degrees at an angular velocity of 2000 degrees per second with the exception of 12 seconds between contractions. Five minutes following the last eccentric contraction, peak isometric torque was re-assessed.

Ex Vivo Assessments of EDL Muscle Contractility

Contractile measurements of isolated EDL muscles included peak twitch force, time-to-peak twitch force, twitch one-half relaxation time, peak isometric tetanic force (P_o), maximal rates of tetanic force production and relaxation, peak eccentric force, and percent decline in isometric tetanic force following eccentric contraction-induced injury [60]. Eccentric contraction-induced injury consisted of five eccentric contractions with 3 minutes in between contractions. Eccentric contractions were evoked by passively shortening the EDL muscle from its anatomical muscle length (L_o) to $0.95L_o$, and then simultaneously stimulating the muscle for 133 ms as the EDL muscle lengthened to $1.05L_o$ at a rate of $0.75L_o/s$ [53]. EDL muscles were trimmed, blotted dry, and weighed immediately following the measurements. Physiological cross-sectional area was calculated using EDL muscle mass, L_o, and a fiber length-to-muscle length ratio of 0.44 [60,61]. Specific P_o was determined by dividing P_o by the calculated physiological cross-sectional area of the muscle.

Muscle Morphology

Altered vascularity within the soleus muscle has been noted following vibration training [30,50], therefore we measured capillary density at the distal end and mid-belly of the soleus muscles. Capillary density was quantified by counting the number of capillaries surrounding a fiber for 200 fibers per muscle following staining by a periodic acid Schiff reaction [50]. Central nucleated fibers (i.e., a marker of muscle damage and regeneration) were also assessed at the distal end and mid-belly of the soleus muscle as well as the mid-belly of the TA muscle. The number of central nucleated fibers present per 300 fibers was counted in each of these regions from hematoxylin and eosin-stained sections [37].

Intramuscular Triglyceride Concentration

Smaller fat pads are consistently reported following long-term vibration training [47,50,54], therefore to extend these results further, we wanted to see what effect vibration has on intramuscular fat. We chose to measure triglyceride concentration within the gastrocnemius muscle, as this method has been previously utilized to assess triglycerides in liver, serum, and fat pads following vibration training [47]. Intramuscular triglycerides were extracted and isolated from gastrocnemius muscles as previously described [62]. Briefly the muscles were homogenized in 20 volumes of a 2:1 chloroform-methanol mixture. The homogenate was vortexed and washed with a volume of saline necessary to obtain an 8:4:3 ratio of chloroform, methanol, and water. The homogenate was centrifuged at 1160 g for 20 minutes to obtain a biphasic separation. A 500-μl sample of the lower phase was removed, transferred to a new tube, dried under nitrogen gas, and resuspended in 250 μL of phosphate-buffered saline containing 1% Triton X-100. Triglyceride concentrations were determined using an enzymatic colorimetric assay employing glycerol-3-phosphate oxidase (Cat. #461-08992; Wako Pure Chemical Industries, Ltd. Richmond, VA). Triglyceride concentration is expressed as milligrams per gram of wet muscle mass.

μCT of Tibial Bone Metaphysis and Mid-diaphysis

A μCT system (Scanco Medical microCT 40, Bruttisellen, Switzerland) was used to quantify trabecular morphometry in the tibial metaphysis as well as cortical bone geometry and volumetric bone density (vBMD) at the tibial mid-diaphysis [34]. Trabecular bone morphometry was assessed in the proximal tibial metaphysis (50 slice region of interest, starting 60 μm distal to the last image containing the growth plate, using 12-μm voxel size) as previously described [34]. Bone volume fraction (BV/TV), trabecular thickness, trabecular number, trabecular separation and trabecular vBMD were determined for each slice and the average value across each of the 50 slices was used for statistical analyses.

The following outcome measures were assessed in the transverse plane on the central 0.8-mm region of the tibial diaphysis: cortical cross-sectional area, cortical thickness, periosteal diameter, cross-sectional moment of inertia (CSMI), and vBMD. CSMI about the anterior-posterior axis corresponds to the CSMI about the bone-bending axis during three-point bending tests. These measures were assessed for each of the 66 slices within the 0.8 mm region of the tibial diaphysis, and the average for all 66 slices was used for statistical analyses. Following the completion of imaging, tibial bones were refrozen in PBS until the time they underwent mechanical testing.

Mechanical Testing of the Tibial Mid-diaphysis

Mechanical testing procedures for assessing the functional capacity of the mouse tibial bone has previously been described in detail [34,63,64,65]. Briefly, the left tibial bone of each mouse was placed on its lateral side in a Mecmesin MultiTest 1-D test machine, and was loaded in three-point bending at the mid-diaphysis using a Mecmesin AFG-25 load cell (Mecmesin, West Sussex, UK). The functional capacity of the tibial bone was quantified by ultimate load, stiffness, and deflection and energy absorbed to ultimate load using custom designed TestPoint software (TestPoint version 7; Measurement Computing Corp.) [34,65].

Trabecular Bone Dynamic Histomorphometry

A subset of the tibiae were dehydrated and embedded without demineralization in methyl-methacrylate (Fisher Scientific, Pittsburgh, PA) as previously described [66]. Briefly, 5-μm thick longitudinal sections were cut on a microtome (Leica, Heidelberg, Germany) and mounted unstained. Fluorochrome labels were visualized at 20x, and dynamic histomorphometric measures were made using OsteoMeasure image analyzer (OsteoMetric, Atlanta, GA) in a region 60 μm distal to the proximal growth plate. Outcome measures of interest include mineralized surface per bone surface, mineral apposition rate, and bone formation rates relative to bone surface or total volume.

Statistical Analyses

Power calculations determined that 10 mice per group were necessary to detect significant group differences with two-way ANOVAs with a minimum power of 80% and α-level of 0.05. The effects of vibration (45 Hz at 0.6 g vs. non-vibrated) and genotype (wildtype vs. mdx) were assessed by two-way ANOVAs with vibration and genotype as the fixed factors. Eccentric contraction data were assessed by three-way repeated measure ANOVAs with vibration, genotype and contraction numbers as the fixed factors. When significant interactions were present, Holm-Sidak post-hoc measures were used to determine differences among the groups. When assumptions of normality or equal variance were violated, Kruskal-Wallis One Way Analysis of Variance on Ranks was performed along with Dunn's post-hoc measures. Dynamic histomorphometry measures of the tibia were only performed on mdx non-vibrated and vibrated mice, and therefore the data were assessed by t-tests. All statistical analyses were carried out using SigmaPlot version 11.0 (Systat Software Inc; Point Richmond, CA).

Results

Body, Muscle and Fat Pad Masses and Intramuscular Triglyceride Content

At the start of the study, body mass did not differ between vibrated and non-vibrated groups (P = 0.654), though mdx mice weighed less than wildtype at 3-wks of age (9.0±0.3 vs 10.7±0.7 g, P = 0.005). Eight weeks later vibrated mice tended to have lower body mass than non-vibrated mice and mdx mice were 12% heavier (Figure 1). Mdx mice also had greater muscle masses than wildtype mice (Table 1). Tibialis anterior, soleus, and gastrocnemius muscle masses were not impacted by vibration (Table 1). Vibrated mice had 5% smaller EDL muscles, primarily due to vibrated mdx mice having 9% smaller EDL muscles compared to non-vibrated mdx mice (Table 1).

Vibrated mice also had significantly smaller subcutaneous fat pads and tended to have lower visceral fat pad masses compared to non-vibrated mice (Figure 1). Main effects of genotype were consistently present for fat pad masses with mdx mice having up to 47% less fat pad masses (Figure 1). Vibrated mice had 26% higher concentrations of triglycerides per gram of wet gastrocnemius muscle mass (Table 1). Triglyceride concentrations were not different between genotypes (Table 1).

Muscle Morphology

Vibration had no impact on capillary density and percentage of centrally-nucleated muscle fibers in either the mid-belly or distal end of the soleus muscle (Table 1). Mdx mice had more centrally-nucleated fibers in soleus and tibialis anterior muscles compared to those of wildtype mice (Table 1).

In Vivo Assessments of Anterior Crural Muscle Functional Capacity

To determine if vibration training affected skeletal muscle tissue in close proximity to the vibrating platform, dorsiflexor torque was assessed. Overall, the contractility measures of anterior crural muscles showed no effect of vibration. Peak isometric dorsiflexor torque and peak isometric torque normalized to body mass were not impacted by vibration (Table 2 and Figure 2A, respectively), indicating that muscle strength was not altered following vibration training. Genotype differences in isometric torque production were only apparent after accounting for the greater body mass of the mdx mice (Figure 2A). Susceptibility to contraction-induced injury, as indicated by the decline in peak eccentric torque over a series of 100 eccentric contractions (Figure 2B) and isometric torque loss (Table 2), was not affected by vibration. Mdx mice had a substantial loss of anterior crural muscle functional capacity following eccentric injury as indicated by a ~70% decline in peak eccentric torque vs. only 34% decline for wildtype mice (Figure 2B), and a larger isometric torque loss (Table 2). These data indicate that lack of dystrophin, but not vibration, is detrimental to muscle function. Similarly, plasma creatine kinase activity did not differ between vibrated and non-vibrated groups (P = 0.974), but was 4-fold higher in mdx than wildtype mice (4507+/−200 U/L vs. 1055+/−210 U/L, P<0.001).

Ex Vivo Assessments of EDL Muscle Contractility

Force-generating capacity of the EDL muscle assessed ex $vivo$ was not affected by 8 weeks of vibration training. Vibration had no impact on peak twitch force, maximal isometric tetanic force, specific P_o, peak eccentric force, and eccentric or isometric force loss following contraction-induced injury (Figure 3 and Table 2). Characteristics relating to speed of EDL muscle contraction, including time-to-peak twitch force, half-relaxation time of twitch force, and maximal rates of tetanic force development and relaxation were also not effected by vibration (Table 2). Most of the EDL contractile measures were different between wildtype and mdx mice, reflecting the expected pathology of the muscle disease (Figure 3 and Table 2).

μCT of Tibial Bone Metaphysis and Mid-diaphysis

μCT was performed to determine the extent to which vibration and genotype influenced trabecular bone morphometry and cortical bone geometry at the proximal metaphysis and mid-diaphysis, respectively. In the proximal metaphysis of the tibia, vibration did not influence trabecular morphometry, though differences between mdx and wildtype were detected (Figure 4). Specifically, bone volume fraction and trabecular thickness, number, and separation did not differ between non-vibrated and vibrated mice (Figure 4). The lack of altered trabecular morphometry in the metaphysis of mdx mice, following vibration, was confirmed by dynamic histomorphometry. Overall, vibration had no impact on bone formation in mdx mice as indicated by the average mineralized surface per bone surface (34.1±1.8% for vibrated mice and 34.1±2.1% for non-vibrated mice, P = 0.989), mineral apposition rate (1.04±0.04 μm/d for vibrated mice and 1.08±0.03 μm/d for non-vibrated mice, P = 0.373), bone formation rate per bone surface (0.36±0.03 $\mu m^3/\mu m^2$/d for vibrated mice and 0.37±0.03 $\mu m^3/\mu m^2$/d for non-vibrated mice, P = 0.633) or bone formation rate per tissue volume

Table 1. Effects of low intensity, high frequency vibration training and genotype on muscle and muscle fiber characteristics.

	Wildtype Non-vibrated	Wildtype Vibrated	Mdx Non-vibrated	Mdx Vibrated	P-values for Two-Way ANOVA		
					Main effect of Vibration	Main effect of Genotype	Interaction (Vibration × Genotype)
Extensor Digitorum Longus Muscle							
Mass (mg)	11.6 (0.3)	11.6 (0.2)	16.2 (0.6)	14.7 (0.4)	0.045	<0.001	0.054
Anatomical Muscle Length (mm)	12.6 (0.1)	12.5 (0.1)	12.7 (0.1)	12.6 (0.1)	0.516	0.411	0.811
Tibialis Anterior Muscle							
Mass (mg)	45.3 (1.2)	48.0 (1.3)	78.1 (2.9)	75.4 (3.3)	0.991	<0.001	0.238
Centrally Nucleated Fibers, Mid-belly (%)	1.6 (0.6)	3.3 (0.9)	71.2 (1.5)	70.3 (1.2)	0.741	<0.001	0.298
Soleus Muscle							
Mass (mg)	6.7 (0.4)	7.0 (0.3)	9.3 (0.6)	8.3 (0.4)	0.405	<0.001	0.183
Centrally Nucleated Fibers, Mid-belly (%)	2.5 (0.4)	5.0 (1.7)	63.6 (2.2)	64.2 (2.7)	0.457	<0.001	0.636
Centrally Nucleated Fibers, Distal (%)	8.7 (7.2)	8.3 (5.5)	62.4 (6.7)	62.7 (3.2)	0.991	<0.001	0.950
Capillaries per Fiber, Mid-belly	2.6 (0.1)	2.9 (0.2)	2.5 (0.2)	2.7 (0.2)	0.174	0.258	0.832
Capillaries per Fiber, Distal	2.5 (0.1)	2.8 (0.2)	2.4 (0.1)	2.5 (0.1)	0.315	0.208	0.598
Gastrocnemius Muscle							
Mass (mg)	125.9 (5.2)	128.6 (4.2)	173.0 (9.0)	164.8 (4.3)	0.654	<0.001	0.363
Triglyceride Concentration (mg/g)	0.889 (0.066)	1.179 (0.116)	0.906 (0.098)	1.085 (0.108)	0.025	0.696	0.577

Values are means (SE).

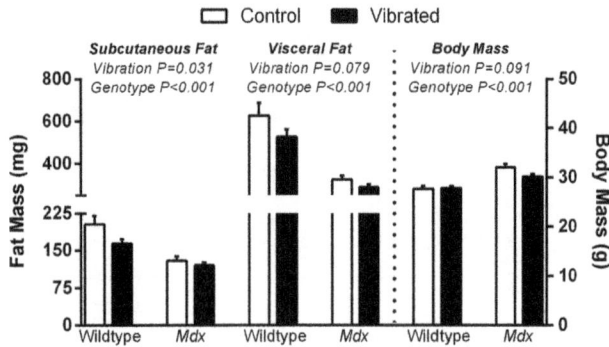

Figure 1. Eight weeks of vibration training affected fat pad masses but not body masses. Vibrated mice had smaller sized subcutaneous fat pads following 8-weeks of training. *Mdx* mice had a larger body mass but smaller fat pad masses compared to wildtype mice following 8-weeks of training. Body masses were not different in mice subject to vibration compared to non-vibrated control mice. Data are means ± SE. P-values associated with the main effects of two-way ANOVAs are indicated above each set of bars. Interactions between vibration and genotype P≥0.056.

(0.192±0.016%/d for vibrated mice and 0.194±0.018%/d for non-vibrated mice, P = 0.908). For the differences in trabecular bone morphometry across genotypes, bone volume fraction showed that *mdx* mice had less bone than wildtype (0.111±0.006 for *mdx* mice and 0.133±0.006 for wildtype mice), which was attributed to having 12% thinner trabeculae (Figure 4B). Trabecular separation and number were not influenced by genotype (Figure 4C and D, respectively).

Neither vibration nor genotype influenced any parameter of cortical bone geometry at the tibial mid-shaft (Table 3 and Figure 5A). These data suggest that the shape of the bone was similar across all groups, despite the tendency of *mdx* mice to have longer tibial lengths (Table 3).

Mechanical Testing of the Tibial Mid-diaphysis

Three-point bending tests were performed at the mid-shaft of the tibial diaphysis to determine if cortical bone mechanical properties were affected, even in the absence of change in cortical bone geometry. The ultimate load and stiffness of tibial bones were not different between vibrated and non-vibrated mice (Figure 5B and C). Energy and deflection to ultimate load were also not different between vibrated and non-vibrated mice (Table 3). Comparisons across genotypes confirmed that mechanical properties of the tibial bone were compromised in *mdx* mice, as indicated by 9% smaller ultimate loads and a trend toward lower tibial stiffness (Figure 5B and C), as well as a significantly lower energy absorption to ultimate load compared to wildtype mice (Table 3).

Overall, vibration had no impact on any measure of intrinsic material properties of the tibia (Table 3). While ultimate stress and modulus of elasticity values were similar across genotypes, μCT revealed differences in vBMD between *mdx* and wildtype mice at both the tibial proximal metaphysis (trabecular) and the tibial mid-diaphysis (cortical) with *mdx* mice having up to 3% lower vBMD (Table 3).

Discussion

Vibration training has been reported to enhance bone and muscle in humans and rodent models in some, but not all studies. Our study failed to show any enhancement in either of these two

tissues. First, 8 weeks of low intensity vibration training did not alter trabecular morphology, cortical bone geometry, or cortical bone mechanical properties in tibia of wildtype mice or mice modeling Duchenne muscular dystrophy. Secondly, vibration did not alter any of our measures of contractile function or histology in lower hindlimb muscles. Despite the lack of benefit, it is noteworthy that muscle function in *mdx* mice was not adversely affected by the vibration training. Lastly, mice that were vibration trained had smaller subcutaneous fat pads and greater intramuscular triglyceride concentrations compared to non-vibrated mice. Combined, these data suggest that vibration training for 15 minutes per day, 5 days per week, for 8 weeks at 45 Hz and 0.6 *g* in rapidly growing mice does not significantly impact musculoskeletal function, but does affect fat.

Trabecular bone

We hypothesized that 8 weeks of low intensity vibration training would improve trabecular morphology. Vibration training, however, did not affect any measure of trabecular morphology or dynamic histomorphometry in the proximal tibial metaphysis of wildtype or dystrophic mice (Figure 4 and Table 3). The anticipation of alterations in trabecular bone morphology was based on several reports of beneficial adaptations to bone in the proximal tibia of mice following vibration training. Specifically, improvements in trabecular thickness [39,67], trabecular number [47], bone volume fraction [39,47,67], dynamic rates of bone formation [48,67], and decreased trabecular separation [47] have been reported in bones of mice in response to 3 to 6 weeks of vibration training that had used similar low intensity parameters. In addition to these beneficial adaptations in healthy mice, vibration has also been shown to preserve or improve trabecular bone in mice modeling disuse [17,68] and in patients with childhood diseases [22,25], thus making vibration training an attractive therapeutic modality for DMD. The lack of vibration-induced alterations in trabecular bone in our study is not alone. Previous studies utilizing mouse models associated with physical inactivity and muscle weakness [41,42,43], as well as an uncontrolled, pilot study assessing the tolerability of high intensity vibration in DMD patients [28], also failed to detect alterations in trabecular or cortical bone density or serum markers of bone formation and metabolism from vibration training.

Cortical Bone

Lower tibial bone ultimate load and stiffness in *mdx* mice compared to wildtype mice (Figure 5) are consistent with previous reports [33,34] and have previously been attributed to altered bone geometry [34]. We hypothesized that 8 weeks of vibration training would improve cortical bone geometry and mechanical properties at tibia mid-diaphysis. Cortical bone, however, was not altered by vibration as indicated by the lack of any differences in cortical bone geometry or mechanical properties between vibrated and non-vibrated groups (Table 3 and Figure 5). These data are corroborated by evidence from others indicating that low intensity vibration did not alter bone geometry at the mid-diaphysis of the tibia [44,45] and femur [14] in mice. Improvements in periosteal bone formation rate and mineral apposition rate at the tibial mid-diaphysis following vibration have been noted [44]. However, this increase in bone growth did not translate to improvements in cortical bone area, ultimate load, or stiffness. Cortical bone dynamic histomorphometry was not measured in the present study due to the lack of observed improvements in cortical bone geometry and mechanical properties.

The lack of an anabolic response in cortical and trabecular bone with vibration training in the present study may be attributed to

Table 2. Muscle contractile measures following 8 weeks of low intensity, high frequency vibration training in wildtype and *mdx* mice.

	Wildtype Non-vibrated	Wildtype Vibrated	*Mdx* Non-vibrated	*Mdx* Vibrated	P-values for Two-Way ANOVA		
					Main effect of Vibration	Main effect of Genotype	Interaction (Vibration×Genotype)
In Vivo **Function of Anterior Crural Muscles**							
Maximal Isometric Torque (N•mm)	2.31 (0.10)	2.42 (0.10)	2.45 (0.16)	2.18 (0.17)	0.526	0.714	0.164
Isometric Torque Loss Following Eccentric Contractions (%)	42.3 (1.9)	41.5 (1.7)	71.9 (5.1)	61.6 (6.9)	0.192	<0.001	0.257
Ex Vivo Function of EDL Muscles							
Peak Twitch Force (mN)	99.3 (3.5)	94.0 (2.2)	94.4 (3.1)	90.2 (3.1)	0.118	0.150	0.860
Time-to-Peak Twitch Force (ms)	19.0 (0.3)	18.8 (0.3)	19.1 (0.5)	20.1 (0.5)	0.276	0.071	0.108
Half-Relaxation Time of Twitch Force (ms)	15.0 (0.4)	13.9 (0.4)	17.8 (0.3)	18.3 (0.7)	0.558	<0.001	0.133
Maximal Rate of Tetanic Force Development (N•s⁻¹)	10.9 (0.4)	10.9 (0.3)	10.2 (0.4)	10.4 (0.5)	0.784	0.167	0.738
Maximal Rate of Tetanic Force Relaxation (N•s⁻¹)	22.6 (0.4)	23.4 (0.7)	18.5 (1.0)	16.5 (1.2)	0.477	<0.001	0.116
Peak Eccentric Force (mN)	639.0 (13.3)	629.6 (15.1)	549.2 (19.2)	517.6 (22.7)	0.253	<0.001	0.533
Isometric Force Loss Following Eccentric Contractions (%)	4.1 (1.0)	3.8 (1.4)	63.1 (5.5)	67.5 (5.7)	0.617	<0.001	0.553

Values are means (SE). Isometric torque loss was calculated as the percent difference between isometric torque measured before and after the 100 eccentric contractions. Isometric force loss was calculated as the percent difference between peak isometric tetanic force measured before and after the 5 eccentric contractions.

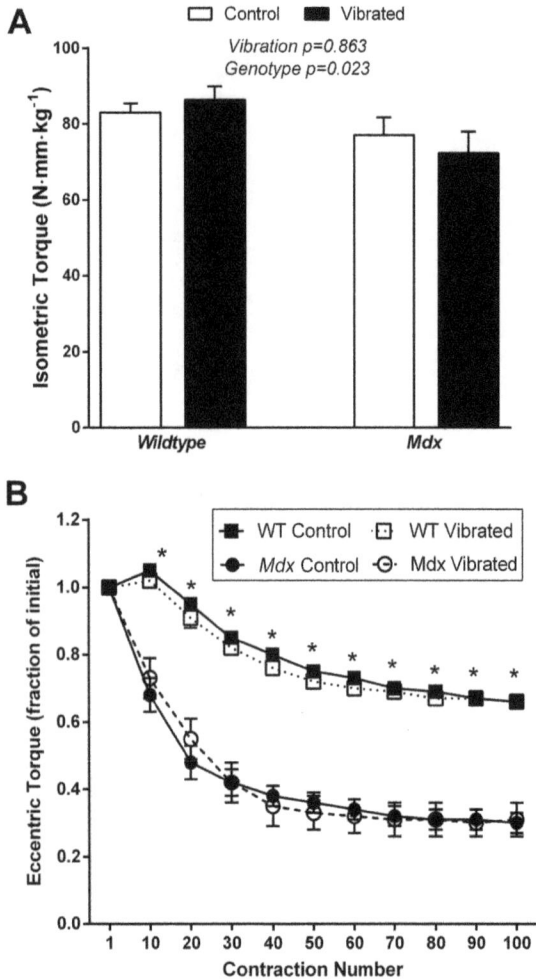

Figure 2. Eight weeks of vibration did not impact *in vivo* muscle strength or susceptibility to injury. A) Maximal isometric torque was not different between vibrated and non-vibrated mice following 8-weeks of training; isometric torque was less in *mdx* than wildtype mice. Interaction between vibration and genotype P≥0.357. **B)** Vibration training for 8 weeks did not alter susceptibility to eccentric contraction-induced injury. As expected, *mdx* mice were more susceptible to eccentric injury relative to wildtype mice. Data are means ± SE. In Panel A, P-values associated with the main effect of two-way ANOVAs are indicated above the bars. In panel B, only a main effect of genotype was present, where * signifies a significant (P<0.05) difference between *mdx* and wildtype mice at that contraction number.

multiple factors including vibration protocol parameters, transmission of the vibration stimuli to the musculoskeletal tissues, or the use of relatively young mice. Bone's response to vibration is not universally effective and has been shown to preferentially respond to certain vibration stimuli [39,40,44,48]. Therefore, it is possible that the vibration parameters utilized in the present study are optimal for eliciting an osteogenic gene expression response after 14 days of training [46], but not optimal for altering tibial bone strength and structure with long-term training. Bone's response to vibration is also dependent upon how well the vibration stimuli are transmitted to the tissues of interest. Skeletal regions closest to the source have more robust responses [40] compared to distal sites where transmission is diminished [56], thus longitudinal growth of the tibia may have altered the magnitude transmission over the 8-week course of the study. Transmission of the stimulus is also

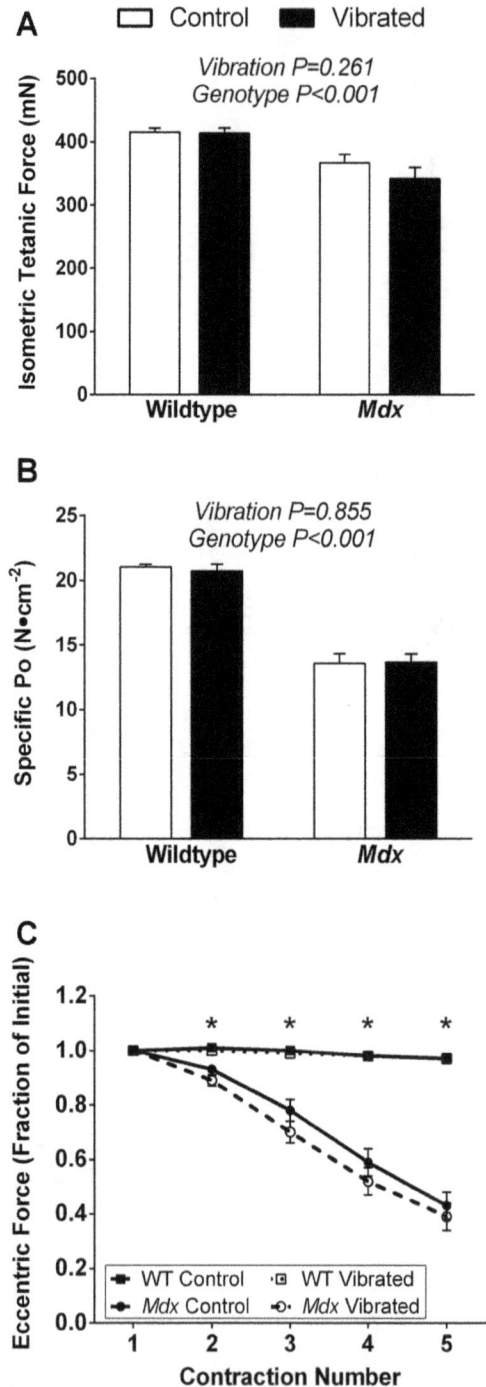

Figure 3. Eight weeks of vibration training did not impact *ex vivo* EDL muscle contractile function. Vibration training for 8 weeks did not influence the following EDL muscle contractile measures: **A)** maximal isometric tetanic force production, **B)** specific force, or **C)** susceptibility to eccentric contraction-induced injury compared to non-vibrated mice. As expected, *mdx* mice had lower values for each of the three measurements of EDL muscle function compared to wildtype mice. Data are means ± SE. P-values associated with the main effects of two-way ANOVAs are indicated above each set of bars in Panel A and B. In panel C, an interaction between genotype and eccentric contraction number was present, where the * signifies a significant (P<0.05) difference between *mdx* and wildtype mice from post-hoc testing. Interactions between vibration and genotype for panels A and B P≥ 0.329.

□ Control   ■ Vibrated

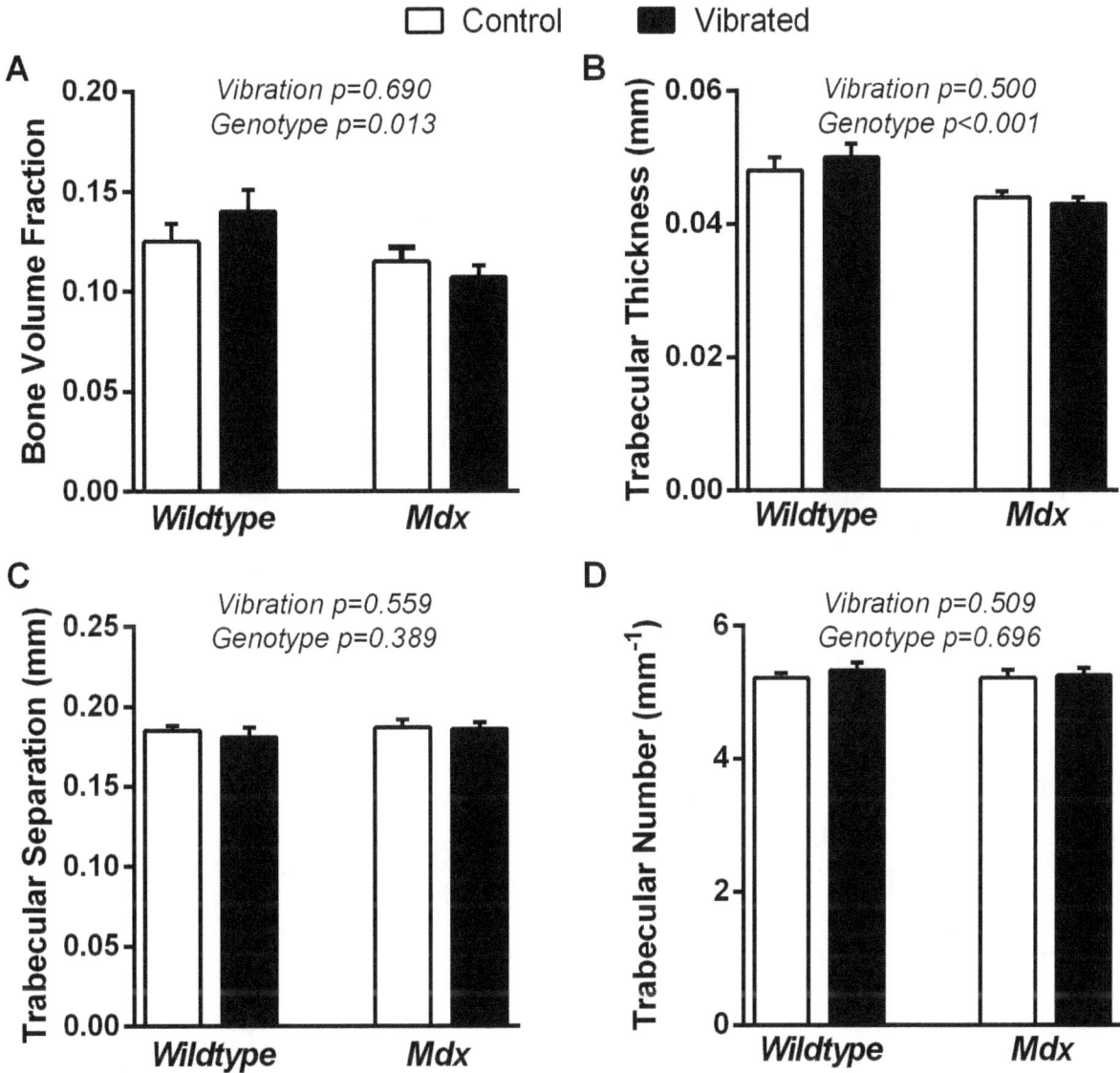

Figure 4. Eight weeks of vibration training did not impact trabecular bone in the tibia. Vibration training for 8 weeks did not influence trabecular bone **A)** volume fraction, **B)** thickness, **C)** separation, or **D)** number. As expected, *mdx* mice had lower values for trabecular bone volume fraction and thickness compared to wildtype mice. Data are means ± SE. P-values associated with the main effects of two-way ANOVAs are indicated above each set of bars. Interactions between vibration and genotype was P≥0.165.

influenced by muscle activation patterns and joint angles [56,69]. These factors were not controlled for in the present study, however mouse behavior and posture while on the platform did not appear to be altered over 8 weeks of training. It is possible that in mice, a higher intensity vibration (i.e., accelerations exceeding 1 *g*) might better amplify transmission and provoke an osteogenic response as previously shown [44]. Lastly, it is plausible that the use of young, growing mice in the present study masked our ability to quantify the efficacy of vibration to improve bone. Between 3–11 weeks of age, the rate of longitudinal bone growth is maximized in mice, and therefore may have a ceiling effect at which the bone becomes unable to respond to additional mechanical stimuli.

Skeletal Muscle

Eight weeks of vibration training did not alter any measure of hindlimb muscle functional capacity or structure (Figures 2 and 3 and Tables 2) and therefore our results do not support the notion

that low intensity vibration is of benefit to muscle. The overall efficacy of low intensity vibration to improve muscle function in humans remains controversial [23,24], with various reports of beneficial effects [21,22,25,26,27] and those reporting lack of alterations [28,29]. Few studies have used mouse models to investigate vibration and skeletal muscle and those reports are also inconsistent in regard to effects on muscle size [30,41,49,50]. The vibration training protocol used in the present study did not improve muscle size or strength in *mdx* or wildtype mice. The lack of vibration-induced improvements in muscle is consistent with results from another study that used botulism toxin to induce muscle weakness [41], but contradicts our previous vibration work in wildtype mice in which muscle strength improved by 10% despite no effect on muscle mass, size, or protein content [50]. Of interest, our previous study on wildtype mice was conducted using the same vibration device except that the vibration parameters were slightly different (1.0 *g* and 45 Hz) and the device was placed

Table 3. Effects of low intensity, high frequency vibration training on tibial bone cortical geometry, mechanical properties, and intrinsic material properties in wildtype and *mdx* mice.

	Wildtype Non-vibrated	Wildtype Vibrated	Mdx Non-vibrated	Mdx Vibrated	P-values for Two-Way ANOVA		
					Main effect of Vibration	Main effect of Genotype	Interaction (Vibration×Genotype)
Tibial Length (mm)	17.73 (0.08)	17.89 (0.07)	17.96 (0.06)	17.92 (0.05)	0.409	0.059	0.154
Cortical Geometric Properties							
Cortical Cross-Sectional Area (mm^2)	0.76 (0.03)	0.79 (0.02)	0.77 (0.02)	0.75 (0.02)	0.818	0.610	0.274
Cortical Thickness (mm)	0.22 (0.01)	0.23 (0.00)	0.23 (0.00)	0.23 (0.00)	0.336	0.495	0.429
Mechanical Functional Properties							
Energy to Ultimate Load (mJ)	4.05 (0.28)	4.26 (0.10)	3.53 (0.11)	3.59 (0.27)	0.521	0.005	0.720
Deflection to Ultimate Load (mm)	0.45 (0.01)	0.48 (0.02)	0.43 (0.01)	0.45 (0.03)	0.148	0.142	0.921
Intrinsic Material Properties							
Ultimate Stress (MPa)	282.8 (5.3)	279.9 (3.5)	273.6 (6.4)	280.0 (5.2)	0.746	0.397	0.385
Modulus of Elasticity (GPa)	10.7 (0.3)	10.6 (0.3)	10.8 (0.4)	10.9 (0.3)	0.945	0.620	0.713
Cortical vBMD (mg·cm^{-3})	1345.1 (17.7)	1351.8 (9.0)	1303.7 (13.9)	1313.2 (10.8)	0.543	0.004	0.918
Trabecular vBMD (mg·cm^{-3})	1095.6 (3.1)	1094.1 (4.1)	1077.8 (3.3)	1079.4 (4.3)	0.986	<0.001	0.679

Values are means (SE) vBMD, volumetric bone mineral density.

A

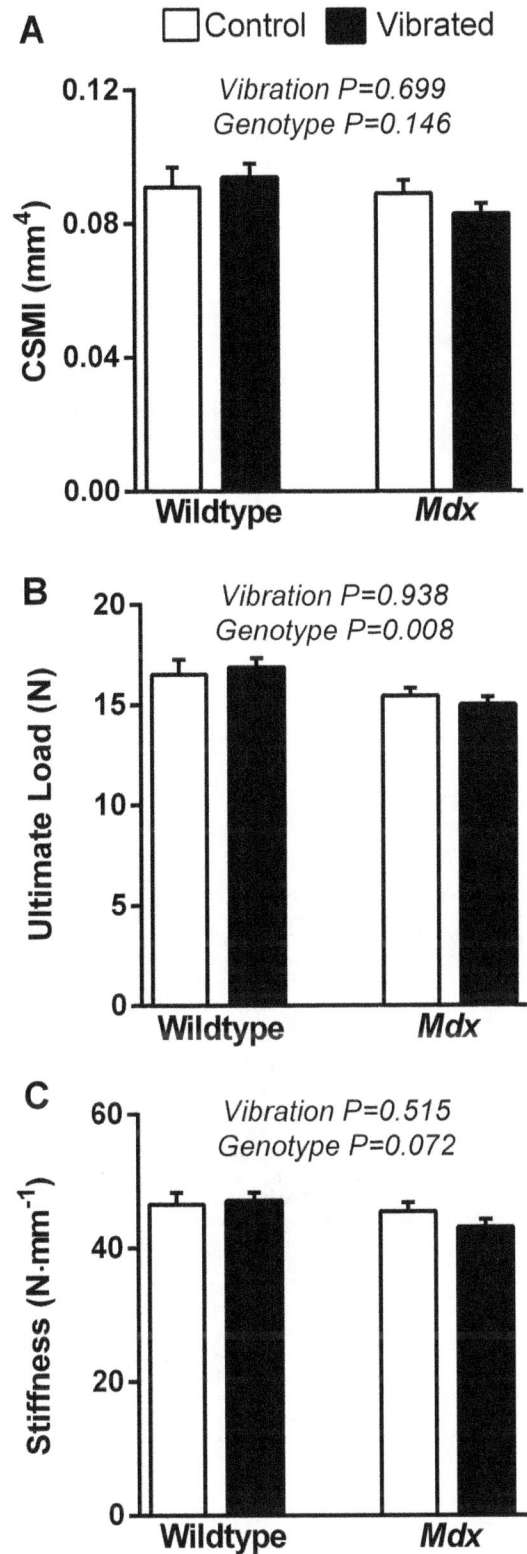

☐ Control ■ Vibrated

Vibration P=0.699
Genotype P=0.146

CSMI (mm⁴) — y-axis values: 0.12, 0.08, 0.04, 0.00

Wildtype   Mdx

B

Vibration P=0.938
Genotype P=0.008

Ultimate Load (N) — y-axis values: 20, 15, 10, 5, 0

Wildtype   Mdx

C

Vibration P=0.515
Genotype P=0.072

Stiffness (N·mm⁻¹) — y-axis values: 60, 40, 20, 0

Wildtype   Mdx

Figure 5. Eight weeks of vibration training did not impact tibial cortical bone. Vibration training for 8 weeks did not influence the following tibial cortical bone properties: **A)** cross-sectional moment of inertia, **B)** ultimate load, or **C)** stiffness. Mdx mice had lower values for ultimate load and trends for lower stiffness compared to wildtype mice. Data are means ± SE. P-values associated with the main effects of two-way ANOVAs are indicated above each set of bars. Interactions between vibration and genotype P≥0.287.

on a bench top [50]. In subsequent studies [46,70] and the current study, our device was mounted on a concrete vibration-isolation base, which reduced the error between actual and target acceleration to ±0.37% [70]. This modification was intended to minimize the variation in acceleration produced by the vibration device. It is possible that the homogenous acceleration stimulus in the present study may be responsible for preventing the improvements in muscle strength we previously observed.

Contraindications of vibration on muscle have been reported [30,31], and due to the high susceptibility of dystrophic muscle to injury, it was necessary to establish that vibration is a safe training modality. Our results show that 8 weeks of low intensity vibration training was not deleterious to any measure of muscle functional capacity (Figures 2–3 and Tables 2). The lack of injury with vibration training corroborates our previous findings in healthy mice [50] and preliminary data in patients [28], and contradicts the two studies which have reported muscle-specific contraindication of vibration (i.e., reduced vascularity in the distal soleus muscle in response to a low intensity vibration [30], and centrally-located nuclei in muscle fibers following relatively high intensity vibration (i.e., accelerations exceeding 1 g) [31]. Our thorough investigation utilized established recommendations for pre-clinical testing in *mdx* mice including a combination of *in vivo* and *ex vivo* assessment of muscle functional capacity providing a comprehensive evaluation of a training modality's efficacy and safety [71]. We further complemented these data with histological analyses and plasma creatine kinase activity to confirm that vibration was not injurious to dystrophic muscle. Our results show that low intensity vibration training does not adversely affect dystrophic mouse muscle.

Fat Pads and Intramuscular Triglyceride Concentration

Vibrated mice had smaller subcutaneous fat pad masses following 8 weeks of training (Figure 1). This vibration-induced reduction in fat mass has been consistently reported in rodents [47,50,54] and vibration training has even been shown to inhibit diet-induced obesity in mice [47]. To determine if vibration training also reduced intramuscular fat, we chose to measure triglyceride concentrations within the gastrocnemius muscle as this is a direct measure of muscle adiposity. The same approach has been utilized to measure triglyceride concentrations in mouse serum, liver and epididymal fat pads following 6 weeks of vibration [47], however we are the first to investigate intramuscular triglycerides. Specifically, we showed that vibration-trained mice had intramuscular triglyceride concentrations that were 26% higher than control mice (Table 1). This finding contrasts the earlier report that triglyceride concentrations were not different in the blood, liver or fat pads [47]. The physiological relevance of the vibration-induced increase in intramuscular triglycerides is not clear. Elevated intramuscular triglyceride concentration has been associated with metabolic disease, however, it also increases in response to exercise training [72]. This latter non-pathological response could potentially be an advantageous adaptation induced by vibration training, but more work will need to be done. Our previous work did show that 8-weeks of vibration-induced reductions in fat were not attributed to alterations in either energy balance (i.e., food intake and physical activity) [46,50] or mitochondrial enzyme activity (i.e., of nicotinamide adenine dinucleotide-tetrazolium reductase reactivity) [50]. An alternative mechanism suggests that vibration may influence bone marrow cells' lineage commitment away from adipocytes toward the osteoblast lineage [16,17,47]. This was based on the finding that vibrated mice had increased expression of the adipogenic gene, PPARγ (27%) and reduced expression of the transcription factor

Runx2 (73%) [47]. Combined, our results indicate that vibration training influences fat distribution in mice.

In conclusion, the present study has established that 8 weeks of low intensity, high frequency vibration training for 15 min per day, 5 days per week at 45 Hz and 0.6 g did not significantly impact trabecular or cortical bone within the tibia of young, growing *mdx* or wildtype mice. Hindlimb muscle functional capacity was also not affected, implying that this type of vibration is safe for dystrophic muscle and would likely not have deleterious effects on disease progression. Vibration training may aid in slowing the acquisition of fat mass and how this could impact the progression of this or other diseases is interesting to consider. Collectively, our results do not support the idea that vibration training could be an effective modality for improving bone or muscle in the context of a muscle disease, but further research is needed to determine if alternative combinations of vibration parameters or a prolonged duration of training, or perhaps using an adult mouse model, could elicit beneficial musculoskeletal functional responses.

Supporting Information

Video S1 Vibration stimulus was well tolerated by 3-week old *mdx* and wildtype mice. Behaviors, ambulation patterns and activities were indistinguishable between 3-week old wildtype mice (n = 2 mice on the left) and *mdx* mice (n = 2 mice on the right).

Video S2 Vibration stimulus was well tolerated by 11-week old *mdx* and wildtype mice. Behaviors, ambulation patterns and activities were indistinguishable between 11-week old wildtype mice (n = 2 mice on the left) and *mdx* mice (n = 2 mice on the right).

Acknowledgments

The authors would like to thank the Biomaterials Characterization & Quantitative Histomorphometry Core at the Mayo Clinic in Rochester, MN for their quantification of our dynamic histomorphometry measurements.

Author Contributions

Conceived and designed the experiments: SAN GLW DAL. Performed the experiments: SAN GLW DAL TLM AGG ASL REG. Analyzed the data: SAN GLW DAL TLM AGG ASL REG. Contributed reagents/materials/analysis tools: GLW DAL ASL REG. Wrote the paper: SAN GLW DAL.

References

1. Cozzi F, Cerletti M, Luvoni GC, Lombardo R, Brambilla PG, et al. (2001) Development of muscle pathology in canine X-linked muscular dystrophy. II. Quantitative characterization of histopathological progression during postnatal skeletal muscle development. Acta Neuropathol 101: 469–478.
2. Blake DJ, Weir A, Newey SE, Davies KE (2002) Function and genetics of dystrophin and dystrophin-related proteins in muscle. Physiol Rev 82: 291–329.
3. Gainer TG, Wang Q, Ward CW, Grange RW (2008) Duchenne Muscular Dystrophy. In: Tiidus PM, editor. Skeletal Muscle Damage and Repair. Champaign, IL: Human Kinetics. 113–124.
4. Larson CM, Henderson RC (2000) Bone mineral density and fractures in boys with Duchenne muscular dystrophy. J Pediatr Orthop 20: 71–74.
5. McDonald DG, Kinali M, Gallagher AC, Mercuri E, Muntoni F, et al. (2002) Fracture prevalence in Duchenne muscular dystrophy. Dev Med Child Neurol 44: 695–698.
6. Bianchi ML, Mazzanti A, Galbiati E, Saraifoger S, Dubini A, et al. (2003) Bone mineral density and bone metabolism in Duchenne muscular dystrophy. Osteoporos Int 14: 761–767.
7. Straathof CS, Overweg-Plandsoen WC, van den Burg GJ, van der Kooi AJ, Verschuuren JJ, et al. (2009) Prednisone 10 days on/10 days off in patients with Duchenne muscular dystrophy. J Neurol 256: 768–773.
8. Moens P, Baatsen PH, Marechal G (1993) Increased susceptibility of EDL muscles from mdx mice to damage induced by contractions with stretch. J Muscle Res Cell Motil 14: 446–451.
9. Petrof BJ, Shrager JB, Stedman HH, Kelly AM, Sweeney HL (1993) Dystrophin protects the sarcolemma from stresses developed during muscle contraction. Proc Natl Acad Sci U S A 90: 3710–3714.
10. Eagle M (2002) Report on the muscular dystrophy campaign workshop: exercise in neuromuscular diseases Newcastle, January 2002. Neuromuscul Disord 12: 975–983.
11. King W, Landoll J, Matkovic V, Kissel J (2009) Volumetric radial and tibial bone mineral density in boys with Duchenne muscular dystrophy; 72.
12. Landoll J, King W, Kissel J, Matkovic V (2008) Forearm pQCT measurements in males with Duchenne muscular dystrophy; 23.
13. Soderpalm AC, Magnusson P, Ahlander AC, Karlsson J, Kroksmark AK, et al. (2007) Low bone mineral density and decreased bone turnover in Duchenne muscular dystrophy. Neuromuscul Disord 17: 919–928.
14. Rubin C, Turner AS, Mallinckrodt C, Jerome C, McLeod K, et al. (2002) Mechanical strain, induced noninvasively in the high-frequency domain, is anabolic to cancellous bone, but not cortical bone. Bone 30: 445–452.
15. Qin YX, Rubin CT, McLeod KJ (1998) Nonlinear dependence of loading intensity and cycle number in the maintenance of bone mass and morphology. J Orthop Res 16: 482–489.
16. Ozcivici E, Luu YK, Adler B, Qin YX, Rubin J, et al. (2010) Mechanical signals as anabolic agents in bone. Nat Rev Rheumatol 6: 50–59.
17. Ozcivici E, Luu YK, Rubin CT, Judex S (2010) Low-level vibrations retain bone marrow's osteogenic potential and augment recovery of trabecular bone during reambulation. PLoS One 5: e11178.
18. Slatkovska L, Alibhai SM, Beyene J, Cheung AM (2010) Effect of whole-body vibration on BMD: a systematic review and meta-analysis. Osteoporos Int 21: 1969–1980.
19. Prisby RD, Lafage-Proust MH, Malaval L, Belli A, Vico L (2008) Effects of whole body vibration on the skeleton and other organ systems in man and animal models: what we know and what we need to know. Ageing Res Rev 7: 319–329.
20. Rubin C, Turner AS, Bain S, Mallinckrodt C, McLeod K (2001) Anabolism. Low mechanical signals strengthen long bones. Nature 412: 603–604.
21. Blottner D, Salanova M, Puttmann B, Schiffl G, Felsenberg D, et al. (2006) Human skeletal muscle structure and function preserved by vibration muscle exercise following 55 days of bed rest. Eur J Appl Physiol 97: 261–271.
22. Ward K, Alsop C, Caulton J, Rubin C, Adams J, et al. (2004) Low magnitude mechanical loading is osteogenic in children with disabling conditions. J Bone Miner Res 19: 360–369.
23. Lau RW, Liao LR, Yu F, Teo T, Chung RC, et al. (2011) The effects of whole body vibration therapy on bone mineral density and leg muscle strength in older adults: a systematic review and meta-analysis. Clin Rehabil 25: 975–988.
24. Mikhael M, Orr R, Fiatarone Singh MA (2010) The effect of whole body vibration exposure on muscle or bone morphology and function in older adults: a systematic review of the literature. Maturitas 66: 150–157.
25. Reyes ML, Hernandez M, Holmgren LJ, Sanhueza E, Escobar RG (2011) High-frequency, low-intensity vibrations increase bone mass and muscle strength in upper limbs, improving autonomy in disabled children. J Bone Miner Res 26: 1759–1766.
26. Muir J, Judex S, Qin YX, Rubin C (2011) Postural instability caused by extended bed rest is alleviated by brief daily exposure to low magnitude mechanical signals. Gait Posture 33: 429–435.
27. Gilsanz V, Wren TA, Sanchez M, Dorey F, Judex S, et al. (2006) Low-level, high-frequency mechanical signals enhance musculoskeletal development of young women with low BMD. J Bone Miner Res 21: 1464–1474.
28. Soderpalm AC, Kroksmark AK, Magnusson P, Karlsson J, Tulinius M, et al. (2013) Whole body vibration therapy in patients with Duchenne muscular dystrophy - A prospective observational study. J Musculoskelet Neuronal Interact 13: 13–18.
29. Torvinen S, Kannus P, Sievanen H, Jarvinen TA, Pasanen M, et al. (2003) Effect of 8-month vertical whole body vibration on bone, muscle performance, and body balance: a randomized controlled study. J Bone Miner Res 18: 876–884.
30. Murfee WL, Hammett LA, Evans C, Xie L, Squire M, et al. (2005) High-frequency, low-magnitude vibrations suppress the number of blood vessels per muscle fiber in mouse soleus muscle. J Appl Physiol 98: 2376–2380.
31. Necking LE, Lundstrom R, Lundborg G, Thornell LE, Friden J (1996) Skeletal muscle changes after short term vibration. Scand J Plast Reconstr Surg Hand Surg 30: 99–103.
32. Anderson JE, Lentz DL, Johnson RB (1993) Recovery from disuse osteopenia coincident to restoration of muscle strength in mdx mice. Bone 14: 625–634.

33. Nakagaki WR, Bertran CA, Matsumura CY, Santo-Neto H, Camilli JA (2011) Mechanical, biochemical and morphometric alterations in the femur of mdx mice. Bone 48: 372–379.

34. Novotny SA, Warren GL, Lin AS, Guldberg RE, Baltgalvis KA, et al. (2011) Bone is functionally impaired in dystrophic mice but less so than skeletal muscle. Neuromuscul Disord 21: 183–193.

35. Rufo A, Del Fattore A, Capulli M, Carvello F, De Pasquale L, et al. (2011) Mechanisms inducing low bone density in Duchenne muscular dystrophy in mice and humans. J Bone Miner Res 26: 1891–1903.

36. Montgomery E, Pennington C, Isales CM, Hamrick MW (2005) Muscle-bone interactions in dystrophin-deficient and myostatin-deficient mice. The anatomical record Part A, Discoveries in molecular, cellular, and evolutionary biology 286: 814–822.

37. Landisch RM, Kosir AM, Nelson SA, Baltgalvis KA, Lowe DA (2008) Adaptive and nonadaptive responses to voluntary wheel running by mdx mice. Muscle Nerve 38: 1290–1303.

38. Rittweger J (2010) Vibration as an exercise modality: how it may work, and what its potential might be. Eur J Appl Physiol 108: 877–904.

39. Christiansen BA, Silva MJ (2006) The effect of varying magnitudes of whole-body vibration on several skeletal sites in mice. Ann Biomed Eng 34: 1149–1156.

40. Judex S, Lei X, Han D, Rubin C (2007) Low-magnitude mechanical signals that stimulate bone formation in the ovariectomized rat are dependent on the applied frequency but not on the strain magnitude. J Biomech 40: 1333–1339.

41. Manske SL, Good CA, Zernicke RF, Boyd SK (2012) High-Frequency, Low-Magnitude Vibration Does Not Prevent Bone Loss Resulting from Muscle Disuse in Mice following Botulinum Toxin Injection. PLoS One 7: e36486.

42. Lee BJ, Judex S, Luu K, Thomas J, Gilsanz V, et al. (2007) Potential mitigation of the skeletal complications of Duchenne's muscular dystrophy with vibration. IEEE. 35–36.

43. Brouwers JE, van Rietbergen B, Ito K, Huiskes R (2009) Effects of vibration treatment on tibial bone of ovariectomized rats analyzed by in vivo micro-CT. J Orthop Res 28: 62–69.

44. Oxlund BS, Ortoft G, Andreassen TT, Oxlund H (2003) Low-intensity, high-frequency vibration appears to prevent the decrease in strength of the femur and tibia associated with ovariectomy of adult rats. Bone 32: 69–77.

45. Xie L, Jacobson JM, Choi ES, Busa B, Donahue LR, et al. (2006) Low-level mechanical vibrations can influence bone resorption and bone formation in the growing skeleton. Bone 39: 1059–1066.

46. Novotny SA, Eckhoff MD, Eby BC, Call JA, Nuckley DJ, et al. (2013) Musculoskeletal response of dystrophic mice to short term, low intensity, high frequency vibration. J Musculoskelet Neuronal Interact 13: 418–429.

47. Luu YK, Capilla E, Rosen CJ, Gilsanz V, Pessin JE, et al. (2009) Mechanical stimulation of mesenchymal stem cell proliferation and differentiation promotes osteogenesis while preventing dietary-induced obesity. J Bone Miner Res 24: 50–61.

48. Garman R, Gaudette G, Donahue LR, Rubin C, Judex S (2007) Low-level accelerations applied in the absence of weight bearing can enhance trabecular bone formation. J Orthop Res 25: 732–740.

49. Xie L, Rubin C, Judex S (2008) Enhancement of the adolescent murine musculoskeletal system using low-level mechanical vibrations. J Appl Physiol 104: 1056–1062.

50. McKeehen JN, Novotny SA, Baltgalvis KA, Call JA, Nuckley DJ, et al. (2013) Adaptations of Mouse Skeletal Muscle to Low-Intensity Vibration Training. Med Sci Sports Exerc 45: 1051–1059.

51. Iwaniec UT, Wronski TJ, Liu J, Rivera MF, Arzaga RR, et al. (2007) PTH stimulates bone formation in mice deficient in Lrp5. J Bone Miner Res 22: 394–402.

52. Moran AL, Warren GL, Lowe DA (2005) Soleus and EDL muscle contractility across the lifespan of female C57BL/6 mice. Exp Gerontol 40: 966–975.

53. Baltgalvis KA, Call JA, Nikas JB, Lowe DA (2009) Effects of prednisolone on skeletal muscle contractility in mdx mice. Muscle Nerve 40: 443–454.

54. Maddalozzo GF, Iwaniec UT, Turner RT, Rosen CJ, Widrick JJ (2008) Whole-body vibration slows the acquisition of fat in mature female rats. Int J Obes (Lond) 32: 1348–1354.

55. Kiiski J, Heinonen A, Jarvinen TL, Kannus P, Sievanen H (2008) Transmission of vertical whole body vibration to the human body. Journal of bone and mineral research: the official journal of the American Society for Bone and Mineral Research 23: 1318–1325.

56. Rubin C, Pope M, Fritton JC, Magnusson M, Hansson T, et al. (2003) Transmissibility of 15-hertz to 35-hertz vibrations to the human hip and lumbar spine: determining the physiologic feasibility of delivering low-level anabolic mechanical stimuli to skeletal regions at greatest risk of fracture because of osteoporosis. Spine (Phila Pa 1976) 28: 2621–2627.

57. Garlich MW, Baltgalvis KA, Call JA, Dorsey LL, Lowe DA (2010) Plantarflexion contracture in the mdx mouse. Am J Phys Med Rehabil 89: 976–985.

58. Ingalls CP, Warren GL, Lowe DA, Boorstein DB, Armstrong RB (1996) Differential effects of anesthetics on in vivo skeletal muscle contractile function in the mouse. J Appl Physiol 80: 332–340.

59. Call JA, Eckhoff MD, Baltgalvis KA, Warren GL, Lowe DA (2011) Adaptive strength gains in dystrophic muscle exposed to repeated bouts of eccentric contraction. J Appl Physiol 111: 1768–1777.

60. Warren GL, Hayes DA, Lowe DA, Williams JH, Armstrong RB (1994) Eccentric contraction-induced injury in normal and hindlimb-suspended mouse soleus and EDL muscles. J Appl Physiol 77: 1421–1430.

61. Brooks SV, Faulkner JA (1988) Contractile properties of skeletal muscles from young, adult and aged mice. J Physiol 404: 71–82.

62. Folch J, Lees M, Sloane Stanley GH (1957) A simple method for the isolation and purification of total lipides from animal tissues. J Biol Chem 226: 497–509.

63. Warren GL, Moran AL, Hogan HA, Lin AS, Guldberg RE, et al. (2007) Voluntary run training but not estradiol deficiency alters the tibial bone-soleus muscle functional relationship in mice. Am J Physiol Regul Integr Comp Physiol 293: R2015–2026.

64. Warren GL, Lowe DA, Inman CL, Orr OM, Hogan HA, et al. (1996) Estradiol effect on anterior crural muscles-tibial bone relationship and susceptibility to injury. J Appl Physiol 80: 1660–1665.

65. Novotny SA, Warren GL, Lin AS, Guldberg RE, Baltgalvis KA, et al. (2012) Prednisolone treatment and restricted physical activity further compromise bone of mdx mice. J Musculoskelet Neuronal Interact 12: 16–23.

66. Lotinun S, Evans GL, Turner RT, Oursler MJ (2005) Deletion of membrane-bound steel factor results in osteopenia in mice. J Bone Miner Res 20: 644–652.

67. Judex S, Donahue LR, Rubin C (2002) Genetic predisposition to low bone mass is paralleled by an enhanced sensitivity to signals anabolic to the skeleton. Faseb J 16: 1280–1282.

68. Rubin C, Xu G, Judex S (2001) The anabolic activity of bone tissue, suppressed by disuse, is normalized by brief exposure to extremely low-magnitude mechanical stimuli. Faseb J 15: 2225–2229.

69. Ritzmann R, Gollhofer A, Kramer A (2012) The influence of vibration type, frequency, body position and additional load on the neuromuscular activity during whole body vibration. Eur J Appl Physiol.

70. Novotny SA, Mehta H, Lowe DA, Nuckley DJ (2013) Vibration Platform for Mice to Deliver Precise, Low Intensity Mechanical Signals to the Musculoskeleton. J Musculoskelet Neuronal Interact 13: 412–417.

71. Grounds MD, Radley HG, Lynch GS, Nagaraju K, De Luca A (2008) Towards developing standard operating procedures for pre-clinical testing in the mdx mouse model of Duchenne muscular dystrophy. Neurobiol Dis 31: 1–19.

72. Koves TR, Sparks LM, Kovalik JP, Mosedale M, Arumugam R, et al. (2013) PPARgamma coactivator-1alpha contributes to exercise-induced regulation of intramuscular lipid droplet programming in mice and humans. Journal of lipid research 54: 522–534.

13

Altered Composition of Bone as Triggered by Irradiation Facilitates the Rapid Erosion of the Matrix by Both Cellular and Physicochemical Processes

Danielle E. Green[1], Benjamin J. Adler[1], Meilin Ete Chan[1], James J. Lennon[1], Alvin S. Acerbo[1,2], Lisa M. Miller[1,2], Clinton T. Rubin[1]*

1 Department of Biomedical Engineering, Stony Brook University, Stony Brook, New York, United States of America, 2 Photon Sciences Directorate, Brookhaven National Laboratory, Upton, New York, United States of America

Abstract

Radiation rapidly undermines trabecular architecture, a destructive process which proceeds despite a devastated cell population. In addition to the 'biologically orchestrated' resorption of the matrix by osteoclasts, physicochemical processes enabled by a damaged matrix may contribute to the rapid erosion of bone quality. 8w male C57BL/6 mice exposed to 5 Gy of Cs^{137} γ-irradiation were compared to age-matched control at 2d, 10d, or 8w following exposure. By 10d, irradiation had led to significant loss of trabecular bone volume fraction. Assessed by reflection-based Fourier transform infrared imaging (FTIRI), chemical composition of the irradiated matrix indicated that mineralization had diminished at 2d by $-4.3\pm4.8\%$, and at 10d by $-5.8\pm3.2\%$. These data suggest that irradiation facilitates the dissolution of the matrix through a change in the material itself, a conclusion supported by a $13.7\pm4.5\%$ increase in the elastic modulus as measured by nanoindentation. The decline in viable cells within the marrow of irradiated mice at 2d implies that the immediate collapse of bone quality and inherent increased risk of fracture is not solely a result of an overly-active biologic process, but one fostered by alterations in the material matrix that predisposes the material to erosion.

Editor: Ryan K. Roeder, University of Notre Dame, United States of America

Funding: Use of the National Synchrotron Light Source, Brookhaven National Laboratory, was supported by the U.S. Department of Energy, Office of Science, Office of Basic Energy Sciences, under Contract No. DE-AC02-98CH10886. Funding for this research was also provided by National Institutes of Health AR43498 and RR23782. The funders had no role in study design, data collection and analysis, decision to publish, or preparation of the manuscript.

Competing Interests: The authors have declared that no competing interests exist.

* E-mail: clinton.rubin@stonybrook.edu

Introduction

Radiation exposure has become a large health concern due to factors including the recent reactor failures at Fukushima Daiichi, the high clinical doses patients receive for radiotherapy, and the exposure astronauts receive during extended space missions [1,2,3]. In addition to radiation's destruction of the bone marrow and the resident hematopoietic and mesenchymal stem cell populations, this exposure – whether intentional or otherwise - leads to the devastation of bone architecture, thereby increasing a person's lifetime risk of fracture [4,5,6]. While the mechanism of bone loss following exposure is presumed to be a biological process mediated by elevated osteoclast activity [7,8], considering the extensive destruction of the precursor population, it is possible that – to some degree – the bone loss is achieved independent of "biology" via an acellular process, perhaps via the physicochemical dissolution of a damaged bone matrix.

The bone matrix is composed of organic components, including collagen type-I and non-collagenous proteins, and an inorganic component comprised of carbonated hydroxyapatite. Damage to either the inorganic or organic constituents of the matrix drastically compromise bone quality, as evidenced by the severe decline in the bone's mechanical properties following irradiation [9,10]. This dose-dependent decline in mechanical properties includes reductions in bone strength, ductility, and fracture resistance, with higher exposure to radiation directly correlating to poorer bone quality [11,12].

As demonstrated *ex vivo*, irradiation also compromises bone strength. In bone allograft transplantation, a method commonly employed in orthopedic bone reconstruction, the bone graft will typically be irradiated at dose exposures greater than 25 kGy to minimize the potential for transmittance of diseases from donor to recipient [13]. Even in acute, *ex vivo* exposures, these high doses directly reduce bone's material properties [14], belying not only a compromised material central to surgery, but suggesting a very real risk that radiation poses to the skeleton during both intentional or unintentional exposures. And while *ex vivo* reductions in material properties occur independent of biologic processes, it is also obvious, but important to point out that this reduction occurs independent of reductions in bone morphology (e.g., bone volume fraction, trabecular number, etc.).

While cancer patients typically receive somewhat lower doses of radiation prior to bone marrow transplantation ($\sim$12 Gy) [15], this exposure can reach as high as $\sim$66 Gy in localized regions targeted to ablate tumors [16], predisposing these specific regions to accelerated bone loss and elevated risk of fracture [4,5,6]. Despite the marked depletion of the bone marrow progenitor population within even two days of irradiation exposure, which include the hematopoietic precursors to osteoclasts [17], bone loss in these clinical cases are typically presumed to result from elevated osteoclast activity. However, if the hematopoietic population is crippled following irradiation, it is difficult to attribute the almost instantaneous decline in bone architecture - realized within 10 days [17]- solely to bio-mediated bone resorption. We propose, therefore, that removal of the matrix following irradiation is facilitated by damage to the bone matrix itself. Testing of this hypothesis has been enabled by recent advances in quantitative microscopy which allow a full characterization of the organic and inorganic constituents of bone, as well as new advances in material property characterization, which allow a full assessment of the mechanics of the matrix.

Fourier transform infrared imaging (FTIRI) can be used to map the chemical composition of bone [18,19,20]. Defined as the phosphate/protein ratio, FTIRI can assess the level of tissue mineralization by integrating the protein amide I peak falling between 1600 cm^{-1}–1700 cm^{-1} and the phosphate peak (v_1,v_3 PO_4^{3-}) falling between 900 cm^{-1}–1200 cm^{-1} (Fig. 1). Any alteration in the content of protein or phosphate will thereby affect the total level of tissue mineralization, a critical correlate to the overall mechanical properties of bone [19,21]. In addition to characterizing key components of tissue mineralization, reflection based FTIRI is capable of identifying independent changes to the bone matrix due to alterations in the organic and inorganic components of the matrix. For instance, a shift in the phosphate intensity infers an alteration to the stoichiometric properties of hydroxyapatite crystals within the mineralized matrix, while a change in amide I intensity implicates a change in the chemical composition of the protein structure, primarily collagen type-I. FTIRI can also be used to assess crystallinity, a measure of the size and shape of crystallites. The spatial resolution of FTIRI mapping facilitates spatial mapping of the chemical composition in different regions of the bone matrix, allowing for the identification of sites most susceptible to radiation-induced destruction.

While FTIRI is a method used to analyze chemical composition, nanoindentation can be used to characterize the surface mechanical properties of materials on the nano-micron scale [22,23]. Typically a diamond probe penetrates into the surface of the tissue, and the material's elastic properties such as hardness and elastic modulus can be determined from the applied load and

depth of sample penetration [23]. A benefit of nanoindentation is its site specificity for indent placement [24], and therefore the mechanical properties of bone in one region can easily be compared to the same region of another sample. Other methods such as whole bone testing often return the bone's bulk material properties which do not necessarily relate to trabecular bone quality, thereby making nanoindentation the most ideal method to determine trabecular bone's elastic properties.

We hypothesize that the rapid bone loss caused by irradiation is a complex product of biological resorption driven by increased osteoclast activity, and physicochemical-based erosion enabled by damage to the matrix that advances independent of the biology of the system. In this study, we investigate the capacity of sub-lethal doses of radiation to alter the chemical and physical properties of the bone matrix in young adult C57BL/6 mice. We attempt to distinguish between the biological and physicochemical processes, although not mutually exclusive, which contribute to bone loss, showing that these processes play not only additive, but independent roles in the destruction of bone quantity and quality.

Materials and Methods

Ethics Statement

All animal procedures were reviewed and approved by Stony Brook University's Institutional Animal Care and Use Committee, ID # 0067.

Irradiating Mice and Tissue Harvest

Six week old male C57BL/6 mice (Jackson Laboratories) were acclimated for 2w prior to the start of the study and fed food and water *ad libitum*. At 8w of age, mice were placed in an irradiation chamber and subjected to 0.6 Gy/min Cs^{137} γ-irradiation for 8.4 min, reaching a 5 Gy cumulative dose. Age-matched control mice were placed into the inactive irradiator for the identical period of time. At 2d, 10d or 8wk following radiation exposure, one third of mice in the irradiated and control groups were anesthetized with isoflurane and sacrificed using cervical dislocation (n = 7–8 per group). The left tibiae were extracted and placed into 70% ethanol and stored at $-20°$C for preservation, and blood was removed from the mice via cardiac puncture at the time of sacrifice. The right tibiae were extracted and bone marrow was flushed with Dulbecco's Modified Essential Medium containing 2% fetal bovine serum (Invitrogen), 10 mM HEPES Buffer (Gibco), and 1% Penicillin & Streptomycin (Gibco). The right humerus was extracted, placed into phosphate buffered saline, and stored at $-20°$C for preservation. Total cell numbers were quantified using a Scepter (Millipore) cell counter following erythrocyte lysis.

Trabecular and Cortical Bone Architecture Using Micro-Computed Tomography (microCT)

In order to analyze the microarchitecture of trabecular and cortical bone following irradiation, the mice tibiae were scanned *ex vivo* using a microCT scanner (μCT 40; Scanco Medical) with a 12 μm isotropic voxel size. The x-ray source voltage was 55 kVp, source current was 145 μA, and integration time was 300 ms. Using well defined automated analysis scripts [25], an 840 μm region of trabecular bone was analyzed approximately 300 μm distal to the growth plate in order to measure trabecular tissue mineral density (Tb.TMD). A 600 μm region of cortical bone was assessed in the midshaft to measure the cortical thickness (Ct.Th), cortical area (Ct.Ar), total area (Tt.Ar), and cortical area fraction (Ct.Ar/Tt.Ar) of the tibiae.

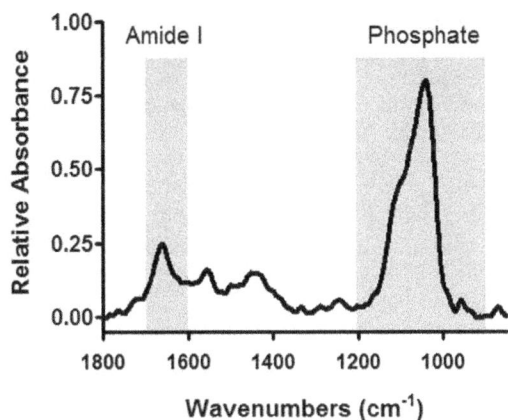

Figure 1. An infrared spectrum of a typical trabecular bone sample collected using Fourier transform infrared microscopy. The highlighted peaks, falling between 900–1200 cm^{-1} and 1600–1700 cm^{-1}, correspond to the phosphate and amide-I peaks, respectively, which are used to calculate the level of tissue mineralization.

Bone Sample Embedding

Before embedding the undecalcified tibiae in poly(methyl methacrylate) (PMMA), the bones were serially dehydrated in 70%, 95%, and 100% isopropanol and cleared in petroleum ether. Samples were then infiltrated with PMMA solution containing 15% n-butyl phthalate (Sigma), 85% methyl methacrylate (Fisher), and 2% w/v benzoyl peroxide (Aldrich). The tibiae were then placed into 20 ml polyethylene scintillation vials containing hardened PMMA, covered with fresh PMMA solution, and stored in a 37°C water bath until the PMMA solidified.

Fourier Transform Infrared Imaging

Fourier transform infrared imaging (FTIRI) provides a high resolution spatial assessment of the mineral composition and collagen cross-linking in the trabecular bone [26]. Cross-sections of bone from irradiated and control mice were first ground with abrasive silicon carbide papers using finer grits with decreasing particle sizes of 600, 800, and 1,200, then polished with a diamond suspension of 3 µm followed by 1 µm, 0.25 µm, and 0.05 µm, and finally sonicated in water to remove any remaining debris. FTIRI data collection were performed on the PMMA embedded and polished tibiae sample blocks using a Hyperion 3000 FTIR microscope (Bruker Optics) equipped with a 64×64 element focal plane array detector at beamline U10B of the National Synchrotron Light Source at Brookhaven National Laboratory. FTIRI data were collected in a reflection geometry over a range of 850–3900 cm^{-1} with a spectral resolution of 8 cm^{-1} and a pixel resolution of 5 µm. A total of 128 scans were collected per pixel, with a reflective gold slide used as a background calibration prior to scanning samples. The resultant reflection spectra were processed and transformed into absorbance spectra using Kramers-Kronig transformation [27].

Using Cytospec 1.4.03 (CytoSpec Inc.), the absorbance spectra of the trabecular bone in the metaphysis first underwent a linear baseline correction from 900 cm^{-1} to 1800 cm^{-1}. The protein of amide I (1600 cm^{-1}–1700 cm^{-1}), amide II (1510 cm^{-1}–1595 cm^{-1}), crystallinity (1033 cm^{-1}–1037 cm^{-1})/(1023 cm^{-1}–1027 cm^{-1}), and phosphate (900–1200 cm^{-1}) peaks were integrated and the mineral/protein ratio was calculated [19,28,29]. Amide I and amide II are both characteristic spectral regions for collagen type-I. To avoid any confounding contribution of a PMMA artifact into the bone spectra, as well as pixels on the bone surface with only partial bone contribution, a PMMA mask was set in place, and pixels with the characteristic PMMA absorbance peak between 1710 cm^{-1} and 1775 cm^{-1} were removed from the analysis. This mask removes 10 µm − 15 µm of data from the bone surface.

To determine the difference in phosphate, protein, and mineralization at the periphery of each trabecular element compared to the average content throughout the strut, all FTIRI data were removed from each independent trabeculae with the exception of the outermost pixel thereby leaving only an outline of the trabeculae, which represents phosphate, protein, or mineralization of the bone surface.

Nanoindentation

Following FTIRI, load controlled nanoindentation (Hysitron Triboindenter) was performed using a Berkovich tip in the same trabecular region analyzed for the bone's chemical composition. The tip shape function was calibrated using fused silica, which is elastically isotropic. A three segment load function was used for indenting, starting with a constant loading rate of 67 µN/s for 15 s, followed by a 10s hold period, and completed by a 67 µN/s unloading rate for 15s, thereby having a 1000 µN peak force, and

20 indentations were made per sample. These 20 points were selected using an optical microscope with a precision stage (500 nm accuracy). The elastic properties were calculated between 50% and 95% of the initial unloading curve and the Oliver-Pharr method was used to calculate elastic modulus and hardness [22,30]. Elastic modulus (E) and hardness (H) of the bone were defined as:

$$E = \frac{1-v^2}{\frac{1}{E_r} - \frac{1-v_i^2}{E_i}} \quad \text{and} \quad H = \frac{P_{max}}{A}$$

where v_i and v are the Poisson's ratio of the indenter and trabecular bone assumed to be 0.07 and 0.25, respectively. E_i is the reduced elastic modulus of the indenter assumed to be 1140 GPa and E_r is the reduced elastic modulus calculated as $E_r = \frac{\sqrt{\pi}}{2} \frac{S}{\sqrt{A}}$ where S is the contact stiffness, A is the contact area at maximum load, and P_{max} is the maximum indentation load.

Mechanical Loading

The right humeri of 10d mice were thawed to room temperature, fixed horizontally within the supports of the MTS, and three point bending strength was assessed. The supports spanned 6 mm. A 0.1 N of static preload was applied to the humeri, and then they were loaded to failure at 0.10 mm/s. The ultimate force and stiffness were evaluated from the force-displacement curves.

Bone Marrow Histology

Bones embedded in PMMA were sectioned using a microtome (Leica) at 6 µm, the PMMA sections were deplasticized in acetone for 1 hr, and stained with Modified Wright Giemsa (Sigma) for 1 min, rinsed with water, and then imaged under a light microscope (Zeiss). The percentage of the bone marrow occupied by adipocytes in the proximal tibiae below the growth plate was quantified using ImageJ software (NIH).

Cell Proliferative Ability Using Colony Forming Cell Assay

The harvested bone marrow from the right tibiae of control and irradiated mice were cultured at 1×10^4 cells per 35 cm plate in methylcellulose based media, and at 14d following the sacrifice, the total number of colonies formed was quantified according to the manufacturer's protocol (R&D Systems). The total number of colonies was identified based on morphology as either colony forming unit-granulocyte macrophage (CFU-GM) or colony forming unit-granulocyte, erythrocyte, macrophage, megakaryocyte (CFU-GEMM).

Osteoclast Activity

The level of osteoclast activity was assayed by the serum concentration of tartrate-resistant acid phosphatase (TRAP5b) assay (MouseTRAP; Immunodiagnostic Systems) according to the manufacturer's protocols.

Collagen I Breakdown

The breakdown of the organic matrix was quantified using plasma samples diluted in PBS. The type-I Collagen was quantified using the Mouse Cross Linked C-telopeptide of Type-I Collagen (CTX-I) Elisa kit according to the manufacturer's protocol (MyBioSource.com). The average of the optical density values was fit to the standard curve and multiplied by the dilution factor to determine the average concentration of CTX-I in the plasma.

Statistics

All data are presented as mean ± standard deviation. Differences between control and irradiated mice are determined using an unpaired sample t-test, where p<0.05 is considered statistically significant. To determine the difference between the mineralization in the center of a trabecular strut as compared to its surface, a paired sample t-test is used, where p<0.05 is considered significant.

Results

Rapid Destruction of Trabecular Bone Architecture Following Radiation Exposure but Cortical Bone Remained Uncompromised

As we previously reported, radiation exposure led to significant loss of trabecular bone as early as 10d following irradiation, as shown by the −41±12% and −33±4% decline in bone volume fraction and trabecular number, and 52±8% increase in trabecular separation compared to the control [17]. By 8w, bone volume fraction and trabecular number showed no evidence of improved trabecular morphology, despite the trend of a 2.7±3.5% increase in Tb.TMD (p = 0.09; Table 1). In contrast, no bone loss was evident in cortical bone at any point measured throughout the study as seen by no differences in Tt.Ar, Ct.Ar, Ct.Th, and Ct.Ar/Tt.Ar between the irradiated and control mice (Table 1).

Decline in Trabecular Bone Mineralization Following Irradiation

Evident at 2d, there is a trend toward a −4.3±4.8% decline in mineralization of the trabecular struts in the irradiated mice as compared to control (p = 0.07), which further falls by 10d to −5.8±3.2% (p<0.01; Fig. 2). A decline in the mineral/matrix ratio appears to occur through a decrease in the phosphate content within the bone matrix (Fig. 2). Irradiated mice show a −4.2±7.8% (NS) and −6.8±8.8% (p = 0.10) decline in phosphate content at 2d and 10d compared to age-matched control, whereas no significant changes were evident in protein level between the irradiated mice and their control at either 2d

or 10d, respectively. Even with the decline in mineralization, crystallinity, a measure of apatite size and shape, shows no differences between the irradiated mice and control at 2d (Control = 1.48±0.07, Irradiated = 1.58±0.16; p = 0.12) or 10d (Control = 1.53±0.11, Irradiated = 1.59±0.14; p = 0.40). By 8w, although the trabecular BV/TV in the tibiae remained −45±9% below age-matched control [17], the level of mineralization in the irradiated mice tibiae had fully recovered, and were 5.8±5.1% (p<0.05) greater than the control, a finding consistent with the increase in Tb.TMD seen using micro-CT. Crystallinity of the irradiated mice though was −12±7% lower than the age-matched control (p<0.05).

In contrast to those measures made in the trabecular compartment, at 10d post-irradiation, FTIRI showed no change in the chemical composition of the cortical bone, confirming the micro-CT analysis which indicated no loss of bone in this region (Table 1). At 10d, mineralization in cortical bone was 4.3±0.2 and 4.4±0.2 in the control and irradiated bones, respectively (p = 0.47). These data also indicated that bone in the trabecular compartment of the control was −11±3% less mineralized than the cortical bone of the control, with phosphate and protein levels −16±3% and −5±3% lower than cortical bone, respectively.

Decline in Phosphate and Protein Following Irradiation was Uniform Across the Trabecular Strut Rather than Localized to the Surface

The level of phosphate and protein in a healthy control mouse at 10d varied across the trabecular bone surface, measuring −7.7±2.9% and −6.9±2.7% lower at the peripheral edges, as compared to the total phosphate and protein through the bone closer to the center (p<0.05). Further, mineralization at these edges was −3.2±1.0% lower than the average level throughout the bone strut (p<0.05). Even following irradiation at 10d, the discrepancy in the phosphate, protein, and level of mineralization between the trabecular surface and the body of each bone strut remained similar, with a −6.6±5.0%, −6.5±3.0%, and −3.7±2.1% lower phosphate, protein, and

Table 1. Trabecular and cortical bone parameters in the tibiae of irradiated mice compared to the age-matched control at 2d, 10d, and 8w.

		Tb.TMD (mg/cc)	Ct.Th (mm)	Tt.Ar (mm²)	Ct.Ar (mm²)	Ct.Ar/Tt.Ar
2 Day	Control (n = 7)	769±16	0.19±0.02	0.57±0.07	0.57±0.07	0.994±0.001
	Irradiated (n = 8)	766±15	0.18±0.01	0.54±0.07	0.53±0.07	0.993±0.002
	% Difference	−0.3±2.0	−6.6±7.0	−5.6±12.5	−5.7±12.5	−0.12±0.22
	P-value	0.79	0.22	0.40	0.55	0.19
10 Day	Control (n = 8)	766±14	0.19±0.01	0.56±0.04	0.55±0.04	0.992±0.004
	Irradiated (n = 8)	773±11	0.19±0.01	0.57±0.06	0.57±0.06	0.994±0.000
	% Difference	1.0±1.4	3.3±7.0	2.5±10.4	2.7±10.4	0.13±0.04
	P-value	0.24	0.33	0.58	0.56	0.40
8 Week	Control (n = 8)	801±8.9	0.20±0.01	0.59±0.04	0.58±0.04	0.994±0.001
	Irradiated (n = 7)	823±28	0.21±0.01	0.60±0.02	0.60±0.02	0.994±0.001
	% Difference	2.7±3.5	3.2±4.6	2.4±3.5	2.4±3.5	−0.05±0.06
	P-value	0.09	0.15	0.38	0.39	0.13

By 10d following irradiation, although there was a deficit to the trabecular bone architecture compared to non-irradiated control, no significant differences were apparent in the cortical bone between the irradiated mice and the age-matched control.

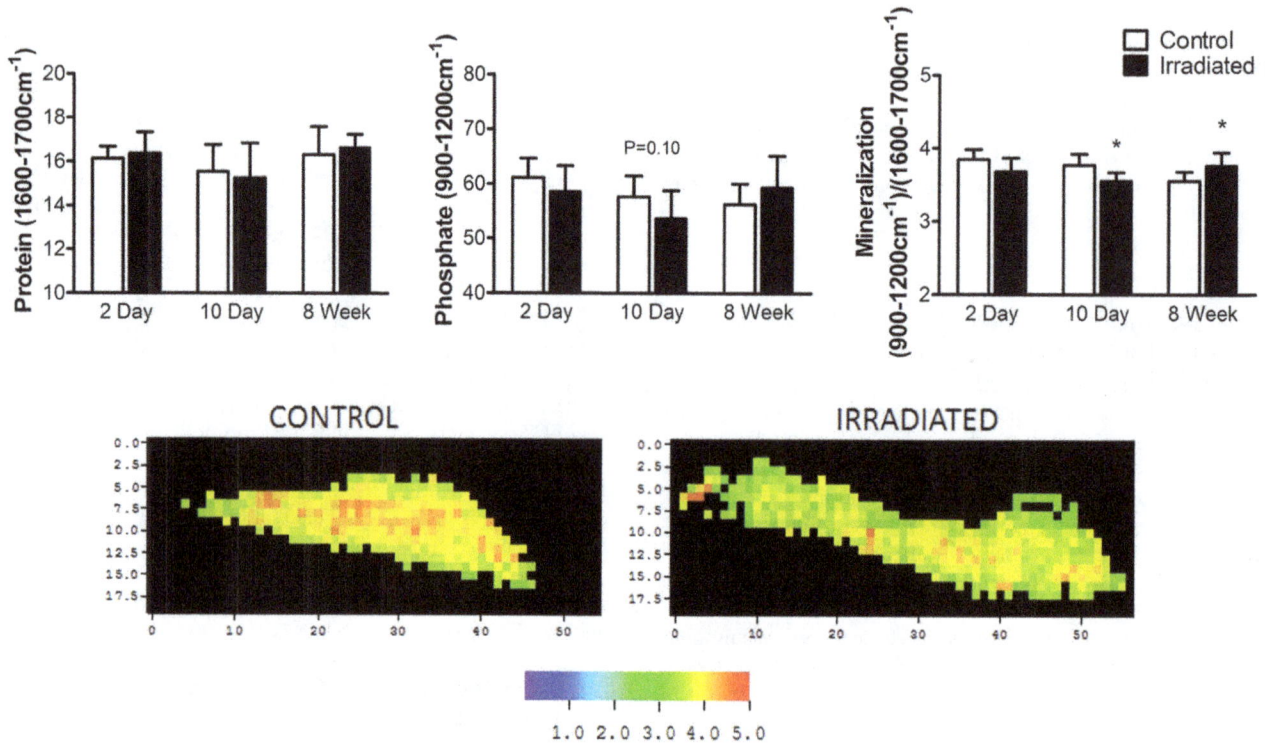

Figure 2. Trabecular bone chemical composition using FTIRI. A) Level of mineralization as a proportion of the inorganic to organic matrix, in the trabecular bone of irradiated mice compared to control at 2d, 10d, and 8w following irradiation. *p<0.05 compared to age-matched control. B) A typical FTIRI heat map of the level of mineralization in a control (left) and an irradiated (right) trabecular strut at 2d. Increased intensity corresponds to increased degree of mineralization.

mineralization on the periphery as compared to the total bone, respectively (p<0.05).

Alteration in Mechanical Properties of Trabecular Bone Following Radiation

As early as 2d following radiation exposure, the elastic modulus of trabecular struts from irradiated mice were 13.7±4.5% higher than control (Irradiated: 15.9±0.6 GPa; Control: 13.9±1.2 GPa), while hardness was 11.5±6.6% higher (Irradiated: 0.78±0.05 GPa; Control: 0.70±0.05 GPa) as compared to control (p<0.05). Even though there was significant trabecular bone loss at 10d, the mechanical properties of the trabeculae from irradiated mice were no longer different from the control at 8w (Fig. 3). Whole bone testing though showed no differences in ultimate force and stiffness between irradiated and control mice at 10d.

Collagen Type-I Release into Circulation

As early as 2d following irradiation, there is a 123% increase in CTX-I in the plasma of irradiated mice compared to control, with 4.1 ng/ml and 9.2 ng/ml of CTX-I in the control and irradiated mice, respectively (Average OD: Irradiated = 0.48±0.28, Control = 0.92±0.05; p<0.05). By 10d, the level of CTX-I in the plasma for irradiated mice remained 100% elevated, at 5.0 ng/ml and 10 ng/ml of CTX-I in the control and irradiated mice, respectively (Irradiated = 0.45±0.25 OD; Control = 0.78±0.19 OD; p<0.05).

Increased Osteoclast Activity Following Irradiation Contributes to Bone Loss Even After Marked Depletion and Proliferative Capability of the Bone Marrow Progenitor Pool

At 2d following irradiation, the total number of cells in the bone marrow of irradiated mice declined by 65±11%, and remained 64±9% depleted at 10d [17]. This cellular depletion at 10d was accompanied by adipocyte infiltration into the marrow, with 6.2±2.9% of the bone marrow compartment occupied by adipocytes compared to only 0.58±0.40% in the control (p<0.05; Fig. 4). By 8w, the adipocytes in the bone marrow remained elevated and had 4.2±0.9% of their bone marrow composed of adipocytes compared to only 1.7±1.1% in the control (p<0.05). The total number of colony forming cells in irradiated mice was also markedly suppressed, showing an −86±8% and −66±9% decline at 2 & 10d relative to control (Fig. 5), indicating a marked reduction in the ability of cells to proliferate. At 2d, the number of CFU-GEMM and CFU-GM were −92±8% and −84±10% lower than control, whereas at 10d, there were −62±11% and −68±9% fewer CFU-GEMM and CFU-GM than control, demonstrating that at 10d the cells remaining were depressed relative to control, but more capable of proliferating than those measured at 2d following irradiation. By 8w, there were still −30±18% fewer CFU-GM cells than control, but the CFU-GEMM were no longer different (p = 0.32). Unlike the decline seen in these hematopoietic progenitor cells, at 2d TRAP5b concentrations in the plasma, used as a measure to quantify osteoclast activity and bone resorption [31], rose 43±35% in the irradiated mice compared to the control

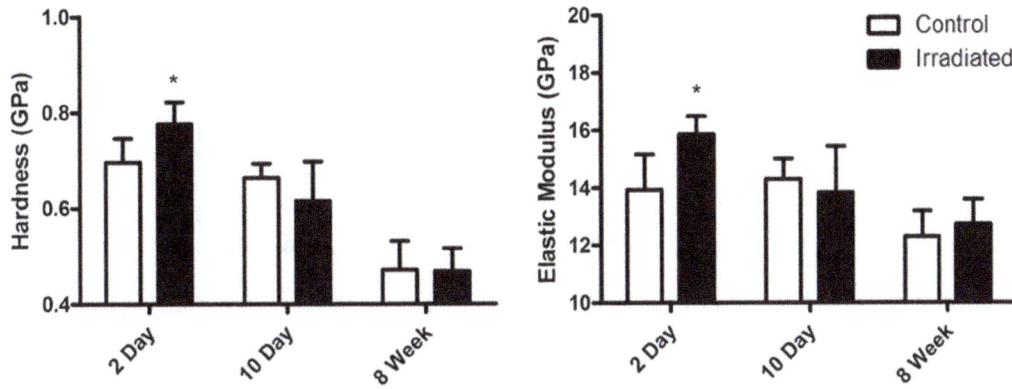

Figure 3. Mechanical properties of trabecular bone at 2d, 10d, and 8w following irradiation. As early as 2d following irradiation, alterations in the trabecular bone led to an increase in hardness and elastic modulus which was no longer present at 10d. *p<0.05 compared to control.

(p<0.05) [17]. The TRAP5b levels remained elevated by 24±33% at 10d (p = 0.06).

Discussion

The rapid and extensive damage to bone as caused by irradiation, markedly diminishing bone quantity and quality and therefore elevating risk of fracture, appears to be the product of two distinct pathways: that driven by biologic processes orchestrated by elevated osteoclastic activity, and that facilitated by physicochemical processes enabled by compositional and material alterations in the composition of the matrix. The physicochemical processes described here refer to the physical and chemical changes to the bone matrix caused directly by the γ-rays. While two days post-irradiation revealed no measurable change in bone quantity as measured by microCT, a −4.3±4.8% (p = 0.07) trend towards a decline in the degree of mineralization of the trabecular bone, defined as the ratio of the phosphate/protein components, was already apparent. Analyzing the protein and phosphate peaks by FTIRI analysis showed this decline in mineralization to be due primarily to a decrease in the inorganic matrix component, representing hydroxyapatite, emphasizing a critical alteration to the material composition immediately following radiation exposure.

Despite a decrease in mineralization, change in the bone's matrix properties as defined by nanoindentation revealed a 13.7±4.5% *increase* in elastic modulus, and 11.5±6.7% *elevation* in hardness in irradiated versus control mice at 2d. These changes occurred prior to visual bone loss. However, considering the rapid decline in bone quantity and mineralization seen at 10d, the elevation in material properties observed as early as 2d suggests that alterations in the matrix composition – specifically the relative increase of collagen relative to hydroxyapatite - makes it more susceptible to erosion, either through resorption achieved by biologically driven (cellular) processes, or by physical attrition, achieved independent of cellular actions. While certainly an extrapolation, perhaps the increase in collagen relative to mineral, as is evident in diseases such as osteomalacia [32], imply why that bone, too, is more susceptible to rapid loss of matrix.

Collagen has been shown to be more vulnerable to irradiation as compared to the inorganic phase of the matrix [14,33], with gamma radiation compromising the collagen structure first by radiolysis of water molecules, which then leads to free radicals and increased cross-linking within the collagen fibrils [14]. The destruction of the organic phase – while contributing to a rise in measured modulus and hardness - somehow makes the matrix more susceptible to erosion *despite* the parallel devastation to the

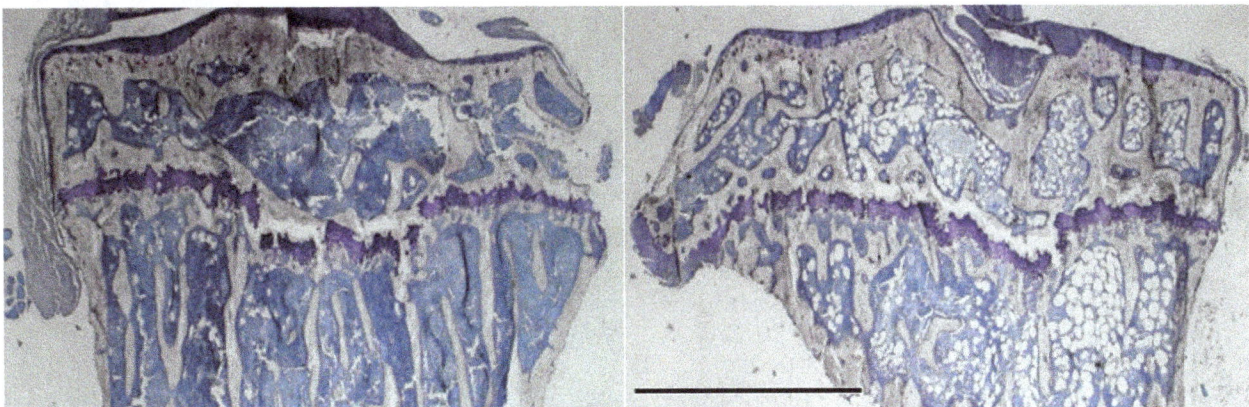

Figure 4. Light microscope image of mouse tibiae stained with Modified Wright Giemsa at 10d following irradiation. Image on the left is a control, and on the right is an irradiated mouse. The empty spaces in the marrow of the irradiated bone correspond to the rapid infiltration of fat cells, a consequence of which is that there is less space for immune cells to occupy the bone marrow space. Scale bar represents 1 mm.

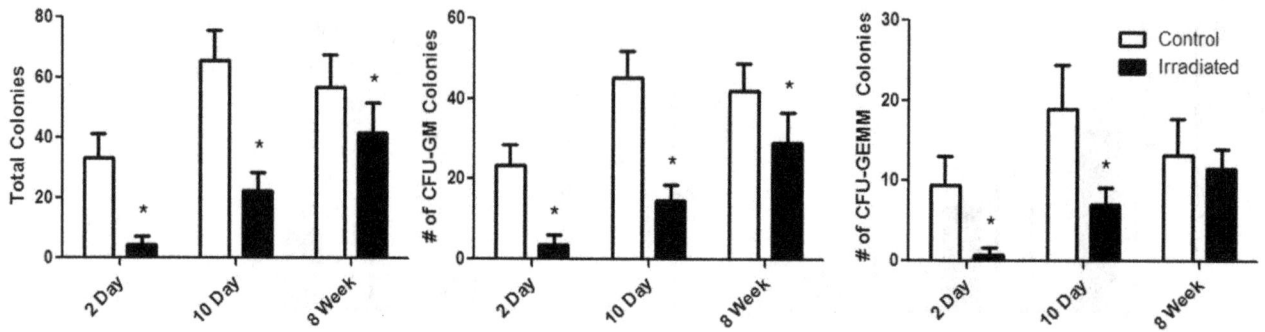

Figure 5. Total number of colony forming cells. CFU-GM, and CFU-GEMM in irradiated mice and age-matched control. Although the cell populations in irradiated mice remain depleted at 10d, they are more proliferative than irradiated cells at 2d and showed greater improvement by 8w. *p<0.05 compared to age-matched control.

osteoclast precursor pool. Perhaps this is enabled by ionizing radiation mediated protein side chain decarboxylation from the phosphate groups bound to the hydroxyapatite crystals [34]. While we were unable to determine the actual physicochemical process of dissolution that were measured here, it does suggest a need to consider chemical – in concert with biologic – processes as contributing to the bone loss, and thus chemical – as well as biologic – strategies to slow this erosion.

By 10d, not only was there a significant depletion of trabecular bone in the irradiated mice, the level of tissue mineralization had declined by −5.8±3.2% relative to control. A question remained, however, whether the decline in tissue mineralization was uniform across the trabecular bone, or if only the periphery of the trabeculae was that which bore the brunt of the irradiation injury. Using the capacity of FTIRI to spatially map mineralization throughout a trabecular strut [35], it became evident that the decline was seen throughout the matrix at both 2d and 10d, implicating the ability of ionizing radiation to instigate physical damage through the trabecular bone struts, and providing further evidence that the increase in osteoclast activity at the bone surface was not the sole contributor to a decline in apparent bone density. However, the PMMA mask created was a limitation to this analysis because it excludes a ∼10 μm region of the trabecular bone surface, which is the region of greatest PMMA infiltration.

It is also important to consider that the bone marrow hematopoietic populations were largely extinguished at 2d, showing little recovery at 10d, reinforcing a conclusion that the loss of bone quantity was not achieved strictly by a biologic process alone. The CFU assay also quantifies the multipotential hematopoietic progenitor cells, and even after bone marrow cells were plated at equal concentrations, there were −86±8% and −66±9% fewer total CFUs at 2d and 10d compared to the control, exemplifying that hematopoietic cells of irradiated mice had inferior proliferative capabilities. The damage to the bone marrow was further exacerbated by immense adipocyte infiltration into the marrow seen at 10d, leading to the inhibition of hematopoietic stem cells necessary to maintain the bone marrow environment [36]. These adipocytes, derived from mesenchymal stem cells, were most likely capable of populating the bone marrow because of the massive deficit of other hematopoietic cell phenotypes [17]. TRAP5b expression though was markedly elevated at 2d, indicating increased osteoclast activity in the irradiated mice as compared to control even as the hematopoietic cells were depleted. Nevertheless, 2d is likely an insufficient period of time for osteoclasts to initiate bone destruction, with at least a

7d period required to increase osteoclast activity and numbers and to induce visible bone resorption in the mouse [37].

FTIRI analysis shows that the inorganic matrix in trabecular bone is more susceptible to radiation damage than the organic matrix following sub-lethal exposure, yet the cortical bone remains intact. The morphology of the two structures is very different, with more surface to volume in trabeculae, while the chemical composition of trabecular and cortical bone are also distinct, as cortical bone is much more mineralized, primarily due to the greater amount of the inorganic matrix. Therefore, it is likely that trabecular bone is more susceptible to radiation induced damage because of its greater surface area and less mineralized properties, leading to a surface more susceptible to dissolution. On the other hand, since the cortical bone is not as chemically sensitive to irradiation compared to trabecular bone, three point bending did not show differences in stiffness and ultimate force between irradiated and control mice at 10d, as these mechanical tests are primarily calculating the mechanical strength of the cortical bone.

While bone quantity had not recovered by 8w, likely due to the inability of the bone marrow to repopulate cellular populations (Fig. 5), irradiated mice exhibited some repair of the quality of the matrix as seen by chemical composition. Indeed, FTIRI analysis showed a 5.8±5.1% *increase* in the level of tissue mineralization of irradiated mice as compared to control (p<0.05), indicating that those trabeculae that remained had restored tissue mineralization properties to the point that they actually exceeded a healthy age-matched control mouse. However the structural properties within the remaining trabeculae were not identical between control and irradiated mice, as seen with lower crystallinity in irradiated mice. The smaller crystals in irradiated mice may be attributed to their more rapid bone turnover rates [17], leaving an insufficient period for their crystals to fully mature. By 8w, the hardness and elastic moduli were reduced for even the control mice, likely due to an age-related decline in mechanical properties [38,39]. No differences in mechanical properties, however, were evident between the irradiated mice and age-matched control by 8w. This reparative process is perhaps best explained by the partial reconstitution of the bone marrow cell populations, as well as the increases in mineral apposition rate and bone formation rate of irradiated mice compared to control at 5w following radiation exposure [17]. This implies that osteoblasts are active in order to compensate for bone deterioration, but are incapable of restoring the structural damage.

Numerous studies have shown the deleterious impact of irradiation on bone [11,40], but this critical outcome has most typically been presumed a result of elevated osteoclast activity.

Here, however, we show that a 5 Gy dose of γ-irradiation, a relatively low dose when compared to those used for implant sterilization, can quickly alter both the chemical and physical makeup of the bone. These changes are likely the result of both biological and physicochemical contributions from osteoclasts and the radiation source. It is important to point out that those patients undergoing tumor ablation – who may receive localized doses greater than 60 Gy [16,41], and bone marrow transplant patients subject to radiation doses greater than 10 Gy [15], are at far greater lifetime risk of fractures due to their bone deterioration. Perhaps the use of pharmacologic agents, such as bisphosphonates, which increase the total mineralization across the bone tissue [42], prior to radiation exposure could help protect bone quantity and quality, but non-intuitively through a process related to protecting the mineral rather than defeating the osteoclast.

Conclusion

A rapid decline in tissue mineralization was seen throughout trabecular morphology of irradiated mice, which preceded a marked collapse of bone quantity. A parallel *increase* in matrix modulus and hardness, when considered in light of the devastation of the cellular population, suggests that bone loss is somehow facilitated by physicochemical changes in the composition of the matrix independent of a hampered biologic state. Ultimately, this study suggests that the bone loss following sub-lethal doses of irradiation, whether it is from intentional exposure used for medical treatments or accidental exposure from environmental disasters, is a result not only from biological processes, but by, chemical and physical changes to the composition of the matrix.

Acknowledgments

We thank all of those who have contributed to this study including the technical assistance of Liangjun Lin, Yi-Xian Qin, Kofi Appiah-Nkansah, Gabriel Pagnotti, Nirukta Patri, Andrea Kwaczala, Alyssa Tuthill, Steven Tommasini, Randy Smith, and Eli Stavitski.

Author Contributions

Conceived and designed the experiments: DEG CTR. Performed the experiments: DEG BJA MEC JJL ASA. Analyzed the data: DEG BJA MEC JJL ASA LMM CTR. Contributed reagents/materials/analysis tools: DEG BJA MEC JJL ASA LMM CTR. Wrote the paper: DEG BJA MEC JJL ASA LMM CTR.

References

1. Cucinotta FA, Durante M (2006) Cancer risk from exposure to galactic cosmic rays: Implications for space exploration by human beings. Lancet Oncol 7: 431–435.
2. Christodouleas JP, Forrest RD, Ainsley CG, Tochner Z, Hahn SM, et al. (2011) Short-term and long-term health risks of nuclear-power-plant accidents. N Engl J Med 364: 2334–2341.
3. Hall EJ, Wuu CS (2003) Radiation-induced second cancers: The impact of 3d-crt and imrt. Int J Radiat Oncol Biol Phys 56: 83–88.
4. van der Sluis IM, van den Heuvel-Eibrink MM, Hählen K, Krenning EP, de Muinck Keizer-Schrama S (2002) Altered bone mineral density and body composition, and increased fracture risk in childhood acute lymphoblastic leukemia. J Pediatr 141: 204–210.
5. Savani BN, Donohue T, Kozanas E, Shenoy A, Singh AK, et al. (2007) Increased risk of bone loss without fracture risk in long-term survivors after allogeneic stem cell transplantation. Biol Blood Marrow Transplant 13: 517–520.
6. Halton JM, Atkinson SA, Fraher L, Webber C, Gill GJ, et al. (1996) Altered mineral metabolism and bone mass in children during treatment for acute lymphoblastic leukemia. J Bone Miner Res 11: 1774–1783.
7. Kondo H, Searby ND, Mojarrab R, Phillips J, Alwood J, et al. (2009) Total-body irradiation of postpubertal mice with cs-137 acutely compromises the microarchitecture of cancellous bone and increases osteoclasts. Radiat Res 171: 283–289.
8. Willey JS, Lloyd SAJ, Robbins ME, Bourland JD, Smith-Sielicki H, et al. (2009) Early increase in osteoclast number in mice after whole-body irradiation with 2 gy×rays. Radiat Res 170: 388–392.
9. Currey JD, Foreman J, Laketić I, Mitchell J, Pegg DE, et al. (1997) Effects of ionizing radiation on the mechanical properties of human bone. J Orthop Res 15: 111–117.
10. Hamer AJ, Strachan JR, Black MM, Ibbotson CJ, Stockley I, et al. (1996) Biomechanical properties of cortical allograft bone using a new method of bone strength measurement - a comparison of fresh, fresh-frozen and irradiated bone. J Bone Joint Surg Br 78B: 363–368.
11. Barth HD, Zimmermann EA, Schaible E, Tang SY, Alliston T, et al. (2011) Characterization of the effects of x-ray irradiation on the hierarchical structure and mechanical properties of human cortical bone. Biomaterials 32: 8892–8904.
12. Fideler BM, Vangsness CT, Lu B, Orlando C, Moore T (1995) Gamma irradiation: Effects on biomechanical properties of human bone-patellar tendon-bone allografts. Am J Sports Med 23: 643–646.
13. Nguyen H, Morgan D, Forwood M (2007) Sterilization of allograft bone: Is 25 kgy the gold standard for gamma irradiation? Cell Tissue Bank 8: 81–91.
14. Akkus O, Belaney RM, Das P (2005) Free radical scavenging alleviates the biomechanical impairment of gamma radiation sterilized bone tissue. J Orthop Res 23: 838–845.
15. Inagaki J, Nagatoshi Y, Sakiyama M, Nomura Y, Teranishi H, et al. (2011) Tbi and melphalan followed by allogeneic hematopoietic sct in children with advanced hematological malignancies. Bone Marrow Transplant 46: 1057–1062.
16. Holt GE, Griffin AM, Pintilie M, Wunder JS, Catton C, et al. (2005) Fractures following radiotherapy and limb-salvage surgery for lower extremity soft-tissue sarcomas. J Bone Joint Surg Am 87: 315–319.
17. Green DE, Adler BJ, Chan ME, Rubin CT (2012) Devastation of adult stem cell pools by irradiation precedes collapse of trabecular bone quality and quantity. J Bone Miner Res 27: 749–759.
18. Boskey AL, Mendelsohn R (2005) Infrared spectroscopic characterization of mineralized tissues. Vib Spectrosc 38: 107–114.
19. Miller LM, Little W, Schirmer A, Sheik F, Busa B, et al. (2007) Accretion of bone quantity and quality in the developing mouse skeleton. J Bone Miner Res 22: 1037–1045.
20. Paschalis EP, Betts F, DiCarlo E, Mendelsohn R, Boskey AL (1997) Ftir microspectroscopic analysis of normal human cortical and trabecular bone. Calcif Tissue Int 61: 480–486.
21. Busa B, Miller L, Rubin C, Qin YX, Judex S (2005) Rapid establishment of chemical and mechanical properties during lamellar bone formation. Calcif Tissue Int 77: 386–394.
22. Oliver WC, Pharr GM (1992) An improved technique for determining hardness and elastic modulus using load and displacement sensing indentation experiments. J Mater Res 7: 1564–1583.
23. Ozcivici E, Ferreri S, Qin Y-X, Judex S (2008) Determination of bone's mechanical matrix properties by nanoindentation. Methods Mol Biol: 323–334.
24. Zysset PK, Edward Guo X, Edward Hoffler C, Moore KE, Goldstein SA (1999) Elastic modulus and hardness of cortical and trabecular bone lamellae measured by nanoindentation in the human femur. J Biomech 32: 1005–1012.
25. Lublinsky S, Ozcivici E, Judex S (2007) An automated algorithm to detect the trabecular-cortical bone interface in micro-computed tomographic images. Calcif Tissue Int 81: 285–293.
26. Paschalis EP (2009) Fourier transform infrared analysis and bone. Osteoporos Int 20: 1043–1047.
27. Acerbo AS, Carr GL, Judex S, Miller LM (2012) Imaging the material properties of bone specimens using reflection-based infrared microspectroscopy. Anal Chem 84: 3607–3613.
28. Boskey A (2006) Assessment of bone mineral and matrix using backscatter electron imaging and ftir imaging. Curr Osteoporos Rep 4: 71–75.
29. Paschalis EP, Verdelis K, Doty SB, Boskey AL, Mendelsohn R, et al. (2001) Spectroscopic characterization of collagen cross-links in bone. J Bone Miner Res 16: 1821–1828.
30. Tai K, Pelled G, Sheyn D, Bershteyn A, Han L, et al. (2008) Nanobiomechanics of repair bone regenerated by genetically modified mesenchymal stem cells. Tissue Eng Part A 14: 1709–1720.
31. Kirstein B, Chambers TJ, Fuller K (2006) Secretion of tartrate-resistant acid phosphatase by osteoclasts correlates with resorptive behavior. J Cell Biochem 98: 1085–1094.
32. Faibish D, Gomes A, Boivin G, Binderman I, Boskey A (2005) Infrared imaging of calcified tissue in bone biopsies from adults with osteomalacia. Bone 36: 6–12.
33. Hamer AJ, Stockley I, Elson RA (1999) Changes in allograft bone irradiated at different temperatures. J Bone Joint Surg Br 81B: 342–344.
34. Hübner W, Blume A, Pushnjakova R, Dekhtyar Y, Hein HJ (2005) The influence of x-ray radiation on the mineral/organic matrix interaction of bone tissue: An ft-ir microscopic investigation. Int J Artif Organs 28: 66–73.
35. Carden A, Morris MD (2000) Application of vibrational spectroscopy to the study of mineralized tissues (review). J Biomed Opt 5: 259–268.

36. Naveiras O, Nardi V, Wenzel PL, Hauschka PV, Fahey F, et al. (2009) Bone-marrow adipocytes as negative regulators of the haematopoietic microenvironment. Nature 460: 259–263.

37. Boyce BF, Aufdemorte TB, Garrett IR, Yates AJP, Mundy GR (1989) Effects of interleukin-1 on bone turnover in normal mice. Endocrinology 125: 1142–1150.

38. Hamrick MW, Ding K-H, Pennington C, Chao YJ, Wu Y-D, et al. (2006) Age-related loss of muscle mass and bone strength in mice is associated with a decline in physical activity and serum leptin. Bone 39: 845–853.

39. Wang X, Shen X, Li X, Mauli Agrawal C (2002) Age-related changes in the collagen network and toughness of bone. Bone 31: 1–7.

40. Barth HD, Launey ME, MacDowell AA, Ager Iii JW, Ritchie RO (2010) On the effect of x-ray irradiation on the deformation and fracture behavior of human cortical bone. Bone 46: 1475–1485.

41. Longhi A, Ferrari S, Tamburini A, Luksch R, Fagioli F, et al. (2012) Late effects of chemotherapy and radiotherapy in osteosarcoma and ewing sarcoma patients. Cancer 118: 5050–5059.

42. Boivin G, Meunier PJ (2002) Effects of bisphosphonates on matrix mineralization. J Musculoskelet Neuronal Interact 2: 538–543.

14

Bone Plasticity in Response to Exercise Is Sex-Dependent in Rats

Wagner S. Vicente[1], Luciene M. dos Reis[1], Rafael G. Graciolli[1], Fabiana G. Graciolli[1], Wagner V. Dominguez[1], Charles C. Wang[5], Tatiana L. Fonseca[3], Ana P. Velosa[4], Hamilton Roschel[2], Walcy R. Teodoro[4], Bruno Gualano[2], Vanda Jorgetti[1]*

1 Nephrology Division, Medical School, University of São Paulo, São Paulo, Brazil, 2 Department of Sports, School of Physical Education and Sport, University of Sao Paulo, Sao Paulo, Brazil, 3 Department of Anatomy, Institute of Biomedical Sciences, University of Sao Paulo, São Paulo, Brazil, 4 Rheumatology Division, Medical School, University of São Paulo, São Paulo, Brazil, 5 Department of Physiological Sciences, Federal University of São Carlos, São Paulo, Brazil

Abstract

Purpose: To characterize the potential sexual dimorphism of bone in response to exercise.

Methods: Young male and female Wistar rats were either submitted to 12 weeks of exercise or remained sedentary. The training load was adjusted at the mid-trial (week 6) by the maximal speed test. A mechanical test was performed to measure the maximal force, resilience, stiffness, and fracture load. The bone structure, formation, and resorption were obtained by histomorphometric analyses. Type I collagen (COL I) mRNA expression and tartrate-resistant acid phosphatase (TRAP) mRNA expression were evaluated by quantitative real-time PCR (qPCR).

Results: The male and female trained rats significantly improved their maximum speed during the maximal exercise test (main effect of training; $p < 0.0001$). The male rats were significantly heavier than the females, irrespective of training (main effect of sex; $p < 0.0001$). Similarly, both the weight and length of the femur were greater for the male rats when compared with the females (main effect of sex; $p < 0.0001$ and $p < 0.0001$, respectively). The trabecular volume was positively affected by exercise in male and female rats (main effect of training; $p = 0.001$), whereas the trabecular thickness, resilience, mineral apposition rate, and bone formation rate increased only in the trained males (within-sex comparison; $p < 0.05$ for all parameters), demonstrating the sexual dimorphism in response to exercise. Accordingly, the number of osteocytes increased significantly only in the trained males (within-sex comparison; $p < 0.05$). Pearson's correlation analyses revealed that the COL I mRNA expression and TRAP mRNA expression were positively and negatively, respectively, related to the parameters of bone remodeling obtained from the histomorphometric analysis ($r = 0.59$ to 0.85; $p < 0.05$).

Conclusion: Exercise yielded differential adaptations with respect to bone structure, biomechanical proprieties, and molecular signaling in male and female rats.

Editor: Edward E Schmidt, Montana State University, United States of America

Funding: The authors have no support or funding to report.

Competing Interests: The authors have declared that no competing interests exist.

* E-mail: vandajor@usp.br

Introduction

The notion that "senile osteoporosis is a pediatric disease" has been increasingly accepted [1,2,3]. This suggestion relies on the observation that physical inactivity during growth has been related to an increased prevalence of osteoporosis-related fractures [2,4,5,6]. In fact, mechanical strain exerts a pivotal role in skeletal growth and modeling. Both *in vitro* and *in vivo* studies have demonstrated that mechanical strain modifies intracellular bone signaling, partially by enhancing the expression of growth factors [7]. The mechanical stimuli-induced intracellular signaling has been suggested to potentially favor bone proliferation and the formation of bone matrix [7], although the actual mechanisms underlying these responses remain to be fully elucidated.

Exercise has been recognized as one of the most effective strategies to counteract low bone mass. Nonetheless, preliminary evidence suggests that the bone response to exercise may be sex-dependent. Differential bone plasticity in males and females may be rooted in the biologic properties of bone [1,8]. In support of this concept, there is evidence suggesting sex-specific differences in the number of osteoprogenitor cells, hormone responses, and hormone regulation [9,10,11], which could potentially influence the normal growth of bone and its response to a given stimulus (e.g., dietary intervention, pharmacological treatment and physical activity). In humans, sex differences in bone size and strength have been suggested to be established in puberty as a result of the greater endocortical and periosteal expansion during the pre-pubertal years and the minimal endocortical contraction in males compared with the high endocortical contraction and inhibition of periosteal apposition in females after the pubertal growth spurt [1,12,13,14]. As a consequence, volumetric density remains constant during growth and similar in both sexes, whereas the bone mass content (BMC) is approximately 20% higher in males than in females at the end of puberty as their bones are larger [1]. Thus, sex-related differences in bone strength are the result of the differences in shape and geometry [1,15]. Recently, rodent studies have provided a growing body of evidence indicating that sex hormones and their receptors have an impact on the mechanical sensitivity of the growing skeleton. According to Callewaert et al. [8], androgen receptor and estrogen receptor-β signaling may attenuate the osteogenic response to mechanical strain in males and females, respectively, whilst estrogen receptor-α may stimulate the response of bone to mechanical strain in the female skeleton. Altogether, these findings support the hypothesis that skeletal sexual dimorphism in response to exercise may exist.

Thus, the objective of this study was to assess the potential sex-based differential bone plasticity in response to exercise in young rats.

Materials and Methods

Ethics Statement

All of the experimental procedures were approved by the local Research Ethics Committee (CAPPesq, permit number 806/05) and developed in strict conformity with our institutional guidelines and with international standards for the manipulation and care of laboratory animals. All of the surgeries were performed under sodium pentobarbital anesthesia, and all efforts were made to minimize suffering.

Experimental Protocol

Male and female 12-week-old Wistar rats, obtained from our local breeding colony with initial weights ranging from 200 to 300 g, were used in this study. The animals were housed in individual cages in a light-controlled environment (12/12-h light/dark cycle) at a constant temperature (22°C) and humidity (25%), with free access to standard laboratory chow (Nuvital Nutrientes S/A, Curitiba, Brazil). After 7 days of acclimation, the animals were randomly divided into four groups as follows: *i)* female sedentary, *ii)* female trained, *iii)* male sedentary, and *iv)* male trained.

The animals were trained by using a treadmill training protocol adapted from Brooks et al. [16] and Ferreira et al. [17]. At the beginning of the training, all of the animals were submitted to a maximal speed test [16,17]. The test was performed on a treadmill (ESD model 01; Funbec, São Paulo, Brazil) at an initial rate of 6 m/min with no slope. The rate was increased by 3 m/min every 3 min until exhaustion. The animals were judged to be exhausted when they could no longer maintain an upright position in the treadmill or continue running at the required pace. The training protocol intensity was set at 60% of the maximal velocity achieved in the test and was performed 5 times a week for 12 weeks. A mid-trial (i.e., week 6) maximal test was performed to adjust the training load. At the end of 12 weeks, all of the animals were submitted to a final maximal test and then anesthetized and euthanized through aortic exsanguination. The sedentary animals remained cage-confined without any physical training program throughout the protocol.

Biomechanical Characterization

The right femur of each rat was removed, dissected free from the soft tissue and placed in saline 9% to - 80°C until mechanical testing. 12 hours before the test, frozen bones were thawed at room temperature and kept in saline. They were tested mechanically through a three-point bending protocol using an Instron Universal Testing machine, model 4444 (Instron Corp.,

Canton, MA, USA). The span of the supports was 21.7 mm, and the deformation rate was 0.5 cm/min. The load-deformation curve was analyzed, and the maximal force, resilience, stiffness and fracture load were calculated from the plot using specific software [18,19].

Bone Histomorphometry

A fluorochrome bone marker (i.e., oxytetracycline) was injected (25 mg/kg ip) on days 12 and 11 and on days 5 and 4 prior to the euthanization of the animals.

The length of the left femur (free from soft tissue) was measured using a caliper, weighed, immersed in 70% ethanol, and processed as previously described [20]. Using a Polycut S equipped with a tungsten carbide knife (Leica, Heidelberg, Germany), the undecalcified distal femurs were cut into sections of 5 μm and 10 μm in thickness. The 5-μm sections were stained with 0.1% toluidine blue, pH 6.4, and coverslips with mounting medium Entellan® (Merck, Darmstadt, Germany) and at least two nonconsecutive sections were examined for each sample. Static, structural and dynamic parameters of bone formation and resorption were measured at the distal metaphyses (magnification, ×250), 195 μm from the epiphyseal growth plate, in a total of 30 fields using a semi-automatic image analyzer and the software Osteomeasure (Osteometrics, Inc., Atlanta, GA, USA) specific for bone histomorphometry. The static and structural histomorphometric indices included the ratio of trabecular bone volume to total bone volume (BV/TV), the ratio of osteoid volume to total bone volume (OV/BV), and the osteoid thickness (O.Th). The percentage of the total trabecular surface was used to express the areas of eroded surface (ES/BS) and osteoid surface (OS/BS). We also determined the number of osteoblasts (N.Ob/T.Ar) and the number of osteoclasts (N.Oc/T.Ar) per tissue area, the number of osteocyte-occupied lacunae (stained) per bone area (N.Ot/B.Ar), the trabecular thickness (Tb.Th), the trabecular separation (Tb.Sp), and the trabecular number (Tb.N). Moreover, the mineral apposition rate (MAR) was determined by calculating the distance between the two oxytetracycline labels divided by the time interval between the two oxytetracycline administrations. The mineralizing surface (MS/BS), which is the rate of cancellous surface that has been mineralized, was calculated as the double-labeled surface plus one-half of the single-labeled surface. The evaluation of the rate of the total trabecular surface that was double-labeled and the bone formation rate (BFR/BS) completed the dynamic evaluation. All of the data were obtained in a blinded fashion. The histomorphometric indices are reported using the nomenclature recommended by the American Society for Bone and Mineral Research [21].

Real-time PCR (qPCR)

The right tibias of each rat were removed, dissected, and crushed in a steel mortar and pestle set (Fischer Scientific International, Inc., Hampton, NH) that was pre-cooled in dry ice. Total RNA was extracted using Trizol (Invitrogen, Calbard, CA, USA), following the manufacturer's instructions. Total RNA was reverse transcribed using RevertAid-H-Minus M-MuLV Reverse Transcriptase (Fermentas, Hanover, MD, USA) to synthesize the first strand cDNA, which was used as a template. The mRNA expression of proteins was determined by qPCR using the ABI Prism 7500 sequence detector (Applied Biosystems), as previously described [22]. The selected genes were Type I collagen (COL I) and tartrate-resistant acid phosphatase (TRAP). COL I was selected as it is a protein produced by the bone-forming cells, osteoblasts; whereas TRAP is an enzyme produced by the bone-resorbing cells, osteoclasts. The analysis was performed using a total volume of 20 µl containing the template (5 ng for Type I collagen and 10 ng for TRAP) and primers (450 ng). The primers used in this study were designed using the Primer Express software (Applied Biosystems) (Table 1) and synthesized (Integrated DNA Technologies, Coralville, IA, USA) specifically for qPCR. All of the Ct values were normalized using an internal control (β-actin mRNA). The relative gene expression quantification was assessed by the ΔΔCt method, as previously described [23]. The final values for the samples are reported as fold-induction relative to the expression of the control, with the mean control value being arbitrarily set to 1.

Statistical Analysis

The results are expressed as the mean ± SD (unless otherwise noted). A two-way ANOVA was performed for each dependent variable, assuming sex and training as fixed factors and the rats as a random factor (SAS® 9.2, Cary, NC, USA). Whenever a significant F-value was obtained, a *post-hoc* test with a Tukeys adjustment was performed for multiple comparison purposes. The associations between the mechanical tests, histomorphometric parameters, COL I protein expression, and qPCR grouped by gender were estimated using Pearson or Spearman correlations, depending on the distribution of the variables. The significance level was set at $p < 0.05$.

Results

The body weight analysis revealed that the male rats were significantly heavier than the females, irrespective of training (main effect of sex; $p < 0.0001$). Similarly, both the weight and length of the femur were greater in the male rats when compared with the females, irrespective of training (main effect of sex; $p < 0.0001$ and $p < 0.0001$, respectively) (Table 2).

The maximum speed achieved by the animals was similar between the sexes at the beginning of the study. The male and female trained rats significantly improved their maximum speed during the test when compared with both their initial values ($p < 0.0001$) and with their sedentary counterpart values at the end of the study ($p < 0.0001$), demonstrating the efficacy of the training program in improving the exercise capacity (Table 2).

Table 3 describes the results of the mechanical parameters of the femur. The maximal force and stiffness were significantly lower in the female rats than in the males, irrespective of training (main effect of sex; $p < 0.0001$ and $p < 0.0001$, respectively). The resilience was positively affected by exercise in the males ($p = 0.04$; within-sex comparison) but not in the females ($p = 0.89$; within-sex comparison). Conversely, the exercise did not affect the fracture load in any of the groups (female: $p = 0.83$; male: $p = 0.89$; within-sex comparisons). However, the males presented a higher fracture load than the females, irrespective of training (main effect of sex; $p < 0.0001$).

With respect to the structural parameters of the histomorphometric analysis, BV/TV was significantly greater in the females than in the males, irrespective of training ($p < 0.001$; main effect of sex). Importantly, the exercise training was effective in changing this parameter in both sexes (female: $p = 0.005$; male: $p = 0.005$; within-sex comparisons) (Figures 1A–1D). The Tb.N was significantly higher in the females when compared with males, and the exercise training did not affect this parameter (main effect of sex; $p < 0.0001$). Similarly, the Tb.Sp was not affected by training, but a main effect of sex was observed ($p < 0.0001$), with the males presenting higher values than the females. In contrast, Tb.Th was positively changed by exercise in the males ($p = 0.0001$; within-sex comparison) but not in the females ($p = 0.1$; within-sex comparison). N.Ot/B.Ar was positively affected by training only in the male rats ($p = 0.002$ within-sex comparison). The analysis of the formation parameters revealed that the male rats presented a higher MAR compared with the females (main effect of sex; $p = 0.0004$). Exercise training was effective in changing not only the MAR but also the BFR/BS in the males ($p = 0.01$ and $p = 0.03$, respectively; within-sex comparisons) but not in the females ($p = 0.86$ and $p = 0.91$, respectively; within-sex comparisons) (Figures 1E–1H). Finally, the analysis of the resorption parameters showed that a main effect of sex was found for Oc.S/BS, and ES/BS ($p < 0.0001$ and $p < 0.0001$, respectively), with the males presenting higher values for both variables than the females, regardless of training (Table 4).

Corroborating the histomorphometric parameters of BV/TV, we found a significant increase in the mRNA expression of COL I (124%) in the trained male rats compared with their sedentary counterparts ($p < 0.05$) (Figure 2A). We also observed a lower osteoclastic activity in both the female and male trained rats. The mRNA expression of TRAP decreased 35% in the trained females ($p < 0.05$) and 45% in the trained males ($p < 0.05$) (Figure 2B).

The correlation analysis between the mRNA expressions of COL I and TRAP and the histomorphometric parameters revealed some significant associations. Positive associations between COL I mRNA expression and BV/TV ($r = 0.77$, $p = 0.0147$), Tb.Th ($r = 0.79$, $p = 0.0114$), and BFR/BS ($r = 0.85$, $p = 0.0034$) were found in males. Conversely, no associations were observed for any of the parameters in the female rats ($p > 0.05$). Additionally, we observed a negative association between TRAP mRNA expression and BV/TV ($r = -0.59$, $p = 0.0439$) and a positive association with Tb.Sp ($r = 0.75$, $p = 0.0052$) and Tb.N ($r = 0.66$, $p = 0.0207$) in the female rats. The male rats displayed negative associations between the mRNA expression of TRAP and BV/TV ($r = -0.65$, $p = 0.0159$), Tb.Th ($r = -0.71$, $p = 0.0063$), MAR ($r = -0.63$, $p = 0.0220$) and BFR/BS ($r = -0.71$, $p = 0.0063$). Figure 3 depicts a summary of the sex-specific exercise-induced changes in bone-related general characteristics, and mechanical, histomorphometric, and molecular parameters.

Table 1. Description and sequences of the genes selected for the study.

Gene	NM	Forward (F)	Reverse (R)
Type I collagen	053304	GCGAAGGCAACAGTCGAT	CTTGGTGGTTTTGTATTCGATGAC
TRAP	019144	GTTCCAGGAGACCTTTGAGGA	TCCAGCCAGCACGTACCA
β-actin	031144	AAGATTTGGCACCACACTTTCTACA	CGGTGAGCAGCACAGGGT

NM: NCBI accession number; TRAP: tartrate-resistant acid phosphatase.

Table 2. General characteristics of the animals.

Parameter	FEMALE		MALE	
	Sedentary (n = 10)	Trained (n = 11)	Sedentary (n = 10)	Trained (n = 9)
Initial body weight (g)*	224.9±9.9	224.5±10.2	325.9±19.4	323.3±35.1
Final body weight (g)*	252.9±17.2	241.7±10.6	448.8±35.7	441.4±32.4
Left femur body (g/100 g of body weight)*	0.75±0.05	0.78±0.06	1.1±0.09	1.1±0.6
Length of the left femur (cm)*	3.3±0.1	3.3±0.1	4.5±0.4	4.4±0.2
Initial maximum speed (m/min)	23.0±4.9	28.0±2.0	22.5±4.9	27.0±2.7
Final maximum speed (m/min)	27.0±5.8	49.0±4.9#	28.0±3.1	51.5±3.9#

*indicates main effect of sex (p<0.0001);
#indicates p<0.05 for within-sex comparisons with both the initial values and with the final speed of the sedentary animals.

Discussion

Although exercise has been recognized as one of the most effective strategies to promote bone mass accrual, whether sexual dimorphism exists in response to exercise remains unknown. To gather knowledge on this topic, we investigated the effects of an exercise training protocol upon bone plasticity in young male and female rats. Collectively, our data showed that exercise yielded differential adaptations with respect to the bone structure, biomechanical properties, and molecular signaling in the bone of male and female rats.

Skeletal sexual dimorphism in bone metabolism is mainly attributed to stimulatory androgen action as opposed to inhibitory estrogen action on periosteal bone formation [8]. Importantly, skeletal growth is stimulated by mechanical loading [24], which in turn may also be influenced by sex hormones. For example, exercise may stimulate periosteal bone formation, whilst estrogen appears to have an inhibitory effect in female mice [25]. Furthermore, estrogen supplementation was able to suppress the periosteal response to mechanical loading in male rats [26]. There is evidence suggesting that estrogen and mechanical strain may also share common signaling pathways, stimulating proliferation independently in osteoblast-like cells derived from male and female rats as well as in human osteoblast cells [27]. Interestingly, estrogen receptor modulators and antagonists block the increase in proliferation in response to mechanical strain, whereas dihydro-testosterone and androgen receptor activation are apparently not involved. These findings suggest that estrogen, but not androgen receptors, influence the response to strain. Together, these findings suggest that sex hormones and their receptors may account for the difference in the mechanical response to loading between sexes observed in the current study.

Even though our findings revealed that trabecular volume was positively affected by exercise in both male and female rats (main effect of training), the trabecular thickness, resilience, mineral apposition rate, and bone formation rate increased only in the trained males, demonstrating the sexual dimorphism in response to exercise. Corroborating these observations, the number of osteocytes increased significantly only in trained males. Osteocytes are the most frequent bone cells that are able to connect to themselves and others (e.g., osteoblasts and osteoclasts) through dendritic processes, creating an extensive net that can "sense" the mechanical stimuli applied to the bone [28,29,30,31]. Interestingly, mechanical loading has been demonstrated to prevent osteocyte apoptosis, possibly through the Wnt/β-catenin pathway [28]. Supporting this concept, Robling et al. [32] demonstrated that mechanical stimulation reduced the osteocyte expression of sclerostin, a negative regulator of the Wnt/β-catenin pathway, thereby increasing bone mass. You et al. [29] demonstrated that osteocytes stimulated by mechanical strain yielded alterations in the precursors of the bone marrow and in pre-osteoblasts, decreasing the formation of osteoclasts and potentially bone resorption. Therefore, one may infer that exercise-induced osteocyte activation may lead to not only increased bone formation but also decreased bone resorption. Despite the significant changes in the parameters of bone resorption (see Table 4), this speculation is only partially corroborated by our molecular findings (possibly due to large variabilities in the data), which indicated that exercise induced an increase in the COL I mRNA expression (only in males) and a decrease in the TRAP mRNA expression (in both sexes), which are genes involved in bone formation and resorption, respectively. Interestingly, in line with their attributed molecular roles, the COL I mRNA expression and TRAP mRNA expression were positively and

Table 3. Mechanical property parameters of the right femur of the sedentary and trained animals.

Parameter	FEMALE		MALE	
	Sedentary (n = 10)	Trained (n = 11)	Sedentary (n = 10)	Trained (n = 9)
Maximal force (N)*	101.5±10.5	100.9±14.1	134.5±21.2	146.9±23.2
Resilience (10^{-3} Joule)* #	0.036±0.01	0.042±0.01	0.039±0.02	0.062±0.02
Stiffness (10^3 N/m)*	247.5±34.9	249.5±35.3	318.9±26.0	344.5±36.5
Fracture load (k/N)*	79.2±36.5	89.9±20.5	114.3±31.6	105.0±28.9

*indicates main effect of sex (p<0.05) # indicates main effect of training (p=0.01).

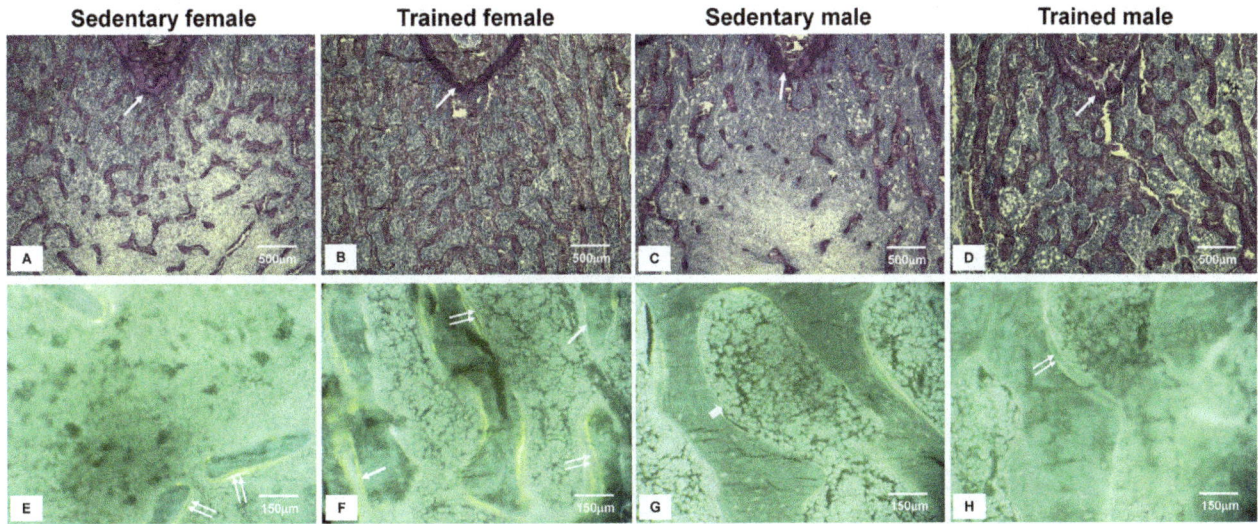

Figure 1. Illustrative bone histological characteristics of sedentary and trained animals. 1A–1D. Undecalcified Bone: Characteristic light microscopy aspects of trabecular bone (femoral metaphysis). Toluidine blue staining showing an increase in the trabecular bone volume (BV/TV) and trabecular thickness (Tb.Th) in the trained animals (B and D) compared with their sedentary counterparts (A and C). The epiphyseal growth plate is indicated by arrows. Histomorphometric analyses were performed at 195 μm under the epiphyseal growth plate (Magnification, x40). 1E–1H. Double oxytetracycline labeling: Characteristic fluorescent light microscopy of undecalcified bone (femoral metaphysis). Unstained bone sections under UV light of the sedentary (E and G) and trained (F and H) animals. Single and double labels are indicated by the single and double arrows, respectively. By quantifying the distance between the oxytetracycline double-labels, we observed that the trained males (H) presented a greater mineral apposition rate (MAR) than the sedentary males (G) and trained females (F). By evaluating the percentage of the trabecular bone surface that was double-labeled, we calculated the bone formation rate (BFR/BS), which was increased only in the trained males (H) (Magnification, x250). Details of the histomorphometric results can be found in Table 4.

Table 4. Histomorphometric analysis of the trabecular bone parameters in the femur.

	FEMALE		MALE	
	Sedentary (n = 10)	Trained (n = 11)	Sedentary (n = 10)	Trained (n = 9)
Structural parameter				
Trabecular Volume (BV/TV, %)*	37.1±3.6	43.5±5.3[#]	30.0±3.0	36.0±4.0[#]
Trabecular Number (Tb.N,/mm)*	6.1±0.3	6.3±0.5	5.1±0.5	4.8±0.7
Trabecular Thickness (Tb.Th, μm)	61.8±8.0	69.0±5.1[#]	59.0±6.2	76.0±11.0[#]
Trabecular Separation (Tb.Sp, μm)*	103.0±5.0	91.0±16.0	141.0±21.0	135.0±23.0
Number of Osteocytes (N.Ot/B.Ar,/mm²)	407.2±116.7	389.2±52.4	486.4±79	552.6±133[#]
Formation parameter				
Osteoid Thickness (O.Th, μm)	1.2±0.1	1.4±0.2	1.2±0.1	1.3±0.2
Osteoid surface (OS/BS, %)	4.8±3.0	6.0±3.4	5.0±2.2	5.6±1.4
Osteoblast surface (Ob.S/BS, %)	3.6±2.4	4.7±2.7	3.9±2.5	4.7±1.1
Osteoblast number (N.Ob/T.Ar,/mm²)	21.0±9.9	25.0±14.0	16.0±9.2	20.0±8.5
Mineralizing surface (MS/BS, %)	3.7±2.0	4.5±1.5	3.7±1.4	4.5±0.9
Mineral apposition rate (MAR, μm/day)*	0.53±0.2	0.59±0.1	0.64±0.1	0.88±0.1[#]
Bone formation rate (BFR/BS, μm³/μm²/day)	0.02±0.02	0.02±0.01	0.02±0.005	0.04±0.008[#]
Resorption parameter				
Eroded Surface (ES/BS, %)*	4.4±1.96	3.4±1.6	6.3±3.0	6.5±1.8
Osteoclast Surface (Oc.S/BS, %)*	0.93±0.41	0.85±0.53	1.56±1.0	1.57±0.46
Osteoclast number (N.Oc/T.Ar,/mm²)	3.62±1.5	3.31±2.1	4.37±2.2	4.32±1.4

*indicates main effect of sex (p<0.05);
[#]indicates p<0.05 for within-sex comparisons (sedentary vs. trained animals).

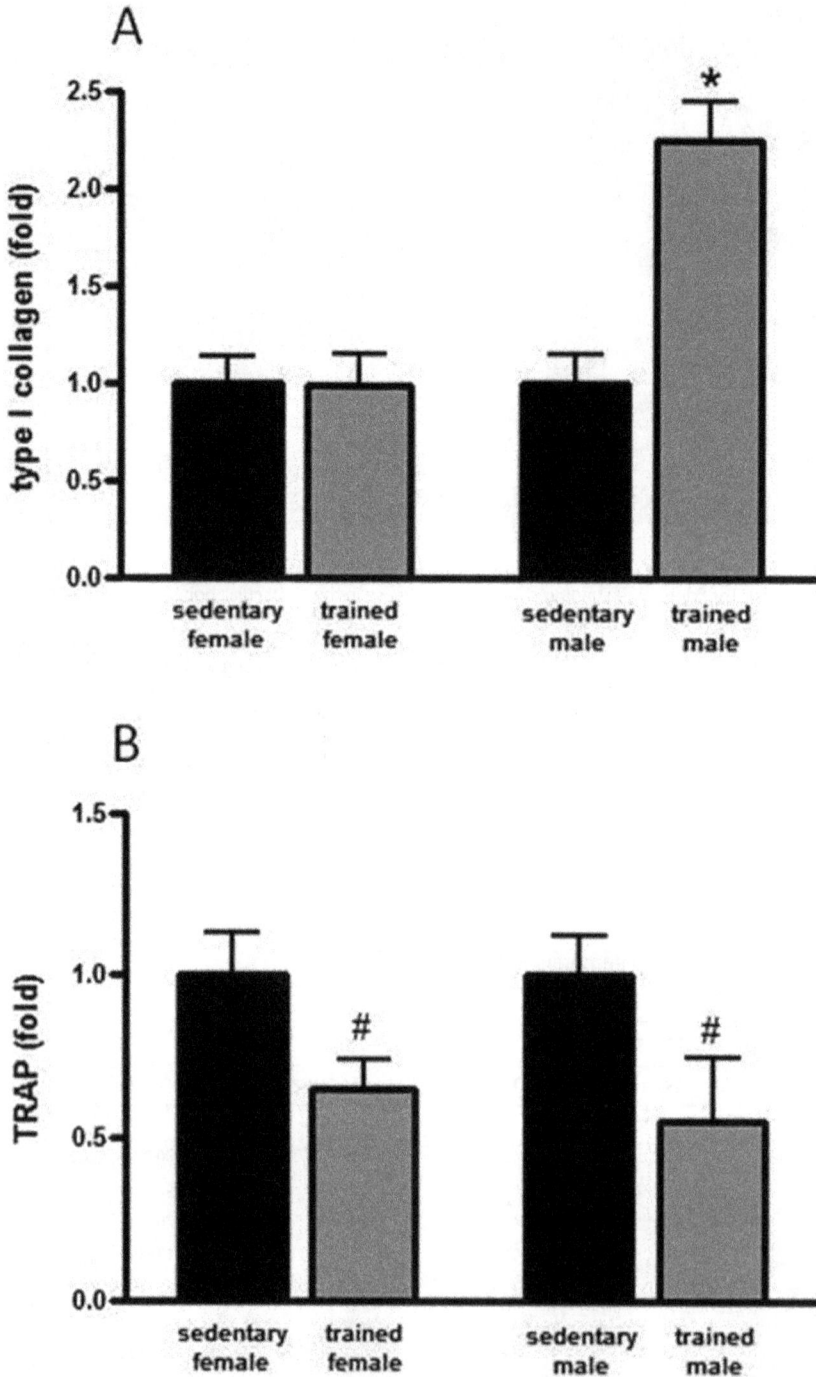

Figure 2. mRNA expression of COL I (panel A) and TRAP (panel B) (data are expressed as the mean and SEM). *indicates $p<0.05$ when compared with their sedentary counterparts (within-sex comparisons).

negatively, respectively, associated with the parameters of bone remodeling (e.g., BV/TV, Tb.N, Tb.Th, MAR, and BFR/BS).

Compared with the female rats, the males had a larger body mass; bigger, thicker and heavier femurs; and greater bone maximal force and stiffness (main effect of sex). In contrast, the trabecular volume was significantly greater in females. The sexual dimorphism related to these bone morphological and biomechanical parameters may be interpreted as a differential developmental feature in our growing rats, reinforcing previous observations [8].

In conclusion, we showed that exercise may elicit differential adaptations with respect to the bone structure, biomechanical proprieties, and molecular signaling in male and female growing rats. Future experimental studies should consider the sexual dimorphism of bone in response to exercise in their study design and data interpretation.

PARAMETER	Sex-specific Exercise-induced Changes			
	Female		Male	
	Within-Sex (sedentary vs. trained)	Between-Sex	Within-Sex (sedentary vs. trained)	Between-Sex
General Characteristics				
Initial Body Weight				↑
Final Body Weight				↑
Left femur body				↑
Length of the left femur				↑
Final Speed			↑	
Mechanical properties				
Maximum Force				↑
Resilience			↑	
Stiffness				↑
Fracture load				↑
Histomorphometric analysis				
Structural				
Trabecular Volume (BV/TV)	↑	↑	↑	
Trabecular Number (Tb.N)		↑		
Trabecular Thickness (Tb.Th)	↑		↑	
Trabecular Separation (Tb.Sp)				↑
Number of Osteocytes (N.Ot/B.Ar)			↑	
Formation				
Mineral Apposition Rate (MAR)			↑	↑
Bone Formation Rate (BFR/BS)			↑	
Resorption				
Eroded Surface (ES/BS)				↑
Osteoclast Surface (Oc.S/BS)				↑
Molecular Biology				
Type I collagen (COL I)			↑	
Tartrate-resistant acid phosphatase (TRAP)	↓		↓	

Figure 3. Summary of the sex-specific exercise-induced changes in bone-related general characteristics, and mechanical, histomorphometric, and molecular parameters. The "within-sex" column indicates the differences induced by exercise training either in female or male rats. The "between-sex" column indicates the differences in sex irrespective of training (i.e. main effect of sex).

Acknowledgments

The authors acknowledge the assistance provided by Cristina Siqueira in the translation and editing of the text and the technical assistance provided by Rosimeire Aparecida Bizerra da Costa.

Author Contributions

Conceived and designed the experiments: WSV BG VJ. Performed the experiments: WSV LMDR RGG FGG CCW TLF APV WRT. Analyzed the data: WSV WVD HR. Contributed reagents/materials/analysis tools: CCW TLF AP. Wrote the paper: WSV BG VJ.

References

1. Vicente-Rodriguez G (2006) How does exercise affect bone development during growth? Sports Med 36(7): 561–569.
2. Martin AD, Mcculloch RG (1987) Bone dynamics: stress, strain and fracture. J Sports Sci 5: 155–163.
3. Peng Z, Tuukkanen J, Zhang H, Jamsa T, Vaananen HK (1994) The mechanical strength of bone in different rats models of experimental osteoporosis. Bone 15: 523–532.
4. American College of Sports Medicine (ACSM) (2009) Exercise and physical activity for older adults. Med Sci Sports Exerc 41(7): 1510–1530.
5. Garber CE, Blissmer B, Desehenes MR, Franklin BA, Lamonte MJ, et al. (2011) Quantity and quality of exercise for developing and maintaining cardiorespiratory, musculoskeletal and neuromotor fitness in apparently healthy adults: guidance for prescribing exercise. Med Sci Sports Exerc 43(7): 1334–1359.
6. Rautava E, Lehtonen VM, Kautiainen H, Kajander S, Heinonen OJ, et al. (2007) The reduction of physical activity reflects on the bone mass among young females: a follow-up study of 142 adolescent girls. Osteoporosis Int 18(1): 1–7.
7. Ehrlich PJ, Lanyon LE (2002) Mechanical Strain and Bone Cell Function: A Review. Osteoporosis Int 13: 688–700.
8. Callewaert F, Sinnesael M, Gielen E, Boonen S, Vanderschueren D (2010) Skeletal sexual dimorphism: Relative contribution of sex steroids, GH-IGF1 and mechanical loading. J Endocrinol 207: 127–134.
9. McMillan J, Fatehi-Sedeh S, Sylvia VL, Bingham V, Zhong M, et al. (2006) Sex-specific regulation of growth plate chondrocytes by estrogen is via multiple MAP kinase signalingpathways. Biochim Biophys Acta 1963: 381–392.
10. Raz P, Nasatzky E, Boyan BD, Ornoy A, Schwartz Z (2005) Sexual dimorphism of growth plate prehypertrophic and hypertrophicchondrocytes in response to testosterone requires metabolism to dihydrotestosterone (DHT) by steroid 5-alpha reductase type 1. J Cell Biochem 95: 108–119.
11. Dy CJ, Lamont LE, Ton QV, Lane JM (2011) Sex and gender considerations in male patients with osteoporosis. Clin Orthop Relat Res 469: 1906–1912.
12. Seeman E (2001) Clinical review 137: sexual dimorphism in skeletal size, density and strength. J Clin Endocrinol Metab 86: 4576–4584.

13. Molgaard C, Thomsen BL, Prentice A, Cole TJ, Michaelsen KF (1997) Whole body bone mineral content in healthy children and adolescents. Arch Dis Child 76: 9–15.

14. Zhang XZ, Kalu DN, Erbas B, Hopper JL, Seeman E (1999) The effects of gonadectomy on bone size, mass, and volumetric density in growing rats are gender-, site-, and growth hormone-specific. J Bone Miner Res 14: 802–809.

15. Schoenau E, Neu CM, Rauch F, Man ZL (2001) The development of bone strength at the proximal radius during childhood and adolescence. J Clin Endocrinol Metab 86: 613–618.

16. Brooks GA, White TP (1978) Determination of metabolic and heart rate responses of rats to treadmill exercise. J Appl Physiol 5: 1009–1015.

17. Ferreira JC, Rolim NP, Bartholomeu JB, Gobatto CA, Kokubun E, et al. (2007) Maximal lactate steady state in running mice: effect of exercise training. Clin Exp Pharmacol Physiol 34(8): 760–765.

18. Engesaeter LB, Ekeland A, Langeland N (1978) Methods for testing the mechanical properties of the rat femur. Acta Orthop Scand 49: 512–518.

19. Peng Z, Tuukkanen J, Zhang H, Jamsa T, Vaananen HK (1994) The mechanical strength of bone in different rats models of experimental osteoporosis. Bone 15: 523–532.

20. Gouveia CH, Jorgetti V, Bianco AC (1997) Effects of thyroid hormone administration and estrogen deficiency on bone mass of female rats. J Bone Miner Res 12: 2098–2107.

21. Parffit AM, Drezner MK, Glorieux FH, Kanis JA, Malluche H, et al. (1987) Bone histomorphometry: stardardization of nomenclature, symbols, and units. Report of the ASBMR Histomorphometry Nomenclature Committee. J Bone Miner Res 2: 595–610.

22. Capelo LP, Beber EH, Huang SA, Zorn TM, Bianco AC, et al. (2008) Deiodinase-mediated thyroid hormone inactivation minimizes thyroid hormone signaling in the early development of fetal skeleton. Bone 43(5): 921–930.

23. Livak KJ, Schmittgen TD (2001) Analysis of relative gene expression data using real-time quantitative PCR and the 2(- Delta Delta C(T)) Method. Methods 25: 402–408.

24. Frost HM (2003) Bone's mechanostat: a 2003 update. Anat Rec a Discov Mol Cell Evol Biol 275(2): 1081–1101.

25. Callewaert F, Venken K, Kopchiek JJ, Torcasio A, Lenthe GHU, et al. (2010) Sexual dimorphism in cortical bone size and strength but not density is determined by independent and time-specific actions of sex steroids and IGF-1: Evidence from pubertal mouse models. J Bone Miner Res 25(3): 617–626.

26. Saxon LK, Tuner CH (2006) Low-dose estrogen treatment suppresses periosteal bone formation in response to mechanical loading. Bone 39(6): 1261–1267.

27. Damien E, Price JS, Lanyon LE (2000) Mechanical strain stimulates osteoblast proliferation through the estrogen receptor in males as well as female. J Bone Miner Res 15(11): 2169–2177.

28. Bonewald LF (2011) The amazing osteocyte. J Bone Miner Res 26(2): 229–238.

29. You L, Temiyasathit S, Lee P, Kim CH, Tummala P, et al. (2008) Osteocytes as mechanosensors in the inhibition of bone resorption due to mechanical loading. Bone 42: 172–179.

30. Cherian PP, Cheng B, Gu S, Sprague E, Bonewald LF, et al. (2003) Effects of mechanical strain on the function of gap junction in osteocytes are mediated through the prostaglandin EP2 receptor. J Biol Chem 278: 43146–43156.

31. Nomura S, Yamamoto TT (2000) Molecular event caused by mechanical stress in bone. Matrix Biol 19: 91–96.

32. Robling AG, Niziolek PJ, Baldridge LA, Condon KW, Allen MR, et al. (2008) Mechanical stimulation of bone in vivo reduces osteocyte expression of Sost/sclerostin. J Biol Chem 283(9): 5866–5875.

15

Effects of Feed Supplementation on Mineral Composition, Mechanical Properties and Structure in Femurs of Iberian Red Deer Hinds (*Cervus elaphus hispanicus*)

Cesar A. Olguin[1,2], **Tomas Landete-Castillejos**[1,2,3]*, **Francisco Ceacero**[4], **Andrés J. García**[1,2,3], **Laureano Gallego**[1,2,3]

1 Animal Science Tech, Applied to Wildlife Management Res.Group, IREC Sec. Albacete, IREC (UCLM-CSIC-JCCM), Campus UCLM, Albacete, Spain, 2 Grupo de Recursos Cinegéticos, Instituto de Desarrollo Regional (IDR), Universidad de Castilla-La Mancha (UCLM), Albacete, Spain, 3 Departamento de Ciencia y Tecnología Agroforestal y Genética, ETSIA, Universidad de Castilla-La Mancha (UCLM), Albacete, Spain, 4 Department of Animal Science and Food Processing in Tropics and Subtropics, Faculty of Tropical AgriSciences – Czech University of Life Sciences, Praha– Suchdol, Czech Republic

Abstract

Few studies in wild animals have assessed changes in mineral profile in long bones and their implications for mechanical properties. We examined the effect of two diets differing in mineral content on the composition and mechanical properties of femora from two groups each with 13 free-ranging red deer hinds. Contents of Ca, P, Mg, K, Na, S, Cu, Fe, Mn, Se, Zn, B and Sr, Young's modulus of elasticity (*E*), bending strength and work of fracture were assessed in the proximal part of the diaphysis (PD) and the mid-diaphysis (MD). Whole body measures were also recorded on the hinds. Compared to animals on control diets, those on supplemented diets increased live weight by 6.5 kg and their kidney fat index (KFI), but not carcass weight, body or organ size, femur size or cortical thickness. Supplemental feeding increased Mn content of bone by 23%, Cu by 9% and Zn by 6%. These differences showed a mean fourfold greater content of these minerals in supplemental diet, whereas femora did not reflect a 5.4 times greater content of major minerals (Na and P) in the diet. Lower content of B and Sr in supplemented diet also reduced femur B by 14% and Sr by 5%. There was a subtle effect of diet only on *E* and none on other mechanical properties. Thus, greater availability of microminerals but not major minerals in the diet is reflected in bone composition even before marked body effects, bone macro-structure or its mechanical properties are affected.

Editor: Brock Fenton, University of Western Ontario, Canada

Funding: CAO was funded by the Postgraduate Scholarship 302198 of the Consejo Nacional de Ciencia y Tecnología (México). FC was supported by a post-doc grant at Czech University of Life Sciences no. 99830/1181/1822 (Czech Republic). This paper has been funded by projects MINECO CGL2011-24811 and AGL2012-38898. The funders had no role in study design, data collection and analysis, decision to publish, or preparation of the manuscript.

Competing Interests: The authors have declared that no competing interests exist.

* E-mail: tomas.landete@uclm.es

Introduction

Bone tissue is the major part of the skeleton and one of its main roles is structural function, such as organ protection, locomotion, muscle activity, load-bearing, and serving as a reserve of minerals [1]. Whole bone mechanical properties depend on factors such as cortical thickness, diameter and quality of material [2]. In long bones, the resistance to flexion increases with cortical thickness [1]. The external diameter of long-bones predicts 55% of variation in resistance to flexion [2]. But bone stiffness also depends on intrinsic material properties (*i.e.*, those independent of size and shape) such as porosity, level of mineralization, crystal size, and properties derived from the organic phase of bone [3,4]. The most widely studied intrinsic mechanical properties include: Young's modulus of elasticity or stiffness (*E*), bending strength (force required to break a sample of bone), and work to fracture (the work required to produce such break) [4,5].

Nutrition is a main factor affecting composition of bone. These in turn affect the degree of mineralization and size of bones, both

of which influence mechanical performance [5]. In addition to the overall effects of the abundance of food, the mineral profile in diet can influence the mechanical performance of bones. This ranges from the more obvious effect of Ca and P [6], to the more subtle effects of minor minerals (*i.e.* Mg, Mn, Cu, S, Zn [1,7,8,9]). Several studies have assessed the importance of almost all minor minerals by examining their relative deficiency in single-mineral studies ([10,11] and references therein). However, several recent studies have calculated the relative importance of these minerals by assessing natural variation of both bone mineral composition and mechanical properties in deer antlers. In antlers the mineral profile differed between different parts, reflecting the size and structural quality of the antler and the adequacy of the diet [12,13,14,15,16]. Furthermore, [13,14] management affected mineral trends along deer antlers, in turn associated with better mechanical performance of bone material in deer with better nutrition. In one case, a change in content of a minor mineral in the diet produced a disproportionate effect in weight, structure and mechanical properties of antlers [9].

Antlers are bones, but differ from ordinary internal bones in that they grow rapidly [17] and are then cut off from the blood supply, and so effectively die, though their function is still to be tested. They show very little remodelling [18]. Thus, whereas antlers may reflect diet in the recent past, internal bones are more likely to reflect diet in the long term.

The aim of this study was to examine the effects of food supplementation on mineral composition, size, structure, and intrinsic mechanical properties of deer femora. In addition, we also aimed to assess variation in mineral profiles among different parts of the femur. Because (in contrast to antlers) nutrition effects on internal bones may constitute a slow process due to remodelling, we studied animals that had been on the same diet from weaning up to 3 years of age. In order to assess the overall importance of the diet for the growth of the animal, we also examined differences in body size, body condition and weight between groups of hinds.

Materials and Methods

Animals and Handling

We studied 26 captive female Iberian red deer (13 with access to supplement food and 13 as control group) from a private game estate in the Ciudad Real province ($38°53'$N, $4°17'$E), Spain. The hinds had been captured as calves at weaning. Ninety percent of Iberian red deer calve over a period of four weeks [19], so that the studied hinds probably differed in age by not more than a few weeks. Animals were kept outdoors in two contiguous fenced enclosures extending over a natural area of 13.5 ha each. All animals were maintained in captivity between 2004 and 2007, when they were hunter harvested (autumn 2007) by gamekeepers at a age of 3.5 years. Because no males lived with them, none of the hinds had been pregnant during the experiment (pregnancy and lactation increase the mobilization and resorption of Ca in the skeleton [20]). No management practice other than daily refilling of feeders was carried out during the experiment.

Ethical Note

We followed Spanish and European (EU Directive 2010/63/ EU for animal experiments) guidelines and laws for the use of animals in research [21]. The experiment was approved by the University Ethical Committee of Universidad de Castilla-La Mancha (no 0610.04).

Protein and Minerals in Diet: Plant and Supplement Feed Analyses

The first group of hinds was supplementally fed with wholemeal feed (pellet feed) while the second group had access only to the natural vegetation present in the area (natural pasture and shrub-steppe in a Mediterranean forest; protein and mineral compositions for wholemeal feed and natural vegetation are shown in Table 1A). The supplemented group of hinds had permanent access to 1 kg day^{-1} animal^{-1} of pelleted food commonly used in deer private game estates. This is more than the deer usually consumed (*i.e.*, in fact they were fed *ad libitum*). In order to estimate overall intake of protein and minerals in the supplemented group, 2.5 kg dry matter intake (DMI) was assumed based on other studies [22,23] and the experience in the experimental deer farm at our university. A mean ingestion of 1 kg of supplementary feed per animal per day would account for 40% of total DMI, whereas natural vegetation would account for the other 60%. Thus, the actual daily intake of protein and minerals could be calculated as 0.4*content in supplementary feed +0.6*content in vegetation (Table 1B).

We collected four samples (at the midpoint of each season) of the supplementary feed offered during the experiment and the natural vegetation. Ten plant species common in the study area and previously reported as preferred by Mediterranean red deer [24] were selected and analysed: strawberry tree (*Arbutus unedo*), gum cistus (*Cistus ladanifer*), rockrose (*C. salviifolius*), gum succory (*Chondrilla juncea*), mock privet (*Phillyrea angustifolia*), mastic tree (*Pistacea lentiscus*), purslane (*Portulaca oleracea*), holm oak (*Quercus ilex*), kermes oak (*Q. coccifera*) and purple vetch (*Vicia benghalensis*). Samples (about 200 g) were harvested in 15 different locations in the study area. Leaves and stems were collected since these are the parts preferred by red deer [25]. The samples were dried in an oven (T-Qtech Model 80L, Barcelona, Spain) at 85°C for 48 h, ground and stored as powder. Finally, 10 g from each sample was mixed for mineral and ash analyses. The data obtained in this way were regarded to reflect the mean yearly mineral content in the diet. Samples of supplementary feed were processed and analyzed in the same way. Crude protein was determined with the Kjeldahl method in a digester Pro-Nitro M (JP Selecta, Barcelona, Spain) and evaluated in a 848 Titrino Plus (Metrohm, Switzerland).

Animal Measurements

The shot animals were transported to a dissection room for data, organ and tissue collection, which took place within 6 h after death.

To assess the effects of diet on body growth and body condition, the following parameters were recorded for each hind: body weight, skin-on carcass weight, kidney weight, kidney fat weight, total body length, femur length, femur cortical thickness (see below for details), chest girth (as described in [26]), and foot length. Kidney fat index (KFI [27]) was calculated as the weight kidney fat divided by kidney weight without fat multiplied by 100. This is an estimator of body condition in deer [28].

Femur Sample Extraction and Specimen Preparation

Left and right femora were removed and stored in a freezer until experimental processing. Each femur was then manually cleaned of adhering soft tissue or other material. Femur length was measured with a digital calliper using standard measurement protocols. The complete femur was cut in 3 parts of similar length with a hand-held drill equipped with a saw blade (Dremel Series 3000, Illinois, USA): upper third or proximal part of diaphysis (PD; Figure 1), central part or mid-diaphysis (MD) and lower third (distal diaphysis). For subsequent analyses we used PD and MD because they probably have slightly different functions and so these sections may have both different mechanical properties and mineral composition. Sawing was performed under running tap water to avoid overheating of the bone tissue.

Following sectioning, cortical thickness was measured using a digital calliper. Cortical thickness of PD and MD was measured at four equally-spaced points of the cross sections (Figure 1). At each point five measurements were performed and the mean of these consecutive measurements was recorded. The single value for cortical thickness (MD, PD) used for statistical analysis was the average of the means obtained for the four sites. Thereafter, we extracted sticks from MD and PD which were used first for mechanical testing and then for chemical analysis. The rough-cut sticks were extracted from the internal parts of left and the right femora. The gross sticks were immersed in Hank's Buffered Salt Solution (HBSS, BioWhittaker) and kept frozen at -20°C until they were processed to produce exact-sized specimens for mechanical testing. The reason we used HBSS is that immersion in non-calcium-buffered saline has been shown to result in a loss of calcium and a 2% reduction in Young's modulus of elasticity E

Table 1. Mineral and protein content of supplemental feed offered to Iberian red deer hinds at 1 kg individual^{-1} day^{-1}, and mean mineral content in main chewed plant species present in the study area.

1A

Nutrient	Feed	Vegetation	Feed/Vegetation ratio
Crude Protein (%)	22.00	9.51	2.3
Calcium (%)	1.69	0.80	2.1
Phosphorus (%)	0.59	0.11	5.5
Magnesium (%)	0.35	0.24	1.5
Potassium (%)	1.00	0.95	1.1
Sodium (%)	0.37	0.02	18.5
Sulfur (mg/kg)	1295.60	899.20	1.4
Copper (mg/kg)	35.50	5.90	6.0
Iron (mg/kg)	475.50	119.20	3.9
Manganese (mg/kg)	467.10	89.80	5.2
Selenium (mg/kg)	1.72	3.72	0.5
Zinc (mg/kg)	401.10	27.60	14.5
Boron (mg/kg)	11.76	26.18	0.4
Strontium (mg/kg)	29.09	49.64	0.6
Silicon (mg/kg)	4100.00	900.00	4.5
Cobalt (mg/kg)	1.21	0.60	2.0
Molybdenum (mg/kg)	4.15	2.45	1.7

1B

Nutrient	Supplemented Group	Control Group	Supplemented/Control diet ratio
Crude Protein (g)	362.7	237.8	1.5
Calcium (g)	28.9	20.0	1.5
Phosphorus (g)	7.6	2.8	2.8
Magnesium (g)	7.1	6.0	1.2
Potassium (g)	24.3	23.8	1.0
Sodium (g)	4.0	0.5	8.0
Sulfur (mg)	2644.4	2248.0	1.2
Copper (mg)	44.4	14.8	3.0
Iron (mg)	654.3	298.0	2.2
Manganese (mg)	601.8	224.5	2.7
Selenium (mg)	7.3	9.3	0.8
Zinc (mg)	442.5	69.0	6.4
Boron (mg)	51.0	65.5	0.8
Strontium (mg)	103.6	124.1	0.8
Silicon (mg)	5450.0	2250.0	2.4
Cobalt (mg)	2.1	1.5	1.4
Molybdenum (mg)	7.8	6.1	1.3

Table 1A shows the composition of the feed and vegetation, as well as their ratio. Table 1B shows overall mean diet composition in both supplemented and control groups, based on 1 kg of supplemental feed +1.5 kg of natural vegetation (*i.e.* 40–60% of diet) in the supplemented group *vs.* a 100% natural vegetation diet in the control group, as well as the intake ratio among groups.

[29,30]. Although no such changes occur when the samples are kept frozen, we nevertheless used immersion in HBBS instead of water, since they were left to thaw for several hours. Specimens were abraded using a semiautomatic polishing equipment (Struers LaboPol-21, Ballerup, Denmark) until they reached dimensions of 45 (length) × 2.5 (depth) × 4.5 (width mm. Samples were kept moist taking care to produce plane-parallel surfaces with a deviation of smaller than 0.01 mm (ACHA, Digital Caliper, Spain). The exact-sized sticks were again immersed in HBBS and kept in a refrigerator until mechanical testing, replacing the solution every week if necessary. The specimen was marked so the orientation was known. Specimens were always loaded with the periosteal side in tension.

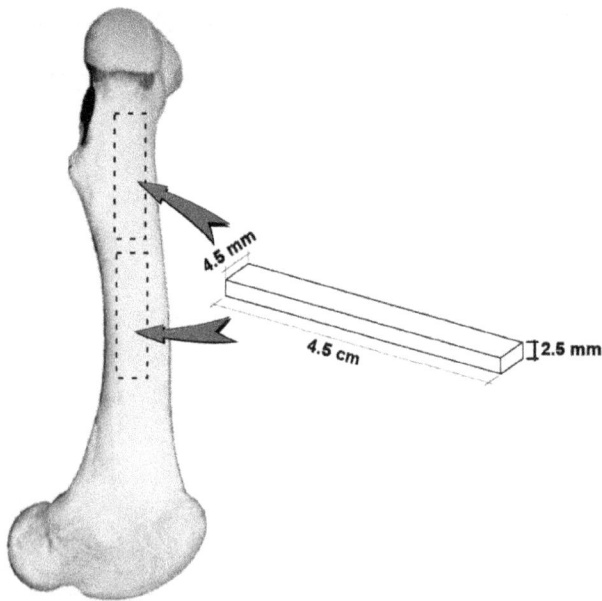

Medial view

Figure 1. Sections of the femur sampled for chemical analysis (arrows) and mechanical testing (femur bars 45 mm×2.5 mm×4.5 mm indicated in the drawing).

Mechanical Testing

The mechanical performance (e.g. resistance to fracture) of a complete bone or bone portion depends on two sets of factors [1]: i) architectural ones, mainly depending on cortical thickness and bone diameter in areas such as the bone shaft [2]; ii) mechanical bone properties that have to be tested in specimens of standardized size. Architectural parameters were determined as detailed above. In addition, the following mechanical properties of the material, that is on the intrinsic mechanical properties [1,3] were determined: stiffness (E), bending strength, and work to facture. We tested exact-sized specimens from MD and PD in a three-point bending test machine (Zwick/Roell 0.5 kN, Ulm, Germany). Span length of the supports was 40 mm and speed of the cross head 32 mm min^{-1}. Because mechanical properties of the femur, antler and other bones differ greatly depending upon the hydration state [4], great care was taken to keep the specimen fully hydrated right up to the start of the mechanical testing. The machine produced an output chart in the software testXpert II (Zwick GmbH & Co, Ulm, Germany).

E was determined from the initial slope of the load-deformation curve between 4 and 10 N, which was usually linear. Bending strength was determined from the maximum load borne. This mechanical property is considered as the relation of bending moment resulting from the mechanical test with the deflection of the sample, multiplied by squared depth and divided by second moment of area [1]. Work to fracture was determined by total work done on the specimen up to the greatest load borne, divided by the 'central' cross-sectional area. It is the amount of energy per unit of area required to break bone material, expressed as J m^{-2} [31].

Chemical Analysis

For chemical analyses, the specimens used in mechanical testing were dried at 60° for 48 h, ground and divided into two subsamples of 0.5 g each (one for assessing ash content and one for mineral content). We used a scale (Gram SR-410M, Barcelona, Spain) with a precision of 0.001 g. Subsamples from natural vegetation and wholemeal were prepared and analysed in the same way.

Samples for ash content were dried in an oven at 105°C for 24 h, weighed, and ashed in a muffle furnace (Hobersal, Model HD-230, Barcelona, Spain) at 550°C for 4 h. Then, samples were cooled and weighed. Ash content was calculated by dividing the ashed weight by the weight of the dry sample, and multiplied by 100.

Samples for mineral content were dissolved with acid solution (12% HCl, 32% HNO$_3$ and 56% H$_2$O). A second wet digestion was carried out in a microwave oven (Perkin-Elmer Multiwave 3000, Boston, USA) under 345 kPa for 30 min. Then, samples were examined with an atomic absorption spectrophotometer (Optima 5300 DV, Perkin-Elmer ICP-OES, Boston, USA). To assess mineral profile, we analyzed thirteen of the most important minerals: Ca, Mg, P, Na, K, S, B, Cu, Fe, Mn, Se, Sr and Zn. We also included Mo, Co and Si for plant and feed analyses, but there were detection problems of these minerals in femurs for an unknown reason, and so they were discarded for bone, but not for comparisons of diet.

Macro-minerals results are expressed as percentages, whilst micro-minerals are expressed as parts per million (mg/kg).

Statistical Analysis

Ratios of mineral availability in both groups were calculated in two ways: i) a simple ratio between content of protein and each mineral in supplemental feed vs. natural vegetation; and ii) a ratio of total protein and mineral content in the diet of experimental vs. control group.

Differences between groups in body weight and body condition (KFI), foot length, body length, thoracic perimeter, femur length, cortical thickness, ash and mineral content, and bone mechanical properties $(E$, bending strength and work to fracture) were examined using one-way ANOVA for those variables which had a single value per hind (i.e. body weight); in the case of KFI, using the mean value between both kidneys; and mean of all sampled sites in the femur variables (although on table 2 both right and left mean values are shown). To avoid excessive degrees of freedom with regard to sample size and the complication of too many potential interactions (two levels of repetition: left-right femurs, and centre-upper shaft within each), an ANOVA was performed to assess if left and right femurs differed in composition and mechanical properties. As no variable showed significant differences between left and right femurs, data of both femurs were aggregated into two mean values for each hind: one for MD and one for PD.

General linear mixed models (GLMM) examined effects of supplementation and femur region (MD vs. PD) on mechanical properties. Because mechanical properties have been usually explained in terms of Ca content, this was included in the model as a covariable, so that the GLMM could evaluate the effects of supplementation independently of the effect of calcium content in the femur. Individual hinds were entered as the subject, and position in the femur as the repeated measure. All analyses were performed using SPSS 18.0 for Windows (SPSS, Chicago, IL, USA).

Results

Table 1A shows protein and mineral composition of supplement food and vegetation, and their ratio. The largest ratio between

Table 2. Differences in body properties measured (Panel A) and femur mechanical properties, ash and minerals between two groups of red deer hinds feeding on natural vegetation (control group) or plants plus 1 kg animal^{-1} day^{-1} of food supplement indicated in Table 1 (food supplemented group).

Parameter	Food supplemented group	Control group	P
A. Body parameters			
Live weight (kg)	90.1±1.4	83.6±1.6	0.005
Carcass weight (kg)	60.1±1.3	56.4±1.7	0.09
Body length (cm)	161±2	160±4	–
Thoracic perimeter (cm)	111.9±1.3	109.0±1.4	–
Hind foot length (cm)	49.19±0.35	49.81±0.40	–
Weight of left kidney (g)	119.6±4.2	109.9±4.4	–
Weight of right kidney (g)	113.9±4.4	109.9±5.7	–
KFI left kidney (%)	121±13	59±7	0.001
KFI right kidney (%)	145±15	72±12	0.002
Left femur length (cm)	27.54±0.13	27.27±0.26	–
Right femur length (cm)	27.46±0.14	27.27±0.20	–
Left femur cortical thickness (mm)	4.97±0.10	4.95±0.18	–
Right femur cortical thickness (mm)	5.07±0.22	4.73±0.13	–
B. Femur composition and mechanical properties			
E (GPa)	22.4±0.3	22.0±0.3	–
Bending strength (MPa)	264.2±6.4	271.6±6.0	–
Work to fracture (kJ m^{-2})	9.4±0.4	9.3±0.4	–
Ash (%)	72.3±0.3	72.5±0.2	–
Calcium (%)	27.5±0.1	27.7±0.1	–
Phosphorus (%)	13.09±0.05	13.05±0.05	–
Magnesium (%)	0.447±0.004	0.451±0.003	–
Potassium (%)	0.0282±0.0003	0.0297±0.0003	0.001
Sodium (%)	0.649±0.004	0.659±0.003	–
Sulfur (mg/kg)	555.4±3.2	554.2±3.9	–
Copper (mg/kg)	0.250±0.007	0.231±0.006	0.048
Iron (mg/kg)	1.4±0.8	1.6±0.5	–
Manganese (mg/kg)	0.32±0.01	0.26±0.01	0.001
Selenium (mg/kg)	0.51±0.05	0.42±0.04	–
Zinc (mg/kg)	63.4±1.2	60.0±0.9	0.024
Boron (mg/kg)	2.06±0.05	2.40±0.06	0.001
Strontium (mg/kg)	238.8±5.1	251.1±3.5	0.050

Means ± SE.

feed and vegetation corresponded to Na and Zn (18.5 and 14.5 times greater in the supplement food, respectively). The greatest ratios after these corresponded to Cu, P, and Mn (6.0, 5.5, and 5.2 times more in feed than vegetation, respectively). Table 1B shows protein and mineral content in the diet in the supplemented group (40% feed +60% vegetation), that in the control group, and the ratio between both. Final ratios in the diet showed: greatest availability in Na and Zn (8.0 and 6.4 times more, respectively), followed by Cu, P and Mn (3.0, 2.8, and 2.7, respectively).

Biometric variables (table 2) showed a significant difference between groups in live weight (supplemented group 90.1±1.4 kg, control group 83.6±1.6 kg; P=0.005) and KFI (supplemented group 131.2±13.7%, control group 65.4±9.9%; P=0.005; table 2 shows left and right values as well as their P), but not in carcass weight, thoracic perimeter, body length, femur length, cortical thickness or foot length (table 2). No differences were found

between left and right values for PD and MD sections of the femur and thus, left and right values for each region were pooled as a single value for further analyses.

Regarding mean mineral composition and mechanical properties, Table 2 also shows differences between femurs from supplemented and control groups. The greater mineral availability (in ratios feed vs. plant content, or between diets) in supplemented group did not affect macromineral femur content (as Na or P). However, greater availability in supplemented diet significantly increased contents of Mn by 23%, Cu by 9% and Zn by 6%. Similarly, the greater content of B and Sr in plants (25% more in the diet based only in plants) may be responsible for the greater content of B and Sr in femurs of hinds under control diet by 14% and 5%, respectively. Finally, despite having the same content of K in both diets, femurs of hinds in the control group had a greater content of K.

No effect of supplementation was found in ANOVAs testing mechanical properties. However, detailed GLMM with repeated measures on mechanical properties including Ca as a covariable showed subtle effects that the ANOVA could not show: 1) the supplemented hinds had bone material with 27.2 GPa additional stiffness (E); 2) there was a Ca effect in the control but not in the experimental group, so that in the control group femurs increased stiffness when the amount of Ca increased in this bone (model intercept = 28.9 ± 9.3; coefficient for control group = -27.2 ± 12.3, $P < 0.05$; interaction coefficient only for control group Ca = 0.73 ± 0.29, $P < 0.05$). No other effect of supplementation was found in mechanical properties. No effect of femur region (MD $vs.$ PD) was found either.

Discussion

The results showed that a greater availability of several major and minor minerals in an enriched diet was reflected in internal bone composition only in the minor minerals with greatest availability, but not in major minerals such as Na or P. Moreover, this effect occurs at a moderate level of supplementation producing only slight changes in live weight and body condition at macrostructural level (probably produced by the greater availability of protein and energy in the supplement), but not changes in body growth or bone structure at largest scale. Thus, long term availability of minerals in the diet seems to be reflected in bone composition only in micro but not macrominerals.

A first step to understand how important the level of supplementation was is to assess its effects on general body weight and size, as well as the effects on internal organs and bone size. Despite being under the nutrition scheme for 3 years, just after weaning, hinds only showed a difference of 6.5 kg between groups and a two-fold difference in KFI, but no difference in body or femur size. The 7.2% difference in weight as a result of food supplement is similar to figures reported by Peek *et al.* [32], who found a mean weight increase of 9.8% in hinds feeding open range with supplements compared with hinds without supplements. In the present study, the differences between groups in KFI showed that supplementation improved body condition significantly, although not markedly to the observer's eye (in contrast to other published studies [33,34,35]). It should be noted that the differences cannot be solely attributable to mineral composition, as most likely they are derived from the greater amount of protein in the supplemented diet, their greater digestibility, as well as fat content, energy, and other nutrients whose measure was beyond the scope of this article.

The weight difference in our study may seem a remarkable difference, but in fact it is not a marked effect for nutrition in deer. Our own experiments showed a 10 kg difference in a group under a 60% restriction in diet compared to a control group after just 10 weeks during lactation [36]. Moreover, unpublished records in our experimental farm under *ad libitum* diets show more than 50 kg difference between adult hinds of the same nutrition level. Indeed, the standard error of the KFI shows an 87% variation with respect to the mean in body condition in the supplemented group, and only 15% in the control group, which also suggest a large variation within the supplemented group. Unfortunately, the experimental set-up with hinds in a nearly free-living situation did not allow us to estimate individual intake of plants (and species composition) or feed. Thus, we could not study further the causes for such large variation of KFI and other variables in the food supplement group. It is even more surprising that no effect was found for growth. Previous studies by our group have shown that a 3-month advance of calf births led to an 11 kg difference between groups one year later despite there being no difference in lactation and

having food *ad libitum* [37]. This is not only a weight difference: differences during lactation at low or standard milk production resulted in significant weaning differences of 7% in thoracic perimeter, 5% in cranial length, and 3% in height at shoulders [38].

Thus, it is remarkable that supplementation affected the mineral composition and to some extent mechanical properties of internal bones, even in a setting that produced slight effects on live weight and body condition, and no effects on growth, cortical femur thickness, its length or ash content. We cannot rule out that further studies analyzing the micro-structure of femurs using micro-CT and back scattered electron microscopy to assess the distribution of mineral at microscopical level, or other fine-detail techniques used in antlers [18,39] could reveal differences. These studies should be very interesting, but they were beyond the scope of our aims. Such studies could also benefit greatly by assessing a wide arrange of nutrients in the diet apart from crude protein (fat content, energy, fibre, etc.).

The results show an increase in mineral content in femur of 23% in Mn, 9% in Cu, and 6% in Zn associated with supplementation. The supplement contained, by order of magnitude, 18.5 times more Na than in natural vegetation, 14.5 for Zn, 6 for Cu, 5.36 for P, and 5.2 for Mn. Considering a 2.5 kg feed intake per animal and day being 40% of the diet, the supplemented group had 8 times more Na, 6.4 times more Zn, 3 times more Cu, 2.8 times more P and 2.7 more Mn. In both ratios the differences found in femur reflected those of the diet except for Na and P. One of the differences between Na or P and the other minerals mentioned is that Na and P are macrominerals. Perhaps microminerals are reflected more or less directly in the bone whilst this is not true for macrominerals. This may point to the role of bone as a store of microminerals. The literature shows contradictory evidence: some early studies have shown that internal bones do reflect the level of Ca, P, and Mg [40,41,42]. However, it has been found in deer that Zn and Mn increased in several tissues in direct proportion to the content in dietary feed [23]. In contrast, Na and P in the diet may not be directly reflected in bone as a result of threshold effects. In fact P and Na, but also Ca and Mg showed a similar concentration in hind femurs as published values for human cortical bone [38]. The greatest availability in supplemented diet after Zn, Cu, and Mn are Ca and Fe. Of these, only Fe is a micromineral and it is not clear why their content in the diet is not reflected, but after these, the following are three minerals showing 25% more availability in the control diet: B, Sr, and Se. Of these, the first two were also reflected in a greater content of B and Sr in the femurs of hinds under the control diet. That is to say, except for Fe and Se, differences in femur micromineral composition reflect their availability in the diet (greater content in femur if the diet has a greater content). Contrary to this, of the major minerals femurs only differed in K, and this despite its content in the diet being the same.

Why did femurs not reflect mineral availability of Na in the diet? Research in other cervids has shown that moose (*Alces alces*) select plants for Na content to meet a Na threshold, and thereafter, the diet is selected for energy content [43]. A similar effect of seeking for Na when it is deficient has been frequently found in red deer [44]. In contrast to our findings here, Na has been found to reflect diet composition in antlers [14,15]. However, in this case, it reflected a deficiency, which suggests that bone Na may reflect diet only in a deficiency situation and only up to the point in which needs have been fulfilled. This would not be surprising, as animals are able to modulate mineral absorption according to their needs, reducing absorption when needs have been met [45]. This, in turn, would also support the hypothesis that internal bones reflect diet only in microminerals as a store to be subsequently liberated by remodeling when they are needed. Recently, a hypothesis has been put forward by our group to explain remodeling as a mechanism to keep flow of microminerals from skeleton to other organs where they are needed [46].

However, the percentage of increase in Mn, Cu and Zn between femurs of supplemented and control groups do not match the ranking in the ratio of these minerals in the diet (in other words, in bones the ratio is similar for Mn and Cu and far less for Zn, whereas it is far greater for Zn in food and similar in the other two). Up to some extent this is not surprising for two reasons: the ratio between content in supplemental diets is based on the assumption of equal intake of all plant species as we did not have information on the percentage of each species in the real hind diet. The second reason is that absorption or bioavailability of a mineral may depend on the interaction between physiological importance of the mineral and its availability in the diet. Thus, animals may store all Mn or Cu they can at relatively low contents, whereas the more concentrated Zn in our setting may have a much lower priority for storing. At least for the case of Cu, its mere addition to an otherwise balanced diet may increase growth in pigs [47], and in fact, it is commonly deficient in ruminants ([48]; Ludek Bartos, John Fletcher and other deer scientists personal communications).

Once discarding Si which unfortunately we could not measure well in bones, the remaining ranking in mineral availability in feed *vs*. plants, corresponds to Fe, Ca, B, Se, Mg and Sr. Of these, Ca and Mg are again two major minerals whose greater availability in the diet are not reflected in femurs. Of the remaining, which are minor minerals, Fe and Se are not reflected in their content in bones, but B and Sr are. B may be absorbed with a greater priority to counteract the lack of other minerals in the bone, as it has been shown to improve bone mechanical properties [49,50]. Some authors reported that B and Sr were associated with increased ash content in bones [23,51], which may explain why hinds in the control group incorporated more of these minerals in their femurs, thus reflecting the diet in contrast to Se, and why we did not find differences in ash content between groups. Se is mainly not stored in bones but in muscles, leading to white-muscle disease if it is deficient [11]. This may be the reason why bone does not reflect Se content of the diet. We do not have an explanation on the lack of effect of Fe in bone.

In contrast, it seems easier to explain why control group had femurs with 5% more K despite having nearly the same availability of K in the diet. The greater K content in femurs of hinds feeding only on plants may indicate that these are under a greater nutritional stress. Studies on antler show that K indicates physiological stress in this kind of bone, as its content is greater in distal parts of the antler which are grown when body stores of minerals are near depletion [13,15]. Similarly, another influence of diet upon antler composition (in this case milk production effects on mineral composition of first deer antler), also pointed to a relation between K and nutrition stress: the lower the milk production by the mothers, the greater the K content in first antlers, whereas the opposite was true for antler ash, Ca and P content [52].

With regard to mechanical properties, we found an effect of supplementation only in stiffness (E), and this could only be assessed after removing the effect of Ca. How can stiffness, usually related to Ca, be increased in the experimental group once the effect of this mineral and femur region is removed? In chemical terms, in the appatite lattice that forms the mineral phase of bone Ca can substitute for Mg or Sr, but also by other minerals with similar valency. This may have happened in our case, or the effect may be caused by other nano-scale factors. To assess further this

question we should test also ratios such as Ca/Mg, Ca/Sr, and other as we did in a previous study [13], but this is a matter for a paper more focused on the effects of chemistry on mechanics at nano-scale, rather than the main effects of diet in bone composition and mechanical properties. The limited effect found of supplementation on bone mechanical properties compared to effects found in antlers [13] may indicate that internal bones, and particularly long bones which sustain the body weight, have more conserved mechanical properties than those antlers. The reason may be that breaking a leg, for example, has more serious consequences for survival than breaking an antler: the former usually ends in death of the animal. However, and as pointed above, we cannot rule out that more subtle effects could be found if other intrinsic mechanical properties are found in test regarding stretching, shear, compression, or hardness tested by micro or nano-indentation. The diet may have affected the distribution of minerals at the microscopical level, which in turn may influence mechanical properties, or may influence porosity [18,39], which is also directly related to some mechanical properties [39]. Thus, further analysis based on these techniques may be very interesting to complete the present study.

If the present results were found in other situations, they may be potentially useful both for deer farming and ecological studies. For example, a chemical analysis of all plants present in an ecosystem may not show gross differences in mineral availability. However if a study finds that two populations of deer feeding on plants of their respective habitats differed, for example, in femur content of Mn, B, Zn or other minor minerals, this would mean that the plants actually ingested by deer do actually differ in mineral availability between populations. Thus, examining the bone composition of a population of deer and comparing it with that of deer under a balanced diet may be a tool to suggest that mineral supplementation should be considered to correct mineral deficiencies. Because threatened deer or indeed other ruminants cannot be killed or caught for detailed physiological studies, this tool could also be used collecting bones from individuals found dead to compare with zoo animals to detect mineral deficiencies in natural populations that may be the cause of the declining of a wild population.

Conclusions

Supplemented food involving greater mineral availability can influence internal bone composition, mostly regarding the microminerals more available in supplement food. This effect occurs even in healthy animals and when supplemented deer only achieved moderate difference in weight or body condition with the control group, but no difference in body growth, femur size, cortical thickness or mechanical properties.

Acknowledgments

This paper is dedicated to the memory of Isidoro Cambronero who helped for years in handling our experimental animals (not those included in this study) until his sudden death in Christmas 2011. The authors wish to thank to Alberto Martínez for the helpful assistance in the chemical analysis. The owner of the game estate, Yolanda Fierro, approved the protocols and helped in shooting the animals, measuring them, and weighing organs. All authors approved the final manuscript. The authors declare that they have no conflict of interest.

Author Contributions

Conceived and designed the experiments: CAO TLC. Performed the experiments: CAO FC AJG. Analyzed the data: CAO TLC. Contributed reagents/materials/analysis tools: AJG LG. Wrote the paper: CAO TLC FC LG.

References

1. Currey JD (2002) Bones: Structure and Mechanics. Princeton: Princeton University Press.
2. Davison SK, Siminoski K, Adachi JD, Hanley DA, Goltzman D, et al. (2006) Bone strength: The whole is greater than the sum of its parts. Semin Arthritis Rheu 36: 22–31.
3. Burr DB (2002) The contribution of the organic matrix to bone's material properties. Bone 31: 8–11.
4. Currey JD, Landete-Castillejos T, Estévez JA, Olguín A, García AJ, et al. (2009) The Young's Modulus and impact energy absorption of wet and dry deer cortical bone. The Open Bone Journal 1: 38–45.
5. Palacios C (2006) The role of nutrients in bone health, from A to Z. CRC Cr Rev Food Sci 46: 621–628.
6. McDowell LR (1996) Feeding minerals to cattle on pasture. Anim Feed Sci Tech 60: 247–271.
7. Corah L (1996) Trace mineral requirements of grazing cattle. Anim Feed Sci Tech 59: 6 1–70.
8. Kotha SP, Guzelsu N (2007) Tensile behavior of cortical bone: dependence of organic matrix material properties on bone mineral content. J Biomech 40: 36–45.
9. Landete-Castillejos T, Currey JD, Estevez JA, Fierro Y, Calatayud A, et al. (2010) Do drastic weather effects on diet influence changes in chemical composition, mechanical properties and structure in deer antlers? Bone 47: 815–825.
10. Spears JW (1996) Organic trace minerals in ruminant nutrition. Anim Feed Sci Tech 58: 151–163.
11. McDowell LR (2003) Minerals in Animal and Human Nutrition. Amsterdan: Elsevier.
12. Landete-Castillejos T, García A, Gallego L (2007) Body weight, early growth and antler size influence antler bone mineral composition of Iberian red deer (Cervus elaphus hispanicus). Bone 40: 230–235.
13. Landete-Castillejos T, Estévez JA, Martínez A, Ceacero F, García A, et al. (2007) Does chemical composition of antler bone reflect the physiological effort made to grow it? Bone 40: 1095–1102.
14. Landete-Castillejos T, Currey JD, Estévez JA, Gaspar-López E, García A, et al. (2007) Influence of physiological effort of growth and chemical composition on antler bone mechanical properties. Bone 41: 794–803.
15. Estévez JA, Landete-Castillejos T, García A, Ceacero F, Gallego L (2009) Antler mineral composition of Iberian red deer (Cervus elaphus hispanicus) is related to mineral profile of diet. Acta Theriol 54: 235–242.
16. Estévez JA, Landete-Castillejos T, García AJ, Ceacero F, Martínez A, et al. (2010) Seasonal variations in plant mineral content and free-choice minerals consumed by deer. Anim Prod Sci 50: 177–185.
17. Gaspar-López E, Landete-Castillejos T, Gallego L, García AJ (2008) Antler growth rate in yearling Iberian red deer (Cervus elaphus hispanicus). Eur J Wildlife Res 54: 753–755.
18. Gomez S, Garcia AJ, Luna S, Kierdorf U, Kierdorf H, et al. (2013) Labeling Studies on Cortical Bone Formation in the Antlers of Red Deer (Cervus elaphus). Bone 52: 50.
19. Kelly R, Whatley J (1975) Observations on the calving of red deer (Cervus elaphus) run in confined areas. App Anim Ethol 1: 293–300.
20. Wysolmerski JJ (2002) The evolutionary origins of maternal calcium and bone metabolism during lactation. J Mammary Gland Biol 7: 267–276.
21. ASAB (2009) Guidelines for the treatment of animals in behavioural research and teaching. Anim Behav 77: I-X.
22. Nicol AM, Brookes IM (2007) The metabolisable energy requirements of grazing livestock. In: Rattray PV, Brookes IM, Nicol AM, editors. Pasture and Supplements for Grazing Animals. Hamilton: New Zealand Society of Animal Production, Occasional Publication No. 14. 151–172.
23. Grace ND, Castillo-Alcala F, Wilson PR (2008) Amounts and distribution of mineral elements associated with liveweight gains of grazing red deer (Cervus elaphus). New Zeal J Agri Res 51: 439–449.
24. Rodríguez-Berrocal J (1978) Introduction to the study and evaluation of red deer (Cervus elaphus L.) food resources in the Sierra Morena: I. Analysis of the deer diet. Arch Zootec 27: 73–82.
25. Minson DJ (1990) Forage in Ruminant Nutrition. San Diego: Academic Press.
26. Cook RC, Cook JG, Irwin LL (2003) Estimating elk body mass using chest-girth circumference. Wildlife Soc B 31: 536–543.
27. Riney T (1995) Evaluating condition on free-ranging red deer (Cervus elaphus). With special reference to New Zealand. New Zeal J Sci 36B: 429–463.
28. Kie JG (1987) Performance in wild ungulates: measuring population density and condition of individuals. General Technical Report PSW-106, 1–17.
29. Gustafson MB, Martin RB, Gibson V, Storms DH, Stover SM, et al. (1996) Calcium buffering is required to maintain bone stiffness in saline solution. J Biomech 29: 1191–1194.
30. Nazarian A, Hermannsson BJ, Muller J, Zurakowski D, Snyder SD (2009) Effect of tissue preservation on murine bone mechanical properties. J Biomech 42: 82–86.
31. Turner CH, Burr DB (1993) Basic biomechanical measurements of bone: A tutorial. Bone 14: 595–608.
32. Peek JM, Schmidt KT, Dorrance MJ, Smith BL (2002) Supplemental feeding and farming of elk. In: Thomas JW, Toweill DE, editors. Elk of North America: Ecology and Management. Mechanicsburg: Stackpole Books. 614–647.
33. Blood DA, Lovaas AL (1966) Measurements and weight relationships in Manitoba Elk. J Wildlife Manage 30: 135–140.
34. Dauphiné TC (1975) Kidney weight fluctuations affecting the kidney fat index in caribou. J Wildlife Manage 39: 379–386.
35. Finger SE, Brsibin IL, Smith MH (1981) Kidney fat as a predictor of body condition in white-tailed deer. J Wildlife Manage 45: 964–968.
36. Landete-Castillejos T, García A, Gómez JA, Laborda J, Gallego L (2002) Effect of nutritional stress during lactation on immunity costs and indices of future reproduction in Iberian red deer (Cervus elaphus hispanicus). Biol Reprod 67: 1613–1620.
37. Gómez JA, Landete-Castillejos T, García AJ, Gallego L (2006) Importance of growth during lactation on body size and antler development in the Iberian red deer (Cervus elaphus hispanicus). Livest Sci 105: 27–34.
38. Hing KA (2004) Bone repair in the twenty-first century: biology, chemistry or engineering? Phil T Roy Soc A 362: 2821–2850.
39. Landete-Castillejos T, Currey JD, Ceacero F, Garcia AJ, Gallego L, et al. (2012) Does nutrition affect bone porosity and mineral tissue distribution in deer antlers? The relationship between histology, mechanical properties and mineral composition. Bone 50: 245.
40. de Sousa JC, Conrad JH, Mendes MO, Blue WG, McDowell LR (1978) Ca, P, Mg and K interrelationships among soil, forage and animal tissues. J Anim Sci 47 (Suppl. 1): 342.
41. Norris LC, Kratzer FH, Lin HJ, Hellewell AB, Beljan JR (1972) Effect of quantity of dietary calcium on maintenance of bone integrity in mature white leghorn male chickens. J Nutr 102: 1085.
42. Varley PF, Callan JJ, O'Doherty JV (2011) Effect of dietary phosphorus and calcium level and phytase addition on performance, bone parameters, apparent nutrient digestibility, mineral and nitrogen utilization of weaner pigs and the subsequent effect on finisher pig bone parameters. Anim Feed Sci Tech 165: 201–209.
43. Belovsky GE (1978) Diet optimization in a generalist herbivore: The moose. Theor Popul Biol 14: 105–134.
44. Ceacero F, Landete-Castillejos T, García AJ, Estévez JA, Martínez A, et al. (2009) Free-choice mineral consumption in Iberian red deer (Cervus elaphus hispanicus) response to diet deficiencies. Livest Sci 122: 345–348.
45. Rodan GA (1998) Bone homeostasis. P Natl Acad Sci USA 95: 13361–13362.
46. Landete-Castillejos T, Molina-Quilez I, Estevez JA, Ceacero F, García AJ, et al. (2012) Alternative hypothesis for the origin of osteoporosis: The role of Mn. Front Biosci E4: 1385–1390.
47. Paik I (2001) Application of chelated minerals in animal production. Asian Australasian J Anim Sci 14: 191–198.
48. Grace ND, Wilson PR (2002) Trace element metabolism, dietary requirements, diagnosis and prevention of deficiencies in deer. New Zeal Vet J 50: 252–259.
49. McCoy H, Kenney MA, Montgomery C, Irwin A, Williams L, et al. (1994) Relation of boron to the composition and mechanical properties of bone. Environ Health Persp 102: 49–53.
50. Naghii MR, Torkaman G, Mofid M (2006) Effects of boron and calcium supplementation on mechanical properties of bone in rats. Biofactors 28: 195–201.
51. Dahl SG, Allain P, Marie PJ, Mauras Y, Boivin G, et al. (2001) Incorporation and distribution of strontium in bone. Bone 28: 446–453.
52. Gómez JA, Landete-Castillejos T, García AJ, Gaspar-López E, Estevez JA, et al. (2008) Lactation growth influences mineral composition of first antler in Iberian red deer Cervus elaphus hispanicus. Wildlife Biol 14: 331–338.

16

Mechanical Influences on Morphogenesis of the Knee Joint Revealed through Morphological, Molecular and Computational Analysis of Immobilised Embryos

Karen A. Roddy[1,2], **Patrick J. Prendergast**[2], **Paula Murphy**[1,2]*

1 Department of Zoology, School of Natural Sciences, Trinity College Dublin, Dublin, Ireland, **2** Trinity Centre for Bioengineering, School of Engineering, Trinity College Dublin, Dublin, Ireland

Abstract

Very little is known about the regulation of morphogenesis in synovial joints. Mechanical forces generated from muscle contractions are required for normal development of several aspects of normal skeletogenesis. Here we show that biophysical stimuli generated by muscle contractions impact multiple events during chick knee joint morphogenesis influencing differential growth of the skeletal rudiment epiphyses and patterning of the emerging tissues in the joint interzone. Immobilisation of chick embryos was achieved through treatment with the neuromuscular blocking agent Decamethonium Bromide. The effects on development of the knee joint were examined using a combination of computational modelling to predict alterations in biophysical stimuli, detailed morphometric analysis of 3D digital representations, cell proliferation assays and in situ hybridisation to examine the expression of a selected panel of genes known to regulate joint development. This work revealed the precise changes to shape, particularly in the distal femur, that occur in an altered mechanical environment, corresponding to predicted changes in the spatial and dynamic patterns of mechanical stimuli and region specific changes in cell proliferation rates. In addition, we show altered patterning of the emerging tissues of the joint interzone with the loss of clearly defined and organised cell territories revealed by loss of characteristic interzone gene expression and abnormal expression of cartilage markers. This work shows that local dynamic patterns of biophysical stimuli generated from muscle contractions in the embryo act as a source of positional information guiding patterning and morphogenesis of the developing knee joint.

Editor: Sudha Agarwal, Ohio State University, United States of America

Funding: This work was supported by a TCD Overhead Investment Plan (OIP) Interdiscipinary Award, Wellcome Trust project grant (083539/Z/07/Z) and Science Foundation Ireland (Programme Award 02/IN1/B267). The funders had no role in study design, data collection and analysis, decision to publish, or preparation of the manuscript.

Competing Interests: The authors have declared that no competing interests exist.

* E-mail: paula.murphy@tcd.ie

Introduction

Each skeletal rudiment and joint of the limb can be identified by its unique, species specific, size and shape. These individual shapes emerge by the local modulation of cellular processes, such as cell proliferation, differentiation, extracellular matrix synthesis, cell shape and size [1], creating complex shapes from relatively simple initial morphologies. Skeletal morphogenesis is regulated by a combination of inductive regulatory signals produced by the constituent tissues [reviewed in 2,3,4]. While such networks of molecular regulatory signals are clearly essential to the correct establishment of spatial patterning, there is evidence that features of the physical environment, such as mechanical forces induced by muscle contraction, contribute to regulatory mechanisms governing morphogenesis. We focus on the developing chick knee joint as a convenient model to investigate how mechanical forces integrate with cellular and molecular events to impact the emerging properties of the skeleton.

We have previously shown that the complex 3D shape of the knee joint emerges following the initiation of muscle contractions, between chick embryonic stages Hamburger and Hamilton (HH)28 and HH34 [5]. The interfacing ends of the cartilaginous rudiments (including the prominent condyles of the distal femur) dictate the shape of the articular surfaces of the knee while other joint structures such as articular cartilages, menisci and the synovium derive from cells in the joint interzone [6,7]. The emergence of knee joint shape and form must therefore involve the local regulation of growth in the cartilaginous rudiments and tissue differentiation within the joint interzone.

Several lines of evidence show that contraction of the developing embryonic musculature is required for normal skeletogenesis. Human congenital malformations [8,9] and animal models where muscle contractions are removed or altered using neuromuscular blocking agents [10,11,12,13], surgery [11] or explant culture [14,15,16] and mouse mutants where no skeletal muscle forms [17,18,19,20], lead to underdeveloped, brittle and in some cases misshapen skeletal elements [11,12,21,22]. Joints appear to be particularly sensitive with immobilisation leading to loss of joint structures such as the cavity, articular surfaces and patella [10,11,12,19,20,21,23]. It is unknown how mechanical stimulation derived from movement can influence rudiment and joint morphogenesis but computational modelling has been used to predict mechanical loads acting on the tissues. Previous studies used Finite Element (FE) modelling to predict mechanical forces in simplified representations of the skeleton to investigate joint formation [24], endochondral ossification [25,26], the emergence of the femoral bicondylar angle [27] and developmental dysplasia of the hip [28]. We previously [29] created a FE model from morphologically accurate 3D data captured from the developing chick tibiotarsus to simulate the dynamic patterns of stimuli generated by muscle contraction. A striking correspondence between the patterns of stimuli and the dynamics of ossification was noted and we further showed that in an altered mechanical environment ossification was reduced and the *in vivo* expression of a number of genes involved in bone formation was altered [13]. More recently we used a similar approach to create a morphologically accurate FE model of knee joint development [30], indicating that the tempero-spatial pattern of mechanical stimuli generated in the distal femur by muscle contraction corresponds with aspects of the pattern of shape changes and with differential rates of cell proliferation in the femoral condyles. This led to the proposal that mechanical forces could act as a physical form of positional information, generating local patterns that modulate cellular events such as cell proliferation and differentiation, thereby guiding tissue morphogenesis.

A number of key molecules regulating cartilage growth and differentiation [31,32,33,34,35,36] and joint formation [32,37,38,39,40,41] have been identified. PTHLP(also known as PTHrP) has been shown to act in a regulatory loop with Ihh to maintain a pool of proliferating chondrocytes at the epiphysis of long bones [34,42]. BMP and FGF family members act in an antagonistic relationship to co-ordinate differentiation and proliferation processes during development of the skeletal rudiment [43]. A large number of genes, including BMP2, FGF2, FGFR2, PTHLH, β1 integrin (ITGB1), CD44 and HAS2, are expressed specifically in the joint region. Several of these gene products regulate chondroctye growth and differentiation while others such as CD44 and HAS2 regulate joint cavitation through the action of hyaluronan [44,45,46]. CD44 encodes one of the major receptors for hyaluronan while HAS2 encodes an enzyme involved in its synthesis. Inhibition of α5β1 integrin leads to ectopic joint formation while missexpression causes the inhibition of joint formation leading to fused long bones [47].

Very little is known about how cells respond to mechanoregulation, especially in an *in vivo* developing system. Mechanoregulation of chondrocyte proliferation and biosynthesis has been extensively studied in a wide range of culture systems, including explants, monolayer and 3D scaffolds [reviewed in 48] where it has been proposed that continuous loads or high frequency, high magnitude

loads inhibit cell matrix synthesis and growth [49,50,51,52] while low magnitude dynamic loading stimulates matrix synthesis [49,50,51,53]. For example, the dynamic compression of cartilage explants by approximately 3% was shown to stimulate matrix synthesis while graded levels of static compression did not [49]. Mechanical forces are known to influence the expression of certain genes in mechanically stimulated cells when compared to non stimulated cells [52,54,55]. Such genes have been called mechanosensitive or mechanoresponsive and evidence exists that several of the molecules involved in the development of the joint and regulation of ossification are mechanosensitive, at least in a cell culture context. Such genes include the previously mentioned BMP2, CD44, β1 integrin subunit, FGF2, FGFR2 and PTHLP (Table 1). A limited number of studies have investigated alterations in the expression patterns of regulatory genes in developing tissues *in vivo* in response to immobilisation, showing for example alteration in FGF2 [37], IHH and COLX [22]. Such *in vivo* studies have the advantage of demonstrating mechanosensitivity within a specific developmental context and also make it possible to relate the changes in gene expression to changes in tissue differentiation and morphogenesis.

In this paper we further explore the link between local patterns of biophysical stimuli generated by embryonic muscle contractions and the generation of shape and structure in the avian knee joint. *In ovo* immobilisation was used to alter the mechanical environment during development and changes in the resulting structure and shape of the knee joint region were revealed. Shape in particular was analysed following 3D imaging of control and immobilised specimens using Optical Projection Tomography (OPT). To explore cellular processes impacted by mechanical stimulation, patterns of cell proliferation in the distal femur were compared between immobilised and control specimens. A FE model of rigid muscle paralysis was used to determine how the local mechanical information produced by muscle contractions would be altered by paralysis and in turn how this compares with the alterations in joint shape and cell proliferation observed. Finally the effect of altered mechanical forces on the expression of regulatory genes was investigated using in situ hybridisation. Genes were chosen for analysis based on previous experimental evidence of a regulatory role in the process of joint formation and an indication of mechanosensitivity in another cellular context (Table 1). The findings support and further the hypothesis that patterns of biophysical stimuli generated by the contracting musculature act as a type of positional information during skeletal morphogenesis, impacting the molecular regulation of cell proliferation and tissue patterning.

Table 1. Summary of regulatory genes selected for analysis based on functional evidence and mechanosensitivity.

Gene	Evidence of skeletal function	Evidence of mechanosensitivity
BMP2	chondrocyte maturation and proliferation [31,35,85],	Distraction osteogenesis (*in vivo*) [86]
CD44	joint cavity formation [44]	in culture [71,87]
HAS 2	joint cavity formation [88]	no evidence
β1 integrin	interzone formation [47]	in explants [89]
WNT9a	interzone specification [90,91]	no evidence
PTHLP	maintains chondrocyte proliferation [31,34,42]	in culture [92]
FGF2	joint cavity formation [93], chondroctye maturation [43,94]	*in vivo* [93]
FGFR2	proliferation of osteoprogenitor cells [95]	in culture [96]
COL2A1	ECM matrix component [97], marker of proliferating chondrocyes	*in vivo* [20]
TNC	Articular cartilage ECM matrix component [98,99]	*in vivo* [98]

Results

Abnormal development of the knee joint following muscle immobilisation

Comparison of embryos immobilised with 0.5% DMB for 4–5 days, commencing on day 4.5 of incubation, and control specimens, revealed a number of consistent abnormalities. Staging of the embryos, using the Hamburger and Hamilton criteria [56], insured that only stage matched specimens were compared. Drug treated embryos showed previously reported effects of immobilisation including spinal curvature and joint contracture (not shown) [10,21]. Specifically in the knee joint, histological sections showed a general reduction in the separation of the rudiments, altered cellular organisation in the interzone with no clear definition of chondrogenous layers and no sign of cavitation in the altered mechanical environment of immobilised specimens (Figure 1A–D and K,L). Additional alterations to knee joint associated tissues were revealed through marker gene expression analysis. Collagen type II alpha1 (COL2A1) gene expression marks the joint capsule, developing ligaments and tendons and initial appearance of the patella, in addition to the cartilaginous rudiments in control specimens at this stage (Figure 1E and G). Tenascin C (TNC) expression also marks the joint capsule and patella and reveals the chondrogenous layers, the perichondrium and the appearance of the menisci in the joint interzone (Figure 1I and K). In immobilised knee joints, expression analysis of these tissue markers

revealed absence of the inter-articular ligaments (Figure. 1F and H), the chondrogenous layers and menisci (Figure 1J and L). The expression of both markers also appeared to be reduced or absent in the joint capsule and patella region (Figure 1F, J) in immobilised joints.

Shape changes in the knee joint

To reveal shape changes in the knee joint following immobilisation, 3D analysis of Alcian blue stained, OPT scanned specimens following 4 or 5 days of immobilisation was carried out (n = 17 and 32 for immobilised specimens on days 4 and 5, n = 16 and 18 for controls). 3D digital representations of each knee joint specimen could be oriented to view comparable sections [57] and take measurements that capture characteristic aspects of shape including the width of the proximal tibiotarsus and fibula, the separation of the tibiotarsus and femur (interzone) and the height and width of the condyles and intercondylar fossa of the distal femur (individual measurements detailed in Figure 2). Statistical analysis of the measurements showed that immobilisation had a significant effect on specific morphological features of the knee (Table 2). Immobilisation caused a significant reduction in the width of the proximal epiphysis of the tibiotarsus and fibula (Table 2). The distal end of the femur, showed a reduction in height of both condyles in the dorso-ventral orientation but no significant reduction in width following 4 days of immobilisation (Table 2), either at the midline or ventral aspect. The same overall

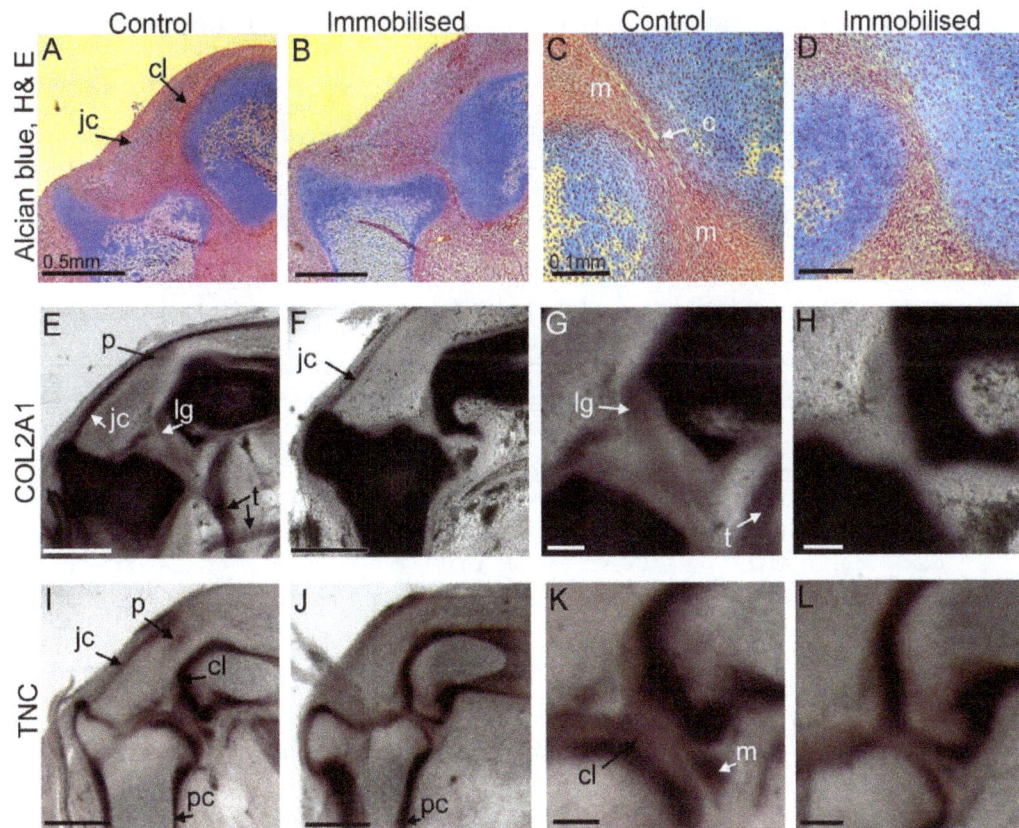

Figure 1. Anatomical changes in the knee joints of immobilised embryos. Longitudinal sections through the chick knee joint of control and immobilised embryos (4.5 days of immobilisation) at low (A, B, E, F, I, J) and high magnification (C, D, G, H, K, L). Histological sections (E–D) were stained using alcian blue and counter stained with haematoxylin and eosin. Other sections show the expression of COL2A1 and TNC mRNA in control and immobilised knee joints. c; cavity, cl; chondrogenous layers, jc; joint capsule, lg ; ligament, m; meniscus, p; patella, pc ; perichondrium, t; tendon. Scale bar 0.5 mm (A, B, E, F, I, J) and 0.1 mm (C, D, G, H, K, L).

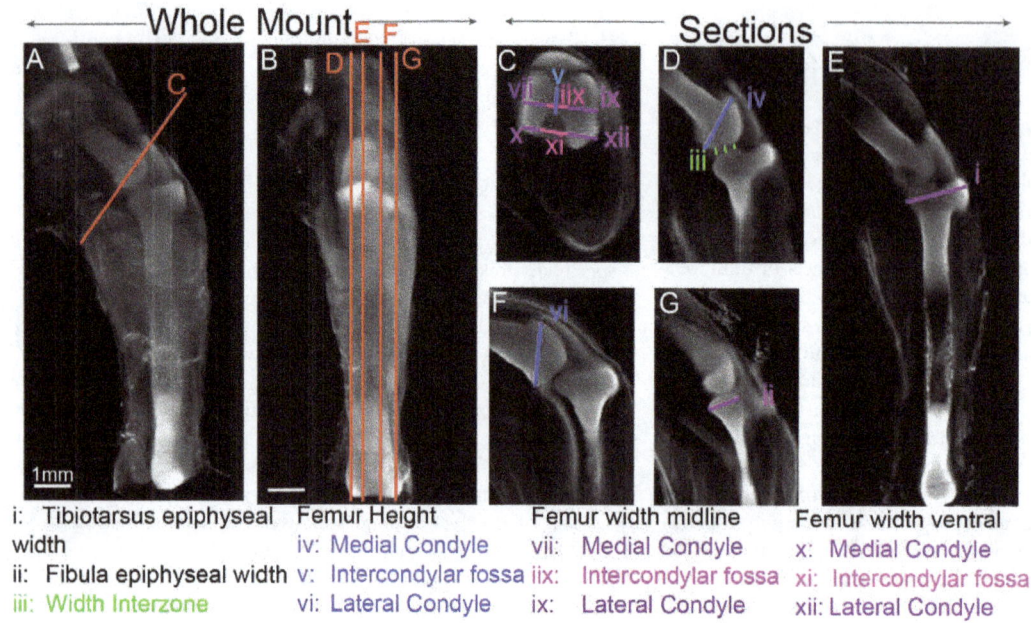

Figure 2. Overview of morphometric analysis of the knee joint. 3D volume representation of the hind limb (A–B) at HH35 with guidelines in red shows the location of consistent virtual sections taken through the hindlimbs of all specimens (C–G). Individual measurements are indicated by lines i-xii. Scale bar 1 mm.

effects on shape were seen following 4 and 5 days of immobilisation except that a reduction in the width of condyles in the ventral aspect became apparent with extended treatment (Table 2, highlighted).

The apparent reduction in the separation of rudiments in the knee joint noted from histological sections was confirmed here through a significant reduction in the size of the interzone separating the femur and the tibiotarsus; 30.1% and 35.6%

following 4 and 5 days of treatments respectively (p<0.001, Table 2).

The strongest and most consistent change in shape was seen in the width of the intercondylar fossa; reduced by 41.6% (P<0.001) at the midline and 44.7% (p<0.001) at the ventral side of the femur following 4 days of immobilisation (Table 2). This reduction in the separation of the femoral condyles is still obvious after 5 days of immobilisation (30.2 and 36%, p<0.001).

Table 2. Comparison of mean morphometric measurements of control and immobilised knee joints following immobilisation for 4 or 5 days.

		Day 4				Day 5			
		Control	Immob	%reduction	significance	Control	Immob	%reduction	significance
Tibiotarsus Epiphyseal width (i)		1.18	0.94	19.65	$F_{(1,29)} = 60.43, <0.001$	1.19	1.01	15.06	$F_{(1,46)} = 62.75, p<0.001$
Fibula Epiphyseal width (ii)		0.55	0.45	18.44	$F_{(1,29)} = 25.61, p<0.001$	0.57	0.46	20.29	$F_{(1,46)} = 90.92, p<0.001$
Interzone (iii)		0.08	0.06	30.12	$F_{(1,29)} = 35.55, p<0.001$	0.10	0.07	35.59	$F_{(1,46)} = 41.61, p<0.001$
Femur Height	Lateral Condyle (iv)	0.96	0.79	18.26	$F_{(1,29)} = 36.40, p<0.001$	1.02	0.88	13.22	$F^{(1,46)} = 58.62, p<0.001$
	Intercondylar fossa (iv)	0.30	0.27	10.33	$F_{(1,29)} = 4.19, p = 0.05$	0.32	0.30	6.64	$F_{(1,46)} = 7.93, p = 0.007$
	Medial Condyle (v)	0.81	0.70	12.98	$F_{(1,29)} = 23.22, p<0.001$	0.82	0.72	12.49	$F_{(1,46)} = 23.37, p<0.001$
Femur Width midline	Lateral Condyle (vi)	0.40	0.39	NS		0.46	0.43	NS	
	Intercondylar fossa (vii)	**0.29**	**0.17**	**41.63**	$F_{(1,29)} = 36.67, p<0.001$	**0.25**	**0.18**	**30.18**	$F_{(1,46)} = 28.51, p<0.001$
	Medial Condyle (iix)	0.30	0.30	NS		0.34	0.32	NS	
Femur Width Ventral	Lateral Condyle (ix)	0.42	0.39	NS		0.50	0.42	15.41	$F_{(1,46)} = 27.46, p<0.001$
	Intercondylar fossa (x)	**0.31**	**0.17**	**44.69**	$F_{(1,29)} = 18.40, p<0.001$	0.31	0.20	35.95	$F_{(1,46)} = 98.37, p<0.001$
	Medial Condyle (xi)	0.29	0.27	NS		0.33	0.28	13.88	$F_{(1,46)} = 24.18, p<0.001$

A particularly interesting pattern with respect to the width of the intercondylar fossa is highlighted in bold. Percentage differences between the mean lengths of the measurements are shown with the associated statistical significance.

This and other detailed aspects of local shape changes in the distal femur were visualised by outlining cartilage (alcian blue stained) in comparable sections of 3D reconstructions. Outlines were generated using the sections shown in Figure 2 C, D and F and physically overlaid so that the medial and lateral sides of the femora in the sections were parallel and the midpoints of the intercondylar fossa were overlapping. Overlaying outlines of this characteristic view of control and immobilised specimens highlighted the effect of rigid paralysis on the emergence of shape in the femoral condyles (Figure 3A–F). Without muscle contraction the general shape of the knee joint is much simpler, joint surfaces are flattened (e.g. flattening of the lateral condyle shown in Figure 3C and F) and functional outgrowths are lost. After 4 days of immobilisation the reduction in width of the intercondylar fossa was very obvious, particularly ventrally (Figure 3A). The surface of

the lateral condyle also appeared to be flattened by immobilisation (Figure 3C). Five days of immobilisation caused greater simplification. In particular note the reduction in height of the medial and lateral condyles in the dorsal aspect (Figure 3D, brackets) and the reduced outgrowths on the ventral aspect of the condyles (arrowheads Figure 3D) in the region of the trochlea fibularis grove where the femur interfaces with the fibula enabling smooth movement in later life. A flattening of the articular surfaces was now apparent in both condyles (Figure 3E,F).

Alteration of cell proliferation patterns in immobilised specimens

A comparison of the proportion of proliferating cells in five selected locations of the distal femur in control and experimentally immobilised embryos (4 days of immobilisation) was performed

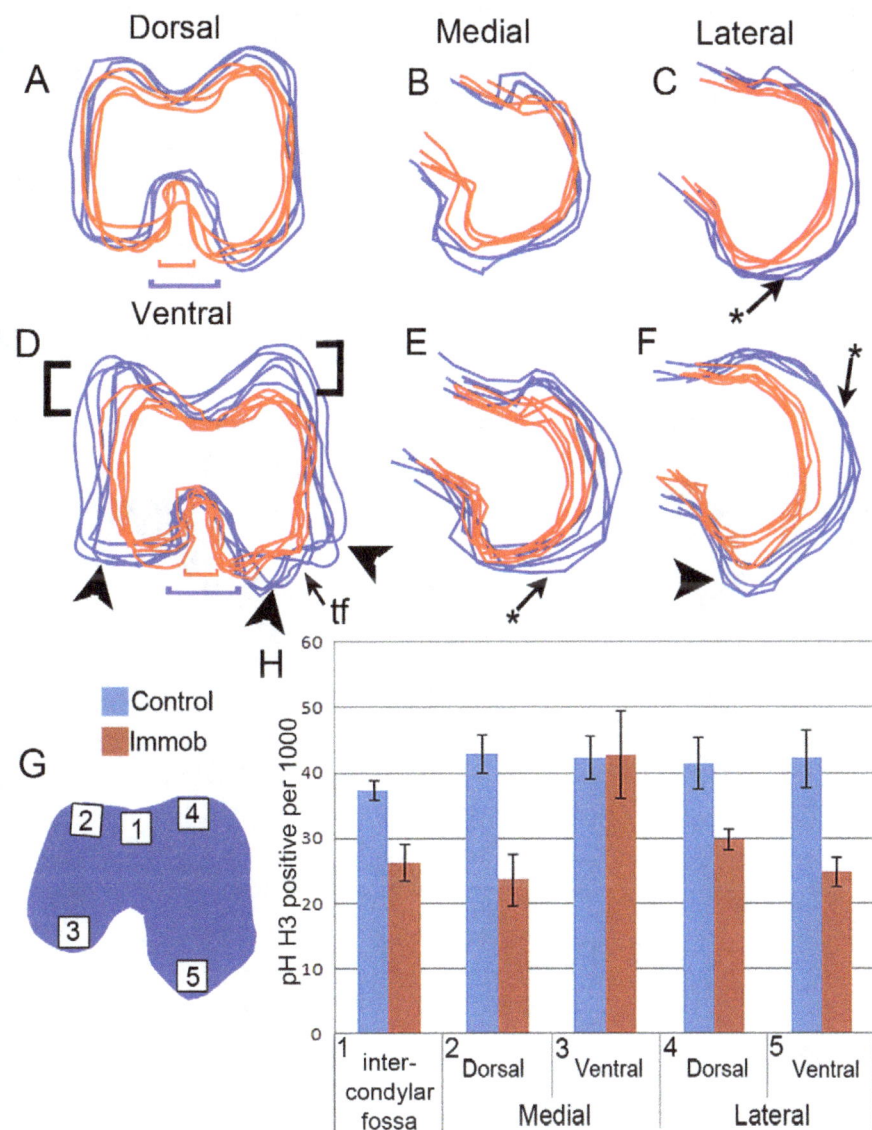

Figure 3. Comparison of cartilage shape and cell proliferation in distal femora of control (blue) and immobilised (red) embryos.
Outlines of the cartilage anlaga (A–F) were extracted from virtual sections (Fig.2, C, D, F). A minimum of four outlines per treatment were overlaid. Arrow heads indicate region of reduced outgrowth in the immobilised animals. *indicates flattening of rudiment surfaces. Black brackets show extent of growth reductions in the dorsal aspect of the condyles. Coloured brackets compare width of the intercondylar fossa of control (red) and immobilised (blue) embryos (A, D). The distribution of proliferating chondroctyes in the femur across 5 regions, represented in G (boxes 1–5), was compared in control (blue) and immobilised (red) embryos (h). tf; trochlea fibularis, IF; intercondylar fossa.

(Figure 3G,H). Looking across the locations in control animals, similar proportions of proliferating cells were observed across the femur head except for a slightly lower proportion in the region adjacent to the intercondylar fossa. The effect of treatment on cell proliferation was investigated using a generalised linear mixed effects model where multiple sections were nested within individuals in order to take account of the nested nature of the data. Combining data across the locations, treatment significantly reduced the proportion of proliferating cells by an average of 11.8/1000 chondrocytes (s.e. = 3.7, df = 4, p = 0.03) (not shown). Examining each location separately, immobilisation caused a significant reduction in the proportion of proliferating cells in four out of the five locations with the ventral aspect of the medial condyle being the only region unaffected (Figure 3H). These changes correspond to the major alterations observed in joint shape following immobilisation with large reductions in the dorsal side of both condyles and no apparent reduction in the ventral aspect of the medial condyle (Figure 3D). Narrowing of the intercondylar fossa was one of the strongest effects of immobilisation revealed by the morphometric study (Table 2) and the proportion of proliferating cells was reduced in the cartilage rudiment adjacent to the dorsal aspect of the fossa.

Effect of immobilisation on patterns of biophysical stimuli in the developing knee

We previously constructed a FE model to predict biophysical stimuli during an extension/flexion cycle of muscle contraction in the developing knee joint region [30]. We adapted this model to represent rigid limb paralysis (simultaneous contraction of all muscles), as induced by DMB, to compare predicted biophysical stimuli in immobilised as compared to normal developing joints. The obvious change to the mechanical environment in the rigid model is the loss of dynamic stimulation associated with a contraction cycle since muscle tension is constant. In addition, changes in the predicted patterns of the local mechanical environment are indicated, including reductions to the magnitude of loads and a general simplification of the spatial pattern of stimuli (Figures 4 and 5).

Focusing on the distal femur (Figure 4), the most obvious pattern of stimuli observed in the normal models was the distinct peak centred on the region adjacent to the intercondylar fossa and a general elevation of stimuli in the dorsal regions of both condyles (Figure 4E–G, Flex, Extend). While this peak was still predicted for patterns of Von Mises stress and Minimum Principal stress (compression) (Figure 4E, G) under rigid paralysis, it is reduced for patterns of Maximum Principal stress (tension) (Figure 4F). Thus, while the normal joint develops under the influence of dynamic patterns of tension and compression (compare Flex and Extend for each) the rigid paralysis model is largely experiencing compression.

Biophysical stimuli patterns in the interzone region correspond with the location of anatomical features that are disrupted following rigid paralysis

The FE simulations of biophysical stimuli in the developing chick knee joint comparing normal muscle contraction and rigid paralysis, as described above, were also used to examine the mechanical environment in the interzone region between the skeletal rudiments (Figure 5). The models were generated using the

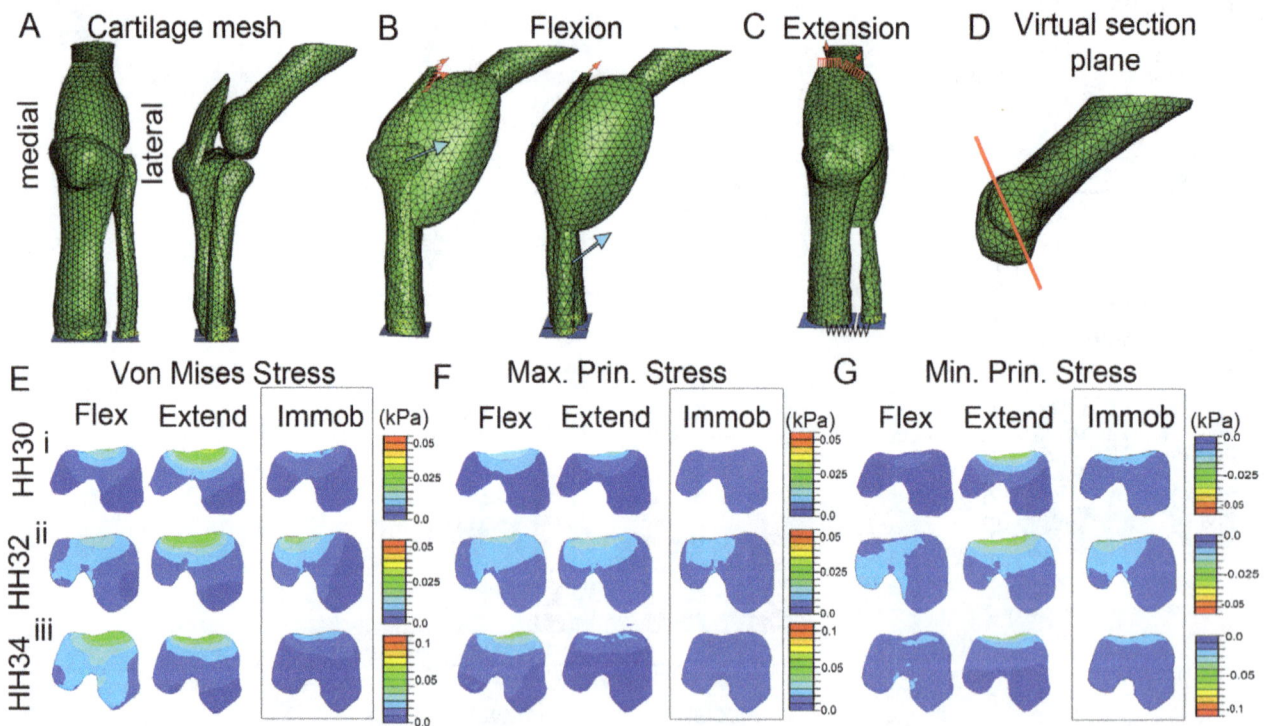

Figure 4. Predicted patterns of biophysical stimuli in FE models of immobilised (boxed, right column of each panel) compared to normal developing femora. Illustrations show the FE mesh (A), the placement of the loads for both flexion (B) and extension (C) and the plane of section (D) shown in E, F and G. The patterns of Von Mises stress (E), Maximum (tension) (F) and Minimum (compression) (G) Principal stress, were captured in equivalent sections through the distal femur (indicated in D) at HH30 (row i), HH32 (ii) and HH34 (iii). Patterns at mid flexion and extension are shown for the normal models whereas the constant pattern during rigid paralysis is shown for immobilised. Stimuli patterns for normal contractions are captured from simulations previously described [30].

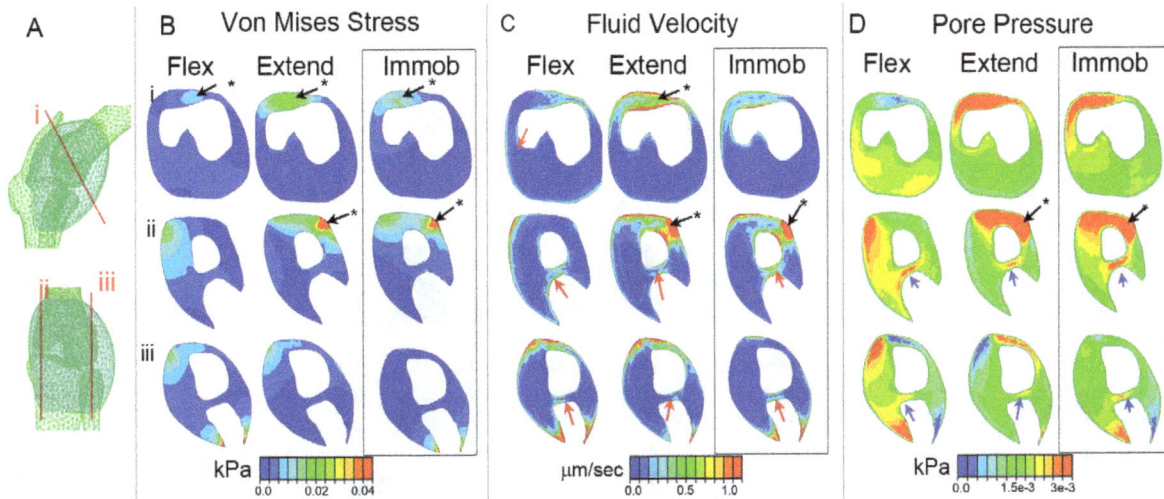

Figure 5. Comparison of the Von Mises Stress (B), Fluid Velocity (C) and Pore Pressure (D) within sections of the developing knee interzone region at HH32 under rigid paralysis (boxed on right) and normal muscle contractions. Location of sections, is indicated by red lines in A. * indicates location of the patella region (B), red arrow indicates pattern of elevated fluid velocity in the presumptive chondrogenous layer (C), blue arrows indicate peak in pore pressure at the intermediate layer (D). Flex; mid flexion, Extend; mid extension, Immob; Immobilisation. The images for normal contractions are captured from simulations previously described [30].

first direct estimates of mechanical properties of cartilage and interzone tissue in the developing chick knee using nanoindentation, as previously described [30]. The general pattern of stimuli did not vary between stages, thus Figure 5 presents the results at HH32 only, the midpoint in the study.

The patterns of stimuli predicted under normal muscle contractions indicated that several territories and tissues developing within the interzone experience specific patterns of stimulation [initially described in 30, Figure 5]. While the patella normally develops under dynamic magnitudes of stress, fluid velocity and pore pressure (Figure 5B–D, indicated by *; compare Flex and Extend), in rigid paralysis, the territory of elevated stimulation is restricted and the dynamic aspect is lost. This is of particular note since the patella fails to appear in immobilised embryos (Figure 1). Appearance of the chondrogenous layers is also lost in immobilised embryos. The chondrogenous layers emerge in a location that experiences dynamic patterns of elevated fluid velocity (Figure 5C. indicated by red arrow) while the intermediate layer, separating the two chondrogenous layers, emerges under a pattern of dynamically elevated pore pressure (Figure 5D, indicated by blue arrow). Under rigid paralysis, elevated fluid velocity in the presumptive chondrogenous layer and elevated pore pressure in the presumptive intermediate layer is still predicted (Figure 5, C,D), although in less extensive territories, but the pattern is no longer dynamic.

The expression of genes that regulate joint morphogenesis is altered in immobilised embryos

Specific changes were observed in the shape of cartilage rudiments and the appearance of tissues associated with the knee joint in immobilised embryos (Figures 1 and 3). To explore the molecular basis of these changes, a number of regulatory genes implicated in cartilage growth or joint cavity formation were selected for expression analysis, comparing immobilised and control embryos. The candidate genes (Table 1) were also selected on the basis of some evidence of mechanosensitivity in another context (e.g. cell culture). Gene expression analysis was carried out on the same embryos used for morphometric analysis above; using

the right hind limb for alcian blue staining/OPT scanning and the left hind limb for gene expression analysis. Three candidate genes, β1 integrin, FGFR2 and WNT9a showed no difference in expression pattern between control and immobilised specimens (not shown). Figure 6 shows differences in the expression observed in control and immobilised embryos on longitudinal sections through the knee joint (plane of section as in Figure 2D).

In the knee joint at this stage PTHLH expression was detected only at the very proximal and distal ends of the cartilaginous anlagen, restricted to the periarticular cartilage where the rudiments oppose (Figure 6A, A'). Immobilisation disrupts the characteristic pattern of PTHLH so that it is no longer restricted to the periarticular regions but is detected across the interzone (Figure 6B, B') (n = 5/5). The expression pattern in immobilised animals appears to have "fuzzy" boundaries compared to the more clearly defined territories in control specimens. This surprising finding of expression across the joint region of a gene normally restricted to cells within cartilage rudiments is complimented by detection of COL2A1 transcripts in some cells spanning the interzone in immobilised individuals (n = 7/7), similar to the expression of PTHLP (Figure 6C, D, C' D').

Expression of FGF2 in control knee joints was detected within the chondrogenous layers of the interzone, part of the forming meniscus, the region of the future patella, prominently within the cranial cnemial crest (Figure 6E) and surrounding the developing tendons (not shown). Immobilisation resulted in a specific alteration to the spatial expression of FGF2 with expression no longer detected in the chondrogenous layers at the point where the femur is closest to the tibiotarsus (n = 5/5) (Figure 6F) i.e. normally continuous expression in the chondrogenous layers is disrupted. Expression in the presumptive patella region and meniscus was also absent in immobilised animals whereas expression in the cranial cnemial crest was unchanged (Figure 6F).

In control specimens BMP2 transcripts were detected in the perichondrium along the length of the rudiments, the intermediate layer of the interzone, the developing patella and joint capsule (Figure 6G). When BMP2 expression was analysed in immobilised embryos, clear, elevated expression was no longer detected in the

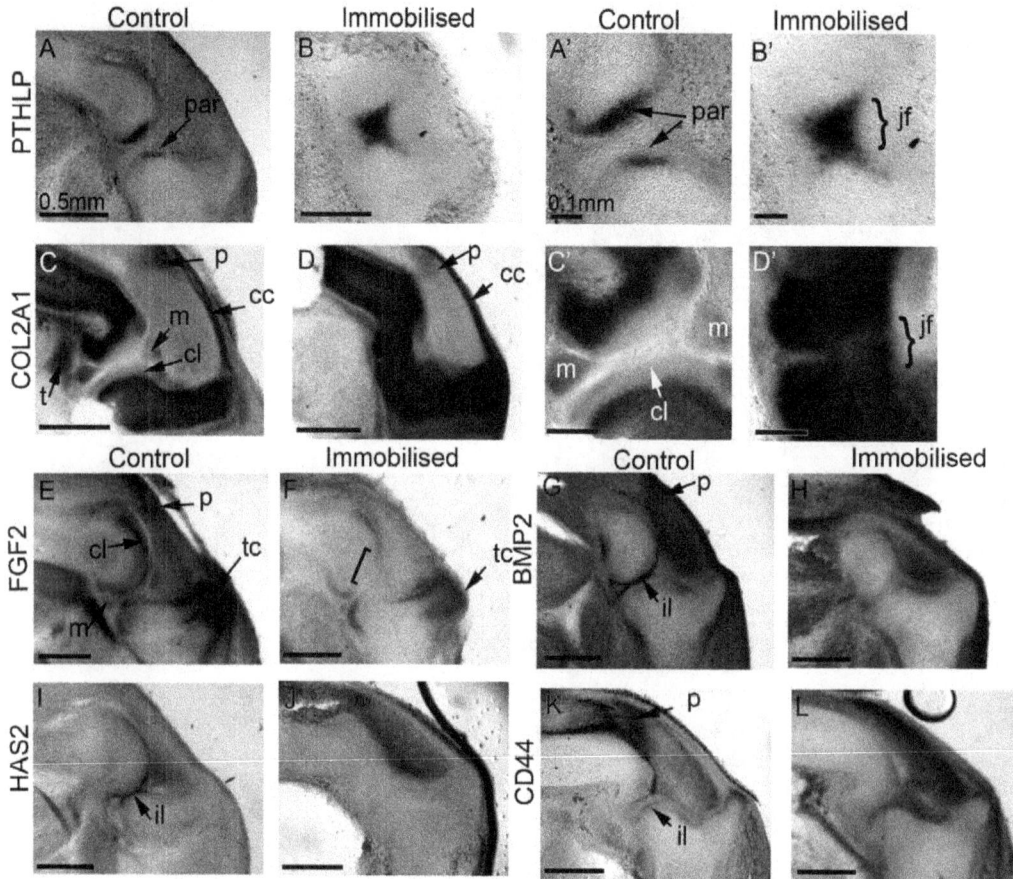

Figure 6. Expression of candidate mechanosensitive genes in control and immobilised specimens on longitudinal sections through the knee joint. Images A'–D' (scale bar 0.1 mm) show the knee region of A to D (scale bar 0.5 mm) at a higher magnification. cc; capsular condensation, cl; chondrogenous layer, iz; intermediate layer, jf; joint fusion, m; meniscus, p; patella, par; periarticular cartilage, t; tendon, tc; tibial crest.

intermediate layer (n = 10/10) (Figure 6H). Expression in the perichondrium and joint capsule remained unchanged and some expression of BMP2 within the patella could be detected although the expression level and size of the expression domain appeared to be reduced (not shown).

Similar to BMP2 expression, CD44 and HAS2 have very defined expression within the presumptive joint line region in the intermediate layer of the interzone (Figure 6I, k). HAS2 is also expressed in the most distal part of the inter-patella-femoral fat pad adjacent to the tibiotarsus (Figure 6I). CD44 also shows additional expression in the region of the future patella (Figure 6K), the muscle blocks and cells surrounding the ligaments (not shown). Expression of both genes was lost from the intermediate layer of immobilised joints (Figure 6J,L). This loss occurred in all specimens analysed (8 assayed for HAS2, 7 for CD44). Specific CD44 expression in the region of the future patella was lost in immobilised specimens (Figure 6L). In addition, both genes showed elevated expression throughout the inter-patella-femoral fat pad of immobilised embryos (Figure 6J,L).

Discussion

Blocking muscle contractions in the chick embryo alters the biophysical environment of the developing musculoskeletal system. In this work we used computational modelling to demonstrate how rigid immobilisation would affect the mechanical stimuli generated

in the developing knee joint and we investigated the impact of such immobilisation on the tissues of the developing joint at morphological and molecular levels. We showed that when dynamic stimulation is removed, patterning of the interzone and joint morphogenesis are altered with very specific changes to the shape of the cartilage rudiments and abnormal definition of tissue territories within the presumptive joint. The altered shape of the cartilaginous rudiments is accompanied by region specific changes in cell proliferation. The altered definition of tissues within the joint interzone is shown not only by the expression of tissue marker genes but also by altered expression of regulatory genes that are involved in steering differentiation and morphogenesis.

The dependence of correct joint development on stimulation from muscle contractions was previously shown by similar immobilisation studies in the chick [10,11,21,23,58] and using genetically altered mice that have absent, reduced or non-contractile muscle [19,20]. However in the altered mechanical environment of mouse models, Nowlan et al [19] showed that forelimbs are more affected than hindlimbs and that while the elbow joint is severely affected, the knee joint appears normal. In the current work we compared the effects of immobilisation on the knee and elbow joints in the chick and found that the elbow joint showed similar but no more severe alterations (not shown) highlighting a clear and intriguing difference between avian and mammalian models. Computational modelling of the mouse model demonstrated that passive displacement of embryonic limbs *in utero* due to movement of the mother and normal littermates would produce greater stimulation of the hindlimbs than the forelimbs providing a possible source of compensation for the reduced stimulation, particularly in the hindlimbs (unpublished data). The *in ovo* situation of the chick embryo means less passive movement from external sources and therefore a greater reliance on muscle contractions to generate mechanical stimuli in the hindlimbs.

Previous studies demonstrated that knee joints in immobilised chick embryos fail to cavitate [10,11,21,23,58,59,60] but there has been very little emphasis on changes to the shape of the joint. Here we used morphometric analysis of 3D digital representations of the specimens to pin point the shape features of the distal femur that are dependent on extrinsically produced stimuli from muscle contractions, showing a link between the shape changes, changes in local cell proliferation and predicted biophysical stimuli. Specific changes included simplification of the shape of the medial and lateral condyles of the distal femur with flattening of the condyles and the absence of characteristic spurs. The strongest effect was seen on the separation of the condyles with a consistent narrowing of the intercondylar fossa. Cell division contributes to morphogenesis of a tissue when proliferation rates differ in a location specific manner and we previously showed that cell proliferation is greater in regions of the distal femur that grow most between stages HH30 and HH34 [30], for example the medial compared to the lateral condyle. Here we show that in immobilised embryos cell proliferation rates in the cartilaginous rudiments are reduced, specifically where dynamic mechanical stimulation is predicted to be strongest (dorsal aspect of both condyles and the region of intercondylar fossa) and where greatest shape changes are observed in immobilised specimens (reduction of height of the condyles and width of the intercondylar fossa). We therefore suggest that local patterns of biophysical stimuli contribute to the regulatory mechanisms controlling local growth. Previous *in vitro* studies have shown that mechanical stimulation can influence cell division and matrix biosynthesis [49,50,51,53] but we show a location specific effect *in vivo*, relevant to morphogenetic changes. The altered mechanical environment of rigid paralysis might resemble that of a statically loaded culture. Static loads have been found to inhibit biosynthesis of articular cartilage while dynamic loading, such as in normal muscle contraction cycles, increases synthesis and proliferation of chondrocytes [50,53]. Comparing the effects of rigid and flaccid paralysis (i.e. static compared to zero load), Osborne et al. [12] found that rigid paralysis led to a greater reduction in the width of the epipheses indicating that static loading may have an inhibitory effect on growth.

Currently much of what is known about cartilage growth in long bones relates to longitudinal growth and the associated regulation of the ossification process; very little is known about the control of local outgrowths and protrusions such as features of the condyles. Changes seen in the expression of PTHLP, FGF2 and BMP2 in immobilised animals have possible implications for the observed altered shape of the femoral condyles due to their roles in the regulation of diaphyseal cartilage growth [31]. PTHLP is known to maintain a pool of proliferating chondrocytes in the rudiments with the rate of chondrocyte proliferation within this pool modulated by BMPs and FGFs as part of a IHH/PTHLP feedback loop. Regional expression of these potential growth modulating molecules could influence local growth patterns within the femoral condyles. The altered expression of the genes encoding these molecules and our previous demonstration of mechanosensitivity of IHH expression in the developing tibiotarsus [29] indicate that mechanisms regulating cartilage growth are affected by immobilisation.

Immobilisation also impacted the process of cell differentiation in the interzone, as indicated by changes in characteristic gene expression patterns and histology. In the absence of normal muscle forces the expression patterns of marker genes Tenascin C and Collagen type II alpha 1 (COL2A1) and regulatory genes PTHLP, BMP2, FGF2, CD44 and HAS2 were altered (summarised in Figure 7). Normal expression of FGF2, BMP2, CD44 and HAS2 was disrupted or lost specifically in the interzone regions of immobilised embryos which acquire cartilage like tissue characteristics as indicated by the inappropriate activation of COL2A1 and PTHLP expression. In addition boundaries of gene expression between cartilage rudiments and the interzone were less distinct suggesting either cell movement and cell mixing between the territories or transdifferentiation of cells in the interzone to a cartilaginous character; the latter interpretation is supported by the findings of Kahn et al [20] of aberrant expression of COL2A1 in lineage labeled interzone cells (descended from Gdf5 expressing cells) in immobile mouse embryo limbs. Finite element models predicted that the patterns of biophysical stimuli created by muscle contractions correspond with the emergence of specific tissues in the joint, suggesting that they could contribute to the patterning of these tissues [30]. For example the chondrogenous layers which ultimately form the articular cartilages are predicted to develop under dynamic elevation of fluid velocity and stress. We propose from these findings that correct differentiation of interzone cells and the maintenance of interzone cell type are dependent on the mechanical environment to which they are exposed. It has been shown that cultured interzone cells initially express the interzone marker Gdf5, however after several days in culture the cells resemble chondrocytes and express markers such as Collagen type II [7]. This strongly supports the conclusion that following initial specification of the interzone, maintenance of the territory and further differentiation toward cell types of particular articular structures is dependent on mechanical stimulation.

Classical descriptions of joint development divide the process into two separate phases: interzone specification and cavitation. Earlier immobilisation studies suggested that only the cavitation phase is sensitive to mechanical stimulation since the interzone forms but fails to cavitate when contractions are altered [reviewed in 61,62]. Our findings show that under rigid paralysis the organisation of joint territories is altered as discussed above. Therefore mechanical stimulation impacts cellular processes involved in the definition of tissues and cellular differentiation prior to cavitation. This shows that mechanical stimulation should not be seen as having a molding effect on intrinsic morphogenetic processes [62,63] but as influencing the processes fundamentally. It also shows the importance of viewing joint development as a series of interlinked events, as argued by Lambe et al [61]; events that are impacted by mechanical stimulation from an early stage.

The data presented here show that multiple aspects of knee joint patterning and morphogenesis are affected when mechanical stimulation is altered (cell proliferation, cell differentiation and tissue boundaries with consequential alterations to the shape of rudiment epiphyses and the structure of the joint and associated tissues). These compound effects support our previously proposed hypothesis [30] that local patterns of biophysical stimuli create a type of positional information that contributes to the correct patterning of emerging tissues in the joint. Finite Element analysis was previously used to predict patterns of biophysical stimuli in the normal developing joint and here we used the same approach to simulate rigid paralysis showing changes in the stimuli patterns that correspond with the major changes observed in immobilised embryos. In rigid paralysis all muscles are in tetanus. Modelling this situation showed a reduction of stimuli in femoral condyles corresponding to sites of reduced proliferation and shape change and the replacement of a complex dynamic pattern of stimuli by a simplified, static environment. The observed changes in growth

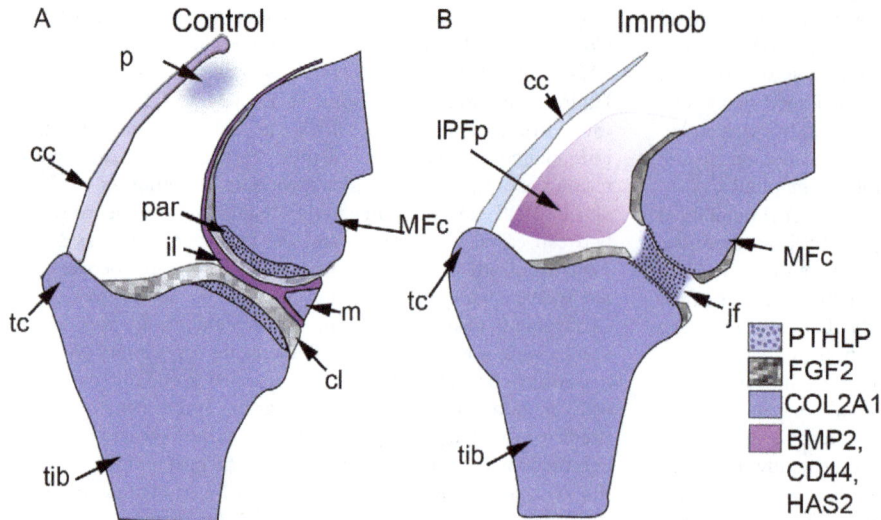

Figure 7. Representation of the altered patterns of regulatory gene expression due to immobilisation. Colour coded expression patterns of markers and regulatory genes of interest in both control and immobilised sections. cc; capsular condensation, cl; chondrogenous layer, il; intermediate layer, IPFp; inter-patella-femoral fat pad, jf; loss of joint line definition, m, meniscus, MFc; medial femoral condyle, p; patella, tc ; tibial crest, tib; tibiotarsus. Note: BMP2 and FGF 2 are expressed in the capsule (not shown in legend).

and proliferation may arise because of either the reduced stimulation or the lack of dynamic stimulation or a combination of both. In the interzone the patterns of stimuli predicted under normal and immobilised situations is similar but of course the forces are static rather than dynamic underlining the importance of dynamic stimuli, also observed in the response of cells in culture [49,50]. However, alterations to the stimuli patterns may be greater than predicted here for a number of reasons. The rigid model was based on normal morphology and assumed normally functioning tendons but tendon development is known to be negatively affected by immobilization [64] so transfer of loads may be compromised. Also the altered shape of the rudiments in immobilized specimens may alter the forces but this is unlikely to have a large effect since we previously found that the general pattern of the forces do not change dramatically with shape changes over time [30]. So our predictions may be conservative and actual changes to the biophysical environment in immobilized embryos may be more extreme than predicted.

Despite numerous examples of mechanoresponsiveness of tissues and cells and a long list of mechanosensitive genes demonstrated in culture, with a growing number demonstrated *in vivo* [reviewed in 65], we know very little about the biological mechanisms that integrate biophysical stimuli with gene regulation. Several possible sensory mechanisms including integrins, stretch activated ion channels and the primary cilium have been indicated in cellular mechanotransduction [66,67,68,69,70]. A number of proteins have been shown to be phosphorylated as a result of mechanical stimulation including MAP kinases like ERK1/2 [71]. Also, mechanical stimulation of human adult articular chondrocytes in culture results in transient tyrosine phosphorylation of the protein kinase pp125FAK, the focal adhesion protein paxillin, and the multifunctional signaling molecule b-catenin [67]. Primary Cilia have also been proposed as a cellular antenna capable of detecting mechanical strains [72,73] and have been identified on both adult and embryonic chondrocytes [74,75,76]. A number of ECM receptors are expressed on cilia [77] leading to the proposal that cilia may transduce mechanical forces from the ECM to the cell. Primary

cilia are particularly interesting in the present context because of the association between cilia and hedgehog signaling [78].

The regulatory genes analysed in this study were previously defined as mechanosensitive based on *in vitro* assays (Table 1). Here we show that spatial restriction of the gene expression patterns of PTHLP, BMP2, FGF2, HAS2, CD44, COL2A1 and TNC is responsive to biophysical stimuli in an *in vivo* context (Figure 7) revealing potential key mediators of mechanical stimulation of joint development that warrant closer analysis. A key question is if the genes respond directly to mechanical stimuli or if they lie downstream of other more direct mediators. If we can demonstrate in an appropriate assay system that the response is direct we open the possibility of revealing the cellular mechanisms that links mechanical stimulation with gene regulation. This is a key focus of future work.

Understanding the input of mechanical signals in the stable differentiation of skeletal tissues is of particular importance in attempts to regenerate tissue for replacement therapies including therapies for patients with articular cartilage defects such as arthritis. A wide range of different stimuli and culture methods have been used to recapitulate the process of articular cartilage development with varying success [69,79,80]. One particular problem is preventing chondrogenic cells from undergoing hypertrophy as they would in endochondral ossification [81]. A better understanding of how the interzone develops and in particular the mechanical requirements for articular cartilage development in embryos provides useful information on the type and magnitude of loads which could be applied to cultures to produce cartilage of the appropriate type and quality for regenerative therapies. The present work indicates that conditions that increase interstitial fluid flow might be beneficial in the regeneration of articular cartilage. To recapitulate stable cartilage differentiation, the process needs to be better understood.

Materials and Methods

In ovo immobilisation

Fertilised chick eggs were purchased from Enfield Broiler Breeders and incubated (Solway Natureform) at 37.5°C and 70%

humidity. The work was approved by the ethics committee Trinity College Dublin. Work on early chick embryos in ovo does not require a license from the Irish Ministry of Health under European Legislation. Immobilisation was induced by the application of the neuromuscular blocking agent Decamethonium bromide (DMB) (Sigma). Following 3 days of incubation, 4 mls of albumen was removed from each egg using a 21 gauge needle. Immobilisation treatments consisted of the application of 100 µl of 0.5% DMB (Sigma) in sterile Hank's Buffered Saline Solution (HBSS) (Sigma) plus 100 units/ml antibiotic/antimycotic (Gibco) were started after 4.5 days of treatment. Controls were treated with 100 µl of sterile HBSS. Treatment was repeated daily until the embryos were harvested after a further 4 or 5 days of incubation. The experiment was repeated independently three times.

At the end of the experiment the embryos were dissected in ice cold Phosphate Buffered saline (PBS) (Sigma), staged using the Hamburger and Hamilton criteria [56] and cut longitudinally down the spine. The right side was fixed overnight in fresh 4% paraformaldehyde (PFA) in PBS at 4°C, dehydrated through a graded series of methanol/PBT (0.1% Triton X100 in PBS; 25%, 50%, 75% methanol; 1×10 minute) washes, followed by 2×10 minutes 100% methanol. After dehydration the embryos were stored in 100% methanol at −20°C until needed. The left hand side of the embryo including the head and neck was fixed in 95% ETOH for 3 days and stained for cartilage using alcian blue [as per 5]. The knee joint region were subsequently imaged using Optical Projection Tomography [as per 5].

In situ hybridisation

Expression probes were prepared from cDNA clones obtained from the Biotechnology and Biological Sciences Research Council (BBSRC) ChickEST Database and its bank of expressed sequence tags (ESTs) [82]. Details of the clones used to produce all probes are given in Table 3. Antisense and sense digoxigenin-labelled RNA was transcribed in vitro from 1 µg of linearized plasmid using T7 and T3 promoter sites (according to insert orientation) in the pBluescript II KS+ vector with all components for in vitro transcription purchased from Roche, Germany. DNA template was degraded by incubation of probes with RNase free DNase (Roche) and probes were purified on G25 columns (Amersham Biosciences, USA) according to the manufacturer's instructions. Probe concentrations were determined by spectophotometry and probes were stored at 20°C.

Limbs fixed in 4%PFA, dehydrated and stored at −20°C were rehydrated through a series of methanol/PBT solutions (75%, 50%, 25%; each 10 minutes) at 4°C and subsequently washed 2×10 minutes in PBT. On rehydration the limbs were further dissected to remove the foot and skin. The knee joint of the specimens were was then embedded in 4% low melting point (LMP) agarose/PBS (Invitrogen, UK) and 100 µm longitudinal sections were cut using a vibrating microtome (VT1000S, Leica). Hybridisation was carried out largely as per Nowlan et al [13].

Histology

Hind limbs of immobilised and control embryos, 4%PFA fixed, dehydrated and stored at −20°C, were embedded in paraffin wax and sectioned as described in Roddy et al 2009, stained using 0.5% Alcian blue (30 min), Harris Haematoxylin (6 min) (Sigma-Aldrich) and counterstained using Eosin. The sections were mounted and photographed using a Nikon Optiphot-2 microscope mounted with a Canon EOS 350D camera.

Determining rates of cell proliferation

The proportion of proliferating cells was determined using the mitosis marker anti-phospho-histone H3 PABs (Millipore). Briefly, the hind limbs of control and immobilised embryos (n = 3) were rehydrated, skinned and embedded in 1.5% agarose, 5% sucrose. The blocks were equilibrated in 30% sucrose solution and frozen over a dry ice bath. 25–30 µm longitudinal sections were collected on BDH superfrost+ slides.

Following heat mediated antigen retrieval (50 µM Tris pH 8 for 35 minutes in a 95°C water bath), sections were washed (3X PBS with 0.5% Triton X-100 and 0.1% Tween 20) and blocked in 5% normal goat serum in 0.5% Triton X-100 and 0.1% Tween 20 for one hour at room temperature. Incubation with the primary antibody (anti-phospho-histone H3 PABs, Millipore P84243), was carried out in blocking solution overnight at 4°. Sections were washed and blocked as before and incubated in secondary antibody Cy3 goat anti-rabbit IGg (1/200, Jackson immuno) at 4°C overnight. Following further washing, the sections were mounted in ProLong Gold anti-fade reagent with DAPI (4′, 6-diamidino 2-phenylindole) (Invitrogen).

The density of proliferating chondrocytes was determined in five cartilage regions (Figure 3g): adjacent to the intercondylar fossa, and the dorsal and ventral portions of the medial and lateral condyles. Each region was imaged separately using an Olympus FV1000 point scanning confocal microscope. The numbers of

Table 3. Summary of in situ probes and elected ChickESTs.

Gene	ChEST reference	Genebank reference	Alignment of ChEST on reference
BMP2	ChEST 367 j4	AY237249.1	66–709
CD44	ChEST 343 m10	XM_001232450.1	1940–1794
HAS2	ChEST 500 e4	NM_204806.1	2507–2025
β1 integrin	ChEST 500 j17	NM_001039254.1	270–1022
WNT9a	ChEST 592 n13*	NM_204891.1	784–1213
PTHLP	ChEST 533 c1	AB175678	68–734
FGF2	ChEST 432 i3	M95706.1	145–445
FGFR2	ChEST 699 I24*	NM_205319.1	1969–2716
COL2A1	ChEST 179 I15	NM204426.1	3887–4689
TNC	ChEST 681 I9	CHKTEN	2809–2075

chondrocytes and proliferating cells were counted within a box 1.44 mm^2 for two independent focal planes on two sections per specimen (n = 3). The effect of treatment on the cell proliferation was statistically analysed using a generalised linear mixed effects model where multiple sections were nested within individuals in order to take account of the nested nature of the data (R statistical package).

Modelling rigid limb paralysis using FE analysis

DMB induces rigid paralysis where the muscles are in continuous contraction [12]. To account for this, the previously described Finite Element models of the developing knee [30] were adjusted to simulate rigid muscle paralysis by applying all muscle forces simultaneously and continuously. In the normal model which represents contractions in the control experimental situation, the muscles attached to the ventral tibiotarsus and fibula were active during the flexion contraction while those attached to the capsular condensation are activate in the extension

contraction. Immobilisation leads to a reduction in muscle size and its ability to transmit forces [83,84]. The magnitudes of the forces applied to the model were therefore adjusted to 75% of the normal estimation [derived from 84]. Reiser et al. [84] measured the forces generated by normal and immobilised embryonic muscles. The same boundary conditions and material properties were used in normal and paralysis models.

Acknowledgments

We thank Dr. Niamh Nowlan, Dr. Suzanne Miller-Delaney, and Mr. Peter Stafford for technical advice and Ms Blaithin Arnold for assistance.

Author Contributions

Conceived and designed the experiments: PM PP KR. Performed the experiments: KR. Analyzed the data: KR PP PM. Contributed reagents/materials/analysis tools: PM PP. Wrote the paper: PM KR.

References

1. Wilsman NJ, Leiferman EM, Fry M, Farnum CE, Barreto C (1996) Differential growth by growth plates as a function of multiple parameters of chondrocytic kinetics. Journal of Orthopaedic Research 14: 927–936.
2. Goldring MB, Tsuchimochi K, Ijiri K (2006) The control of chondrogenesis. Journal of Cellular Biochemistry 97: 33–44.
3. Pacifici M, Koyama E, Iwamoto M (2005) Mechanisms of synovial joint and articular cartilage formation: recent advances, but many lingering mysteries. Birth Defects Res C Embryo Today 75: 237–248.
4. Provot S, Schipani E (2005) Molecular mechanisms of endochondral bone development. Biochemical and Biophysical Research Communications 328: 658–665.
5. Roddy KA, Nowlan NC, Prendergast PJ, Murphy P (2009) 3D representation of the developing chick knee joint: a novel approach integrating multiple components. J Anat 214: 374–387.
6. Pacifici M, Koyama E, Shibukawa Y, Wu C, Tamamura Y, et al. (2006) Cellular and molecular mechanisms of synovial joint and articular cartilage formation. Ann N Y Acad Sci 1068: 74–86.
7. Koyama E, Shibukawa Y, Nagayama M, Sugito H, Young B, et al. (2008) A distinct cohort of progenitor cells participates in synovial joint and articular cartilage formation during mouse limb skeletogenesis. Developmental Biology 316: 62–73.
8. Hammond E, Donnenfeld AE (1995) Fetal akinesia. Obstet Gynecol Surv 50: 240–249.
9. Hall JG (1986) Analysis of Pena Shokeir phenotype. Am J Med Genet 25: 99–117.
10. Persson M (1983) The role of movements in the development of sutural and diarthrodial joints tested by long-term paralysis of chick embryos. J Anat 137(Pt 3): 591–599.
11. Drachman DB, Sokoloff L (1966) The role of movement in embryonic joint development. Developmental Biology 14: 401–420.
12. Osborne AC, Lamb KJ, Lewthwaite JC, Dowthwaite GP, Pitsillides AA (2002) Short-term rigid and flaccid paralyses diminish growth of embryonic chick limbs and abrogate joint cavity formation but differentially preserve pre-cavitated joints. J Musculoskelet Neuronal Interact 2: 448–456.
13. Nowlan NC, Prendergast PJ, Murphy P (2008) Identification of mechanosensitive genes during embryonic bone formation. PLoS Comput Biol 4: e1000250.
14. Mitrovic D (1982) Development of the Articular Cavity in Paralyzed Chick Embryos and in Chick Embryo Limb Buds Cultured on Chorioallantoic Membranes. Cells Tissues Organs 113: 313–324.
15. Lelkes G (1958) Experiments in vitro on the role of movement in the development of joints. J Embryol Exp Morphol 6: 183–186.
16. Fell HB, Canti RG (1934) Experiments on the Development in vitro of the Avian Knee-Joint. Proceedings of the Royal Society of London, Series B, Biological Sciences, 116: 316–351.
17. Rot-Nikcevic I, Reddy T, Downing K, Belliveau A, Hallgrímsson B, et al. (2006) Myf5 −/−:MyoD −/− amyogenic fetuses reveal the importance of early contraction and static loading by striated muscle in mouse skeletogenesis. Development Genes and Evolution 216: 1–9.
18. Gomez C, David V, Peet NM, Vico L, Chenu C, et al. (2007) Absence of mechanical loading in utero influences bone mass and architecture but not innervation in Myod-Myf5-deficient mice. Journal of Anatomy 210: 259–271.
19. Nowlan NC, Bourdon C, Dumas G, Tajbakhsh S, Prendergast PJ, et al. (2010) Developing bones are differentially affected by compromised skeletal muscle formation. Bone 46: 1275–1285.
20. Kahn J, Shwartz Y, Blitz E, Krief S, Sharir A, et al. (2009) Muscle Contraction Is Necessary to Maintain Joint Progenitor. Cell Fate 16: 734–743.
21. Murray PD, Drachman DB (1969) The role of movement in the development of joints and related structures: the head and neck in the chick embryo. J Embryol Exp Morphol 22: 349–371.
22. Rodriguez JI, Garcia-Alix A, Palacios J, Paniagua R (1988) Changes in the long bones due to fetal immobility caused by neuromuscular disease. A radiographic and histological study. J Bone Joint Surg Am 70: 1052–1060.
23. Ruano-Gil D, Nardi-Vilardaga J, Tejedo-Mateu A (1978) Influence of extrinsic factors on the development of the articular system. Acta Anat (Basel) 101: 36–44.
24. Heegaard JH, Beaupré GS, Carter DR (1999) Mechanically modulated cartilage growth may regulate joint surface morphogenesis. Journal of Orthopaedic Research 17: 509–517.
25. Carter DR, Wong M (1988) The role of mechanical loading histories in the development of diarthrodial joints. J Orthop Res 6: 804–816.
26. Stevens SS, Beaupré GS, Carter DR (1999) Computer model of endochondral growth and ossification in long bones: biological and mechanobiological influences. J Orthop Res 17: 646–653.
27. Shefelbine SJ, Carter DR (2002) Development of the femoral bicondylar angle in hominid bipedalism. Bone 30: 765–770.
28. Shefelbine SJ, Carter DR (2004) Mechanobiological predictions of growth front morphology in developmental hip dysplasia. J Orthop Res 22: 346–352.
29. Nowlan NC, Murphy P, Prendergast PJ (2008) A dynamic pattern of mechanical stimulation promotes ossification in avian embryonic long bones. Journal of Biomechanics 41: 249–258.
30. Roddy KA, Kelly GM, van Es MH, Murphy P, Prendergast PJ (2011) Dynamic patterns of mechanical stimulation co-localise with growth and cell proliferation during morphogenesis in the avian embryonic knee joint. Journal of Biomechanics 44: 143–149.
31. Minina E, Wenzel HM, Kreschel C, Karp S, Gaffield W, et al. (2001) BMP and Ihh/PTHrP signaling interact to coordinate chondrocyte proliferation and differentiation. Development 128: 4523–4534.
32. Francis-West PH, Parish J, Lee K, Archer CW (1999) BMP/GDF-signalling interactions during synovial joint development. Cell Tissue Res 296: 111–119.
33. Hilton MJ, Tu X, Cook J, Hu H, Long F (2005) Ihh controls cartilage development by antagonizing Gli3, but requires additional effectors to regulate osteoblast and vascular development. Development 132: 4339–4351.
34. Karp SJ, Schipani E, St-Jacques B, Hunzelman J, Kronenberg H, et al. (2000) Indian hedgehog coordinates endochondral bone growth and morphogenesis via parathyroid hormone related-protein-dependent and -independent pathways. Development 127: 543–548.
35. Duprez D, de H. Bell EJ, Richardson MK, Archer CW, Wolpert L, et al. (1996) Overexpression of BMP-2 and BMP-4 alters the size and shape of developing skeletal elements in the chick limb. Mechanisms of Development 57: 145–157.
36. Bi W, Deng JM, Zhang Z, Behringer RR, de Crombrugghe B (1999) Sox9 is required for cartilage formation. Nat Genet 22: 85–89.
37. Merino R, Macias D, Ganan Y, Economides AN, Wang X, et al. (1999) Expression and function of Gdf-5 during digit skeletogenesis in the embryonic chick leg bud. Dev Biol 206: 33–45.
38. Brunet LJ, McMahon JA, McMahon AP, Harland RM (1998) Noggin, Cartilage Morphogenesis, and Joint Formation in the Mammalian Skeleton. Science 280: 1455–1457.
39. Spagnoli A, O'Rear L, Chandler RL, Granero-Molto F, Mortlock DP, et al. (2007) TGF-{beta} signaling is essential for joint morphogenesis. J Cell Biol 177: 1105–1117.
40. Spater D, Hill TP, O'Sullivan RJ, Gruber M, Conner DA, et al. (2006) Wnt9a signaling is required for joint integrity and regulation of Ihh during chondrogenesis. Development 133: 3039–3049.

41. Hartmann C, Tabin CJ (2001) Wnt-14 plays a pivotal role in inducing synovial joint formation in the developing appendicular skeleton. Cell 104: 341–351.

42. Lanske B, Karaplis AC, Lee K, Luz A, Vortkamp A, et al. (1996) PTH/PTHrP Receptor in Early Development and Indian Hedgehog–Regulated Bone Growth. Science 273: 663–666.

43. Minina E, Kreschel C, Naski MC, Ornitz DM, Vortkamp A (2002) Interaction of FGF, Ihh/Pthlh, and BMP Signaling Integrates Chondrocyte Proliferation and Hypertrophic Differentiation. 3: 439–449.

44. Dowthwaite GP, Edwards JC, Pitsillides AA (1998) An essential role for the interaction between hyaluronan and hyaluronan binding proteins during joint development. J Histochem Cytochem 46: 641–651.

45. Dowthwaite GP, Flannery CR, Flannelly J, Lewthwaite JC, Archer CW, et al. (2003) A mechanism underlying the movement requirement for synovial joint cavitation. Matrix Biol 22: 311–322.

46. Pitsillides AA (2003) Identifying and characterizing the joint cavity-forming cell. Cell Biochem Funct 21: 235–240.

47. Garciadiego-Cazares D, Rosales C, Katoh M, Chimal-Monroy J (2004) Coordination of chondrocyte differentiation and joint formation by alpha5beta1 integrin in the developing appendicular skeleton. Development 131: 4735–4742.

48. McMahon L, Reid A, Campbell V, Prendergast P (2008) Regulatory Effects of Mechanical Strain on the Chondrogenic Differentiation of MSCs in a Collagen-GAG Scaffold: Experimental and Computational Analysis. Annals of Biomedical Engineering 36: 185–194.

49. Buschmann M, Gluzband Y, Grodzinsky A, Hunziker E (1995) Mechanical compression modulates matrix biosynthesis in chondrocyte/agarose culture. J Cell Sci 108: 1497–1508.

50. Davisson T, Kunig S, Chen A, Sah R, Ratcliffe A (2002) Static and dynamic compression modulate matrix metabolism in tissue engineered cartilage. Journal of Orthopaedic Research 20: 842–848.

51. Fukuda K, Asada S, Kumano F, Saitoh M, Otani K, et al. (1997) Cyclic tensile stretch on bovine articular chondrocytes inhibits protein kinase C activity. Journal of Laboratory and Clinical Medicine 130: 209–215.

52. Sironen RK, Karjalainen HM, Elo MA, Kaarniranta K, Torronen K, et al. (2002) cDNA array reveals mechanosensitive genes in chondrocytic cells under hydrostatic pressure. Biochim Biophys Acta 1591: 45–54.

53. Wu QQ, Chen Q (2000) Mechanoregulation of chondrocyte proliferation, maturation, and hypertrophy: ion-channel dependent transduction of matrix deformation signals. Exp Cell Res 256: 383–391.

54. Cillo JE, Gassner R, Koepsel RR, Buckley MJ (2000) Growth factor and cytokine gene expression in mechanically strained human osteoblast-like cells: Implications for distraction osteogenesis. Oral Surgery, Oral Medicine, Oral Pathology, Oral Radiology & Endodontics 90: 147–154.

55. Kanbe K, Inoue K, Xiang C, Chen Q (2006) Identification of clock as a mechanosensitive gene by large-scale DNA microarray analysis: downregulation in osteoarthritic cartilage. Modern Rheumatology 16: 131–136.

56. Hamburger V, Hamilton HL (1951) A series of normal stages in the development of the chick embryo. J Morphol 88: 49–92.

57. Summerhurst K, Stark M, Sharpe J, Davidson D, Murphy P (2008) 3D representation of Wnt and Frizzled gene expression patterns in the mouse embryo at embryonic day 11.5 (Ts19). Gene Expr Pattern doi:101016/jgep200801007.

58. Mikic B, Johnson TL, Chhabra AB, Schalet BJ, Wong M, et al. (2000) Differential effects of embryonic immobilization on the development of fibrocartilaginous skeletal elements. J Rehabil Res Dev 37: 127–133.

59. Hamburger V, Waugh M (1940) The Primary Development of the Skeleton in Nerveless and Poorly Innervated Limb Transplants of Chick Embryos. Physiological Zoology 13: 367–382.

60. Hogg DA, Hosseini A (1992) The effects of paralysis on skeletal development in the chick embryo. Comp Biochem Physiol Comp Physiol 103: 25–28.

61. Lamb KJ, Lewthwaite JC, Bastow ER, Pitsillides AA (2003) Defining boundaries during joint cavity formation: going out on a limb. Int J Exp Pathol 84: 55–67.

62. Thorogood PV (1983) Morphogenesis of cartilage. In: Hall BK, ed. Cartilage. New York: Academic Press.

63. Murray PDF, Selby D (1930) Intrinsic and extrinsic factors in the primary development of the skeleton. Development Genes and Evolution 122: 629–662.

64. Wortham RA, Eastlick HL (1960) Studies on transplanted embryonic limbs of the chick. VI. The development of muscles and tendons in nerveless and weakly innervated chick limb grafts. J Morphol 106: 131–146.

65. Nowlan N, Sharpe J, Roddy K, Prendergast P, Murphy P (2010) Mechanobiology of Embryonic Skeletal Development. Insights from Animal Models Birth Defects Research Part C 90(3): 203–13.

66. Pazour GJ, Witman GB (2003) The vertebrate primary cilium is a sensory organelle. Curr Opin Cell Biol 15: 105–110.

67. Lee HS, Millward-Sadler SJ, Wright MO, Nuki G, Salter DM (2000) Integrin and mechanosensitive ion channel-dependent tyrosine phosphorylation of focal adhesion proteins and beta-catenin in human articular chondrocytes after mechanical stimulation. J Bone Miner Res 15: 1501–1509.

68. Millward-Sadler SJ, Wright MO, Lee HS, Caldwell H, Nuki G, et al. (2000) Altered electrophysiological responses to mechanical stimulation and abnormal signalling through [alpha]5[beta]1 integrin in chondrocytes from osteoarthritic cartilage. Osteoarthritis and Cartilage 8: 272–278.

69. McMahon LA, Campbell VA, Prendergast PJ (2008) Involvement of stretch-activated ion channels in strain-regulated glycosaminoglycan synthesis in mesenchymal stem cell-seeded 3D scaffolds. J Biomech 41: 2055–2059.

70. Whitfield JF (2008) The solitary (primary) cilium-A mechanosensory toggle switch in bone and cartilage cells. Cellular Signalling 20: 1019–1024.

71. Bastow E, Lamb K, Lewthwaite J, Osborne A, Kavanagh E, et al. (2005) Selective activation of the MEK-ERK pathway is regulated by mechanical stimuli in forming joints and promotes pericellular matrix formation. J Biol Chem 280: 11749–11758.

72. Low SH, Vasanth S, Larson CH, Mukherjee S, Sharma N, et al. (2006) Polycystin-1, STAT6, and P100 Function in a Pathway that Transduces Ciliary Mechanosensation and Is Activated in Polycystic Kidney Disease. Developmental Cell 10: 57–69.

73. Nauli SM, Alenghat FJ, Luo Y, Williams E, Vassilev P, et al. (2003) Polycystins 1 and 2 mediate mechanosensation in the primary cilium of kidney cells. Nat Genet 33: 129–137.

74. Jensen CG, Poole CA, McGlashan SR, Marko M, Issa ZI, et al. (2004) Ultrastructural, tomographic and confocal imaging of the chondrocyte primary cilium in situ. Cell Biol Int 28: 101–110.

75. Poole CA, Jensen CG, Snyder JA, Gray CG, Hermanutz VL, et al. (1997) Confocal analysis of primary cilia structure and colocalization with the Golgi apparatus in chondrocytes and aortic smooth muscle cells. Cell Biol Int 21: 483–494.

76. Poole CA, Zhang ZJ, Ross JM (2001) The differential distribution of acetylated and detyrosinated alpha-tubulin in the microtubular cytoskeleton and primary cilia of hyaline cartilage chondrocytes. J Anat 199: 393–405.

77. McGlashan SR, Jensen CG, Poole CA (2006) Localization of Extracellular Matrix Receptors on the Chondrocyte Primary Cilium. J Histochem Cytochem 54: 1005–1014.

78. Rohatgi R, Milenkovic L, Scott MP (2007) Patched1 Regulates Hedgehog Signaling at the Primary Cilium. Science 317: 372–376.

79. Angele P, Yoo JU, Smith C, Mansour J, Jepsen KJ, et al. (2003) Cyclic hydrostatic pressure enhances the chondrogenic phenotype of human mesenchymal progenitor cells differentiated in vitro. Journal of Orthopaedic Research 21: 451–457.

80. Thorpe SD, Buckley CT, Vinardell T, O'Brien FJ, Campbell VA, et al. (2008) Dynamic compression can inhibit chondrogenesis of mesenchymal stem cells. Biochemical and Biophysical Research Communications 377: 458–462.

81. Dickhut A, Pelttari K, Janicki P, Wagner W, Eckstein V, et al. (2009) Calcification or dedifferentiation: requirement to lock mesenchymal stem cells in a desired differentiation stage. J Cell Physiol 219: 219–226.

82. Boardman PE, Sanz-Ezquerro J, Overton IM, Burt DW, Bosch E, et al. (2002) A Comprehensive Collection of Chicken cDNAs. Current Biology 12: 1965–1969.

83. Hall BK, Herring SW (1990) Paralysis and growth of the musculoskeletal system in the embryonic chick. J Morphol 206: 45–56.

84. Reiser PJ, Stokes BT, Walters PJ (1988) Effects of immobilization on the isometric contractile properties of embryonic avian skeletal muscle. Experimental Neurology 99: 59–72.

85. Macias D, Ganan Y, Sampath TK, Piedra ME, Ros MA, et al. (1997) Role of BMP-2 and OP-1 (BMP-7) in programmed cell death and skeletogenesis during chick limb development. Development 124: 1109–1117.

86. Sato M, Ochi T, Nakase T, Hirota S, Kitamura Y, et al. (1999) Mechanical tension-stress induces expression of bone morphogenetic protein (BMP)-2 and BMP-4, but not BMP-6, BMP-7, and GDF-5 mRNA, during distraction osteogenesis. J Bone Miner Res 14: 1084–1095.

87. Dowthwaite GP, Ward AC, Flannely J, Suswillo RF, Flannery CR, et al. (1999) The effect of mechanical strain on hyaluronan metabolism in embryonic fibrocartilage cells. Matrix Biol 18: 523–532.

88. Pitsillides AA, Archer CW, Prehm P, Bayliss MT, Edwards JC (1995) Alterations in hyaluronan synthesis during developing joint cavitation. J Histochem Cytochem 43: 263–273.

89. Lucchinetti E, Bhargava MM, Torzilli PA (2004) The effect of mechanical load on integrin subunits alpha5 and beta1 in chondrocytes from mature and immature cartilage explants. Cell Tissue Res 315: 385–391.

90. Guo X, Day T, Jiang X, Garrett-Bea L, Topol L, et al. (2004) Wnt/beta-catenin signaling is sufficient and necessary for synovial joint formation. Genes Dev 18: 2404–2417.

91. Hartmann C, Tabin CJ (2000) Dual roles of Wnt signaling during chondrogenesis in the chicken limb. Development 127: 3141–3159.

92. Tanaka N, Ohno S, Honda K, Tanimoto K, Doi T, et al. (2005) Cyclic Mechanical Strain Regulates the PTHrP Expression in Cultured Chondrocytes via Activation of the Ca2+ Channel. Journal of Dental Research 84: 64–68.

93. Kavanagh E, Church VL, Osborne AC, Lamb KJ, Archer CW, et al. (2006) Differential regulation of GDF-5 and FGF-2/4 by immobilisation in ovo exposes distinct roles in joint formation. Dev Dyn 235: 826–834.

94. Böhme K, Winterhalter KH, Bruckner P (1995) Terminal Differentiation of Chondrocytes in Culture Is a Spontaneous Process and Is Arrested by Transforming Growth Factor-[beta]2 and Basic Fibroblast Growth Factor in Synergy. Experimental Cell Research 216: 191–198.

95. Yu K, Xu J, Liu Z, Sosic D, Shao J, et al. (2003) Conditional inactivation of FGF receptor 2 reveals an essential role for FGF signaling in the regulation of osteoblast function and bone growth. Development 130: 3063–3074.

96. Li C-F, Hughes-Fulford M (2006) Fibroblast Growth Factor-2 Is an Immediate-Early Gene Induced by Mechanical Stress in Osteogenic Cells. Journal of Bone and Mineral Research 21: 946–955.

97. Craig FM, Bentley G, Archer CW (1987) The spatial and temporal pattern of
     collagens I and II and keratan sulphate in the developing chick metatarsopha-
     langeal joint. Development 99: 383–391.
98. Mikic B, Wong M, Chiquet M, Hunziker EB (2000) Mechanical modulation of
     tenascin-C and collagen-XII expression during avian synovial joint formation.
     J Orthop Res 18: 406–415.

99. Pacifici M, Iwamoto M, Golden EB, Leatherman JL, Lee Y-S, et al. (1993)
     Tenascin is associated with articular cartilage development. Developmental
     Dynamics 198: 123–134.

17

Interaction of Age and Mechanical Stability on Bone Defect Healing: An Early Transcriptional Analysis of Fracture Hematoma in Rat

Andrea Ode[1,2], **Georg N. Duda**[1,2,3]*, **Sven Geissler**[1,2], **Stephan Pauly**[1,3], **Jan-Erik Ode**[1], **Carsten Perka**[1,2,3], **Patrick Strube**[1,3]

1 Julius Wolff Institute, Charité - Universitätsmedizin, Berlin, Germany, 2 Berlin-Brandenburg Center for Regenerative Therapies, Berlin, Germany, 3 Klinik für Orthopädie, Centrum für Muskuloskeletale Chirurgie, Charité - Universitätsmedizin, Berlin, Germany

Abstract

Among other stressors, age and mechanical constraints significantly influence regeneration cascades in bone healing. Here, our aim was to identify genes and, through their functional annotation, related biological processes that are influenced by an interaction between the effects of mechanical fixation stability and age. Therefore, at day three post-osteotomy, chip-based whole-genome gene expression analyses of fracture hematoma tissue were performed for four groups of Sprague-Dawley rats with a 1.5-mm osteotomy gap in the femora with varying age (12 vs. 52 weeks - biologically challenging) and external fixator stiffness (mechanically challenging). From 31099 analysed genes, 1103 genes were differentially expressed between the six possible combinations of the four groups and from those 144 genes were identified as statistically significantly influenced by the interaction between age and fixation stability. Functional annotation of these differentially expressed genes revealed an association with *extracellular space, cell migration* or *vasculature development*. The chip-based whole-genome gene expression data was validated by q-RT-PCR at days three and seven post-osteotomy for MMP-9 and MMP-13, members of the mechanosensitive matrix metalloproteinase family and key players in cell migration and angiogenesis. Furthermore, we observed an interaction of age and mechanical stimuli *in vitro* on cell migration of mesenchymal stromal cells. These cells are a subpopulation of the fracture hematoma and are known to be key players in bone regeneration. In summary, these data correspond to and might explain our previously described biomechanical healing outcome after six weeks in response to fixation stiffness variation. In conclusion, our data highlight the importance of analysing the influence of risk factors of fracture healing (e.g. advanced age, suboptimal fixator stability) in combination rather than alone.

Editor: Dimitrios Zeugolis, National University of Ireland, Galway (NUI Galway), Ireland

Funding: This study was supported partly by the Federal Ministry of Education and Research (BMBF, Grant 0315848A) excellence cluster, Berlin-Brandenburg Center for Regenerative Therapies, and partly by the German Research Foundation (DFG SFB 760). The funders had no role in study design, data collection and analysis, decision to publish, or preparation of the manuscript.

Competing Interests: The authors have declared that no competing interests exist.

* Email: Georg.Duda@charite.de

Introduction

Due to the ageing of the population the high incidents of delayed or mal-unions after fracture trauma develops to a growing concern. The classical boundary conditions that influence healing (mechanics, surgery, accompanying traumata) are overlapped by the age-related changes in regenerative capacity.

Fracture consolidation is significantly influenced by many biological and mechanical factors. An important biological risk factor is the age of the patient. Animal experiments in rats and clinical studies in humans show a delayed course of bone healing with increasing age[1–3]. Possible reasons for this could be a diminished number of mesenchymal progenitor cells, their reduced migration potential and higher susceptibility towards senescence, and reduced local or systemic blood flow in older individuals [4,5]. Inadequate fracture stability - determined by fixation stability - is the principle mechanical factor that leads to a non-union [6,7]. An optimal mechanical stimulus enables successful fracture healing, whereas too little or too much disables

it. Especially the early phase of bone healing seems to be sensitive to mechanical loading conditions [8]. In clinical cases biological and mechanical boundary conditions both jointly interact with each other and influence regeneration. The negative influence of mechanical instability on the biological factor of vascularity, endochondral ossification and maturation is an important example for this [9,10]. Vascularity is not only disturbed by the trauma itself and/or by surgical disruption but also by (initial) instability at the fracture site[6][11]. A failure of angiogenesis is critical, since angiogenesis is not only responsible for the oxygen supply, but also a prerequisite for the resorption of necrotic tissue and recruitment of different cell types including mesenchymal progenitor cells, which is necessary for a mechanically stable repair of the bone defect [12]. However, it remains unclear whether one stressor to regeneration - mechanical stability or age - dominates or how they interact.

In recent animal studies we provided evidence that the healing outcome of bone regeneration depends not only on mechanical stability or age alone, but more importantly, on the overlap of both stressors [1,9]. Our data revealed a statistically significant interaction between the effects of mechanical stability and age on radiological outcome at two and six weeks and on biomechanical callus competence at six weeks post-operative. For example, in young rats, the biomechanical parameters torsional stiffness and maximum torque at failure were improved when bone defects were rigidly fixated, whereas the opposite was true for old rats [1]. However, on a more microscopically level, results were not as explicit. Micro-computed tomography could not reveal an interaction between the effects of mechanical stability and age on callus size, geometry, microstructure, and mineralization. Similar results were observed for histological analysis of vascularity and bone remodelling. Rather, we found a complex mixture of differences in the investigated parameters between the groups. For example, fixator stability influences callus size and geometry, whereas age influences callus strut thickness and perforation within these struts [9].

Gene expression in the early fracture hematoma is also known to be influenced by either age or fixation stability. For example, Meyer *et al.* found significantly lower levels of mRNA levels for Indian hedgehog and bone morphogenetic protein 2 (*BMP2*) in the fracture callus of old rats compared to young ones [13]. Comparative analysis of stabilized and non-stabilized fractures in small animals revealed differences in molecular signals controlling chondrogenesis [14]. In large animals, mRNA expression levels of members of the BMP-, tumor necrosis factor (*TNF*)- and matrix metalloproteinase (*MMP*) families as well as genes involved in bone matrix generation were lower in the critical fixation compared to the rigid fixation group at several time points [10]. However, little is known about the interaction between the effects of mechanical stability and age on gene expression. This is especially important during the early phase of bone healing, which has been shown to be mechanically sensitive and therefore crucial for the healing outcome [8].

In the present study we performed a chip-based whole-genome gene expression analysis of fracture hematoma tissue from young and old rats that underwent rigid and semi-rigid bone defect fixation. Our aim was to identify genes and, through their functional annotation, related biological processes that are influenced by an interaction between the effects of mechanical stability and age. In conclusion, we identified a number of genes and their functional annotation revealed an association with cell migration and blood vessel formation, which is so far unknown.

Materials and Methods

Animals and groups

All animal experiments were carried out according to the policies established by the Animal Welfare Act, the NIH Guide for Care and Use of Laboratory Animals, and the National Animal Welfare Guidelines and were approved by the local legal representative (LAGeSo Berlin, G0190/05).

Operations and postoperative care were performed according to a previously published protocol and employed a standardized biomechanically validated external fixation device [15]. Preoperatively the animal husbandry was performed in large cages (ground area 1800 cm^2, height 19 cm, ground covered with soft-wood granule animal bedding) with a maximum of 6 animals per cage. After operation we changed to single animal husbandry in a smaller cage (ground area 810 cm^2, height 19 cm, soft-wood granule bedding). Housing facility was specific pathogen free with a 12/12 h light/dark rhythm and a room temperature of 24°C. Animals had free access to water and food (pressed diet pellets for

rodents). Preoperatively, as well as before surgical intervention for harvesting the fracture hematoma rectal temperature was measured to detect possible infections (temperature >38 °C). Postoperatively the animals were visited daily and if necessary analgesia was given. The experimental model has been previously described [1] and is briefly summarized here. For gene-chip and q-RT-PCR (day 3) analysis thirty-six and for q-RT-PCR analysis (day 7) twenty female Sprague–Dawley (Sprague-Dawley SD (Aged for old groups) Outbred rats, Harlan Laboratories, Indianapolis, USA) rats were divided into four groups with nine (day 3) respectively five (day 7) animals each group. Groups were defined by variation of fixator stabilities (rigid vs. semi-rigid) and age (12 vs. 52 weeks): young rigid (YR), young semi-rigid (YSR), old rigid (OR), and old semi-rigid (OSR). Weights of the old animals ranged at 313.4±17.4 g whereas that of the young ones was 251.9±15.3 g ($p<0.001$ in t-test). Animals were not restricted in weight bearing. In cases of adverse events (infection, major bleeding, pin loosening, implant failure or complications related to anaesthesia) the animal was sacrificed as described and a new animal was included/operated to gain the planned group sizes. Regarding this, two animals of the YSR group died for unknown reasons during primary surgery in general anaesthesia, and two aged rats (one OR, one OSR) presented with pin loosening prior to harvesting at day 7.

Surgical procedure

Using an anterolateral approach, the left femur was osteotomized at the midshaft, distracted to a gap of 1.5 mm and externally fixated employing a previously described fixation system [1]. The distance between fixator and bone (offset) was set to 7.5 mm in the rigid configuration (leading to a torsional 8.13 Nmm/° and axial 25.21 N/mm fixator stiffness) and 15 mm in the semi-rigid configuration (torsional 6.62 Nmm/°, axial 10.39 N/mm stiffness) (Figure 1). Before sacrifice and under general anaesthesia (see anaesthesia protocol published before [16]) the wound was reopened, inter-fragmentary fracture hematoma was harvested with a sterile forceps, directly transferred into a sterile container and frozen immediately in liquid nitrogen. For the gene-chip-analysis follow-up was three days, for q-RT-PCR analysis follow-up was three (RNA of the gene chip animals was used) and seven days. Animal sacrifice was performed in deep general anaesthesia by intracardial injection of 5 ml potassium chloride (7.45%, B.Braun Melsungen AG, Melsungen, Germany) [16].

RNA Isolation, cDNA Synthesis, and Quantitative Reverse Transcription-Polymerase Chain Reaction

At day three and seven post-OP, total RNA was isolated from fracture hematoma by using Trizol (following the instructions of the manufacturer) starting with an initial stepwise mechanical destruction of the tissue with syringe cannulas of three different diameters until liquid was homogeneous without visible tissue particles. Next, RNA was reversely transcribed to cDNA using iScript cDNA Synthesis kit (Bio-Rad, Munich, Germany) according to the manufacturer's instructions. RNA quality was evaluated by visualizing the 18S/28S rRNA on a 1.5% agarose gel. Quantification of *MMP-2*, *MMP-9*, *MMP-13*, and *TIMP-2* were assessed by quantitative reverse transcription-polymerase chain reaction (q-RT-PCR) using the iQ SYBR Green Supermix and the iQ 5 Multicolor Realtime PCR Detection System and software (Bio-Rad, Munich, Germany) using the delta-Ct-method. The transcript expression was normalized versus the housekeeping gene β-actin (*ACTB*), elongation factor 1-alpha 1 (*EEF1A*), and glyceraldehyde-3-phosphate dehydrogenase (*GAPDH*). The primers used in the real-time PCR assay were commercially purchased

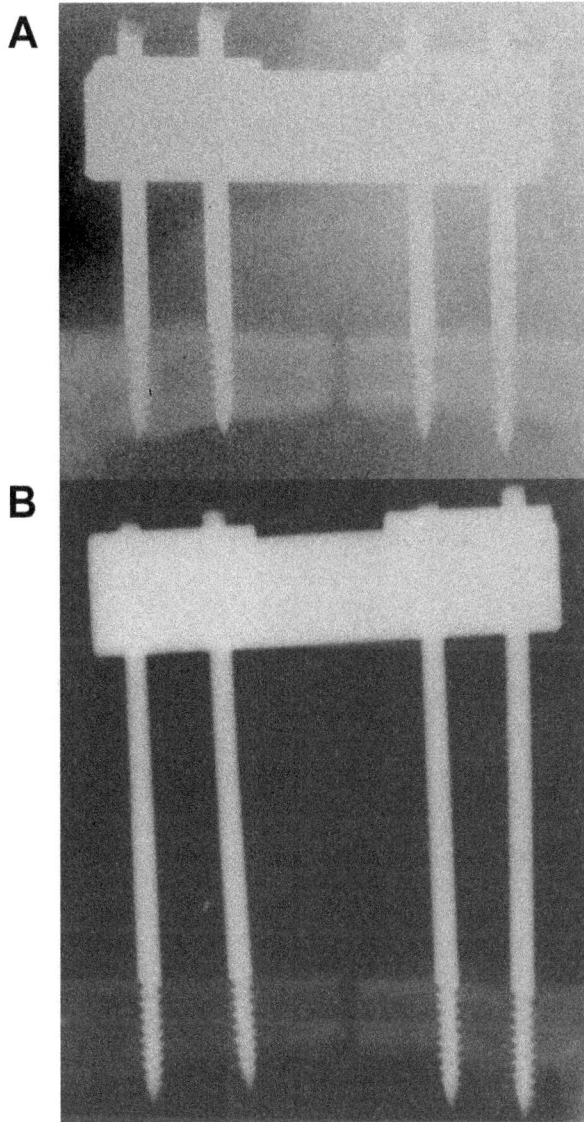

Figure 1. Radiographs of the two fixator configurations varying in the distance between bone and fixator crossbar (offset). (A) Rigid configuration with a 7.5 mm offset. (B) Semi-rigid configuration with a 15 mm offset. Osteotomy gap was set to 1.5 mm.

(Invitrogen, Karlsruhe, Germany; Table 1). Amplification efficiency (E) was assessed to be between 1.9 and 2. At day 7, transcripts from five animals were analyzed. At day 3, analysis was performed with three pools of total RNA from three animals per pool. Each experiment was conducted in triplicates.

Affymetrix gene chip hybridization

The amplification and labeling of the RNA samples, isolated at day three post-OP, were carried out according to the manufacturer's instructions (Affymetrix, Santa Clara, CA). Briefly, total RNA was quantified by UV-spectroscopy and its quality was checked by analysis on a LabChip (BioAnalyzer, AGILENT Technologies, Santa Clara, CA). Between one to three micrograms from each sample were synthesized into double-stranded cDNA using SuperScript transcriptase II (Life Technologies, Inc., Carlsbad, CA) and with an oligo(dT)24 primer containing a T7

RNA polymerase promoter (TIBMOL Biol, Berlin, Germany). After RNAse H – mediated (Roche, Germany) second strand cDNA synthesis, the product was purified and served as template in the subsequent in vitro transcription (IVT) reaction. Labeled complementary RNA (cRNA) was prepared from double-stranded cDNA by in vitro transcription using the GeneChip RNA transcript labeling kit (Affymetrix, Santa Clara, CA). After cleanup (Qiagen, Hilden, Germany), the biotin-labeled cRNA was fragmented by alkaline treatment [40 mmol/L Trisacetate (pH 8.2), 100 mmol//L potassium acetate, and 50 mmol//L magnesium acetate] at 94°C for 35 minutes. 15 µg of each cRNA sample was hybridized for 16 hours at 45°C to an Affymetrix Rat GeneChip Array 230 2.0. Chips were washed and stained with streptavidin-phycoerythrin using a fluidics station according to the protocols recommended by the manufacturer. Finally, probe arrays were scanned at 1.56-µm resolution using the Affymetrix GeneChip System confocal scanner 3000. Raw data were submitted to Affymetrix Expression Console software (v.1.3) to generate probe set summarization (CHP) files from feature intensity (CEL) files using PLIER algorithm. Analysis of differentially expressed genes was performed with Affymetrix Transcriptome Analysis Console (TAC) Software and PASW Statistics 18 (SPSS Inc., Chicago, USA). To conduct functional categorizing, all differentially expressed genes were submitted to the Database for Annotation, Visualization and Integrated Discovery (DAVID) V6.7 (http://david.abcc.ncifcrf.gov/) [17]. P-values were determined using EASE, followed by a Benjamini-Hochberg correction for multiple comparisons. Summary and visualization of Gene Ontology (GO) terms was performed with REVIGO (http://revigo.irb.hr/) [18]. For each condition group (YSR, YR, OSR, OR) three Affymetrix gene chip hybridizations were performed with separate pools of total RNA. Each pool comprised of total RNA from three animals. Thus, the analysis is based on total RNA from nine animals per condition group.

MSC isolation, culture and mechanical stimulation

MSCs were isolated from bone marrow of 12 months old Sprague–Dawley rats selected by plastic adherence (Dobson et al., 1999). Dulbecco's modified Eagle's medium (DMEM) (Gibco, NY, USA) supplemented with 10% fetal calf serum (FCS) (Biochrom AG, Berlin, Germany) and 10 U/ml penicillin plus 100 µg/ml streptomycin was used as expansion medium for MSCs. Only cells from passages 2–4 were used for experiments. The bioreactor system used has been described previously [15]. Briefly, MSCs were trypsinized, and 2×106 cells in 350 µl of bioreactor medium (culture medium containing 2.4% Trasylol [Bayer, Leverkusen, Germany]) were mixed with 300 µl of fibrinogen/bioreactor medium (1:2) mixture and 50 µl of thrombin S/bioreactor medium (1:2) mixture (Tissucol; Baxter, Munich, Germany). This MSC/fibrinogen/thrombin mixture was placed between two spongiosa bone chips and allowed to solidify for 30 minutes at 37°C. The sandwich construct was placed into the bioreactor, and 25 ml of bioreactor medium was added. A strain of approximately 20% at a frequency of 1 Hz was applied in accordance with in vivo measurements of interfragmentary movement [19]. Mechanical loading was carried out for 72 hours. Afterwards, cells within the fibrin construct were isolated by 225 U trypsin/1 ml PBS.

Transwell Migration Assay

Random migration (i.e. equal concentrations of bioactive molecules in both compartments) was measured by a modified Boyden chamber assay (Falk et al., 1980) using polycarbonate filters (8 µm pore size; Nunc, Wiesbaden, Germany) coated with or without Collagen I (100 µg/ml; Pure Col, Inamed Biomaterials,

Table 1. Primer sequences.

Protein	Gene	Primer Sequence (forward/reverse)
matrix metalloproteinase 9	Mmp-9	5′ GTCTGGATAAGTTGGGGCTA 3′
		5′ GCCTTGTCTTGGTAGTGAAA 3′
matrix metalloproteinase 13	Mmp-13	5′ CAGTCTCTCTATGGTCCAGG 3′
		5′ TGGTCAAAAACAGTTCAGGC 3′
actin cytoplasmic 1 (β-actin)	Actb	5′ TGTCACCAACTGGGACGATA 3′
		5′ GGGGTGTTGAAGGTCTCAAA 3′
glyceraldehyde-3-phosphate dehydrogenase	Gapdh	5′ ATGGGAAGCTGGTCATCAAC 3′
		5′ GTGGTTCACACCCATCACAA 3′
elongation factor 1-alpha 1	Eef1a	5′ CCCTGTGGAAGTTTGAGACC 3′
		5′ CTGCCCGTTCTTGGAGATAC 3′

Fremont, U.S.), which is the most abundant extracellular protein of bones (Rossert and de Crombrugghe, 2002). MSCs (4×10^4) were seeded onto the filters and incubated for 5 h at 37°C. Equal cell seeding was validated by an MTS test. Non-migrated cells were removed from the upper side of the filter by scraping, and remaining migrated cells were stained with 10μg/ml Hoechst-33342 (Invitrogen, Karlsruhe, Germany). The average numbers of migrated cells from five microscopic fields (1 mm×0.8 mm) per filter (0.47 cm^2) were analysed using the NIH ImageJ software package (http://rsb.info.nih.gov/nih-image/). MSCs were isolated form three (old MSCs) and five (young MSCs) different animals followed by separate migration assays, which were performed in duplicates, i.e. two wells per group, the mean value being used for statistical analysis.

Statistical analyses

The statistical analysis of q-RT-PCR data was performed using statistics software PASW Statistics 18 (SPSS Inc., Chicago, USA). If not stated otherwise, the influence of age and mechanical stability and their interaction on gene expression and migration were tested with a 2-tailed, 2-way Analysis of Variance (ANOVA) and posthoc Bonferroni correction. The parameters time (gene expression) and coating (migration) were set as covariates. The assumption of normality was tested using the Shapiro-Wilk normality test. For graphical presentation, results are presented in boxplots. The dark line in the middle of the boxes is the median. The box represents the interquartile range (IQR = Q3-Q1). The whiskers indicate 1.5xIQR. Outliers are circles between 1.5xIQR and 3xIQR of the quartiles. Extreme values are stars more than 3xIQR away from quartiles. For statistical analysis of chip-based

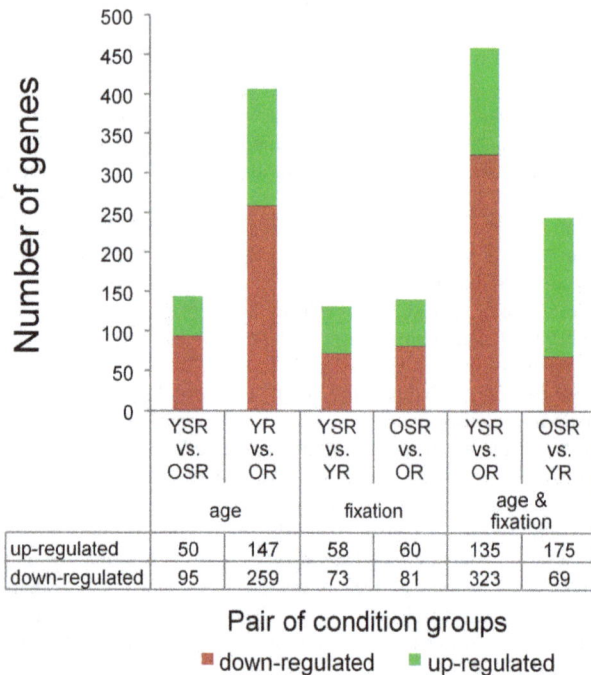

	YSR vs. OSR	YR vs. OR	YSR vs. YR	OSR vs. OR	YSR vs. OR	OSR vs. YR
	age		fixation		age & fixation	
up-regulated	50	147	58	60	135	175
down-regulated	95	259	73	81	323	69

Pair of condition groups

■ down-regulated   ■ up-regulated

Figure 2. Number of differentially expressed genes for each pair of condition groups. With Affymetrix Transcriptome Analysis Console (TAC) a traditional unpaired One-Way Analysis of Variance (ANOVA) for each pair of condition groups was performed (Linear Fold Change <−2 or > 2; ANOVA p-value (condition pair) <0.05).

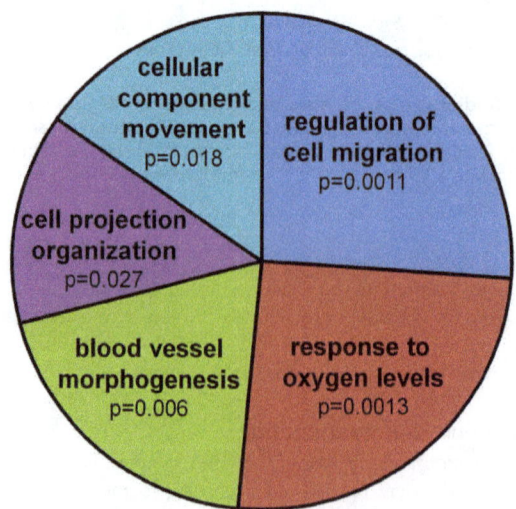

Figure 3. Functional annotation of 144 genes that are affected by a statistically significant interaction between age and fixation stability. Pie Chart view of REVIGO results: The single GO terms (Table S8 in File S3) are joined into clusters of related terms, visualized as sectors with different colours. Size of the sectors is adjusted (log10p-value) to reflect the p-values.

Table 2. Gene list of cluster *Regulation of cell migration.*

Affymetrix ID	Gene Name	Linear Fold Change						p-value
		(YSR vs. YR)	(OSR vs. OR)	(YSR vs. OSR)	(YR vs. OR)	(YSR vs. OR)	(OSR vs. YR)	(2-way ANOVA)
1393403_AT	angiopoietin-like 3	−2.9	2.2	−1.5	4.1	1.4	−1.9	1.87E-02
1369814_AT	chemokine (C-C motif) ligand 20	1.2	−2.6	4.3	1.4	1.7	−3.5	2.35E-02
1370634_X_AT	chemokine (C-X-C motif) ligand 3	−3.0	−1.7	1.2	2.1	−1.4	−3.6	3.50E-02
1388459_AT	collagen, type XVIII, alpha 1	1.1	−1.8	−1.1	−2.2	−2.0	1.2	2.04E-02
1369113_AT	gremlin 1, cysteine knot superfamily, homolog (Xenopus laevis)	−1.2	−2.9	−3.2	−8.1	−9.3	2.8	4.81E-02
1369166_AT	matrix metallopeptidase 9	1.0	−5.0	−2.0	−10.4	−10.1	2.1	3.97E-03
1398275_AT	matrix metallopeptidase 9	2.0	−4.8	−1.8	−17.7	−8.7	3.7	4.67E-03
1370642_S_AT	platelet derived growth factor receptor, beta polypeptide	−1.0	−1.8	−1.1	−2.0	−2.1	1.1	2.35E-02
1393456_AT	podoplanin	−1.2	2.0	1.2	2.8	2.3	−1.4	3.90E-02
1391369_AT	serum response factor (c-fos serum response element-binding transcription factor)	−1.0	2.0	1.1	2.1	2.1	−1.1	1.22E-04
1382685_AT	slit homolog 2 (Drosophila)	1.2	−1.9	−1.1	−2.6	−2.2	1.4	4.26E-02
1392382_AT	transforming growth factor, beta 2	−1.4	1.5	−2.7	−1.2	−1.8	1.9	4.80E-02
1371240_AT	tropomyosin 1, alpha	−1.3	6.9	−1.2	7.4	5.7	−1.1	3.90E-02

Bold: Linear Fold Change <−2 or > 2.

Table 3. Gene list of cluster *Response to oxygen levels*.

Affymetrix ID	Gene Name	Linear Fold Change						p-value
		(YSR vs. YR)	(OSR vs. OR)	(YSR vs. OSR)	(YR vs. OR)	(YSR vs. OR)	(OSR vs. YR)	
1393902_AT	aldo-keto reductase family 1, member C1 (dihydrodiol dehydrogenase 1; 20-alpha (3-alpha)-hydroxysteroid dehydrogenase)	1.9	-3.2	1.9	-3.2	-1.7	1.0	8.82E-03
1398333_AT	endothelial PAS domain protein 1	-1.1	-3.0	-1.1	-3.0	-3.3	1.0	4.07E-02
1370604_AT	leptin receptor	1.0	-3.8	1.0	-3.8	-3.8	1.0	1.24E-02
1388204_AT	matrix metallopeptidase 13	-1.1	-4.2	-1.5	-6.0	-6.4	1.4	4.47E-02
1369166_AT	matrix metallopeptidase 9	1.0	-5.0	-2.0	-10.4	-10.1	2.1	3.97E-03
1398275_AT	matrix metallopeptidase 9	2.0	-4.8	-1.8	-17.7	-8.7	3.7	4.67E-03
1387410_AT	nuclear receptor subfamily 4, group A, member 2	1.6	-1.8	1.4	-2.1	-1.3	1.2	1.30E-03
1370642_S_AT	platelet derived growth factor receptor, beta polypeptide	-1.0	-1.8	-1.1	-2.0	-2.1	1.1	2.35E-02
1393456_AT	podoplanin	-1.2	2.0	1.2	2.8	2.3	-1.4	3.90E-02
1368170_AT	solute carrier family 6 (neurotransmitter transporter, GABA), member 1	1.0	2.8	-2.8	1.0	1.0	2.8	3.70E-05
1392382_AT	transforming growth factor, beta 2	-1.4	1.5	-2.7	-1.2	-1.8	1.9	4.80E-02

Bold: Linear Fold Change < -2 or > 2.

whole-genome gene expression data see section 2.4 Affymetrix gene chip hybridization. The level of significance for all statistical tests was defined $p<0.05$.

Results

The data discussed in this publication have been deposited in NCBI's Gene Expression Omnibus [20] and are accessible through GEO Series accession number GSE53256 (http://www.ncbi.nlm.nih.gov/geo/query/acc.cgi?acc = GSE53256).

In total, 31099 genes were analysed with Affymetrix Transcriptome Analysis Console (TAC). TAC computes and summarizes a traditional unpaired One-Way (single factor) Analysis of Variance (ANOVA) for each pair of condition groups and for all six condition groups. In our study, 1103 genes were differentially expressed between the six possible combinations (Linear Fold Change < -2 or > 2; ANOVA p-value (condition pair) < 0.05). To illustrate the differences between the six possible combinations the numbers of up- and down-regulated genes for each pair of condition groups are displayed in Figure 2 and listed in Table S1–S6 in File S1. Since TAC cannot examine the combined effect of age and fixation stability on gene expression, a two-way ANOVA was conducted with PASW Statistics 18. However, complete analysis of 31099 genes would be computationally intensive. Therefore, 521 genes were pre-selected for analysis under the following conditions: (1) ANOVA p-value (All conditions) < 0.05 and (2) Linear Fold Change < -2 or > 2 in at least one of the six condition pairs. Statistical analysis of these 521 genes revealed a statistically significant interaction between age and fixation stability on gene expression of 144 genes (2-way ANOVA $p< 0.05$). A complete list of these genes, the pre-selection criteria, Linear Fold Change in all six condition pairs and the results of 2-way ANOVA analysis can be found in Table S7 in File S2.

To identify biological processes related to these differently expressed genes, functional categorizing was conducted using DAVID v6.7. Functional annotation of these differentially expressed genes resulted in a list of Gene Ontology terms related to biological processes (Table S8 in File S3). Single GO terms were then joined into five clusters of related terms using REVIGO; the two dominant clusters, based on p-values, being *Regulation of cell migration* and *Response to oxygen levels* (Figure 3 and Table S9–S11 in File S4). The respective genes that were functionally categorized to these clusters by the functional annotation tool DAVID v6.7 are listed in Table 2 and 3.

Interestingly, among the genes listed in Table 2 and 3 with a noticeable Linear Fold Change in gene expression are members of the family of matrix metalloproteinases (MMPs): MMP-9 and MMP-13. To validate the results of Affymetrix chip-based whole-genome gene expression analyses, the expression level of MMP-9 and -13 were analysed via qRT-PCR at day 3 and also day 7. There was also a statistically significant interaction between the effects of fixation and age on gene expression of MMP-9 ($p = 0.009$) and MMP-13 ($p = 0.016$). MMP-9 and -13 expression is also influenced by time (MMP-9, $p = 0.001$; MMP-13, $p = 0.007$). Thus results are presented separately for day 3 and day 7 (Figure 4). Post-hoc inter-group comparison revealed that MMP-9 expression is significantly higher in YSR than in OSR ($p = 0.019$) at day 7; for MMP-13 a trend was observed between YSR and OSR ($p = 0.057$).

Functional annotation clustering of the 144 genes that were influenced by a significant interaction between age and fixation stability revealed the cluster *Regulation of cell migration* having the most significant p-value. To validate whether this biological process is influenced by a significant interaction between age and

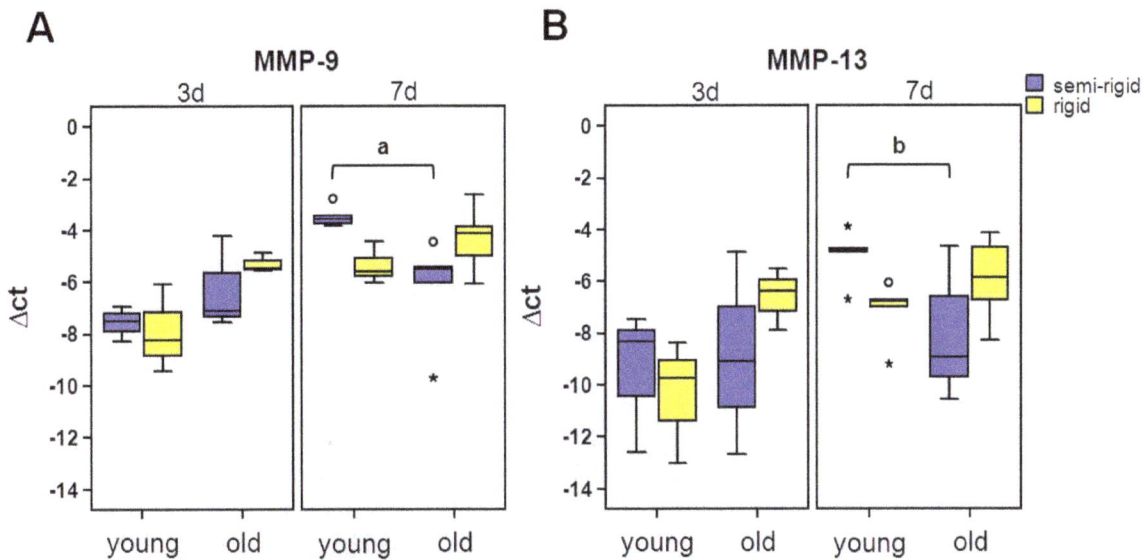

Figure 4. Gene expression of MMP-9 and -13 were significantly affected by the interaction of fixation and age. The expression of mRNA of YR, YSR, OR and OSR was evaluated by quantitative qRT-PCR, normalized for the housekeeping genes (HKG) *Actb*, *Gapdh*, and *Eef1a* and quantified using the delta-Ct-method: Δct = ct (geo mean HKG) - ct (gene of interest). (day 3, n = 3 RNA pools á 3 animals; day 7, n = 5; a, *p* = 0.019; b, *p* = 0.057; °, outlier; *, extreme value)

mechanical stability, we investigated this parameter in a simplified *in vitro* model. In order to do so, we mimicked the early phase fracture gap conditions by embedding mesenchymal stromal cells (MSCs) in fibrin, the major extracellular matrix of the hematoma, and by stimulating these cells with cyclic-compressive loading. MSCs were chosen as cell source, because they are known to be mechanosensitive and key players in bone regeneration, and they are present in the fracture hematoma by day three post-fracture [21–23], the same day we harvested the hematoma tissue for gene expression analyses. The MSCs were isolated from young (10–12 weeks; data published in [24]) and old (12 months) rats (yMSCs and oMSCs, respectively) in accordance with the *in vivo* experiments. To apply two different mechanical loading regimes to MSCs, similar to the ones *in vivo* (semi-rigid and rigid fixation allowing more or less interfragmentary movement, respectively), the cells underwent cyclic compression in a bioreactor (loaded) and were compared to nonloaded controls. Although the loading regimes *in vivo* (more or less movement) and *in vitro* (cyclic or no compression) did not perfectly match, the migratory behaviour of MSCs was found to be influenced by the interaction between age and mechanical stimulation (p = 0.007). Coating of the migration filter with collagen I had no influence on the results compared to no coating (p = 0.942). Thus results are displayed together (Figure 5). Inter-group comparison revealed that migration of nonloaded yMSCs is significantly higher compared to loaded yMSCs (p<0.001), nonloaded oMSCs (p = 0.005) and loaded oMSCs (p = 0.003). No statistical significance was observed between the latter three groups. A summary of the work flow in this study is given in Figure 6.

Discussion

This study is a follow-up of two previous ones [1,9]. Our aim in this study was to compare gene expression in fracture hematoma tissue of young and old rats that underwent either rigid or semi-rigid fixation. By using Affymetrix chip-based whole-genome gene expression analyses our aim was to identify genes and related

biological processes that are influenced by an interaction between the effects of mechanical stability and age. By this we wanted to gain insights into the early hematoma's biological processes that led to the previously reported biomechanical long-term outcomes in bone defect healing, which were influenced by age and varying fixator configurations, i.e. fixation stiffness. In total, we had four experimental groups: young semi-rigid (YSR), young rigid (YR), old semi-rigid (OSR) and old rigid (OR).

The majority of fractures heal by secondary, or indirect, fracture healing involving callus formation [22]. This process is both spatially and temporally regulated [25]. By using a model of experimental fracture healing in the rat the healing cascade has been elucidated [21]. Compared to human fractures, the rat fracture healing cascade proceeds at about twice the speed [23]: Initially, bone fracture is accompanied by disruption of bone marrow, bone matrix, blood vessels, and surrounding soft tissue. Within the first 24 hours, bleeding of these tissues and releasing of bone marrow into the fracture gap give rise to the initial hematoma [26]. Degranulating platelets and inflammatory cells release cytokines and growth factors that induce migration of MSCs from bone marrow and acute inflammatory cells and further aggregation of platelets. From day two to six, in the area between the cortices, soft callus begins to form via endochondral ossification, where MSCs begin to proliferate by day three [21–23].

In our study fracture hematoma tissue was harvested at three days post-osteotomy. From 31099 analysed genes, 1103 genes were differentially expressed between the six possible combinations and from those 1103 genes 521 genes were selected to be analysed with a two-way ANOVA. In total, 144 genes were identified as statistically significantly influenced by the interaction between age and fixation stability. Functional annotation of these genes revealed an involvement in cell migration. Thus far, there is no data published that report on the interaction of age and mechanical stimulation on cell migration *in vivo*.

We have two possible explanations for the observations in our study: either (1) the cell type composition in the fracture

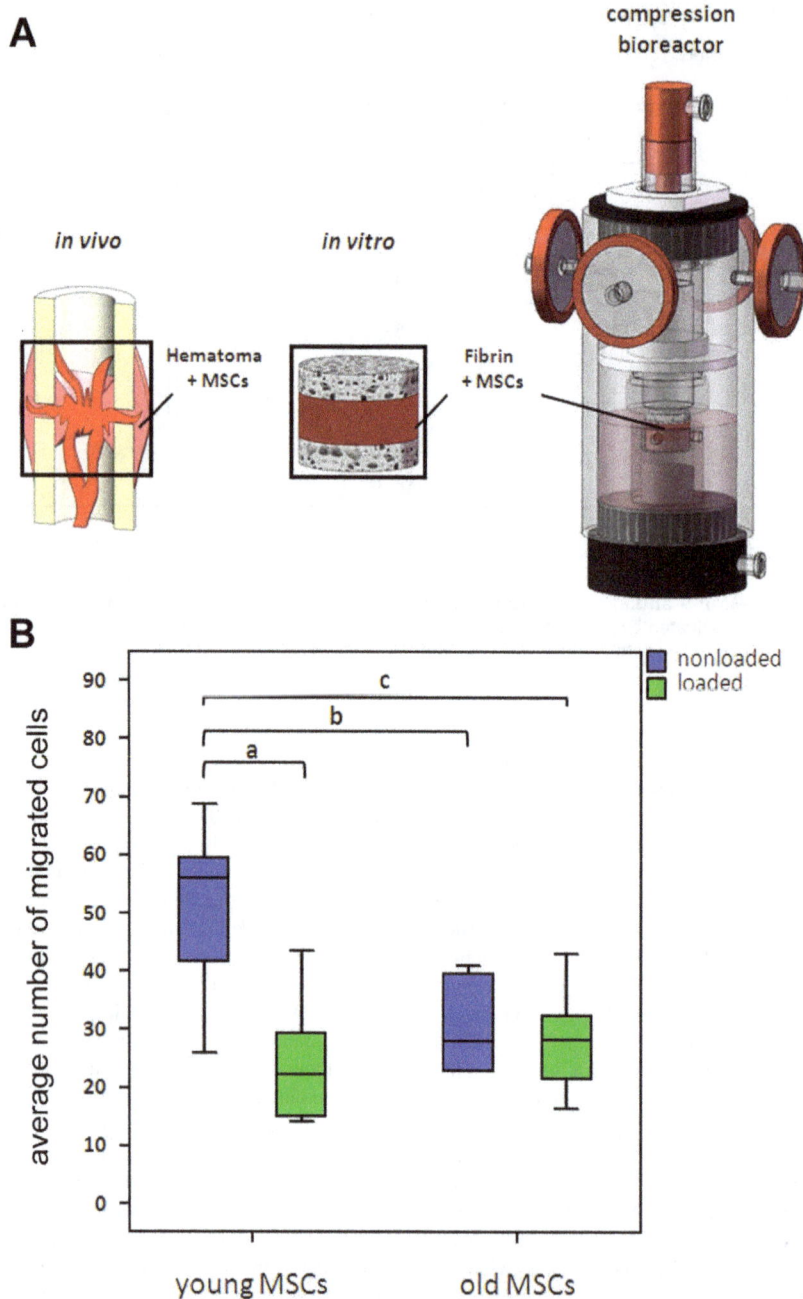

Figure 5. Migration of mechanically stimulated young and old MSCs. (A) Experimental set-up to investigate the effect of mechanical loading of MSC/fibrin constructs. MSCs were embedded in fibrin, placed between two cancellous bone chips and mechanically stimulated. (B) MSC migration was investigated in a modified Boyden-Chamber assay. The average number of migrated cells from five microscopic fields per filter was analysed using NIH ImageJ software. (young MSCs, n = 10; old MSCs, n = 6, a, p<0.001; b, p = 0.005; c, p = 0.003).

hematoma of young and old animals varies so that different cells respond differently to mechanical stimulation or (2) the cell type composition is similar, but the ability of these cells to sense and adapt to mechanical stimulation has changed during aging. We recently reported that migration of young MSCs, key players in bone regeneration and present in the fracture hematoma, is reduced *in vitro* if the cells underwent mechanical stimulation compared to non-stimulated controls [24]. Therefore, we now compared migration of young and old MSCs in response to cyclic-compressive loading. Interestingly, a statistically significant inter-

action between the effects of mechanical stimulation and age on MSC migration could also be observed in this setting. These results point towards a reduced ability of MSCs to sense and/or adapt to mechanical stimulus with advanced age. Based on several studies, it was proposed that fracture hematoma and bone tissue of old individuals is less responsive to mechanical stimulation than that of young ones. For example, an obvious growth of the loaded tibia was reported in young but not in old animals [27]. And a higher mechanical loading threshold was needed for initiation of bone growth during remodelling in old compared to young rats

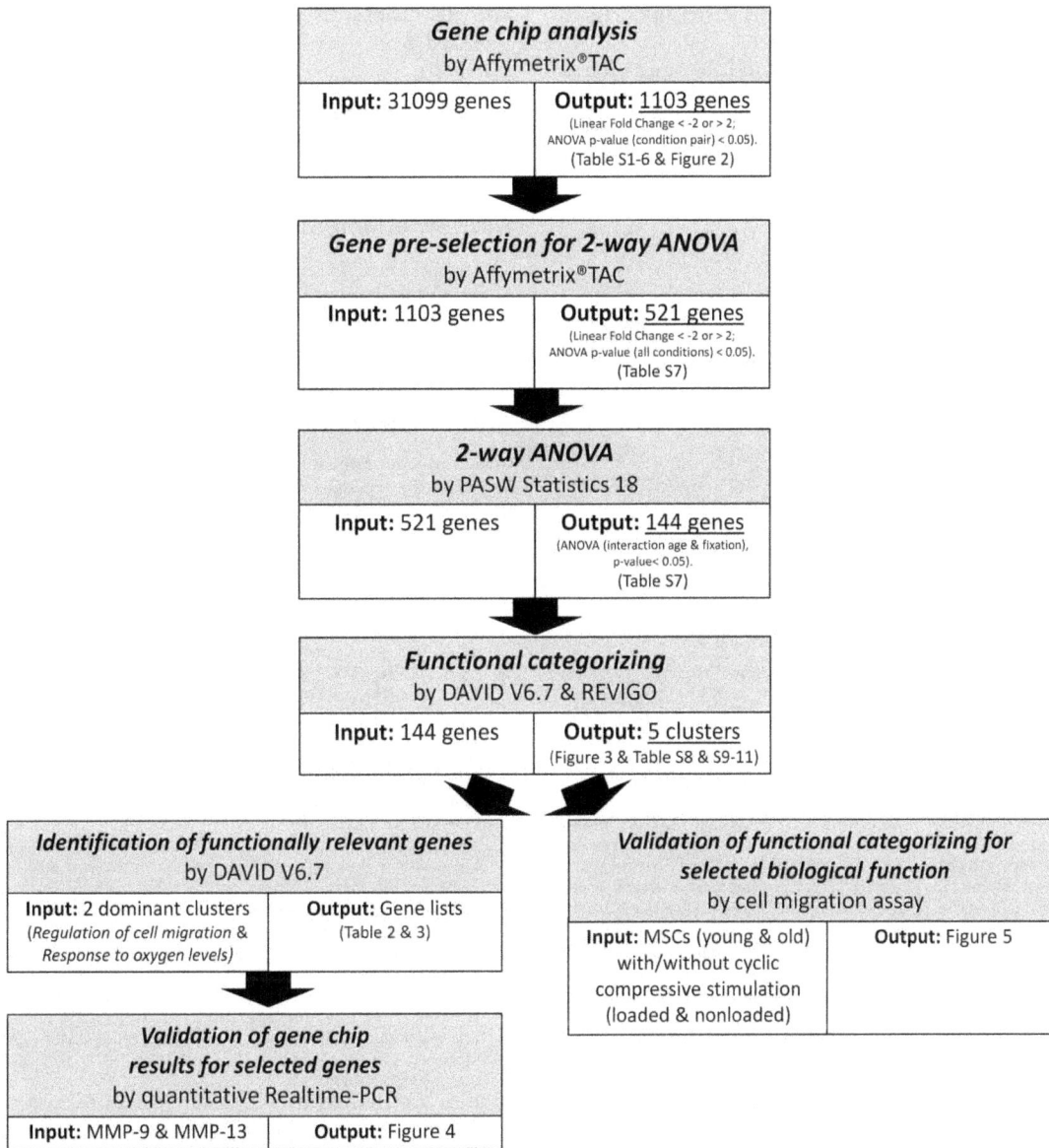

Figure 6. Summary and overview of the work flow in this study.

[28]. We recently reported that mechanical loading *in vitro* stimulates the paracrine pro-angiogenic capacity of MSCs and human fracture hematoma [29,30]. Interestingly, in the latter study, the angiogenic regulator *vascular endothelial growth factor* (VEGF) was up-regulated in hematoma of young but not old patients in response to mechanical loading.

The chip-based whole-genome gene expression analysis revealed that members of the family of matrix metalloproteinases (MMPs), MMP-9 and -13, were strongly influenced by an interaction of the effects of age and mechanical stability. Therefore, these results were further validated by q-RT-PCR with similar outcome. MMPs degrade most components of the extracellular matrix (ECM), such as aggrecan, collagens, elastin, or vitronectin, as well as many non-ECM molecules. Thereby, MMPs allow cell migration, participate in cleavage or release of biologically active molecules and regulate cellular behaviour such as cell attachment, growth, differentiation, and apoptosis [31,32]. During successful enchondral ossification, MMP-9 and -13 play an

important role in the fracture callus [33]. They coordinate not only cartilage matrix degradation, but also the recruitment and differentiation of endothelial cells, osteoclasts, chondroclasts and osteoprogenitors. Lack of MMP-9 in mice results in non-unions and delayed unions of their fractures caused by persistent cartilage at the injury site [34]. MMP-13-null mice showed profound defects in growth plate cartilage [35]. Several studies provided evidence for the regulation of MMP mRNA by mechanical loading *in vitro*. For example, mRNA level of MMP-9 increased as early as 3-6 h after the application of cyclic tensile load on cultured chondrocytes isolated from young rabbits [36]. MMP-13 mRNA up-regulation has been described after stretching of murine osteoblasts [37]. This is the first study that describes the simultaneous influence of age and mechanical stimuli on their expression. However, it is known that MMP function is spatially and temporally regulated at transcriptional, post-transcriptional, and post-translational levels via MMP controlled activation, inhibition and cell surface localization. Therefore, it is now crucial to validate these results also on protein level.

In summary, our results indicate that cellular migration is differently affected by fixator stability in young and old rats three days post-osteotomy possibly leading to the previously reported biomechanical long-term outcomes in bone defect healing. In conclusion, our data highlight the importance of analysing the influence of risk factors of fracture healing (e.g. advanced age, suboptimal fixator stability) in combination rather than alone.

Supporting Information

File S1   Table S1–S6. Up- and down-regulated genes for each pair of condition groups identified by Affymetrix Transcriptome Analysis Console (TAC)

File S2   Table S7. Genes that are influenced by the interaction between age and fixation stability identified by 2-way ANOVA.

File S3   Table S8. Functional annotation of differentially expressed genes using DAVID v6.7.

File S4   Table S9–S11. Single GO terms were joined into five clusters of related terms using REVIGO.

Acknowledgments

The authors would like to thank Dr. Ute Ungethuem from the Laboratory of Functional Genome Research Charité Core Facility for performing Affymetrix hybridization. We are grateful to Dr. Nicola Ott, Martin Textor, Liliya Schumann, Annett Kurtz, Marcel Gaetjen, Delia Koennig, and Claudia Schaar for excellent technical support.

Author Contributions

Conceived and designed the experiments: AO GND SG SP JEO CP PS. Performed the experiments: AO SG JEO PS. Analyzed the data: AO GND SG SP JEO CP PS. Wrote the paper: AO GND JEO PS.

References

1. Strube P, Sentuerk U, Riha T, Kaspar K, Mueller M, et al. (2008) Influence of age and mechanical stability on bone defect healing: Age reverses mechanical effects. Bone 42: 758–764. doi:10.1016/j.bone.2007.12.223.
2. Meyer RA, Tsahakis PJ, Martin DF, Banks DM, Harrow ME, et al. (2001) Age and ovariectomy impair both the normalization of mechanical properties and the accretion of mineral by the fracture callus in rats. J Orthop Res 19: 428–435. doi:10.1016/S0736-0266(00)90034-2.
3. Skak SørV, Jensen TT (1988) Femoral shaft fracture in 265 children: Log-normal correlation with age of speed of healing. Acta Orthop Scand 59: 704–707.
4. Kasper G, Mao L, Geissler S, Draycheva A, Trippens J, et al. (2009) Insights into Mesenchymal Stem Cell Aging: Involvement of Antioxidant Defense and Actin Cytoskeleton. STEM CELLS 27: 1288–1297. doi:10.1002/stem.49.
5. Bloomfield SA, Hogan HA, Delp MD (2002) Decreases in bone blood flow and bone material properties in aging Fischer-344 rats. Clin Orthop Relat Res: 248–257.
6. Panagiotis M (2005) Classification of non-union. Injury 36: S30–S37. doi:10.1016/j.injury.2005.07.019.
7. Hayda RA, Brighton CT, Esterhai JL (1998) Pathophysiology of delayed healing. Clin Orthop Relat Res: S31–40.
8. Klein P, Schell H, Streitparth F, Heller M, Kassi J, et al. (2003) The initial phase of fracture healing is specifically sensitive to mechanical conditions. Journal of Orthopaedic Research 21: 662–669. doi:10.1016/S0736-0266(02)00259-0.
9. Mehta M, Strube P, Peters A, Perka C, Hutmacher D, et al. (2010) Influences of age and mechanical stability on volume, microstructure, and mineralization of the fracture callus during bone healing: Is osteoclast activity the key to age-related impaired healing? Bone 47: 219–228. doi:10.1016/j.bone.2010.05.029.
10. Lienau J, Schmidt-Bleek K, Peters A, Weber H, Bail HJ, et al. (2010) Insight into the Molecular Pathophysiology of Delayed Bone Healing in a Sheep Model. Tissue Engineering Part A 16: 191–199. doi:10.1089/ten.tea.2009.0187.
11. Claes L, Blakytny R, Göckelmann M, Schoen M, Ignatius A, et al. (2009) Early dynamization by reduced fixation stiffness does not improve fracture healing in a rat femoral osteotomy model. Journal of Orthopaedic Research 27: 22–27. doi:10.1002/jor.20712.
12. Carano RA, Filvaroff EH (2003) Angiogenesis and bone repair. Drug Discov Today 8: 980–989. doi:10.1016/S1359-6446(03)02866-6.
13. Meyer RA Jr, Meyer MH, Tenholder M, Wondracek S, Wasserman R, et al. (2003) Gene expression in older rats with delayed union of femoral fractures. J Bone Joint Surg Am 85-A: 1243–1254.
14. Le AX, Miclau T, Hu D, Helms JA (2001) Molecular aspects of healing in stabilized and non-stabilized fractures. Journal of Orthopaedic Research 19: 78–84. doi:10.1016/S0736-0266(00)00006-1.
15. Strube P, Mehta M, Putzier M, Matziolis G, Perka C, et al. (2008) A new device to control mechanical environment in bone defect healing in rats. Journal of Biomechanics 41: 2696–2702. doi:10.1016/j.jbiomech.2008.06.009.
16. Kaspar K, Schell H, Toben D, Matziolis G, Bail HJ (2007) An easily reproducible and biomechanically standardized model to investigate bone healing in rats, using external fixation/Ein leicht reproduzierbares und biomechanisch standardisiertes Modell zur Untersuchung der Knochenheilung in der Ratte unter Verwendung eines Fixateur Externe. bmte 52: 383–390. doi:10.1515/BMT.2007.063.
17. Huang DW, Sherman BT, Lempicki RA (2008) Systematic and integrative analysis of large gene lists using DAVID bioinformatics resources. Nat Protocols 4: 44–57. doi:10.1038/nprot.2008.211.
18. Supek F, Bošnjak M, Škunca N, Šmuc T (2011) REVIGO Summarizes and Visualizes Long Lists of Gene Ontology Terms. PLoS ONE 6: e21800. doi:10.1371/journal.pone.0021800.
19. Claes LE, Heigele CA, Neidlinger-Wilke C, Kaspar D, Seidl W, et al. (1998) Effects of mechanical factors on the fracture healing process. Clin Orthop Relat Res: S132–147.
20. Edgar R, Domrachev M, Lash AE (2002) Gene Expression Omnibus: NCBI gene expression and hybridization array data repository. Nucl Acids Res 30: 207–210. doi:10.1093/nar/30.1.207.
21. Einhorn TA (1998) The cell and molecular biology of fracture healing. Clin Orthop Relat Res: S7–21.
22. Dimitriou R, Tsiridis E, Giannoudis PV (2005) Current concepts of molecular aspects of bone healing. Injury 36: 1392–1404. doi:10.1016/j.injury.2005.07.019.
23. Phillips AM (2005) Overview of the fracture healing cascade. Injury 36: S5–S7. doi:10.1016/j.injury.2005.07.027.
24. Ode A, Kopf J, Kurtz A, Schmidt-Bleek K, Schrade P, et al. (2011) CD73 and CD29 concurrently mediate the mechanically induced decrease of migratory capacity of mesenchymal stromal cells. Eur Cell Mater 22: 26–42.
25. Gerstenfeld LC, Cullinane DM, Barnes GL, Graves DT, Einhorn TA (2003) Fracture healing as a post-natal developmental process: Molecular, spatial, and temporal aspects of its regulation. Journal of Cellular Biochemistry 88: 873–884. doi:10.1002/jcb.10435.
26. Barnes GL, Kostenuik PJ, Gerstenfeld LC, Einhorn TA (1999) Growth Factor Regulation of Fracture Repair. Journal of Bone and Mineral Research 14: 1805–1815. doi:10.1359/jbmr.1999.14.11.1805.
27. Rubin CT, Bain SD, McLeod KJ (1992) Suppression of the osteogenic response in the aging skeleton. Calcif Tissue Int 50: 306–313.
28. Turner CH, Takano Y, Owan I (1995) Aging changes mechanical loading thresholds for bone formation in rats. Journal of Bone and Mineral Research 10: 1544–1549. doi:10.1002/jbmr.5650101016.
29. Groothuis A, Duda GN, Wilson CJ, Thompson MS, Hunter MR, et al. (2010) Mechanical stimulation of the pro-angiogenic capacity of human fracture haematoma: Involvement of VEGF mechano-regulation. Bone 47: 438–444. doi:10.1016/j.bone.2010.05.026.
30. Kasper G, Dankert N, Tuischer J, Hoeft M, Gaber T, et al. (2007) Mesenchymal Stem Cells Regulate Angiogenesis According to Their Mechanical Environment. STEM CELLS 25: 903–910. doi:10.1634/stemcells.2006–0432.
31. Ortega N, Behonick D, Stickens D, Werb Z (2003) How Proteases Regulate Bone Morphogenesis. Annals of the New York Academy of Sciences 995: 109–116. doi:10.1111/j.1749-6632.2003.tb03214.x.
32. Sternlicht MD, Werb Z (2001) How matrix metalloproteinases regulate cell behavior. Annu Rev Cell Dev Biol 17: 463–516. doi:10.1146/annurev.cellbio.17.1.463.
33. Uusitalo H, Hiltunen A, Söderström M, Aro HT, Vuorio E (2000) Expression of Cathepsins B, H, K, L, and S and Matrix Metalloproteinases 9 and 13 During Chondrocyte Hypertrophy and Endochondral Ossification in Mouse Fracture Callus. Calcified Tissue International 67: 382–390. doi:10.1007/s002230001152.
34. Colnot C, Thompson Z, Miclau T, Werb Z, Helms JA (2003) Altered fracture repair in the absence of MMP9. Development 130: 4123–4133.
35. Stickens D, Behonick DJ, Ortega N, Heyer B, Hartenstein B, et al. (2004) Altered endochondral bone development in matrix metalloproteinase 13-deficient mice. Development 131: 5883–5895. doi:10.1242/dev.01461.

36. Fujisawa T, Hattori T, Takahashi K, Kuboki T, Yamashita A, et al. (1999) Cyclic mechanical stress induces extracellular matrix degradation in cultured chondrocytes via gene expression of matrix metalloproteinases and interleukin-1. J Biochem 125: 966–975.

37. Yang C-M, Chien C-S, Yao C-C, Hsiao L-D, Huang Y-C, et al. (2004) Mechanical Strain Induces Collagenase-3 (MMP-13) Expression in MC3T3-E1 Osteoblastic Cells. J Biol Chem 279: 22158–22165. doi:10.1074/jbc.M401343200.

18

Changes in the Mechanical Properties and Composition of Bone during Microdamage Repair

Gang Wang[1,2¶], **Xinhua Qu**[1¶], **Zhifeng Yu**[1]*

1 Shanghai Key Laboratory of Orthopaedic Implants, Department of Orthopaedic Surgery, Shanghai Ninth People's Hospital, Shanghai Jiaotong University School of Medicine, Shanghai, China, 2 Department of Orthopedic Surgery, the Second Affiliated Hospital of Nanjing Medical University, Nanjing, China

Abstract

Under normal conditions, loading activities result in microdamage in the living skeleton, which is repaired by bone remodeling. However, microdamage accumulation can affect the mechanical properties of bone and increase the risk of fracture. This study aimed to determine the effect of microdamage on the mechanical properties and composition of bone. Fourteen male goats aged 28 months were used in the present study. Cortical bone screws were placed in the tibiae to induce microdamage around the implant. The goats were euthanized, and 3 bone segments with the screws in each goat were removed at 0 days, 21 days, 4 months, and 8 months after implantation. The bone segments were used for observing microdamage and bone remodeling, as well as nanoindentation and bone composition, separately. Two regions were measured: the first region (R1), located 1.5 mm from the interface between the screw hole and bone; and the second region (R2), located >1.5 mm from the bone-screw interface. Both diffuse and linear microdamage decreased significantly with increasing time after surgery, with the diffuse microdamage disappearing after 8 months. Thus, screw implantation results in increased bone remodeling either in the proximal or distal cortical bone, which repairs the microdamage. Moreover, bone hardness and elastic modulus decreased with microdamage repair, especially in the proximal bone tissue. Bone composition changed greatly during the production and repair of microdamage, especially for the C (Carbon) and Ca (Calcium) in the R1 region. In conclusion, the presence of mechanical microdamage accelerates bone remodeling either in the proximal or distal cortical bone. The bone hardness and elastic modulus decreased with microdamage repair, with the micromechanical properties being restored on complete repair of the microdamage. Changes in bone composition may contribute to changes in bone mechanical properties.

Editor: Damian Christopher Genetos, University of California Davis, United States of America

Funding: This research was supported by grants from the National Natural Science Foundation for Youth (11002090), the Shanghai Natural Science Foundation (10ZR1417900), and the Program for Key Disciplines of the Shanghai Municipal Education Commission (J50206). The funders had no role in study design, data collection and analysis, decision to publish, or preparation of the manuscript.

Competing Interests: The authors have declared that no competing interests exist.

* Email: zhifengyu@gmail.com

❥ These authors contributed equally to this work.

¶ GW and XQ are co-first authors on this work.

Introduction

The skeleton is the main weight-bearing structure in humans. Under normal conditions, loading activities result in microdamage in the living skeleton, which is repaired by bone remodeling [1]. Microdamage is primarily caused by excessive continuous fatigue loading and mechanical loading, such as that generated by implantation [2]. Compared with the microdamage caused by fatigue loading, the microdamage resulting from implant surgery has been the focus of a limited number of studies, suggesting that the implantation procedure may cause extensive microdamage to the peri-implant bone [3]. Moreover, microdamage accumulation may impair the strength of the bone-implant interface, potentially causing implant loosening; compromise the mechanical properties of bone; and increase the risk of fracture [4].

Until recently, most studies have investigated the effect of prevalent microdamage on the mechanical properties of bone. To our knowledge, there is limited data regarding the restoration of bone quality after microdamage repair. Apparently, extensive microdamage often occurs in peri-implant bone, which is affected mainly by diffuse damage. Moreover, microdamage is removed by bone remodeling and—possibly—the deposition of mineralized tissue [5,6], which may change the structure and mechanical properties of local bone. The present study aimed to determine the effect of microdamage on the mechanical properties and composition of bone. Overall, we were able to confirm our hypothesis that the accumulation and repair of microdamage changes the composition of local bone, which decreases the mechanical properties of bone.

Materials and Methods

Screw implantation and groups

Fourteen skeletally mature Chinese mountain goats with a body weight ranging from 27–32 kg were used in the present study. The goats are all male and aged 28 months. Skeletal maturity was confirmed by radiography showing closure of the distal femoral and proximal tibia growth plates [7]. The goats were kept on a farm and cared for by a qualified veterinarian during the entire study. Animal Research Ethics approval was obtained from the Research Ethics Committee of the Shanghai Ninth People's Hospital. Under pentobarbital sodium (50 mg/kg, IV) anesthesia, a longitudinal incision was made along the cranial-lateral aspect of the leg. In each tibia, 5 holes were drilled into the cranial-lateral diaphysis with the centered hole located at the midline of the tibia diaphysis. The screws only passed through one side of the cortex, and the distance between neighboring screws was 20 mm. After drilling and tapping, titanium cortical bone screws measuring 2 mm in diameter and 10 mm in length were inserted into each hole with a torque of 40 N·cm. The incisions were closed in layers using 4.0 vicryl sutures. All animals were injected with tetracycline (20 mg/kg, IV), 14 days before being euthanized, for 2 consecutive days. Seven days prior to necropsy, all animals were injected with calcein (15 mg/kg, IV), for 2 consecutive days. The schedules for screw implantation in each tibiae and the time of euthanasia are presented in Fig. 1. The bone segments with screws were removed and numbered according to the longitudinal direction. Segment 2 was stained with basic fuchsin, embedded in polymethyl methacrylate (PMMA) and cut cross-sectionally for the observation of microdamage and bone remodeling. Segment 4 was embedded in PMMA without basic fuchsin stain for nanoindentation testing. Segment 5 was used for bone composition testing. The samples taken at 0 days and 21 days were prepared for short-term observation, and the 4-month and 8-month samples were prepared for long-term observation. Fourteen goats were allocated to short-term (n = 7) and long-term (n = 7) observation separately (Fig. 1).

Microdamage observation and histomorphometry measurements

After euthanasia, the bilateral tibiae of the goats were removed, and bone segments implanted with screws were prepared for bulk staining in basic fuchsin for microdamage analysis [8]. After basic fuchsin stained, screws were removed, bone samples of 2 cm in length were embedded in PMMA and cut into slices with a thickness of 50 μm using a hard tissue slicer (Leica SP1600, Germany). Stained sections were observed and measured under a fluorescence microscope. The Bioquant Image Analysis system (Bioquant OSTEO II VB.10.20, USA) was used to measure the porosity of the whole segments and the level of microdamage in the following 2 regions: the R1 region located 1.5 mm from the

interface between the screw hole and the bone, and the R2 region located >1.5 mm (1.5~3) from the bone-screw interface (Fig. 2) [9]. Diffuse damage often occurred in a narrow area adjacent to the bone-screw interface. In diffuse damage microcracks were too small to be discriminated by light microscopy. Therefore, pooled staining was present in bone with diffuse damage. In cross-hatched damage microcracks were vaguely distinguishable under fluorescence microscope [10]. The microdamage measurements included the damaged area in diffuse microdamage area (Dif.Dx.Ar = the area of diffuse microdamage/bone area) and the crack density (Cr.Dn = the number of cracks/bone area), mean crack length (Cr.Le), and crack surface density (Cr.S.Dn = the mean crack length/bone area) in the linear cracks [11]. The variables related to bone remodeling were measured in the same region of microdamage measurement. Moreover, the densities of the cortical porosity (Ct.Po = the number of resorption cavities/total bone area), mineralized surface (Md.S = the mineralized surface/bone surface) and mineral apposition rate (MAR = interlabel distance/label interval) were calculated.

Nanoindentation measurement

The polished PMMA embedded bone blocks were examined using nanoindentation techniques as reported previously [12]. The elastic modulus (E) and contact hardness (H) were calculated using the Oliver-Pharr method [13], where the elastic modulus is a function of the unloading stiffness of the force-displacement curve, under the assumption that the unloading is elastic. Contact hardness is calculated as the maximum load divided by the projected area of the indenter tip at maximum load. A Nano Indenter XP system (MTS Nanoindenter XP, Oak Ridge, USA) was employed to measure the force and displacement during the indentation of the polished bone specimen. For each specimen, the sites selected for nanoindentation were consistent with the areas for microdamage measurement. Sixty-four points were selected per region. The measurement areas were determined using an optical microscope at 50× magnification. A Berkovich shape

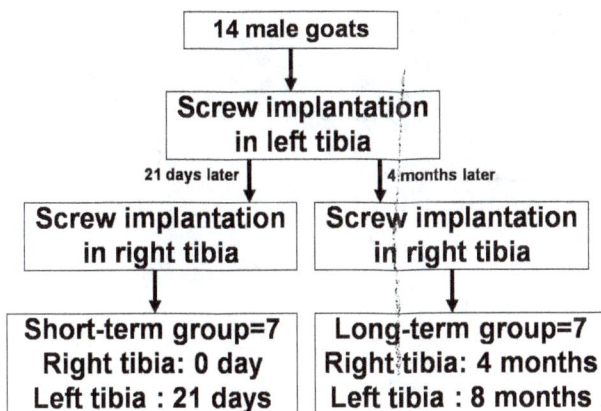

Figure 1. Schedule for screw implantation in goat tibiae in both groups.

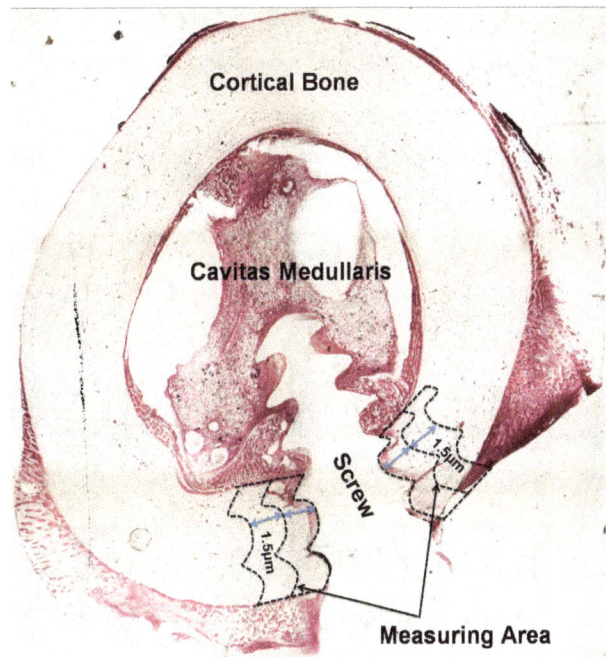

Figure 2. Schematic diagram of the measuring area.

diamond indenter tip (Ei = 1141 Gpa, vi = 0.07) was used to perform the nanoindentation tests at each site. The indentation procedure was under displacement control. After the surface was identified, the indenter was advanced to 500 nm at a speed of 10 nm/s to avoid the effect of bone surface roughness. A typical indentation load-displacement curve includes a loading segment, a 10-s holding period at maximum load, an unloading segment, and a 50-s holding period for thermal drift measurement at 10% of maximum load.

Bone composition measurement

JEOL JXA-8100 electron microprobe wavelength dispersive system (WDS) was employed to detect the kinds of elements contained in the bone segment by analyzing the wavelength (or energy) of characteristic X ray. Backscattered electron (BSE) images were produced by Oxford Instruments INCA energy dispersive system (EDS). Then through quantitative analysis, by comparing the diffraction intensity with standard sample, quantitative result can be got from the testing sample.

Statistical analyses

All data have been tested by SPSS Statistics 18.0 (SPSS Inc., Chicago, USA) and demonstrated to have normal distributions with homogeneous variances. We used two-way analysis of variance to compare differences in microdamage, bone remodeling, nanoindentation, and bone composition between different time points in different regions. Differences between the R1 and R2 regions were analyzed by the paired sample t-test. P values of <0.05 were considered significant.

Results

Microdamage morphology around the implant

Under fluorescence microscopy, areas of microdamage dyed with basic fuchsin were observed around the implant in both regions. Diffuse microdamage and linear cracks were found in the cortical bone around the implants immediately after screw implantation (0 days), and diffuse microdamage was more easily observed than linear cracks in the R1 region (Fig. 3). Moreover, Fig. 3 shows that the diffuse microdamage and linear cracks were more pronounced in the R1 region.

Region and time differences in the microdamage observed around the implant

We used a two-way analysis of variance to determine the region and time differences between microdamage parameters. The Dif.Dx.Ar, Cr.Dn, Cr.Le, and Cr.S.Dn in the R1 region were significantly higher than those in the R2 region (Fig. 4), suggesting that screw implantation can produce more diffuse microdamage and more linear cracks around the screw. Interestingly, the values of the linear cracks, especially Cr.Dn and Cr.S.Dn, are relatively higher in the R1 region than the R2 region 21 days after surgery— 5-fold and 10-fold higher in the R1 region, respectively— indicating that the linear cracks are more likely to be observed proximal to the screw.

Repair mode for microdamage

By comparing the results between different regions, we found no significant difference between the different regions at 0 days. At 21 days and 4 months, the labeled surface and bone porosity areas were significantly higher in the R1 than in the R2 region. However, at 8 months, although the bone porosity was higher in the R1 region, the labeled surface area was not different from that in the R2 region. Mineral apposition rate (MAR) changed greatly after surgery, especially in the R1 region. MAR in R1 region is higher than it in the R2 region after 21 days and 4 months of surgery. (Fig. 5).

Change in micromechanical properties

Nanoindentation testing showed that at different time points, the hardness and elastic modulus from 0 days, 21 days to 4 months showed a downward trend in both regions. Both of the hardness and elastic modulus began to increase after 4 month in R1 and R2 region. The hardness and elastic modulus of the bone tissue adjacent to the screw thread was significantly lower in the R1 than in the R2 region 4 months after surgery. After 8 months, the bone quality recovered slightly (Fig. 6). There was significant time difference among the 4 time points in each group (P<0.01).

Change in bone composition

We measured 4 elements, namely, C (Carbon), O (Oxygen), P (Phosphorus), and Ca (Calcium), in different regions. There was no difference between the 4 elements 0 days after surgery (P>0.05). At 21 days, significant differences were observed in the C, P and Ca elements between the R1 and R2 regions (P<0.05), with C

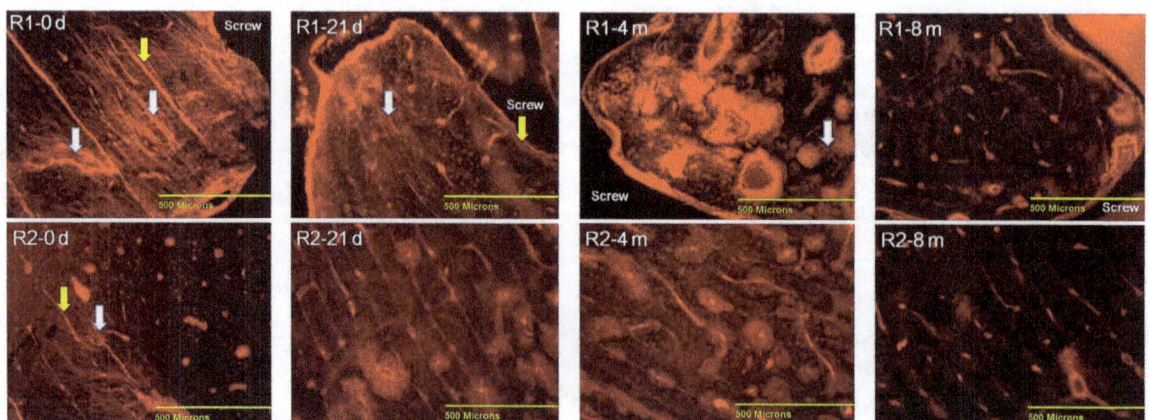

Figure 3. Microdamage accumulation and associated changes at 0 days, 21 days, 4 months, and 8 months after surgery (40×). White arrow referred to diffuse microdamage, yellow arrow referred to linear microcrack.

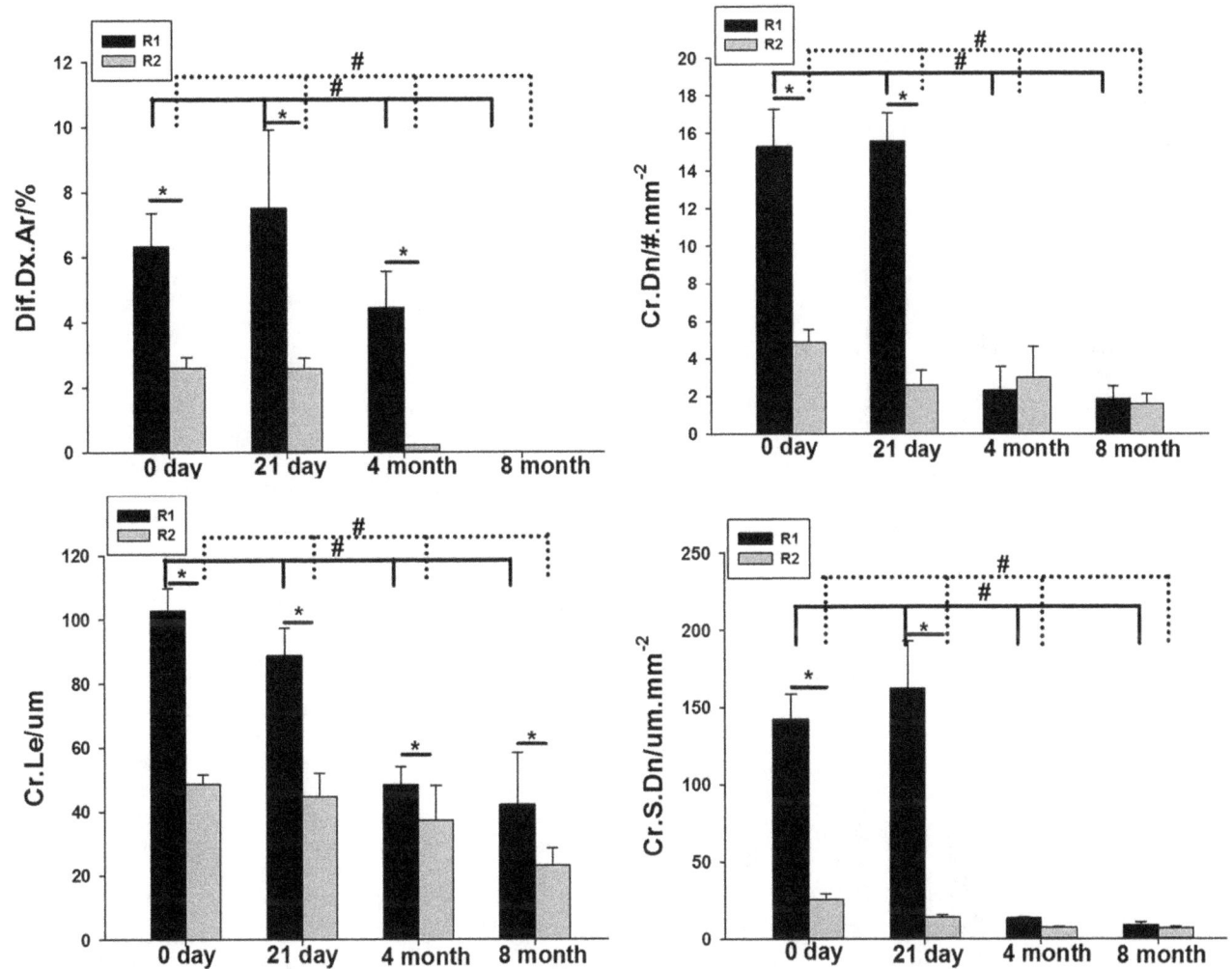

Figure 4. Production and repair of microdamage after screw implantation. * means difference between R1 and R2 region at different time point. # means difference between four time point in each region, solid line referred to R1 region, dotted line referred to R2 region.(*<0.05, #< 0.05).

being higher in the R1 region, and P and Ca being higher in the R2 region. At 4 months, C increased to peak levels in the R1 region, whereas Ca decreased to a minimum. At 4 months and 8 months, the C and Ca composition was significantly different between the R1 and R2 regions (P<0.01), whereas the O and P composition was not significantly different between groups (P< 0.05, Fig. 7).

Discussion

Although microdamage accumulation and increased bone remodeling are clearly evidenced in peri-screw bone [10,14], the mechanisms of microdamage repair around the implant and change in the mechanical quality and composition of bone after screw insertion remain unclear. Previous studies suggest that the activation of bone resorption is considerably less effective in diffuse than in linear damage, suggesting that the response to diffuse and linear damage is different [15,16]. However, the large-size linear cracks are known to cause severe damage to the bone matrix, leading to osteocyte apoptosis, which initiates bone remodeling [16] and acts as the primary factor contributing to targeted bone remodeling [16]. Overall, the mechanical properties of bone and

mode of microdamage repair seem to depend on the different types of microdamage.

With regard to the location of the microdamage, Wang et al. [10] reported that diffuse damage tended to appear in the area adjacent to the bone-screw interface, which seems unlikely. In the present study, we frequently observed a combination of diffuse damage and linear cracks in peri-screw bone, which was referred to as complex microdamage. However, the diffuse microdamage decreased as the distance to the implant increased, and the linear cracks were still observed. As confirmed by O'Brien et al. [17], implant surgery on the cortical bone results in increased micro-damage produced near the implant, owing to the greater damage to the proximal bone. Moreover, the complex microdamage is likely to be caused by the strong impact—the drilling and cutting forces in implant surgery result in rapid damage to bone. After surgery, implant micromotion is likely to increase microdamage accumulation.

Microdamage has been postulated as a contributing factor for bone remodeling, which replaces the damaged bone with new bone. Thus, the rate of bone remodeling increased in the area with microdamage, with Wang et al. [10] which suggested that bone resorption becomes considerably active within 1 month after screw

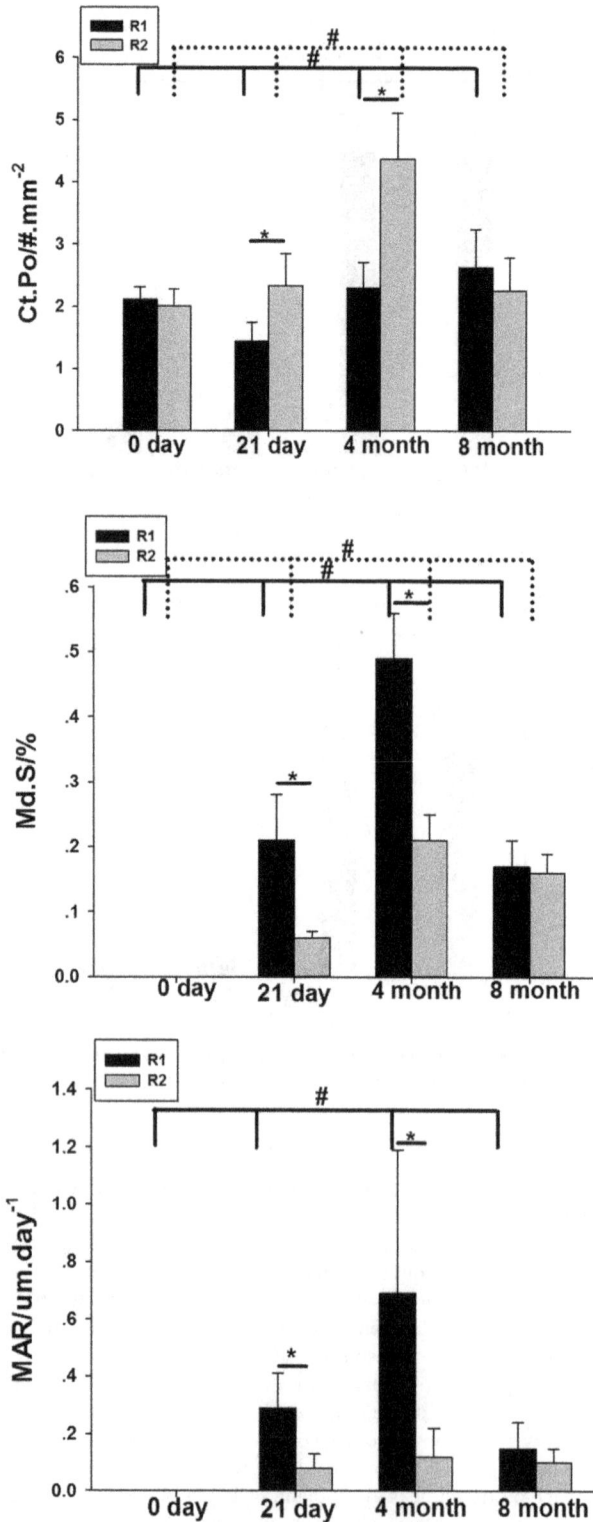

Figure 5. Comparison of bone remodeling parameters in different regions at different time points. * means difference between R1 and R2 region at different time point. # means difference between four time point in each region, solid line referred to R1 region, dotted line referred to R2 region.(*<0.0.5, #<0.05).

implantation. Moreover, increased bone resorption was observed which may decreased diffuse damage and linear cracks. Although screw implantation can result in different types of microdamage in cortical bone, the type of microdamage associated with activation of bone remodeling remains unclear. A previous study by Bentolila et al. [15] reported that intracortical resorption was preferentially associated with linear-type cracks, with a 40% reduction in linear microcracks occurring 10 days after fatigue loading. In contrast, diffuse microdamage did not show a significant decrease at that time. In our study, we found that the bone resorption cavity and mineralized surface increased to peak levels 4 months after surgery, while the linear crack decreased rapidly compared with that 21 days after surgery, suggesting that linear microdamage is the main contributor to activation of bone remodeling. By using bisphosphonate coating screw, Agholme et al [18] found bisphosphonate coating leads to increased bone volume around screws. This maybe caused by preventing resorption of the tissue with microdamage and provide a new way to increase implant stability. But how much bisphosphonate were released to the bone and the change of bone remodeling should be confirmed, because bisphosphonate could increases microdamage accumulation during early treatment [19].

Moreover, the accumulation of microdamage has been associated with a reduction in the elastic modulus and strength of bone tissue, and an increase in energy dissipation when the bone is loaded [1,4,17,20–22]. Although the effect of diffuse damage on bone strength remains unclear [17], the accumulation of microcracks at lower stress levels is believed increase the strength of cortical bone [23]. In fact, extensive microdamage often occurs in peri-implant bone, which is affected mainly by diffuse damage. Moreover, microdamage is removed by bone remodeling and—possibly—the deposition of mineralized tissue [5,6], which may both change the structure and mechanical properties of local bone. In the early stage of repair, multiple resorption cavities impair bone structure and act as stress concentrators, both of which reduce the mechanical properties of bone [6]. During that period, the ideal osteointegration of the implant has not been established. Our results suggest that the hardness and elastic modulus did not change significantly during the early stage. However, after 4 months, with an increase in bone remolding and decrease of microdamage, the mechanical properties of local bone began to decrease. The area located near the implant—where significant microdamage occurred—was characterized by a greater decrease in the mechanical properties of bone than the region located far away from the implant. Both of elastic modulus and hardness declined until 4 months after surgery. At the same time, bone remodeling results show that they increased significantly after surgery. The decreased bone mechanical property may be caused by newly formed bone tissue being present following resorption.

The material properties of bone are determined by the matrix composition, including minerals, collagen and water—each of which can influence the mechanical properties of bone [24]. Micropetrotic bone is characterized by the loss of osteocytes, decreased water content and high mineralization [5,25,26]. However, bone remodeling—although increased—is not sufficient to rapidly repair all microdamage, particularly for extensive diffuse damage. The coexistence of lower mineralized new osteonal bone and higher mineralized micropetrotic bone would further compromise the mechanical properties of bone. Accordingly, the bone quality around the endosseous implant may not return to normal immediately after microdamage repair. Moreover, increased diffuse damage may change the normal repair mode, which may compromise the restoration of bone quality after microdamage repair. By using BSE [27] and Fourier transform infrared (FTIR)

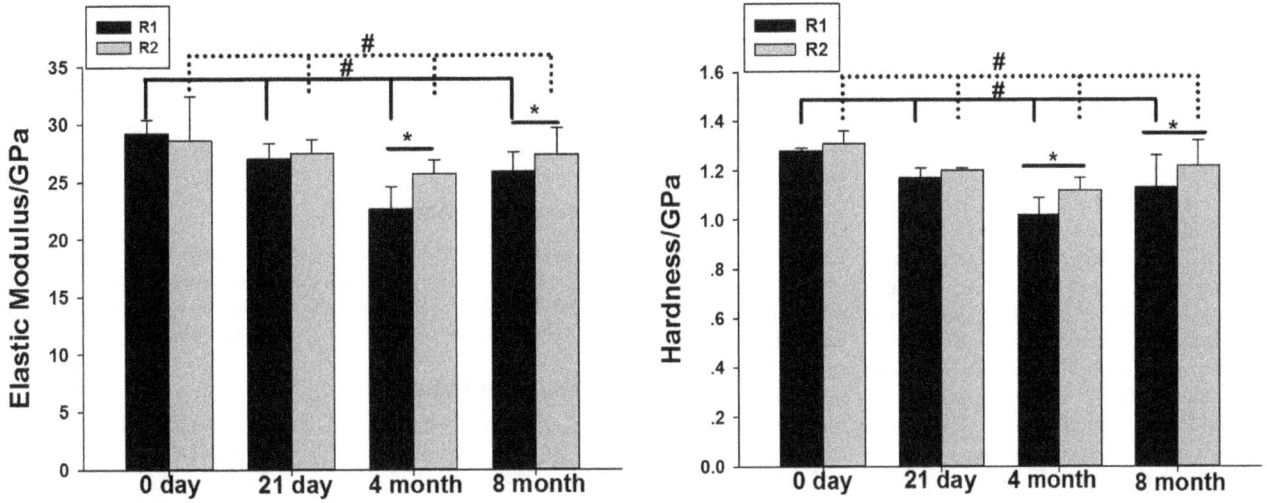

Figure 6. Comparison of nanoindentation data in different regions at different time points. * means difference between R1 and R2 region at different time point. # means difference between four time point in each region, solid line referred to R1 region, dotted line referred to R2 region.(*<0.0.5, #<0.05).

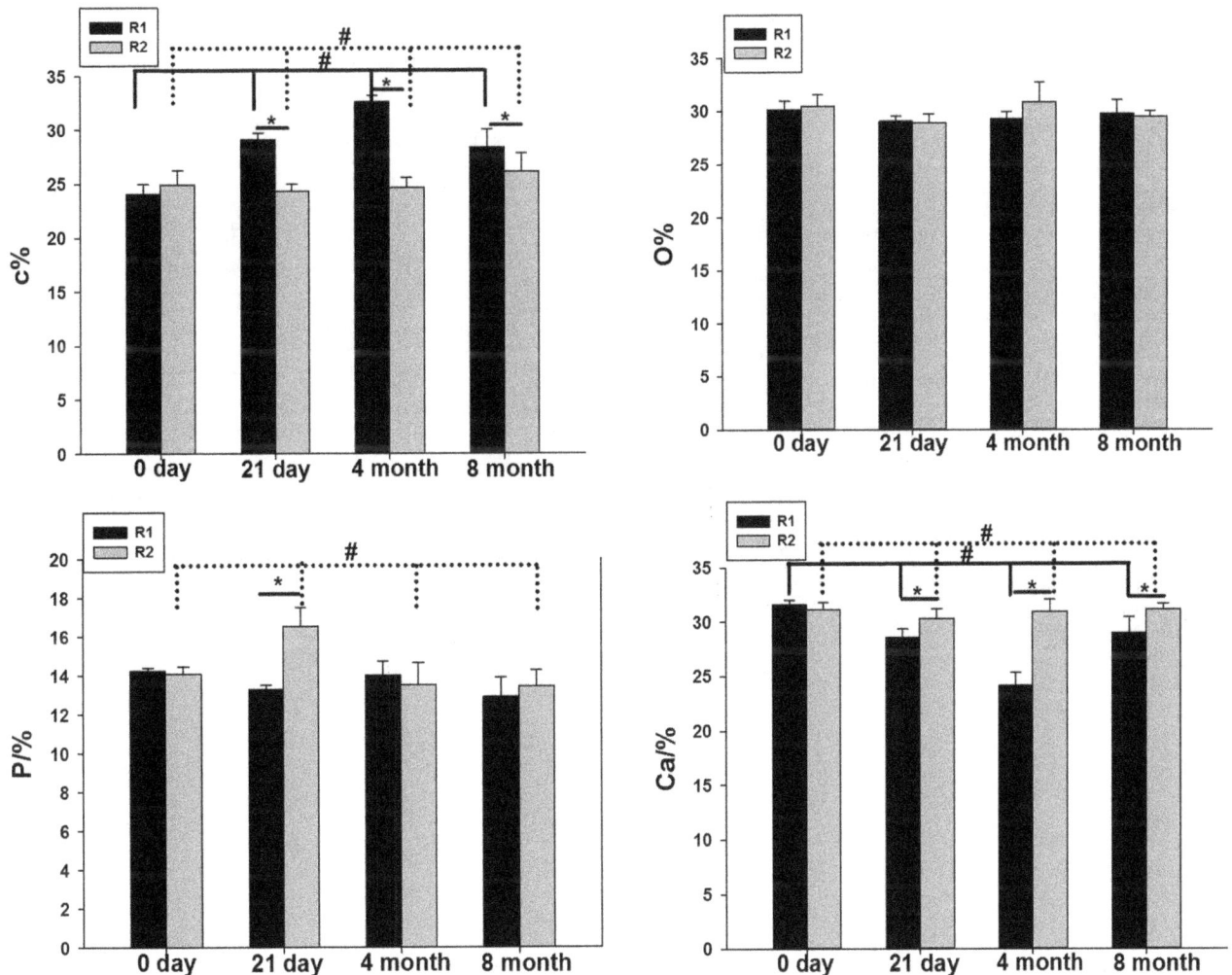

Figure 7. Comparison of bone composition (C, O, P, and Ca) in different regions at different time points. * means difference between R1 and R2 region at different time point. # means difference between four time point in each region, solid line referred to R1 region, dotted line referred to R2 region. (*<0.0.5, #<0.05).

[28] imaging, previous studies have found that calcium and collagen contribute to the mechanical properties of bone. In the present study, although we did not measure changes in collagen, we did measure changes in C, O, P, and Ca. We found that 4 elements—especially C and Ca—changed greatly in the region surrounding the screw during the production and repair of microdamage. The C levels were highest 4 months after surgery, while the Ca levels were lowest at that time. Moreover, the above results suggest that bone remodeling increased to a maximum 4 months after surgery. The variation trend of bone composition may caused by bone remodeling during repair of microdamge. Change of Ca was the same as elastic modulus and hardness, but because the sample size is too low, we cannot find strong correlation between Ca and bone mechanical property.

Overall, both diffuse microdamage and linear cracks were found around the implant and accumulated during the early stage after surgery, and these types of damage may impair the initial stability of the implant. The decreased microdamage at later stages after surgery indicates that the implant-induced microdamage was repaired. The repair of microdamage by bone remodeling can change the structure and mechanical properties of local bone. Accordingly, the bone will return to normal after microdamage repair, through further remodeling activated by mechanical stimulation or other factors, which may take a relatively long time.

Acknowledgments

The present research was supported by grants from the National Natural Science Foundation for Youth (11002090), the Shanghai Natural Science Foundation (10ZR1417900), and the Program for Key Disciplines of the Shanghai Municipal Education Commission (J50206).

Author Contributions

Conceived and designed the experiments: ZY. Performed the experiments: GW XQ ZY. Analyzed the data: XQ ZY. Contributed reagents/materials/analysis tools: XQ ZY. Wrote the paper: ZY.

References

1. Martin RB (2003) Fatigue microdamage as an essential element of bone mechanics and biology. Calcif Tissue Int 73: 101–107.
2. Donahue SW, Sharkey NA, Modanlou KA, Sequeira LN, Martin RB (2000) Bone strain and microcracks at stress fracture sites in human metatarsals. Bone 27: 827–833.
3. Huja SS, Katona TR, Burr DB, Garetto LP, Roberts WE (1999) Microdamage adjacent to endosseous implants. Bone 25: 217–222.
4. Burr D (2003) Microdamage and bone strength. Osteoporos Int 14 Suppl 5: 67–72.
5. Boyde A (2003) The real response of bone to exercise. J Anat 203: 173–189.
6. Schaffler MB (2003) Role of bone turnover in microdamage. Osteoporos Int 14 Suppl 5: 73–80.
7. Leung KS, Siu WS, Cheung NM, Lui PY, Chow DH, et al. (2001) Goats as an osteopenic animal model. J Bone Miner Res 16: 2348–2355.
8. Burr DB, Hooser M (1995) Alterations to the en bloc basic fuchsin staining protocol for the demonstration of microdamage produced in vivo. Bone 17: 431–433.
9. Wang L, Shao J, Ye T, Deng L, Qiu S (2012) Three-dimensional morphology of microdamage in peri-screw bone: a scanning electron microscopy of methyl-methacrylate cast replica. Microsc Microanal 18: 1106–1111.
10. Wang L, Ye T, Deng L, Shao J, Qi J, et al. (2014) Repair of microdamage in osteonal cortical bone adjacent to bone screw. PLoS One 9: e89343.
11. Mashiba T, Turner CH, Hirano T, Forwood MR, Jacob DS, et al. (2001) Effects of high-dose etidronate treatment on microdamage accumulation and biomechanical properties in beagle bone before occurrence of spontaneous fractures. Bone 29: 271–278.
12. Wang X, Sudhaker Rao D, Ajdelsztajn L, Ciarelli TE, Lavernia EJ, et al. (2008) Human iliac crest cancellous bone elastic modulus and hardness differ with bone formation rate per bone surface but not by existence of prevalent vertebral fracture. J Biomed Mater Res B Appl Biomater 85: 68–77.
13. Oliver WC, Pharr GM (1992) An Improved Technique for Determining Hardness and Elastic-Modulus Using Load and Displacement Sensing Indentation Experiments. Journal of Materials Research 7: 1564–1583.
14. Yu Z, Wang G, Tang T, Fu L, Yu X, et al. (2014) Production and repair of implant-induced microdamage in the cortical bone of goats after long-term estrogen deficiency. Osteoporos Int 25: 897–903.
15. Bentolila V, Boyce TM, Fyhrie DP, Drumb R, Skerry TM, et al. (1998) Intracortical remodeling in adult rat long bones after fatigue loading. Bone 23: 275–281.
16. Herman BC, Cardoso L, Majeska RJ, Jepsen KJ, Schaffler MB (2010) Activation of bone remodeling after fatigue: differential response to linear microcracks and diffuse damage. Bone 47: 766–772.
17. O'Brien FJ, Brennan O, Kennedy OD, Lee TC (2005) Microcracks in cortical bone: how do they affect bone biology? Curr Osteoporos Rep 3: 39–45.
18. Agholme F, Andersson T, Tengvall P, Aspenberg P (2012) Local bisphosphonate release versus hydroxyapatite coating for stainless steel screw fixation in rat tibiae. J Mater Sci Mater Med 23: 743–752.
19. Allen MR, Burr DB (2007) Three years of alendronate treatment results in similar levels of vertebral microdamage as after one year of treatment. J Bone Miner Res 22: 1759–1765.
20. Norman TL, Yeni YN, Brown CU, Wang Z (1998) Influence of microdamage on fracture toughness of the human femur and tibia. Bone 23: 303–306.
21. Burr DB, Turner CH, Naick P, Forwood MR, Ambrosius W, et al. (1998) Does microdamage accumulation affect the mechanical properties of bone? J Biomech 31: 337–345.
22. Schaffler MB, Radin EL, Burr DB (1989) Mechanical and morphological effects of strain rate on fatigue of compact bone. Bone 10: 207–214.
23. Sobelman OS, Gibeling JC, Stover SM, Hazelwood SJ, Yeh OC, et al. (2004) Do microcracks decrease or increase fatigue resistance in cortical bone? J Biomech 37: 1295–1303.
24. Currey JD (2002) The mechanical properties of bone. In: Currey JD, editor. Bones: Structure and mechanics. Princeton and Oxford: Princeton University Press. pp. 54–123.
25. Frost HM (1960) Micropetrosis. J Bone Joint Surg 42A: 144–150.
26. Qiu S, Sudhaker Rao D, Fyhrie DP, Palnitkar S, Parfitt AM (2005) The morphological association between microcracks and osteocyte lacunae in human cortical bone. Bone 37: 10–15.
27. Smith IJ, Schirer JP, Fazzalari NL (2010) The role of mineral content in determining the micromechanical properties of discrete trabecular bone remodeling packets. J Biomech 43: 3144–3149.
28. Brennan O, Kuliwaba JS, Lee TC, Parkinson IH, Fazzalari NL, et al. (2012) Temporal changes in bone composition, architecture, and strength following estrogen deficiency in osteoporosis. Calcif Tissue Int 91: 440–449.

19

Raloxifene Prevents Skeletal Fragility in Adult Female Zucker Diabetic Sprague-Dawley Rats

Kathleen M. Hill Gallant[1,2]*, Maxime A. Gallant[1], Drew M. Brown[1], Amy Y. Sato[1], Justin N. Williams[1], David B. Burr[1]

1 Department of Anatomy and Cell Biology, Indiana University School of Medicine, Indianapolis, Indiana, United States of America, 2 Department of Nutrition Science, Purdue University, West Lafayette, Indiana, United States of America

Abstract

Fracture risk in type 2 diabetes is increased despite normal or high bone mineral density, implicating poor bone quality as a risk factor. Raloxifene improves bone material and mechanical properties independent of bone mineral density. This study aimed to determine if raloxifene prevents the negative effects of diabetes on skeletal fragility in diabetes-prone rats. Adult Zucker Diabetic Sprague-Dawley (ZDSD) female rats (20-week-old, n = 24) were fed a diabetogenic high-fat diet and were randomized to receive daily subcutaneous injections of raloxifene or vehicle for 12 weeks. Blood glucose was measured weekly and glycated hemoglobin was measured at baseline and 12 weeks. At sacrifice, femora and lumbar vertebrae were harvested for imaging and mechanical testing. Raloxifene-treated rats had a lower incidence of type 2 diabetes compared with vehicle-treated rats. In addition, raloxifene-treated rats had blood glucose levels significantly lower than both diabetic vehicle-treated rats as well as vehicle-treated rats that did not become diabetic. Femoral toughness was greater in raloxifene-treated rats compared with both diabetic and non-diabetic vehicle-treated ZDSD rats, due to greater energy absorption in the post-yield region of the stress-strain curve. Similar differences between groups were observed for the structural (extrinsic) mechanical properties of energy-to-failure, post-yield energy-to-failure, and post-yield displacement. These results show that raloxifene is beneficial in preventing the onset of diabetes and improving bone material properties in the diabetes-prone ZDSD rat. This presents unique therapeutic potential for raloxifene in preserving bone quality in diabetes as well as in diabetes prevention, if these results can be supported by future experimental and clinical studies.

Editor: Carlos M. Isales, Georgia Regents University, United States of America

Funding: This project was funded by a National Institutes of Health grant (AR047838) to DBB. The funders had no role in study design, data collection and analysis, decision to publish, or preparation of the manuscript.

Competing Interests: Raloxifene was provided by Eli Lilly and Co., Indianapolis, IN. DBB has received grants/research support from Eli Lilly and Co., and Amgen, served as a consultant/scientific advisor for PharmaLegacy, Wright Medical, Agnovos, and AbbieVie, and served as a speaker for the Japan Implant Practice Society. AYS has a family member who is a retired employee of Eli Lilly and Co. KMHG, MAG, DMB and JNW have no competing interests to declare.

* Email: hillgallant@purdue.edu

Introduction

People with type 2 diabetes mellitus have a greater risk for bone fragility fractures compared with healthy adults, despite normal or higher bone mineral density [1–5]. This suggests that bone quality, not quantity, is responsible for the increase in fracture risk in diabetes. Raloxifene is a selective estrogen receptor modulator (SERM) used clinically in women to treat post-menopausal osteoporosis. Our group has previously shown that dogs treated with raloxifene have greater femoral and vertebral toughness, despite no significant effect on bone mineral density [6,7]. Similarly, in post-menopausal women, raloxifene decreases risk of fracture with little effect on bone mineral density [8–10]. This indicates that raloxifene improves bone resistance to fracture by affecting bone quality, and may therefore be an agent with potential to improve bone properties in diabetes where fracture risk is higher apparently due to reduced bone quality rather than reduced bone mass.

The Zucker Diabetic Sprague-Dawley (ZDSD) rat is a recently developed rodent model of type 2 diabetes crossbred from the diet-induced-obesity CD (Sprague-Dawley-derived) and lean Zucker Diabetic Fatty rats (ZDF$^{fa/+}$) [11]. Unlike the diabetic obese ZDF$^{fa/fa}$ rats, ZDSD rats do not have a leptin receptor mutation, and both sexes develop a type 2 diabetes phenotype of polygenic origin more gradually with age or by induction with a high-fat diet, thus reflecting more closely the pathogenesis of human type 2 diabetes [11,12]. This study aimed to test the effects of raloxifene on bone quality and strength in adult female ZDSD rats. Although we have shown positive effects of raloxifene on bone material properties in normoglycemic animals, no studies have been performed in a model subject to diabetes to determine whether raloxifene in a hyperglycemic environment will prevent increased skeletal fragility.

Materials and Methods

Animals and Experimental Design

Twenty-week-old female (n = 24) Zucker Diabetic Sprague Dawley (ZDSD) rats (PreClinOmics, Indianapolis, IN) were randomized (n = 12/group) to receive daily subcutaneous injections of raloxifene (0.5 mg/kg, Eli Lilly Co., Indianapolis, IN) or vehicle (10% cyclodextrin, Sigma-Aldrich) for 12 weeks, and all rats were fed a diabetogenic high-fat diet (48% fat; 5SCA, TestDiet, Richmond, IN) for the duration of the study. The high-fat diet is used to synchronize diabetes induction. Additionally, in contrast to male ZDSD rats that will develop diabetes with age even while on a normal rat diet[11], female ZDSD rats are more resistant to developing diabetes and require the high-fat diet for diabetes induction and to maintain the diabetic state. Blood glucose was measured weekly by glucometer (AlphaTRAK, Abbott Laboratories, Abbott Park, IL) and diabetes was defined as blood glucose $\geq$ 250 mg/dL for 2 consecutive weeks. Whole blood and serum samples were collected at baseline and sacrifice. Glycated hemoglobin (HbA1c,%) was measured in whole blood by immunological assay (Daytona Chemistry Anlayzer, Randox Laboratories, Kearneysville, WV). Serum insulin was measured by ELISA (Mercodia Inc., Winston Salem, NC) and serum triglycerides were measured by colorimetric assays (Daytona Chemistry Analyzer, Randox Laboratories, Kearneysville, WV). Serum c-telopeptide of type I collagen was measured by ELISA (Biotang Inc., Lexington, MA). Prior to sacrifice, rats were double-labeled by intraperitoneal injections of calcein (5 mg/kg; Sigma-Alrich, St. Louis, MO) with a 7-day interlabel period and a 3-day period for incorporation and washout (i.e. 1-7-1-3). Bones (femora, lumbar vertebrae) were collected at the time of sacrifice when rats were 32-weeks-old. Femora and L4 vertebrae were wrapped in saline-soaked gauze and frozen at −20°C for storage prior to bone imaging and mechanical testing; L5 vertebrae were cleaned of soft tissue and fixed in 10% phosphate-buffered formalin for 48 h, then transferred to 70% ethanol, dehydrated in a graded series of ethanol from 70–100%, then embedded (undecalcified and unstained) in methyl-methacrylate with 3% dibutyl phthalate (Sigma-Aldrich, St. Louis, MO) for dynamic histomorphometry. This protocol was approved by the Indiana University Animal Care and Use Committee, and all institutional and national guidelines for the care and use of laboratory animals were followed.

Bone Imaging

Dual-energy x-ray absorptiometry (DXA, GE Lunar PixiMus, Madison, WI) was performed on excised right femora and L4 vertebrae for measures of areal bone mineral density (aBMD g/cm^2), bone mineral content (BMC, g) and area (cm^2). Peripheral quantitative computed tomography (pQCT, XCT Research SA+, Stratec Medizintechnik GmbH, Pforzheim, Germany) was performed on the right femur midshaft for cortical bone morphometric properties (volumetric BMD (vBMD), BMC, cortical area and thickness, periostal and endosteal circumferences and x-axis cross-sectional moment of inertia). Micro-computed tomography (μCT, Brucker Skyscan 1172, Kontich, Belgium) was performed on L4 vertebral bodies and the right distal femur for cancellous bone morphometric properties. Scans were done at 8 μm resolution, 65 kV and 120 μA using a 0.7° rotation step. Reconstructed μCT images (NRecon software) were analyzed using CT Analyzer software (Skyscan, Kontich, Belgium). The same parameters/thresholds were used for each site for reconstruction and analysis. Outcome measurements included whole vertebral body bone volume (BV, mm^3), trabecular bone volume fraction (BV/TV,% [where TV is tissue volume]), trabecular number (Tb.N, mm^{-1}), trabecular thickness (Tb.Th, mm), trabecular separation (Tb.Sp, mm), connectivity density (Conn.D, mm^{-3}), and structural model index (SMI).

Mechanical Testing

Mechanical properties of the femur mid-diaphysis were determined by three-point bending using standard methods [13]. Briefly, bones were thawed to room temperature, and placed posterior side down on the bottom support (18 mm wide) of a servohydraulic test system (100P225 Modular Test Machine, TestResources, Shakopee, MN), so that the descending probe contacted the central anterior surface. Bones were loaded to failure using a displacement rate of 2 mm/min. Force vs. displacement data was collected at 10 Hz. Material properties were calculated based on standard equations using structural mechanical properties and geometric measures from pQCT [13]. Reduced platen compression (RPC) was used to determine mechanical properties of cancellous bone on a 2 mm thick slab of distal femur (100P225 Modular Test Machine, TestResources, Shakopee, MN). For RPC, platen size was set at 70% of the maximum circle diameter to include only cancellous bone [6], which was determined by uCT scanning of the samples prior to mechanical testing. Tests were performed at 0.5 mm/min and data collected at 2 Hz until sample failure.

Mechanical properties of L4 vertebrae were determined by axial compression after removal of vertebral processes using a dremel tool with a minisaw attachment, and removal of the cranial and caudal endplates parallel to each other using a low-speed bone saw (Isomet, Buehler, Lake Bluff, IL). L4 vertebral bodies (+/- 3.5 mm height) were loaded at a rate of 0.5 mm/min until failure (100P225 Modular Test Machine) and data were collected at 10 Hz.

Structural mechanical properties of femoral cortical bone, L4 vertebrae and cancellous bone from the RPC testing were determined from the load-deformation curves using standard definitions. Material properties were calculated based on standard equations using structural mechanical properties and geometric measures from caliper measurements and pQCT (cortical bone) or μCT [13].

Bone Turnover

Bone turnover was measured by serum C-terminal telopeptides of type I collagen (Ctx) by EIA (RatLaps™, IDS, Inc.), and by dynamic histomorphometry of L5 vertebrae. Thin sections (approximately 6 μm) of the L5 vertebra were cut longitudinally with a microtome (Reichert-Jung SuperCut). Approximately 5 mm^2 of cancellous bone tissue 0.5 mm from the endocortical surface was analyzed from one section. Measurements were made at 200x magnification using a fluorescent microscope (Nikon Optiophot 2, Nikon, Inc., Garden City, NY) and images were analyzed using the Bioquant system (R&M Biometrics, Nashville, TN). All measurements and calculations were performed following the guidelines of the American Society for Bone and Mineral Research Histomorphometry Nomenclature Committee [14]. Parameters measured included single-label perimeter (sL.Pm), double-label perimeter (dL.Pm), and interlabel width (Ir.L.Wi). From these primary measurements, the following outcome parameters were calculated: mineral apposition rate (MAR = Ir.L.Wi/7 days [μm/day]); mineralizing surface (MS/BS = (0.5*sL.Pm + dl.Pm)/B.Pm*100 [%]); and bone formation rate (BFR/BS = MAR*MS/BS*365 [μm^3/μm^2/year]).

Statistical analyses

The planned two-way analysis of variance (diabetes and raloxifene as factors) was not possible because none of the raloxifene-treated rats became diabetic. Thus, one-way analysis of variance with Tukey's posthoc analysis was used to detect differences in means among the three groups: raloxifene-treated (RAL), vehicle-treated non-diabetic (VEH-ND), and vehicle-treated diabetic rats (VEH-D). Body weight was tested as a covariate for all measures and used for DXA variables and L4 BV/TV. Diabetes induction was analyzed by log-rank test of Kaplan-Meier survival curves and Fisher's exact tests. Statistical analyses were performed using SAS 9.2 software (Cary, NC) and significance was set at α 0.05. Values are presented as least squares means $\pm$ SEM unless otherwise noted.

Results

Rats randomized to receive raloxifene injections and vehicle injections had similar baseline body weight (mean $\pm$ SD: 342$\pm$14 and 348$\pm$23 g, respectively), blood glucose (mean $\pm$ SD: 114$\pm$7 and 112$\pm$12 mg/dL, respectively), HbA1c (mean $\pm$ SD: 4.5$\pm$0.3 [n = 11] and 4.5$\pm$0.1% [n = 9], respectively), serum insulin (mean $\pm$ SD: 0.50$\pm$0.26 [n = 6] and 0.47$\pm$0.36 [n = 10], respectively) and serum triglycerides (mean $\pm$ SD: 3.37$\pm$1.39 [n = 7] and 2.73$\pm$1.01 [n = 9]). After 12 weeks, none of the 12 rats treated with raloxifene became diabetic whereas 4 out of 12 rats that received vehicle injections became diabetic. By Fisher's exact test, the difference in diabetes frequency did not reach statistical significance (p = 0.09). However, by survival analysis of Kaplan-Meier curves, raloxifene significantly reduced diabetes induction (p = 0.03) (**Fig. 1a**).

At the time of sacrifice, vehicle-treated non-diabetic (VEH-ND) rats weighed more than both vehicle-treated diabetic (VEH-D) (p<0.0001) and raloxifene-treated (RAL) (p<0.0001) rats (**Table 1**).VEH-D rats had higher blood glucose over the course of the study, as determined by area-under-the-curve (AUC) compared with RAL or VEH-ND rats (p<0.0001) (**Fig. 1b,c**). Additionally, RAL-treated rats had lower cumulative blood glucose over the course of the study (AUC) than VEH-ND rats (p = 0.048) (**Fig. 1c**), but endpoint values were not significantly different between RAL-treated and VEH-ND rats (**Table 1**). At sacrifice, HbA1c was higher in VEH-D (p<0.0001) compared with RAL and VEH-ND rats. Serum insulin tended to be lower in the VEH-D rats compared with the VEH-ND and RAL-treated rats, but this was not significant. Serum triglycerides were higher in VEH-D rats compared with the RAL-treated rats (p = 0.02) (**Table 1**). Bone resorption measured by serum Ctx was similar among VEH-ND, VEH-D, and RAL-treated rats. However, dynamic histomorphometry showed non-significant trends for lower MS/BS (−28%) and BFR/BS (−26%) but higher MAR (+ 10%) in RAL-treated rats versus VEH-ND rats. Additionally, diabetic rats (VEH-D) had significantly lower MAR and non-significant trends for lower MS/BS and BFR/BS compared to the non-diabetic animals (VEH-ND or RAL-treated) (**Table 1**).

Areal bone mineral density and bone mineral content of the whole femur were lower in VEH-D compared with the other two groups (**Table 2**). There were no significant differences among groups for pQCT measures of the femoral midshaft. In the distal femur, bone volume normalized to tissue volume, trabecular thickness, and trabecular number were lower, and trabecular separation was higher in VEH-D rats compared with the other two groups, and structure model index was higher (more rod-like) in VEH-D compared with RAL rats (**Table 2**). There were no significant differences among groups for DXA or μCT measures of L4 vertebrae (**Table 2**).

RAL-treated rats had greater energy to failure and post-yield energy to failure in femoral cortical bone compared with VEH-ND and VEH-D rats (**Table 3**). Correspondingly, the material-level properties of femoral toughness and post-yield toughness were also higher in RAL-treated rats (**Table 3**). There were no differences among groups in structure-level or material-level mechanical properties from vertebral axial compression (**Table 3**). RPC of the distal femur cancellous bone revealed greater energy to ultimate force in RAL versus VEH-D rats, and non-significant trends for greater toughness in RAL rats versus VEH-ND (p = 0.07) and VEH-D (p = 0.06) rats. Ultimate stress was significantly greater in RAL rats compared with VEH-D rats (**Table 3**). VEH-D rats had lower ultimate force and stiffness compared with RAL and VEH-ND, but the corresponding material property of modulus was not different among groups.

Discussion

This study showed that raloxifene treatment in female ZDSD rats improved blood glucose levels and showed a trend for prevention of type 2 diabetes while imparting a beneficial effect on bone material properties. While the frequency of diabetes between vehicle and raloxifene treated animals was not statistically different by Fisher's exact test, there was a significant difference by survival analysis using Kaplan-Meier curves. This is not conclusive but at least suggestive of a benefit of raloxifene for prevention of diabetes. The finding that raloxifene might prevent the onset of type 2 diabetes in ZDSD rats was an unexpected outcome of this study. A randomized controlled trial [15] found that raloxifene did not improve insulin sensitivity or glycemic control in postmenopausal women who had type 2 diabetes, and a post-hoc analysis [16] of the Multiple Outcomes of Raloxifene Evaluation (MORE) trial found no effect of raloxifene on glycemic control in postmenopausal women with or without diabetes, although a beneficial effect was found on serum lipids. However, these two studies did not evaluate the effect of raloxifene on diabetes onset. Conversely, our results are supported by experimental evidence on the effect of raloxifene on glucose homeostasis and diabetes: it has been shown that estradiol prevents pancreatic β-cell failure in diabetic rats fed a high-fat diet by suppressing fatty acid synthesis and accumulation within the β-cells through estrogen receptor signaling [17], and the same research group found similar results with raloxifene in an *in vitro* study [18]. Therefore, a potential mechanism by which raloxifene could prevent the onset or slow the progression of diabetes is by preventing pancreatic β-cell failure. Additionally, two recent clinical studies [19,20] found a beneficial effect of raloxifene on serum lipids in women with type 2 diabetes, further supporting a role beyond bone for raloxifene to improve health in people with diabetes.

The fact that none of the raloxifene treated animals became diabetic, while an interesting outcome in itself, was a limitation of this study as it precluded our ability to analyze the effects of raloxifene on bone in rats with established diabetes. However, we were able to show a benefit of raloxifene on bone toughness in a diabetes-prone rat model. While we have previously reported a positive effect of raloxifene on bone toughness in non-diabetic canines [6,7], this effect has not been previously shown in bones of normal rats [21,22]. It is possible the predisposition to diabetes in the ZDSD rats creates a therapeutic window for an effect of raloxifene on bone toughness that is not present in normal rats.

Another possible limitation of our study is that we did not include a ZDSD group on a normal diet. However, this would not

A

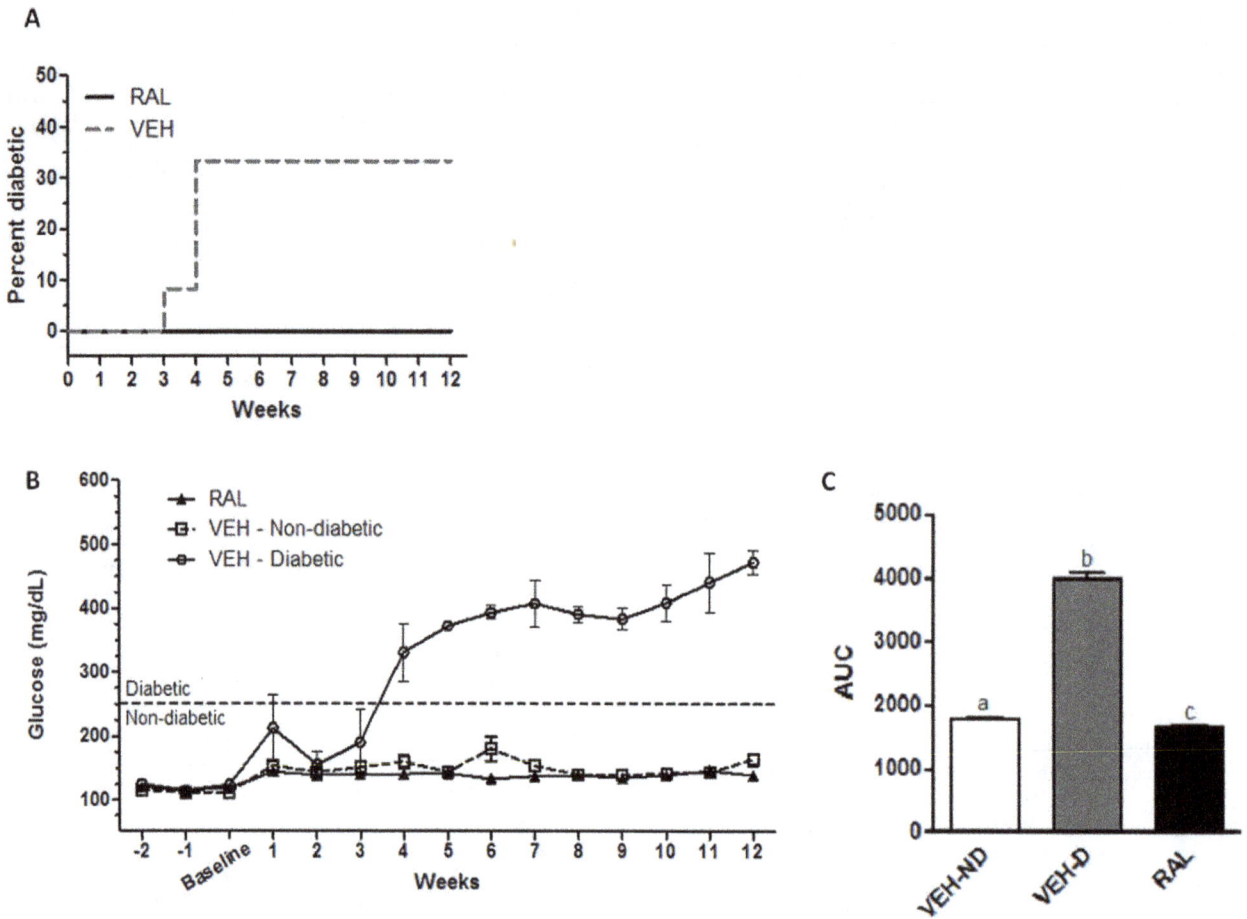

B

C

Figure 1. Diabetes Incidence and Glucose Levels in Raloxifene and Vehicle-Treated Rats. Panel A) Female ZDSD rats treated with raloxifene (RAL) had lower incidence of diabetes compared with vehicle treated rats (VEH) by survival analysis (p = 0.03) (shown), but by Fisher's exact test, the frequency of diabetes in VEH and RAL treated rats did not reach statistical significance (p = 0.09). Panels B,C) Over the course of the study, blood glucose was lowest in raloxifene treated rats (RAL), and highest in vehicle-injected rats that became diabetic (VEH-D), as assessed by area-under-the-curve (AUC). Different letters indicate differences in means with p<0.05.

Table 1. Body weight, metabolic parameters and bone turnover at end of study[a].

	VEH-ND (n = 8)		VEH-D (n = 4)[b]		RAL (n = 12)	
Body weight, g	532.5 (10.9)	a	411.0 (15.4)	b	417.8 (8.9)	b
Serum glucose, mg/dL	163.8 (8.0)	a	472.3 (11.3)	b	138.9 (6.5)	a
Blood HbA1c,%	4.8 (0.2)	a	10.3 (0.2)	b	4.7 (0.1)	a
Serum triglycerides, mg/dL	5.7 (0.7)	ab	7.7 (1.0)	a	4.4 (0.6)	b
Serum insulin, µg/L	3.6 (0.6)	a	1.8 (0.9)	a	3.2 (0.5)	a
Serum Ctx, ng/mL	19.6 (5.0)	a	34.9 (7.0)	a	23.9 (4.1)	a
L5 Histomorphometry						
MAR, µm/day	1.03 (0.11)	a	0.43 (0.18)	b	1.13 (0.10)	a
MS/BS,%	4.60 (0.85)	a	1.07 (1.39)	a	3.33 (0.76)	a
BFR/BS, µm³/µm²/year	18.27 (3.78)	a	2.13 (6.18)	a	13.52 (3.38)	a

[a]Different letters in each row indicate differences among groups by Tukey's post-hoc comparisons, p<0.05.
[b]n = 3 for VEH-D for the L5 histomophometry measures due to unavailable sample from 1 rat.

Table 2. Bone mass and microarcitecture of the femur and L4 vertebra from female ZDSD rats[a].

	VEH-ND (n = 8)		VEH-D (n = 4)		RAL (n = 12)	
Total Femur DXA						
aBMD, g/cm²	0.249 (0.005)	a	0.224 (0.004)	b	0.246 (0.003)	a
BMC, g	0.626 (0.010)	a	0.574 (0.009)	b	0.619 (0.006)	a
Area, cm²	2.51 (0.03)	a	2.56 (0.03)	a	2.51 (0.02)	a
Midshaft femur pQCT						
Ct. vBMD, mg/cm³	1473 (2)	a	1467 (3)	a	1475 (2)	a
Ct. BMC, mg/mm	10.2 (0.1)	a	10.1 (0.1)	a	10.2 (0.1)	a
Ct.Ar, mm³	6.90 (0.06)	a	6.89 (0.08)	a	6.90 (0.05)	a
Ct.Th, mm	0.86 (0.01)	a	0.85 (0.01)	a	0.85 (0.01)	a
Periosteal Circumference, mm	10.7 (0.08)	a	10.9 (0.11)	a	10.7 (0.07)	a
Endosteal Circumference, mm	6.02 (0.08)	a	5.99 (0.11)	a	6.01 (0.06)	a
Distal Femur μCT						
BV/TV,%	42.2 (2.2)	a	27.9 (3.2)	b	42.3 (1.8)	a
Tb.Th, mm	0.112 (0.003)	a	0.092 (0.005)	b	0.110 (0.003)	a
Tb.Sp, mm	0.170 (0.008)	a	0.206 (0.011)	b	0.171 (0.006)	a
Tb.N, #	3.75 (0.13)	a	3.02 (0.19)	b	3.83 (0.11)	a
Conn.Dn, #/mm³	126.9 (6.4)	a	116.7 (9.1)	a	135.7 (5.3)	a
SMI, units	0.50 (0.20)	ab	1.32 (0.28)	a	0.39 (0.16)	a
L4 DXA						
aBMD, g/cm²	0.127 (0.004)	a	0.125 (0.003)	a	0.129 (0.002)	a
BMC, g	0.025 (0.002)	a	0.024 (0.002)	a	0.025 (0.001)	a
L4 μCT						
BV/TV,%	38.6 (2.5)	a	40.7 (2.2)	a	42.6 (1.5)	a
Tb.Th, mm	0.102 (0.001)	a	0.097 (0.002)	a	0.102 (0.001)	a
Tb.Sp, mm	0.197 (0.008)	a	0.197 (0.011)	a	0.194 (0.006)	a
Tb.N, #	4.04 (0.14)	a	4.06 (0.19)	a	4.08 (0.11)	a
Conn.Dn, #/mm³	97.2 (7.1)	a	96.8 (10.0)	a	100.4 (5.8)	a
SMI, units	0.22 (0.10)	a	0.22 (0.14)	a	0.18 (0.08)	a

[a]Different letters in each row indicate differences among groups by Tukey's post-hoc comparisons, p<0.05.

have been a true control for the diabetes-prone rats because the effects of the different dietary composition on bone's material properties are not known. Moreover, we did not use CD rats which are sometimes used as non-diabetic controls. CD rats are not prone to diabetes even on a high fat diet, but are prone to obesity, and would introduce additional weight-related variables that could affect BMD or other mechanical properties of bone.

The effect of raloxifene on bone toughness was significant only for the femur, representing an effect on cortical bone, but a near-significant trend for greater toughness with raloxifene treatment was also observed for the distal femur by RPC, indicating a possible effect on cancellous bone as well. Our previous canine studies showed a beneficial effect of raloxifene on toughness in both cortical and cancellous bone [6,7]. Because cortical bone turnover is relatively slow [23], the effect of raloxifene on cortical bone toughness implies a direct effect of raloxifene on the existing bone material, rather than on newly formed bone. Furthermore, intracortical remodeling does not normally occur in rats and does not occur in the ZDSD rats. One mechanism by which raloxifene may improve toughness is by altering the hydration of the bone. We have shown that bone beams carved from human and dog cortical bone, when soaked in a raloxifene solution, have greater

toughness and that this is associated with higher water content of the bone [24]. Greater toughness and hydration were also observed in cortical bone beams from dogs treated *in vivo* with raloxifene for 1 year [24].

Additionally, our previous canine study showed no effect of raloxifene on BMD, which corresponds to clinical data from raloxifene trials in which fracture risk is reduced with little change in BMD [8−10]. Similarly, in the present study of diabetes-prone ZDSD rats, treatment with raloxifene resulted in greater femoral toughness without an effect on BMD, suggesting that raloxifene affects bone strength by improving bone quality rather than quantity. Because people with type 2 diabetes often have normal or high bone mineral density, the increased fracture risk observed in these patients appears to be due to impaired bone quality rather than reduced bone quantity. However, in this study, diabetic rats actually had lower bone density and mass compared with non-diabetic rats. This difference between human type 2 diabetes and the ZDSD rat model might be explained as follows: in humans with type 2 diabetes, overweight and obesity often persist after the onset of diabetes, and excess body weight may be protective of bone mass through mechanical loading or the positive effects of leptin and estrogen produced by adipose tissue. Conversely,

Table 3. Structure-level and material-level mechanical properties of femoral cortical and cancellous bone and L4 cancellous bone from female ZDSD rats[a].

	VEH-ND (n = 8)[b]		VEH-D (n = 4)		RAL (n = 12)[c]	
Femur 3-point bending (cortical bone)						
Structure-level						
Ultimate Force, N	135.1(2.7)	a	135.6 (3.6)	a	142.7 (2.1)	a
Stiffness, N/mm	342.4 (11.1)	a	346.8 (14.7)	a	342.4 (8.5)	a
Energy to Failure, mJ	46.0 (2.9)	a	43.7 (3.8)	a	57.4 (2.2)	b
Post-Yield Energy to Failure, mJ	26.5 (3.0)	a	23.5 (4.0)	a	36.6 (2.3)	b
Post-Yield Displacement, mm	0.210 (0.023)	a	0.187 (0.030)	a	0.275 (0.017)	a
Material-level						
Ultimate Stress, MPa	61.4 (1.6)	a	61.3 (2.1)	a	65.0 (1.2)	a
Elastic Modulus, MPa	2683 (92)	a	2724 (122)	a	2691 (70)	a
Toughness, mJ/m^3	1.21 (0.08)	a	1.13 (0.10)	a	1.51 (0.06)	b
Post-Yield Toughness, mJ/m^3	0.69 (0.07)	a	0.61 (0.10)	a	0.96 (0.06)	b
Distal femur RPC (cancellous bone)[b]						
Structure-level						
Ultimate Force, N	21.5 (2.3)	a	7.3 (3.1)	b	23.7 (1.9)	a
Stiffness, N/mm	239.4 (16.8)	a	142.6 (22.3)	b	249.1 (14.1)	a
Energy to Ultimate Force, mJ	1.41 (0.89)	ab	0.26 (1.18)	a	4.16 (0.75)	b
Material-level						
Ultimate Stress, MPa	20.6 (2.5)	ab	13.6 (3.4)	a	27.6 (2.1)	b
Modulus, MPa	400.7 (51.6)	a	474.5 (68.3)	a	534.8 (43.2)	a
Toughness, mJ/m^3	0.68 (0.55)	a	0.25 (0.73)	a	2.40 (0.46)	a
L4 axial compression (cancellous bone)						
Structure-level						
Ultimate Force, N	369.9 (18.9)	a	317.3 (25.0)	a	350.4 (15.1)	a
Stiffness, N/mm	1739 (117)	a	1476 (155)	a	1793 (93)	a
Energy to Ultimate Force, mJ	46.6 (3.2)	a	40.9 (4.2)	a	41.1 (2.5)	a
Material-level						
Ultimate Stress, MPa	2.27 (0.12)	a	2.11 (0.16)	a	2.22 (0.10)	a
Modulus, MPa	1175 (93)	a	1060 (123)	a	1269 (74)	a
Toughness, mJ/mm^3	2.51 (0.13)	a	2.28 (0.17)	a	2.31 (0.10)	a

[a]Different letters in each row indicate differences among groups by Tukey's post-hoc comparisons, p<0.05.
[b]n = 7 for VEH-ND for distal femur RPC, L4 axial compression, and femur 3-point bending measures, due to specimens breaking during preparation or unavailable sample.
[c]n = 10 for RAL for the distal femur RPC measures and n = 11 for RAL for L4 axial compression measures, due to specimens breaking during preparation.

ZDSD rats gain weight with the high fat diet until the onset of diabetes, after which they begin to lose body weight due to the catabolic state produced by the diabetes. Indeed, diabetic rats in the present study had significantly lower body weight at the time of sacrifice compared with non-diabetic animals, and this may be associated with their lower BMD.

The results did not show lower bone resorption as measured by serum CTX in the raloxifene treated rats. However, the numerically lower BFR/BS and MS/BS with raloxifene treatment (−26% and −27% respectively in RAL versus VEH-ND rats) supports that the raloxifene treatment had an effect on reducing bone turnover. The non-significant differences are not surprising given this study was not powered to detect differences in these outcomes, and that raloxifene is a relatively weak antiresorptive agent [25]. However, the magnitude of the difference in BFR/BS with raloxifene treatment is similar to what we previously observed in dogs [7]. Additionally, the rats in this study were not

ovariectomized, which also may have reduced our ability to detect a significant antiresorptive effect of raloxifene. The rats that became diabetic (VEH-D) had a lower bone formation rate, which is consistent with reduced bone observed in humans and animals with diabetes [26]. Despite the lack of significant differences in bone turnover measures with raloxifene treatment, our results show that raloxifene improves bone material properties, potentially through direct action of raloxifene on the bone matrix, and may prevent the induction of diabetes in female ZDSD rats. The risk of diabetes increases with age [27], as does the risk for bone fragility fractures [28]. If these results are supported by future experimental and clinical studies, they suggest that raloxifene could be a useful drug to prevent skeletal fragility in diabetes with an added benefit of ameliorating the diabetic condition.

Author Contributions

Conceived and designed the experiments: KMGH MAG DBB. Performed the experiments: KMHG MAG DBB AYS JNW. Analyzed the data: KMGH MAG DBB. Contributed reagents/materials/analysis tools: DBB. Wrote the paper: KMHG MAG DMB AYS JNW DBB.

References

1. Janghorbani M, Feskanich D, Willett WC, Hu F (2006) Prospective study of diabetes and risk of hip fracture: the Nurses' Health Study. Diabetes Care 29: 1573–1578.
2. Janghorbani M, Van Dam RM, Willett WC, Hu FB (2007) Systematic review of type 1 and type 2 diabetes mellitus and risk of fracture. Am J Epidemiol 166: 495–505.
3. Strotmeyer ES, Cauley JA, Schwartz AV, Nevitt MC, Resnick HE, et al. (2005) Nontraumatic fracture risk with diabetes mellitus and impaired fasting glucose in older white and black adults: the health, aging, and body composition study. Arch Intern Med 165: 1612–1617.
4. Yamamoto M, Yamaguchi T, Yamauchi M, Kaji H, Sugimoto T (2009) Diabetic patients have an increased risk of vertebral fractures independent of BMD or diabetic complications. J Bone Miner Res 24: 702–709.
5. Schwartz AV, Sellmeyer DE, Ensrud KE, Cauley JA, Tabor HK, et al. (2001) Older women with diabetes have an increased risk of fracture: a prospective study. J Clin Endocrinol Metab 86: 32–38.
6. Allen MR, Hogan HA, Hobbs WA, Koivuniemi AS, Koivuniemi MC, et al. (2007) Raloxifene enhances material-level mechanical properties of femoral cortical and trabecular bone. Endocrinology 148: 3908–3913.
7. Allen MR, Iwata K, Sato M, Burr DB (2006) Raloxifene enhances vertebral mechanical properties independent of bone density. Bone 39: 1130–1135.
8. Delmas PD, Ensrud KE, Adachi JD, Harper KD, Sarkar S, et al. (2002) Efficacy of raloxifene on vertebral fracture risk reduction in postmenopausal women with osteoporosis: four-year results from a randomized clinical trial. J Clin Endocrinol Metab 87: 3609–3617.
9. Ettinger B, Black DM, Mitlak BH, Knickerbocker RK, Nickelsen T, et al. (1999) Reduction of vertebral fracture risk in postmenopausal women with osteoporosis treated with raloxifene: results from a 3-year randomized clinical trial. Multiple Outcomes of Raloxifene Evaluation (MORE) Investigators. Jama 282: 637–645.
10. Siris ES, Harris ST, Eastell R, Zanchetta JR, Goemaere S, et al. (2005) Skeletal effects of raloxifene after 8 years: results from the continuing outcomes relevant to Evista (CORE) study. J Bone Miner Res 20: 1514–1524.
11. Reinwald S, Peterson RG, Allen MR, Burr DB (2009) Skeletal changes associated with the onset of type 2 diabetes in the ZDF and ZDSD rodent models. Am J Physiol Endocrinol Metab 296: E765–774.
12. Fajardo RJ, Karim L, Calley VI, Bouxsein ML (2014) A review of rodent models of type 2 diabetic skeletal fragility. J Bone Miner Res 29: 1025–1040.
13. Turner CH, Burr DB (1993) Basic biomechanical measurements of bone: a tutorial. Bone 14: 595–608.
14. Dempster DW, Compston JE, Drezner MK, Glorieux FH, Kanis JA, et al. (2013) Standardized nomenclature, symbols, and units for bone histomorphometry: a 2012 update of the report of the ASBMR Histomorphometry Nomenclature Committee. J Bone Miner Res 28: 2–17.
15. Andersson B, Johannsson G, Holm G, Bengtsson BA, Sashegyi A, et al. (2002) Raloxifene does not affect insulin sensitivity or glycemic control in postmeno-pausal women with type 2 diabetes mellitus: a randomized clinical trial. J Clin Endocrinol Metab 87: 122–128.
16. Barrett-Connor E, Ensrud KE, Harper K, Mason TM, Sashegyi A, et al. (2003) Post hoc analysis of data from the Multiple Outcomes of Raloxifene Evaluation (MORE) trial on the effects of three years of raloxifene treatment on glycemic control and cardiovascular disease risk factors in women with and without type 2 diabetes. Clin Ther 25: 919–930.
17. Tiano JP, Delghingaro-Augusto V, Le May C, Liu S, Kaw MK, et al. (2011) Estrogen receptor activation reduces lipid synthesis in pancreatic islets and prevents beta cell failure in rodent models of type 2 diabetes. J Clin Invest 121: 3331–3342.
18. Tiano J, Mauvais-Jarvis F (2012) Selective estrogen receptor modulation in pancreatic beta-cells and the prevention of type 2 diabetes. Islets 4: 173–176.
19. Matsumura M, Monden T, Nakatani Y, Shimizu H, Domeki N, et al. (2010) Effect of raloxifene on serum lipids for type 2 diabetic menopausal women with or without statin treatment. Med Princ Pract 19: 68–72.
20. Mori H, Okada Y, Kishikawa H, Inokuchi N, Sugimoto H, et al. (2013) Effects of raloxifene on lipid and bone metabolism in postmenopausal women with type 2 diabetes. J Bone Miner Metab 31: 89–95.
21. Diab T, Wang J, Reinwald S, Guldberg RE, Burr DB (2011) Effects of the combination treatment of raloxifene and alendronate on the biomechanical properties of vertebral bone. J Bone Miner Res 26: 270–276.
22. Sato M, Bryant HU, Iversen P, Helterbrand J, Smietana F, et al. (1996) Advantages of raloxifene over alendronate or estrogen on nonreproductive and reproductive tissues in the long-term dosing of ovariectomized rats. J Pharmacol Exp Ther 279: 298–305.
23. Burr DB, Diab T, Koivunemi A, Koivunemi M, Allen MR (2009) Effects of 1 to 3 years' treatment with alendronate on mechanical properties of the femoral shaft in a canine model: implications for subtrochanteric femoral fracture risk. J Orthop Res 27: 1288–1292.
24. Gallant MA, Brown DM, Hammond M, Wallace J, Du J, et al. (2014) Bone cell-independent benefits of raloxifene on the skeleton: A novel mechanism for improving bone material properties. Bone 61: 191–200.
25. Sambrook PN, Geusens P, Ribot C, Solimano JA, Ferrer-Barriendos J, et al. (2004) Alendronate produces greater effects than raloxifene on bone density and bone turnover in postmenopausal women with low bone density: results of EFFECT (Efficacy of FOSAMAX versus EVISTA Comparison Trial) International. J Intern Med 255: 503–511.
26. Vestergaard P (2011) Risk of newly diagnosed type 2 diabetes is reduced in users of alendronate. Calcif Tissue Int 89: 265–270.
27. Narayan KM, Boyle JP, Thompson TJ, Sorensen SW, Williamson DF (2003) Lifetime risk for diabetes mellitus in the United States. Jama 290: 1884–1890.
28. Melton LJ, 3rd, Kan SH, Wahner HW, Riggs BL (1988) Lifetime fracture risk: an approach to hip fracture risk assessment based on bone mineral density and age. J Clin Epidemiol 41: 985–994.

20

Low Intensity Pulsed Ultrasound Enhanced Mesenchymal Stem Cell Recruitment through Stromal Derived Factor-1 Signaling in Fracture Healing

Fang-Yuan Wei[1], Kwok-Sui Leung[1,2], Gang Li[1], Jianghui Qin[1], Simon Kwoon-Ho Chow[1], Shuo Huang[1], Ming-Hui Sun[1], Ling Qin[1,2], Wing-Hoi Cheung[1,2]*

1 Department of Orthopaedics and Traumatology, Clinical Sciences Building, The Chinese University of Hong Kong, Shatin, New Territories, Hong Kong SAR, China, **2** Translational Medicine Research & Development Center, Institute of Biomedical and Health Engineering, Shenzhen Institutes of Advanced Technology, Chinese Academy of Sciences, Shenzhen, China

Abstract

Low intensity pulsed ultrasound (LIPUS) has been proven effective in promoting fracture healing but the underlying mechanisms are not fully depicted. We examined the effect of LIPUS on the recruitment of mesenchymal stem cells (MSCs) and the pivotal role of stromal cell-derived factor-1/C-X-C chemokine receptor type 4 (SDF-1/CXCR4) pathway in response to LIPUS stimulation, which are essential factors in bone fracture healing. For *in vitro* study, isolated rat MSCs were divided into control or LIPUS group. LIPUS treatment was given 20 minutes/day at 37°C for 3 days. Control group received sham LIPUS treatment. After treatment, intracellular CXCR4 mRNA, SDF-1 mRNA and secreted SDF-1 protein levels were quantified, and MSCs migration was evaluated with or without blocking SDF-1/CXCR4 pathway by AMD3100. For *in vivo* study, fractured 8-week-old young rats received intracardiac administration of MSCs were assigned to LIPUS treatment, LIPUS+AMD3100 treatment or vehicle control group. The migration of transplanted MSC to the fracture site was investigated by *ex vivo* fluorescent imaging. SDF-1 protein levels at fracture site and in serum were examined. Fracture healing parameters, including callus morphology, micro-architecture of the callus and biomechanical properties of the healing bone were investigated. The *in vitro* results showed that LIPUS upregulated SDF-1 and CXCR4 expressions in MSCs, and elevated SDF-1 protein level in the conditioned medium. MSCs migration was promoted by LIPUS and partially inhibited by AMD3100. *In vivo* study demonstrated that LIPUS promoted MSCs migration to the fracture site, which was associated with an increase of local and serum SDF-1 level, the changes in callus formation, and the improvement of callus microarchitecture and mechanical properties; whereas the blockade of SDF-1/CXCR4 signaling attenuated the LIPUS effects on the fractured bones. These results suggested SDF-1 mediated MSCs migration might be one of the crucial mechanisms through which LIPUS exerted influence on fracture healing.

Editor: Amarjit S. Virdi, Rush University Medical Center, United States of America

Funding: This research project was supported by an AO Grant (Ref: S-11-10C) and partially by OTC Foundation Research Fund (Ref: 2009-WHLG). The funders had no role in study design, data collection and analysis, decision to publish, or preparation of the manuscript.

Competing Interests: The authors have declared that no competing interests exist.

* Email: louis@ort.cuhk.edu.hk

Introduction

Millions of fractures occur annually as a result of traumatic injuries or pathological conditions. Although most fractures will successfully heal within a few months, a considerable proportion of fracture cases still result in delayed healing [1], which may prolong treatment period and increase morbidity of the patients.

Mesenchymal stem cells (MSCs) are multipotent stromal cells able to differentiate into many cell types and contribute to the regeneration of musculoskeletal tissues such as bone, cartilage, tendon, adipose, and muscle [2–4]. It is widely accepted that MSCs are normally retained in the special niches of different adult tissues. In stressful situations such as injury, when there is a need for tissue repair and to maintain tissue homeostasis, MSCs can be recruited to the site of injury and contribute to the repair process. When bone integrity is disrupted after fracture, the bone tissue

would enter a healing process that is generally divided into three overlapping phases including the inflammation, soft and hard callus formation, and the callus remodeling [5]. The damage of blood vessel and other tissues lead to local tissue bleeding and hypoxia. This process will trigger the inflammatory cascade [6]. In this early inflammatory phase of fracture healing, many types of cytokines, such as interleukin 6 (IL-6) and stromal cell-derived factor-1 (SDF-1), released from the damaged bone facilitate the egress of MSCs from the periosteum and bone marrow into the blood stream, which rapidly accumulate and engraft at fracture site, and initiate bone regeneration process [3,7,8]. Although the interactions between cytokines and MSCs in bone repair remain controversial, many studies found that MSCs expressed both SDF-1 and CXCR4 genes [9–11], and SDF-1/CXCR4 signaling is critical for the recruitment of MSCs to the fracture site during fracture healing. Granero-Molto et al. found that implanted MSCs were recruited to the fracture site in an exclusively CXCR4-dependent manner [12]. Kitaori et al. showed that SDF-1 level was elevated in the periosteum of injured bone, which recruited MSCs homing to the graft bone at the fracture site and promoted endochondral bone formation [8].

Low intensity pulsed ultrasound (LIPUS) has been reported to be effective in promoting fracture healing in both animal models and clinical trials [13–16]. In brief, the beneficial effects of LIPUS on fracture healing include the decrease in healing time at the tissue level, and the increase in the cellular responses including osteogensis-related gene expression [17], protein synthesis and cell proliferation [18]. The mechanical stimulation produced by the pressure waves of LIPUS on bone can result in series of biochemical events at cellular level [19,20]. However, the detailed mechanism through which LIPUS stimulates tissues remains unclear. Although osteocytes have been considered as the primary mechanosensors in bone, convincing data show that MSCs also have the ability to sense and respond to physical stimuli [21–23]. To date, very little is known about how physical stimuli affect MSCs mobilization. One possible mechanism through which LIPUS enhances fracture healing is through the enhancement in MSC recruitment. A recent report has demonstrated that LIPUS was able to enhance MSC recruitment from a parabiotic source at the fracture site in a surgically conjoined mice pair model. The report also suggested the involvement of SDF-1/CXCR4 signaling pathway by an apparent increase immuno-detection of the two proteins [24].

In this study, we attempted to investigate that under LIPUS treatment, (a) the migration of MSCs to the fracture site; (b) the role of SDF-1/CXCR4 in regulating the recruitment of MSCs; (c) the MSCs engraftment and fracture healing. The aim of the first part of this study was to evaluate the direct influence of LIPUS on MSCs migration and intracellular SDF-1/CXCR4 signaling in vitro. The second part was to investigate the effects of LIPUS on MSCs recruitment and femoral fracture repair in a rat model, with or without blocking SDF-1/CXCR4 pathway.

Materials and Methods

2.1. MSCs Isolation and Identification

2.1.1. MSCs Isolation. All experiments were approved by the Animal Experimentation Ethics Committee (AEEC) of the authors' institution (Ref: 10/007/GRF-5). MSCs were isolated from two 8-week female Sprague-Dawley (SD) rats, following protocol previously established in our laboratory [25]. Briefly, intact tibiae and femora were collected from euthanized healthy 8-week SD rats and carefully dissected free of muscles in the Petri dish containing sterile phosphate-buffered saline (PBS) and 1% penicillin-streptomycin (Invitrogen Corporation, Carlsbad, California, US). The bones were rinsed once in sterile 1×PBS before being transferred to the biosafety cabinet hood in the culture

room. After rinsed once with sterile 1×PBS, the bone ends were cut with bone clipper. With the cut surface facing the bottom of the centrifuge tube, the tube was spun at 2000 rpm for 1 minute (most of the bone marrow (BM) was collected at the bottom of the tube). Mononuclear cells were then isolated by density gradient centrifugation (850 g, 30 minutes) using Lymphoprep (1.077 g/ml; AXIS-SHIELD PoC AS, Oslo, Norway), and re-suspended in complete culture medium containing alpha minimum essential medium (α-MEM) (Gibco, Grand Island, NY, US), 10% fetal bovine serum (FBS) (Gibco, Grand Island, NY, US), 100 U/ml penicillin, 100 µg/ml streptomycin and 2 mM L-glutamine (Invitrogen, Carlsbad, California, US). These mononuclear cells were plated at an optimal low cell density (10^5 cells/cm^2) to isolate stem cells and cultured at 37°C, 5% CO_2/20% O_2 to form colonies. When colonies reached 80–90% confluence, the MSCs were sub-cultured and re-plated for further expansion. Medium was changed every three days. Cells at passage 3 were used for all the experiments. The surface marker expression and multi-lineage differentiation potential of MSCs were characterized before being used for the further experiments.

2.1.2. MSCs Characterization. The methods of the characterization of MSCs in this study were mainly based on the minimal criteria for human MSCs suggested by the Mesenchymal and Tissue Stem Cell Committee of the International Society for Cellular Therapy [26].

Flow Cytometry Assay. The surface marker expression of MSCs isolated from healthy rats, including CD90, CD44, CD45, CD31 and CD34, was analyzed by flow cytometry assay as described in the previous study [27]. Briefly, the MSCs at passage 3 were harvested by trypsinization, washed twice in PBS, then pelleted by centrifugation at 350 g for 5 minutes at room temperature and re-suspended in the staining buffer (Becton Dickson, Franklin Lakes, NJ, US) at $2×10^6$/ml for 15 minutes at 4°C. One-hundred microliters cell suspension was incubated with primary antibodies against rat CD90 and CD44 (Abcam, Cambridge, UK) conjugated with phycoerythrin (PE), CD31 (Abcam, Cambridge, UK) and CD34 (Santa Cruz Biotechnology, Santa Cruz, CA, US) conjugated with fluorescein isothiocyanate (FITC), CD45 (Abcam, Cambridge, UK) without conjugation for 15 minutes at 4°C. Unbound antibodies were washed away by adding ice-cold staining buffer. The cell pellet was re-suspended in the staining buffer containing goat anti-rabbit immunoglobulin G (IgG) conjugated with FITC (Santa Cruz Biotechnology, Santa Cruz, CA, US) for CD45 detection for at least 15 minutes at 4°C. The cells were washed with ice-cold PBS containing 2% bovine serum albumin (BSA) before analysis using the LSRFortessa flow cytometer (Becton Dickinson, San Jose, CA, US). PE-conjugated isotype-matched mouse IgG1 (R&D systems Inc, Minneapolis, MN, US) was used as isotype control for both CD90 and CD44; FITC-conjugated isotype-matched mouse IgG1 (R&D systems, Inc, Minneapolis, MN, US) was used as isotype control for both

Table 1. Primer sequences of the target genes.

Gene	Primer nucleotide sequence	Product Size (bp)	Ta (°C)
SDF-1	Forward: 5'-TTGCCAGCACAAAGACACTCC-3' Reverse: 5'-CTCCAAAGCAAACCGAATACAG-3'	225	58
CXCR4	Forward: 5'-TCCGTGGCTGACCTCCTCTT-3' Reverse: 5'-CAGCTTCCTCGGCCTCTGGC-3'	210	56
GAPDH	Forward: 5'-AACTCCCATTCCTCCACCTT-3' Reverse: 5'-GAGGGCCTCTCTCTTGCTCT-3'	200	57

GAPDH = glyceraldehide-3-phosphate dehydrogenase.

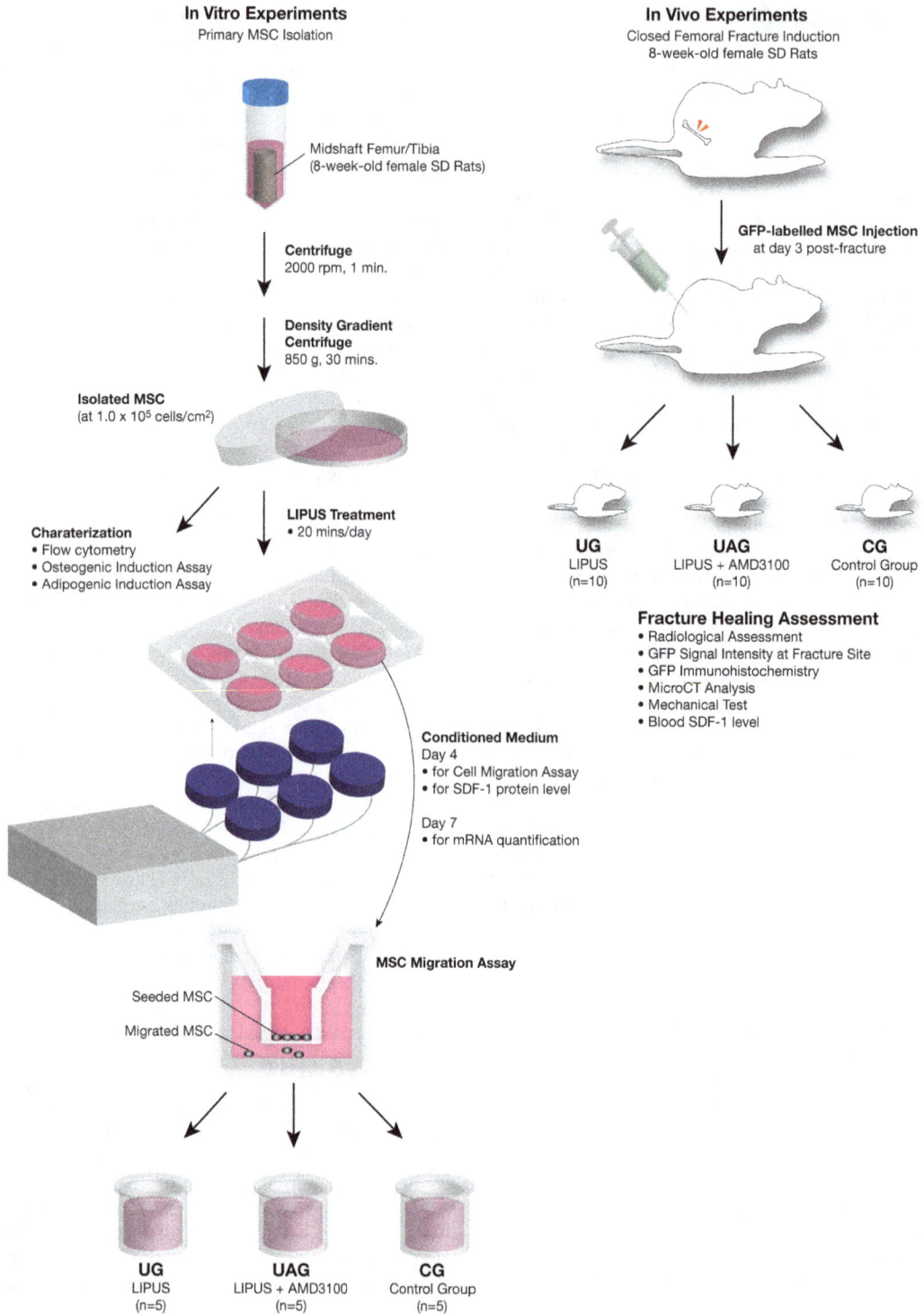

Figure 1. Flowchart of the study design. MSCs were isolated from two 8-week-old female SD rats and characterized by flow cytometry assay, osteogenic induction assay and adipogenic induction assay. In the in vitro experiments (n = 5), the SDF-1 protein and mRNA expression levels in condition medium were compared between control (CG) and LIPUS treatment (UG) groups, the MSCs migration ability was compared among CG, UG and LIPUS plus AMD3100 treatment (UAG) groups. In the in vivo experiments, closed femoral fractured rats were randomly divided into CG, UG and UAG groups (n = 10). GFP-labeled MSCs were intracardiac injected to all the rats on day 3 post-fracture and recruitment effects by LIPUS were compared among groups.

CD31 and CD34; and rabbit polyclonal IgG (Epitomics, Burlingame, CA, US) was used as isotype control for CD45.

Osteogenic Induction Assay and Alizarin Red S Staining. The methods have been described previously [25]. Briefly, MSCs at passage 3 were trypsinized and re-plated in a six-

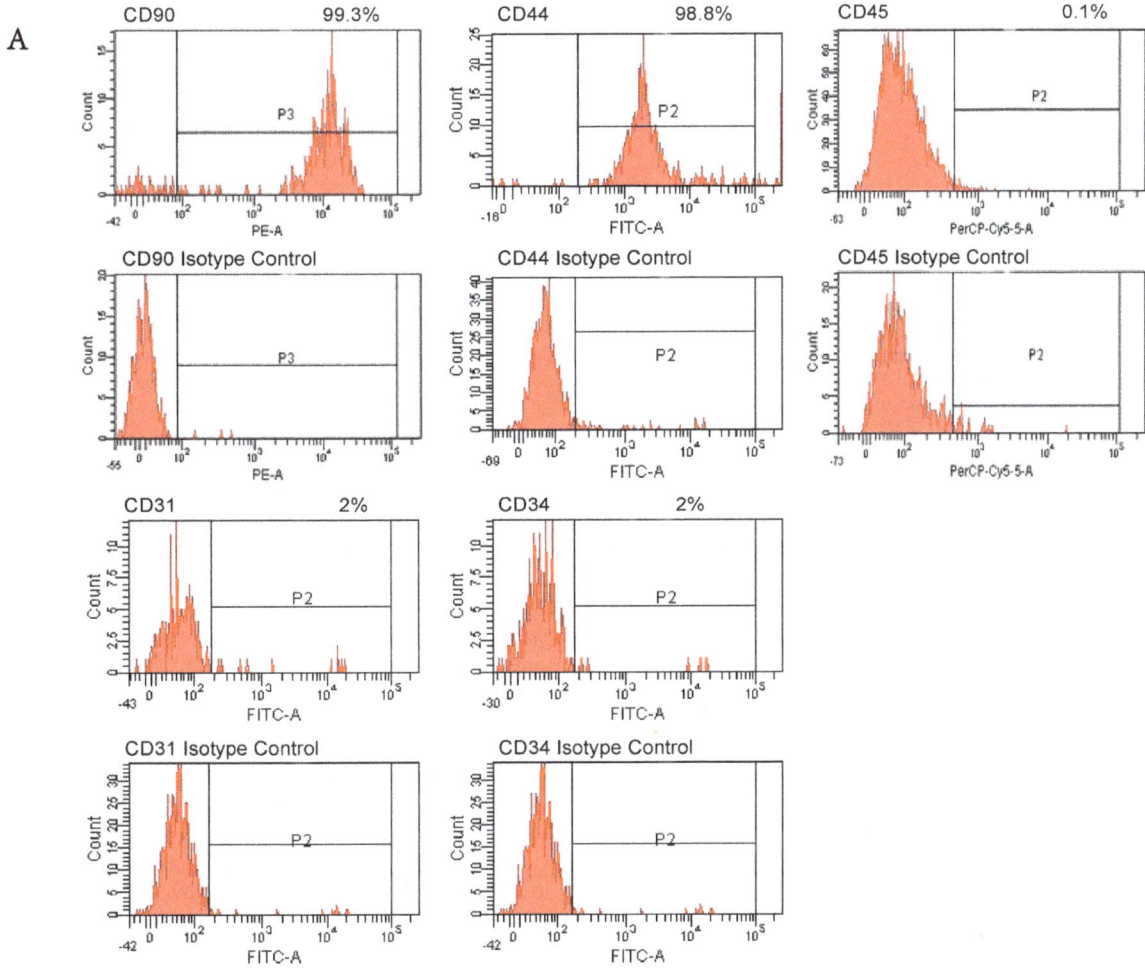

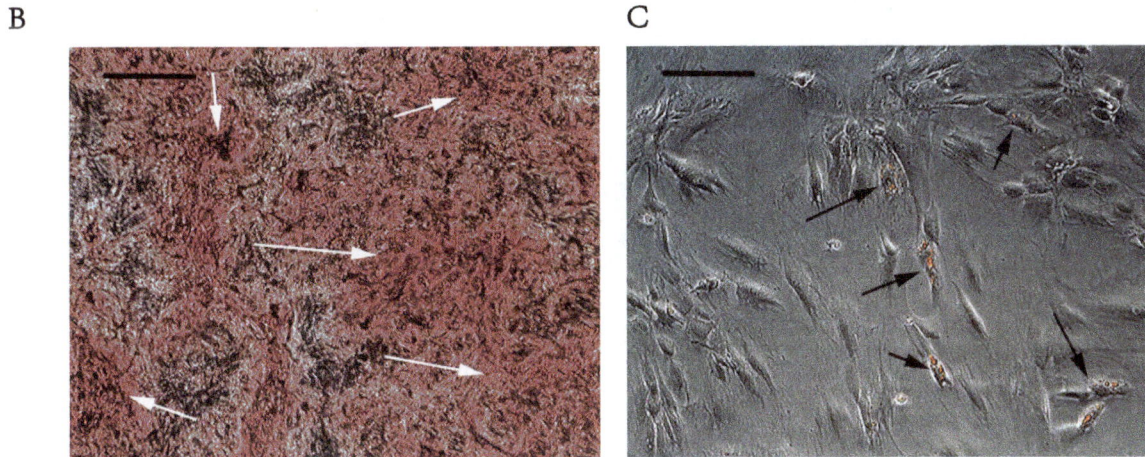

Figure 2. MSCs characterization. (**A**) The expressions of selected surface markers on the isolated cells from BM of SD rats. This figure shows the expressions of mesenchymal stem cell markers (CD90 and CD44), endothelial cell marker (CD31) and hematopoietic cell markers (CD34 and CD45) on the isolated cell colonies. (**B**) Representative microphotograph of Alizarin Red S-stained cells isolated from BM of SD rats. White arrows indicate the obvious calcium deposition areas in the matrix. Scale bar = 100 μm. (**C**) Representative microphotograph of Oil Red O-stained cells isolated from BM of SD rats. The black arrows indicate the adipose-differentiated cells. The small red bubbles in cells are lipids. Scale bar = 100 μm.

well plate at a concentration of 1×10^5 cells per well, and cultured in complete culture medium for three days. These cells were then incubated in osteogenic induction medium (OIM) containing 100 nmol/L dexamethasone, 10 mmol/L beta-glycerophosphate, and 0.05 mmol/L L-ascorbic acid-2-phosphate. The OIM was changed every three days for 21 days. Cell and matrix layer was washed with PBS, fixed with 70% ethanol for 10 minutes, and

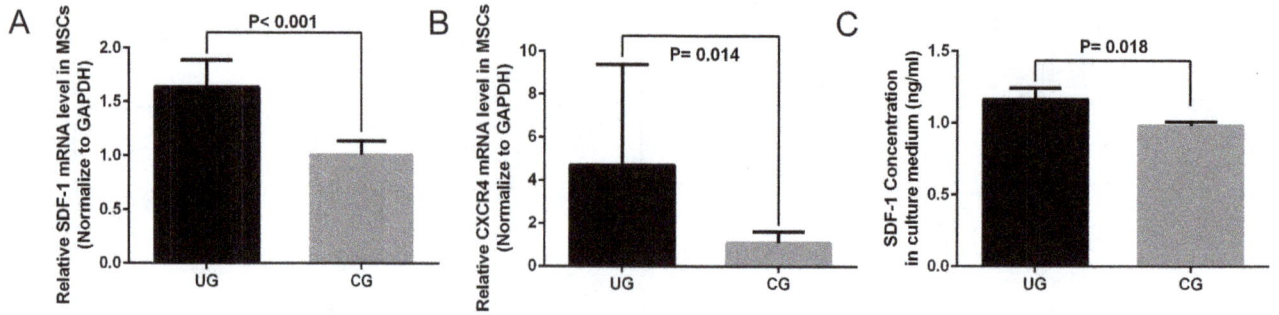

Figure 3. Target genes quantification and SDF-1 protein measurement. (**A**) LIPUS enhanced SDF-1 gene expression in MSCs significantly (p<0.0001), and (**B**) LIPUS enhanced CXCR4 gene expression in MSCs significantly (p = 0.014). (**C**) ELISA assay showed increased SDF-1 protein level in the culture medium of MSCs treated with LIPUS (p = 0.018).

stained with 0.5% Alizarin Red S (pH 4.1, Sigma, St. Louis, MO, US) for 30 minutes.

Adipogenic Induction Assay and Oil Red-O Staining. The method has been described previously [25]. MSCs were trypsinized and re-plated in a six-well plate at the same concentration as that for the osteogenic assay. The cells were cultured in the complete culture medium for three days, and were then incubated in adipogenic induction medium (AIM) containing 10% FBS, 1 μM dexamethasone, 10 μg/ml insulin, 50 μM indomethacin, and 0.5 mM isobuthyl-methylxanthine. The cells were cultured for an additional 21 days for assessment of the presence of oil droplets, which was confirmed by staining the cells with 0.3% fresh Oil Red-O solution (Sigma-Aldrich, St. Louis, MO, US) for 2 hours after fixation with 70% ethanol for 10 minutes.

2.2. LIPUS Treatment of MSCs

The isolated rat MSCs were divided into control (CG) or LIPUS treatment (UG) group (n = 5). The procedure of LIPUS treatment on cells was based on our previous protocol [19,28]. Briefly, LIPUS was provided by a Sonic Accelerated Fracture Healing System (Smith & Nephew, Memphis, TN, US) for cell culture. The 6-well culture plate (Corning, Lowell, MA, US) was placed on the matched ultrasound transducers with a thin layer of coupling gel. The ultrasound energy was calibrated by the manufacturer and tested with the output checker before use. For UG, LIPUS treatment (unfocused plane waves, frequency 1.5 MHz, duty cycle 1:4, spatial average-temporal average intensity 30 mW/cm^2, pulse repetition frequency 1 kHz for pulse duration of 200 μs) was given from the bottom of the culture plate for 20 minutes/day in open air at 37°C for 3 days. CG received sham LIPUS treatment, with

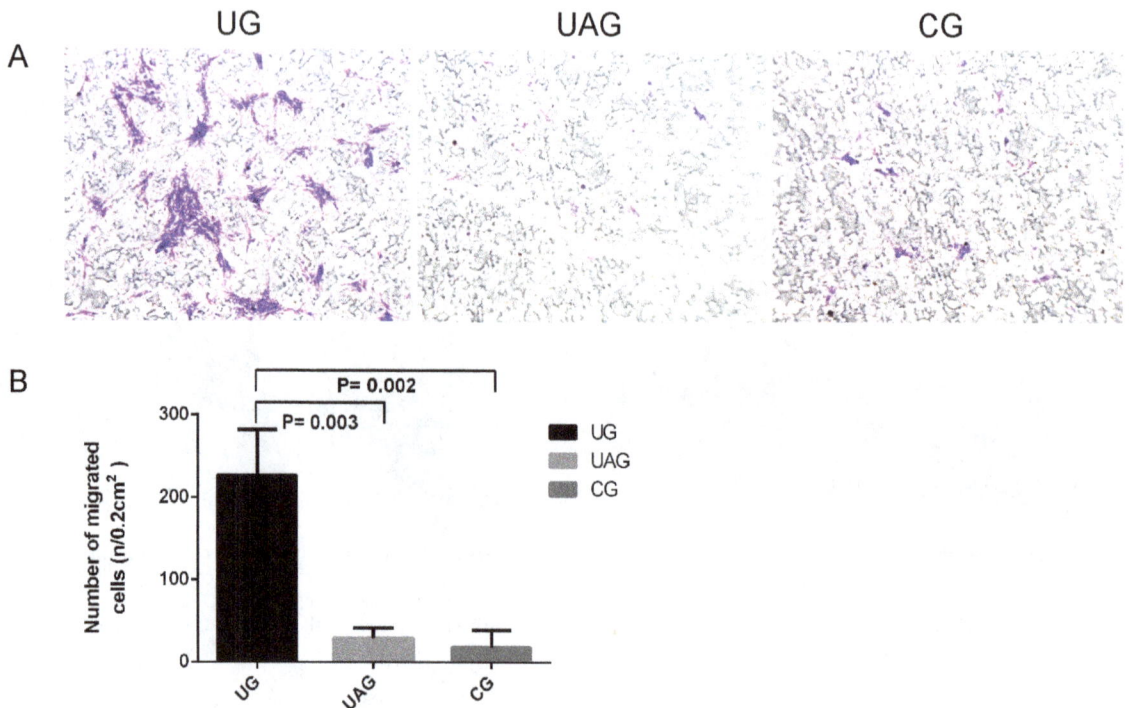

Figure 4. MSCs migration assay. (**A**) The migrated MSCs on the exterior of the insert in UG (left) were remarkably increased compared with UAG (middle) and CG (right) at ×100 magnification. (**B**) The number of migrated cells in UG was increased 12.1 times than in CG (p = 0.002); the number of migrated cells in UAG was decreased by 87.2% as compared to UG (p = 0.003).

Figure 6. *Ex vivo* **GFP intensity measurement. (A)** On the representative image of each group, blue circles indicate the region of interest (ROI) for fluorescent imaging analysis. GFP signals in UG (left) was much higher than in UAG (middle) and CG (right). **(B)** Semi-quantitative GFP intensity of fracture callus in UG was increased 2.66 times than in UAG (p = 0.014), although no significant differences were found among other groups.

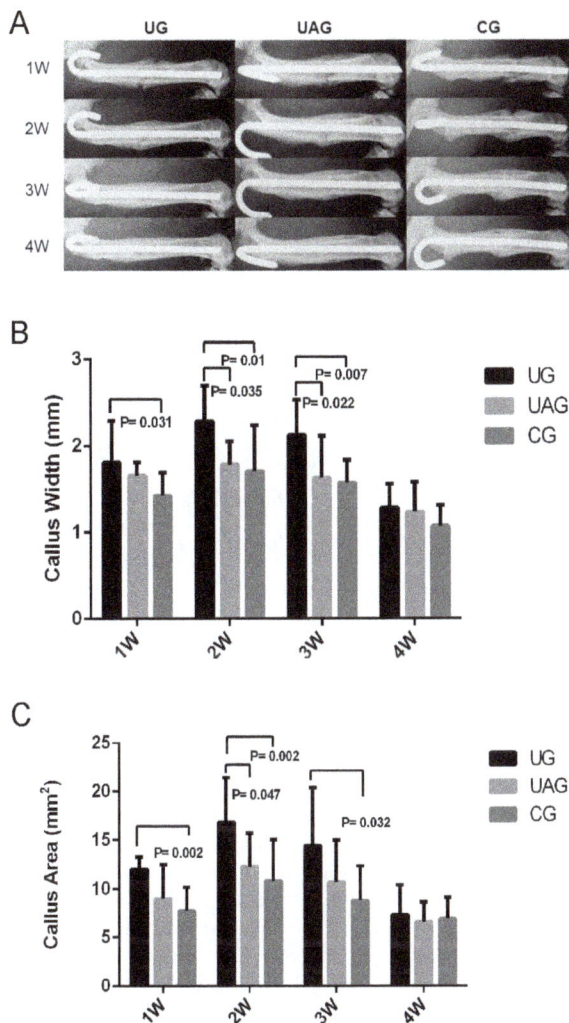

Figure 5. Radiographic analysis of fracture healing in young rats. (A) Series of representative radiographies showed better callus bridging in UG and UAG, compared with in CG at different time points. **(B)** The measurement of callus width (CW) showed: CW in UG was larger by 26.8% at week 1 (p = 0.031), by 33.6% at week 2 (p = 0.01) and by 35.0% at week 3 (p = 0.007) post-fracture than in CG, and by 27.8% at week 2 (p = 0.035) and by 30.0% at week 3 (p = 0.022) post-fracture than in UAG. **(C)** The quantitative measurement of callus area (CA) showed: CA was significantly larger in UG by 55.1% at week 1 (p = 0.002), by 55.5% at week 2 (p = 0.002), by 64.0% at week 3 (p = 0.032) than in CG, and was significantly larger by 37.7% at week 2 than in UAG (p = 0.047).

the culture plate placed on the transducers (with coupling gel) yet without ultrasound. On day 4, after changing the medium, the old conditioned medium from each group was collected for SDF-1 protein level analysis and cell migration assay; on day 7, the MSCs were harvested for real-time RT-PCR analysis.

2.3. Real-time RT-PCR Analysis

After 6 days of treatment, total RNA was isolated based on the established protocol [17]. The mRNA was then reverse-transcribed and amplified (Applied Biosystems, CA, USA). The synthesized cDNA was used as the template to quantify the relative content of mRNA by using LightCycler Real-Time PCR System (Roche Diagnostics, Penzberg, Germany). The glyceraldehyde 3-phosphate dehydrogenase (GAPDH) was used as the

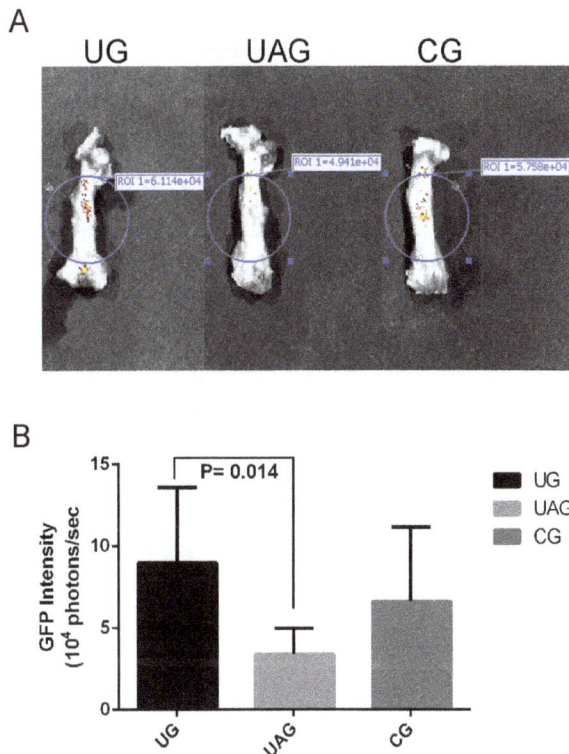

internal control. Primer sequences used in the experiments are summarized in Table 1 (all from Invitrogen, Carlsbad, CA, USA).

2.4. SDF-1 Protein Level Measurement

The culture medium was collected from UG and CG after 3 days of treatment. Protein level of SDF-1α was quantified using Quantikine SDF-1α enzyme immunoassay kit (R&D System,

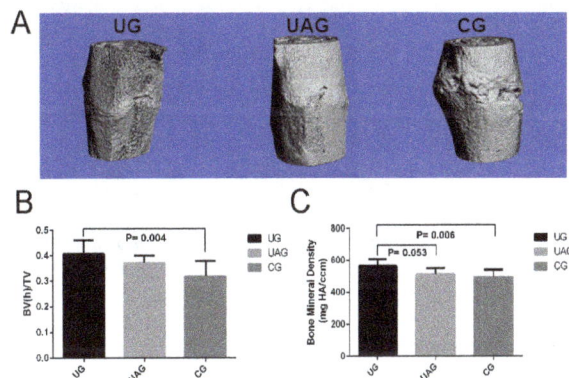

Figure 7. µCT measurement. (A) Representative 3D reconstructed micro-CT images of the three groups at week 4 post-fracture showed improved fracture healing in UG and UAG. **(B)** BVh/TV in UG was increased by 27.9% than in CG (p = 0.004). **(C)** BMD in UG was higher by 10% than in UAG (p = 0.053) and by 14.5% than in CG (p = 0.006).

Table 2. Mechanical properties of the femurs of the three groups at week 4 post-fracture.

Parameters	UG	UAG	CG
Ultimate Load (N)	126.2±7.0[a, b]	96.9±10.4[c]	62.9±26.9
Stiffness (N/mm)	53.5±7.2[b]	44.9±12.4	38.7±10.9
Energy to Failure (j)	0.05±0.02	0.03±0.01	0.03±0.02

UG, LIPUS treatment group; UAG, LIPUS+AMD3100 treatment group; CG, control group received sham treatment.
[a]p<0.05 between UG and UAG;
[b]p<0.05 between UG and CG;
[c]p<0.05 between UAG and CG.

Minneapolis, Minnesota, USA). Colorimetric density of the developed plates was determined using BioTek µQuant Microplate Spectrophotometer (Bio-Tek Instruments Inc, Winooski, VT, US) at 450 nm wavelength. The enzyme-linked immunosorbent assay (ELISA) assay was performed in duplicate. A standard curve was constructed by plotting the mean absorbance for each standard against the concentration.

2.5. Cell Migration Assay

QCM 24-Well colorimetric cell migration assay (Millipore, MA, US) was used to study the migration of MSCs under LIPUS treatment, with or without the presence of AMD3100 (Sigma, St Louis, MO, US), in comparison with sham control [29]. AMD3100 is a specific antagonist of CXCR4 and not cross-reactive with other chemokine receptors [30]. MSCs at passage 3 were starved in serum-free medium for 1 day, washed twice with PBS and incubated in Harvesting Buffer at 37°C for 15 minutes. 20 ml Quenching Medium was added and cells were centrifuged at 1500 rpm for 5 minutes. The pellet was re-suspended in Quenching Medium, and brought to a final concentration at 5.0×10^5 cells/ml. 300 µl cell suspension were added to each insert (8 µm pore size). [31] These inserts then were randomly divided into 3 groups (n = 5), including LIPUS treatment group (UG), LIPUS plus AMD3100 treatment group (UAG) and sham control group (CG). For UAG, an additional 1 µM AMD3100 was added to the insert. LIPUS treatment was applied with the same custom-built platform designed for 6-well plated as described above, then 500 µl of old serum-free conditioned medium collected after 3 days of LIPUS treatment were added to the lower chambers in UG and UAG; 500 µl of old serum-free conditioned medium collected after 3 days of sham treatment was added to the lower chamber in CG. MSCs were incubated at 37°C in 5% CO_2/20% O_2 for 18 hours. The remaining cell suspension was removed; the migration insert was placed into a clean well containing 400 µl of Cell Stain and incubated for 20 minutes at room temperature. The insert was rinsed in water several times, and non-migratory

cells were carefully removed from the interior of the insert. The migrated cells on the exterior of the insert were counted manually under the microscope (Leica DM IRB, Heerbrugg, Switzerland), and the images were taken at ×100 magnification.

2.6. Animal Model, Groupings and LIPUS Treatment

Thirty 8-week-old female SD rats were obtained from the Laboratory Animal Services Center of the Chinese University of Hong Kong. Closed femoral fractures were created at femur shaft based on our established protocol [32,33]. Postoperative radiographies were used to confirm the quality of fracture.

On day 3 post-fracture, the rats were anaesthetized by intraperitoneal injection of a mixture of ketamine (50 mg/kg) and xylazine (10 mg/kg). 1.0×10^6 GFP-labelled MSCs (RASMX-01101, Cyagen, Guangzhou, China) in 500 µl PBS were transplanted by intracardiac injection as previous described [34,35]. Briefly, the GFP-labeled MSCs were sub-cultured to the fifth passage, and then were trypsinized and diluted in 0.9% normal saline. Under anaesthesia, the rat was placed supine, kept firmly on the desk holding the chest between the thumb and forefinger. After removing the air in the liquid, a 23G needle was inserted through the thoracic wall at a point left to the sternum on a line connecting the left axillary pivot with the caudal tip of the sternum. In the meantime, the syringe containing MSCs suspension was aspirated; the injection of the MSCs suspension was performed by gently pressing the piston of the syringe. The position of needle was confirmed by an ultrasound system (Vevo 770, VisualSonics, ON, Canada) under B-mode (brightness mode).

After MSCs injection, the rats were randomly assigned to the following groups: 1) LIPUS group (UG, n = 10) in which LIPUS treatment (Exogen 3000+, Smith & Nephew Inc, Memphis, TN, USA) was applied 20 minutes/day, 5 days/week, after the cell injection. During treatment, the rats were laid on the ventral side under general anaesthesia and a 2.5 cm diameter ultrasound transducer was placed on the lateral side of the fracture site. The ultrasound signal consisting of a 200 µs burst of 1.5 MHz sine

Table 3. Histomorphometric analysis of femoral microarchitecture at week 4 post-fracture.

Parameters	UG	UAG	CG
Cl.Ar (mm^2)	5.0±0.6	4.0±1.9	5.3±2.7
Cg.Ar (mm^2)	0.0±0.0	0.1±0.0	0.9±0.9
Cg.Ar/Cl.Ar (%)	0.0±0.0[a]	0.0±0.0	0.1±0.1

UG, LIPUS treatment group; UAG, LIPUS treatment plus AMD3100 treatment group; CG, control group received sham treatment, Cl.Ar, total callus area; Cg.Ar, cartilage area; Cg.Ar/Cl.Ar, the percentage of cartilage area.
[a]p = 0.038 between UG and CG.

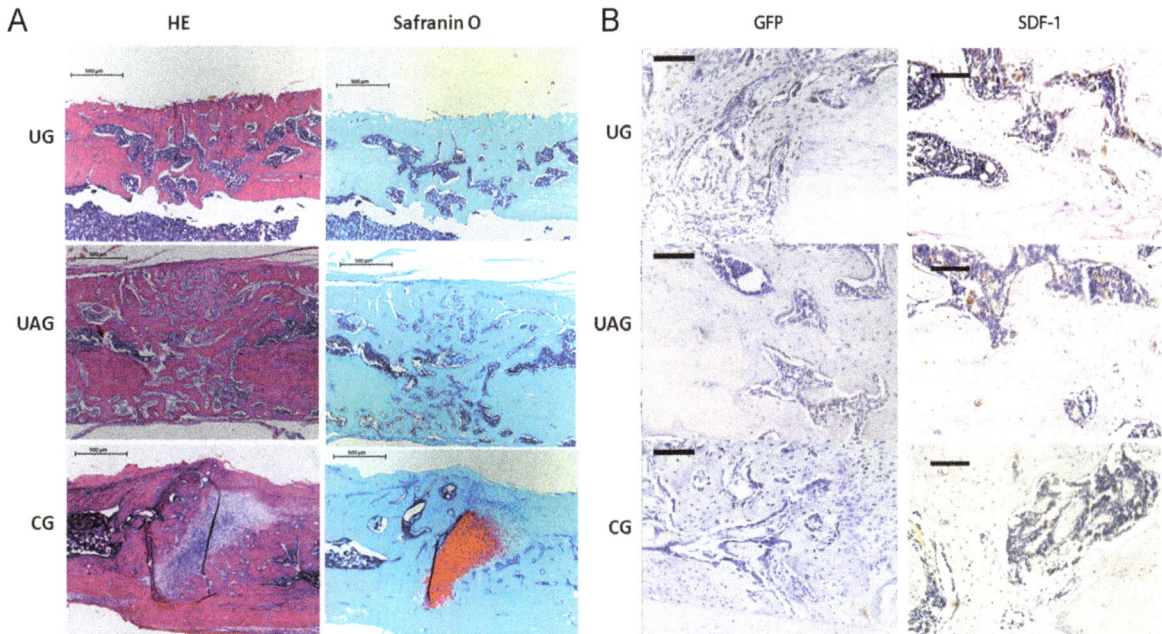

Figure 8. Histological and immunohistochemical results. (A) Representative H&E and safranin-O/fast green staining showed that in UG and UAG, there were large amounts of woven bone formation in the fracture areas, with almost no chondroid tissues (stained red by safranin-O), whereas lots of chondroid and fibrous tissues still existed at the fracture site in CG at week 4. Scale bars, 50 µm. **(B,** left) Representative immunohistochemistry for GFP showed that a large number of the GFP positive cells engrafted in the fracture area in UG, whereas fewer GFP positive cells were detected in CG, and even fewer GFP positive cells were found in UAG. Scale bars at ×100, 200 µm. **(B,** right) Representative images of SDF-1 staining of callus at week 4 post-fracture for young rats. Brown color indicates positive staining. SDF-1 was located mainly in the blood vessels or sinusoid-like regions. Scale bars at ×200, 100 µm.

wave repeating at 1.0 kHz with 30.0 ± 5.0 mW/cm^2 spatial average and temporal average incident intensity was given. 2) LIPUS+AMD3100 group (UAG, n = 10) in which daily LIPUS treatment was applied (same as UG), AMD3100 was resolved in saline to a final concentration of 1 mg/ml for injection and administered (1 mg/kg/day, intraperitoneal) [36] 30 minutes before LIPUS treatment. 3) Sham control group (CG, n = 10) in which the daily sham treatment (LIPUS machine turned off) was applied. Both UG and CG groups received vehicle injections of 0.9% normal saline (1 ml/kg/day, intraperitoneal). Animals were allowed full weight bearing, free cage activity, and food and water ad libitum. At week 4 post-fracture, animals were euthanized by overdosed sodium pentobarbital; the femora and blood were collected for the end-point assessments.

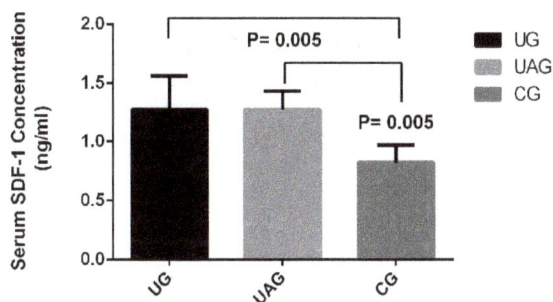

Figure 9. Serum SDF-1 protein concentration. The serum SDF-1 protein level measured by ELISA assay in UG was increased by 1.55 times than in CG (p = 0.005); the serum SDF-1 level in UAG was increased by 1.55 times than in CG (p = 0.005).

2.7. Radiological Assessment

Two-dimensional digital radiographs (MX-20, Faxitron, Lincolnshire, IL, USA) were taken weekly post-fracture to confirm the quality and degree of fracture healing. The quantified temporal changes of callus morphology were evaluated by using the Metamorph Image Analysis System (Universal Imaging Corporation, Downingtown, PA, USA) according to our previous established protocol [37], where callus width (CW) was defined as the maximal outer diameter of the mineralized callus (d2) minus the outer diameter of the femur (d1); and callus area (CA) was calculated as the sum of the areas of the external mineralized callus.

2.8. Ex Vivo GFP Signal Intensity Analysis

Animals were euthanized at week 4 post-fracture. The *ex vivo* fluorescent images were taken by the Xenogen Imaging System (IVIS 200; Caliper Life Sciences, Hopkinton, MA, USA), immediately after the femur was harvested with removal of soft tissues and K-wire. The GFP signals at callus area were acquired and measured by using the live image 2.5 software of Xenogen Imaging System with the settings of exposure time at 5 seconds, binning at 8 and f/stop at f/16 [12,35]. A standard circular region of interest (ROI) with a diameter of 1 cm at the callus area was selected for the measurements. The fluorescent image data was displayed in units of photons. The tissue autofluorescence was subtracted by using GFP background filter with the excitation passband at 440 nm, and emission passband at 550 nm.

2.9. µCT Analysis

Ex vivo micro-computed tomography scans (µCT40, Scanco Medical, Brüttisellen, Switzerland) were performed at 4 weeks post-fracture based on our established protocol [32,38]. The femora were positioned vertically with normal saline-soaked gauze in the sample holder during scanning. The newly formed bone (low-density bone, threshold = 165–350) and highly mineralized bone (high-density bone, threshold = 350–1000) were reconstructed separately [16,32]. The ratios of low-density bone volume over total tissue volume (BVl/TV), high-density bone volume over total tissue volume (BVh/TV), total bone volume fraction (BVt/TV) and bone mineral density (BMD) were calculated.

2.10. Mechanical Testing

A complete healing of closed femoral fracture in young rat was reported taking place around week 4 post-surgery [39]. After μCT scanning, the fractured femora were subjected to mechanical testing as previously described [32,37]. The ultimate load (UL), stiffness, and the energy to failure were calculated from load displacement curves using built-in software (QMAT Professional Material testing software).

2.11. Histomorphometric and Immunohistochemical Analysis

The harvested samples were performed hematoxylin–eosin (H&E) and safranin-O/fast green staining for histomorphometric analysis. The images were taken at ×50 magnification under microscope (Leica DMRB DAS; Leica, Heerbrugg, Switzerland). Quantitative assessment of the safranin-O-positive cartilage at the region of 1.5 mm proximal and distal to the fracture line was performed by using ImageJ (NIH, MD, USA). Cartilage area (Cg.Ar), callus area (Cl.Ar) and their ratio (Cg.Ar/Cl.Ar) were measured.

Immunohistochemical staining was carried out on deparaffinized sections using a rabbit ABC staining system (Santa Cruz Biotechnology, Santa Cruz, CA, USA). The sections were incubated overnight at 4°C in 1:500 rabbit anti-GFP polyclonal antibody (Abcam, Cambridge, MA, USA) or 1:200 rabbit anti-SDF-1 polyclonal antibody (Abcam, Cambridge, MA, USA), followed by incubation with biotinylated secondary antibody and color development according to the manufacturer's instructions. Images were captured using bright field microscopy at ×100 and ×200 magnification (Leica DMRB DAS; Leica, Heerbrugg, Switzerland).

2.12. Blood Collection and Serum SDF-1 Analysis

Five milliliter of blood was collected by cardiac puncture shortly before the animals were euthanized. The blood was centrifuged at 1,800 g for 10 minutes, and the separated serum samples were then stored at −80°C until analysis. The level of SDF-1 in the serum was measured by using the same SDF-1α ELISA kit and microplate reader settings used for culture medium testing as described above. All the serum samples were run in duplicate. A flowchart of the study design was shown in Fig. 1.

2.13. Statistical Analysis

All quantitative data were expressed as mean ± standard deviation and analyzed with SPSS version 20.0 software (IBM, NY, USA). Independent student's t test and one-way analysis of variance (ANOVA) with Tukey's post-hoc test were used for two-group or multiple-group comparisons respectively, as time since fracture induction was not considered as independent variable due to known temporal changes for our measured parameters. Statistical significance was set at $p < 0.05$.

Results

3.1. MSCs Identification

The results of flow cytometry demonstrated that over 99.3% and 98.8% of the mononucleated cell colonies isolated from BM of SD rats were positive for the fibroblastic marker CD90 and MSC marker CD44 respectively. They were negative for the endothelial stem cell marker CD31 and negative for the hematopoietic lineage markers, including CD34 and CD45 (Fig. 2A). Osteogenic induction assay showed abundant calcium deposits in the cell culture (Fig. 2B). Adipogenic induction assay indicated that a number of isolated cells were positively stained by Oil Red O (Fig. 2C).

3.2. Real-time RT-PCR and SDF-1 Protein Level Measurement

Real-time RT-PCR and ELISA results demonstrated that gene expression of SDF-1 and CXCR4 were significantly upregulated in UG, as compared with CG ($p < 0.0001$ and $p = 0.014$, respectively) (Fig. 3A, B). The SDF-1 and CXCR4 mRNA levels were increased 1.6 times and 4.3 times in UG than in CG respectively. SDF-1 protein level in the culture medium was also significantly increased in UG than CG ($p = 0.018$) (Fig. 3C).

3.3. Cell Migration Assay

Under light microscopy, abundant MSCs were found on the exterior of the inserts in UG (Fig. 4A, left); whereas only a small number of MSCs were observed on the exterior of the inserts in both UAG (Fig. 4A, middle) and CG (Fig. 4A, right).

Quantitatively, the number of migrated MSCs in UG was significantly increased than in UAG and CG ($p = 0.003$, $p = 0.002$, respectively) (Fig. 4B). No significant difference was observed in the number of migrated cells between UAG and CG.

3.4 Radiological Assessment

Radiographic analysis demonstrated that both UG and UAG showed earlier fracture healing than CG, as indicated by earlier callus bridging occurred at week 2, whereas callus bridging started from week 3 in CG (Fig. 5A). Quantitative measurement of callus morphometry showed that CW was significantly larger in UG than in CG at week 1, 2 and 3 ($p = 0.031$ for week 1, $p = 0.01$ for week 2, $p = 0.007$ for week 3, respectively); CW was significantly larger in UG than in UAG at week 2 ($p = 0.035$) and week 3 ($p = 0.022$) (Fig. 5B). CA was significantly larger in UG than in CG at week 1, 2 and 3 ($p = 0.002$ for week 1, 2; $p = 0.032$ for week 3); CA was significantly larger in UG than in UAG at week 2 ($p = 0.047$) (Fig. 5C).

3.5. Ex Vivo GFP Signal Intensity Analysis

GFP intensity in UG was significantly higher than in UAG ($p = 0.014$). Higher GFP intensity was found in UG as compared with CG, and in CG as compared with UAG, but no significance was found between these groups (Fig. 6).

3.6. μCT Analysis

3D reconstructed μCT images of the three groups at week 4 post-fracture showed that the fracture healing in UG and UAG were much faster than in CG, as indicated by early closure of fracture gap (Fig. 7A). The BVh/TV in UG was significantly higher than in CG ($p = 0.004$) (Fig. 7B); BMD in UG was higher than in UAG ($p = 0.053$) and CG ($p = 0.006$) (Fig. 7C). For BVl/TV and BVt/TV, no significant differences were found among groups.

3.7. Mechanical Testing

Femora in UG were significantly stronger than in other two groups (Table 2). The ultimate load of the fractured femur in UG was significantly greater than those in UAG and CG (p<0.05, p< 0.05 respectively). The stiffness of the fractured femur in UG was also marginally higher than that of CG (p = 0.065).

3.8. Histomorphometric and Immunohistochemical Analysis

Histological evaluation using hematoxylin–eosin (H&E) and safranin-O/fast green staining demonstrated enhanced fracture healing in UG and UAG at week 4, as reflected by newly formed woven bone with better bridging of fracture gap, whereas newly formed woven bone was observed to some extent in CG at week 4 but there were still many chondroid tissues in the fracture area (Fig. 8A). Quantitative analysis showed that Cg.Ar/Cl.Ar (%) in UG was significantly lower than in CG (Table 3).

Immunohistochemical analysis of GFP and SDF-1 demonstrated that at week 4 post-fracture, many GFP positive cells could be found in the callus area in all three groups, where the number of GFP positive cells in UG was remarkably increased than those in UAG and CG. In UAG, only a scarce number of GFP-positive cells could be found in the fracture area, as compared with UG and CG (Fig. 8B, left). SDF-1 protein in the callus area of UG and UAG was higher than in CG respectively (Fig. 8B, right).

3.9. Serum SDF-1 Protein Level Measurement

At week 4, the serum SDF-1 protein levels in UG and UAG were significantly increased, as compared with CG (p = 0.005 for both), while the level of SDF-1 in serum was comparable between UG and UAG (Fig. 9).

Discussion

Fracture healing is a complex and well-orchestrated biological process composed of three phases: inflammation, repair and remodeling. In the inflammation and repair phases, SDF-1/ CXCR4 signaling participates in bone repair mainly by acting as a regulator of MSCs trafficking to fracture site [40]. In this study, we confirmed that LIPUS applied on the fractured bone promoted MSCs recruitment and that SDF-1/CXCR4 played a very important role in this process. Blocking of SDF-1 signaling with AMD3100 inhibited the migration of MSCs, and also reduced the promoting effect of LIPUS on fracture healing. The present study demonstrates that the enhanced MSCs migration mediated by SDF-1/CXCR4 pathway may be one of the crucial mechanisms through which LIPUS promotes fracture healing.

A major finding of the current study is that physical stimulation in form of LIPUS, can promote MSCs migration *in vitro* and during bone fracture healing *in vivo*. Mechanical stimuli are very important for the development and maintenance of bone [41]. Recently, increasing studies have shown that mechanical stimulation is crucial for regulating MSCs activities during bone repair. Luu *et al.* reported that low magnitude mechanical signals could significantly increase the proliferation and osteogenic differentiation of MSCs in the bone marrow of male C57BL/6J mice [42]; Lai *et al.* further demonstrated that LIPUS could increase osteogenic differentiation of human MSCs [43]. However, the role of mechanical signals in regulating MSCs migration is not reported. Our results revealed that mechanical signals might work in several ways to regulate MSCs behavior and functions. From *in vitro* study, the spontaneous migration capacity of the isolated rat MSCs in CG was observed. Similar phenomenon was observed by Adriana *et al.* in an *in vitro* migration study, which

demonstrated a low spontaneous migration capacity of BM-derived MSCs in the presence of medium alone (without growth factors or chemokines), after overnight incubation of the transwells at 37°C, 5%CO$_2$ [44]. It was most likely that the conditioned medium in the lower chamber of CG contained many bio-active factors that served as chemoattractant and induced the spontaneous migration of MSCs [45–48]. Our data found that LIPUS stimulation enhanced the migration of cultured MSCs, as indicated by the increased number of migrated cells at the exterior of inserts from UG, compared with those in CG. The effect of LIPUS on cell migration has been reported by several studies. Takao *et al.* studied the migration of osteoblast-like cell (MC3T3-E1 cells) under LIPUS treatment by using a wound healing assay. They found that after 20 minutes LIPUS treatment, the migration of MC3T3-E1 osteoblastic cells was significantly increased than the control group, as indicated by the distance between the wound line and the migration front at 6 h, 12 h and 20 h after wounding [49]. The upregulated SDF-1 and CXCR4 expression by LIPUS may be responsible for the enhanced motility of MSCs observed in cell migration assay in two possible ways. First, the enhanced CXCR4 expression of MSCs could lead to the increased MSCs migration. Shyam *et al.* isolated and cultured MSCs from healthy volunteers and transduced them with a retroviral vector containing either CXCR4 and GFP or GFP alone. They used a transwell migration system to study MSC migration to SDF-1, and found that MSCs transduced with CXCR4 showed significantly more migration toward SDF-1, with 3-fold greater at 3 h and more than 5-fold greater at 6 h [50]. Second, the upregulated SDF-1 expression, especially the increased SDF-1 protein level in the conditioned medium, promoted MSCs migration. Previously Son *et al.* used a chemoinvasion assay to evaluate the ability of MSCs to cross the reconstituted basement membrane Matrigel. After 24 hours of incubation at 37°C, 5% CO$_2$, the number of migrated MSCs toward the lower compartment containing SDF-1 was significantly higher than that in control [51]. Our results indicated that the enhanced cell migration may be due to the combined effects as described above. BM-derived MSCs are able to express CXCR4 and secrete SDF-1 simultaneously. The present study further demonstrated that after treating MSCs with AMD3100, the antagonist of SDF-1/CXCR4 pathway, the migration of MSCs under LIPUS treatment was strikingly reduced. This indicates the LIPUS-induced MSC migration is CXCR4-dependent. Although we still found a very small number of migrated cells in the combined treatment group (UAG), when compared with that in CG, there was no statistical significance. It suggested that AMD3100 might almost completely block the effect of LIPUS on MSC migration, since SDF-1 may not be the only chemokine in the conditioned medium of MSCs. Other bioactive factors secreted by MSCs may also influence MSC migration to some extent [45–48]. *In vivo*, the transplanted GFP-labeled MSCs were found to migrate to the callus, as indicated by the GFP intensity measured by fluorescent imaging and immunohistochemistry. Our findings confirmed that in young rat model, after 4 weeks of LIPUS treatment, both the serum and local SDF-1 protein levels in the callus of UG were increased than in CG, together with the higher GFP intensity from *ex vivo* fluorescent imaging, and increased GFP-positive cells in the callus of UG as detected by immunohistochemistry. Our findings were also substantiated by a similar report by Kumagai *et al.* demonstrating a positive effect in the recruitment of GFP-positive cells from a parabiotic mouse to the fracture site of another surgically conjoined mouse [24]. Therefore, there are sufficient evidence to conclude that there exist a strong relationship between LIPUS stimulation and MSCs migration.

This study demonstrated LIPUS treatment could activate SDF-1/CXCR4 pathway, which were substantiated by a few previous studies. Carlet et al. observed an intense expression of SDF-1 in the compression side of periodontal ligament during orthodontic tooth movement [52]; Li et al. also reported that cyclic stretch could upregulate SDF-1/CXCR4 axis in human saphenous vein smooth muscle cells [53]. Kumagai et al. also detected increased protein expression of both SDF-1 and CXC-R4 in the fracture site in the LIPUS treatment group as compared to the control group [24]. However, the mechanisms responsible for mechanical stimuli induced SDF-1/CXCR4 signaling in these cells are largely unknown. Integrins are the main receptors that connect the cytoskeleton to the extracellular matrix (ECM) [54]. They transmit mechanical stresses across the plasma membrane that enables the tractional forces developed in the cytoskeleton to be conveyed to the ECM [55]. Integrins also regulates signaling pathways [56,57]. Many recent studies have demonstrated the important interactions between SDF-1/CXCR4 pathway and integrins in regulating cellular activities in different cell types, including MSCs [58–61]. Thus, the effect we observed might be the downstream of LIPUS's interactions with the transmembrane integrins on the MSCs.

The direct regulatory effect of LIPUS on MSCs found in vitro may not fully reflect the complexity of the in vivo situation. During fracture repair, SDF-1 was found not only in MSCs, but also in other cell types, including endothelial cells [62,63], periosteal cells [28,64], chondrocytes [65], and osteoblasts [66,67] etc. Many previous studies have shown these cells might also secrete SDF-1 and contribute to the recruitment of MSCs. Given the fact that mechanical signals provided by LIPUS could be sensed and transduced by many cell types, which has been extensively studied in the past [68–73], it is most likely that LIPUS may act on these cells through different ways. LIPUS might promote SDF-1 secretion from different cells through physical interactions [74,75] and integrins signal transductions [76–79] in the site of injury; simultaneously, LIPUS enhances CXCR4 expression on the surface of MSCs from circulation or the adjacent BM, thus promoting these cells to migrate toward the SDF-1 gradient and engraft in the fracture site.

In the rat model, LIPUS promoted early callus formation, as indicated by weekly radiographic analysis. In all the groups, the temporal changes of CW and CA followed the similar pattern, i.e. from week 1 to week 2, both CW and CA increased gradually; from week 2 to week 4, both CW and CA decreased rapidly. The largest callus size was generally found at week 2, which indicated the most active callus formation; whereas the smallest callus size was found at week 4, which represented the callus remodeling. Significantly increased callus size in UG, in comparison with CG, was observed from week 1 to week 3 post-fracture, which reflected the promoting effect of LIPUS on callus formation in the early phase of fracture healing. Another finding of the radiographic analysis was that the callus bridging in UG was accelerated, which started at week 2, in contrast to CG at week 3. The present findings were consistent with many previous researches. In 1983, Dyson et al. applied therapeutic ultrasound on the complete bilateral transverse fibular fractures in adult female Wistar rats, and found that ultrasound therapy was most effective during the first two weeks after injury [80]. Later, by using bilateral closed femoral shaft fracture rat model, Wang et al. [81] and Yang et al. [82] demonstrated increased callus size after 7 daily 15-minute exposures to LIPUS treatment, compared with the contralateral controls. Our recent studies on LIPUS effects in rat closed fracture

healing substantiated these early works and demonstrated the promoted early callus formation [17,35,38,74].

LIPUS was also found to promote callus mineralization and remodeling in rats. As described earlier, from the radiographic analysis, the callus size of all the groups decreased rapidly, after reaching the peak value at week 2. It represented the start of the remodeling phase, overlapping significantly with the reparative phase, characterized by the slow modeling and remodeling of the fracture callus from woven to mature lamellar bone, and ultimately, the restoration of the bone to normal or near normal morphology and mechanical strength [83]. Among all the groups, the callus size in UG reduced more rapidly from week 2 to week 4 post-fracture, which suggested an accelerated remodeling process. The radiographic findings were supported by the results from µCT, which reflected the microstructure and mineralization of the callus. At week 4 post-fracture, we found BVh/TV and BMD in UG was significantly increased than in CG. Both BVh/TV and BMD are important tools for reflecting the degree of mineralization in callus. BMD of the callus has been used for the quantitative evaluation of the mechanical properties of healing bone [84–86]. The present results suggested LIPUS might accelerate the maturation process of callus. We also observed the apparent trend in the reduction of TV, BVl and BVl/TV in the callus of UG at week 4, as compared to in CG, although the differences were not statistically significant. These parameters reflected the reduced callus size and unmineralized tissue in the callus of UG, which was consistent with the radiographic analysis. The mechanical testing results further demonstrated the improved mechanical properties of callus under LIPUS treatment, as indicated by the higher UL and increased stiffness in UG at week 4. These findings were substantiated by the histological analysis, which showed the enhanced endochondral ossification in UG, in which fracture gap was better bridged at week 4 and filled with woven bone, whereas chondroid tissues were still present in the fracture area in CG.

Disrupting SDF1/CXCR4 signaling pathway by daily administration of AMD3100 on the rats receiving LIPUS treatment resulted in the significantly reduced GFP-positive cells in the fracture area, reduced callus size and less cartilage volume during fracture healing. These findings were supported by several previous studies. Toupadakis et al. examined the effect of AMD3100 on bone repair by using a murine fracture model. They found that the administration of AMD3100 led to a significantly reduced hyaline cartilage volume, callus volume and mineralized bone volume, associated with reduction of genes expression related to endochondral ossification [40]. Recently, Zhou found that in a traumatic brain injury/closed femoral fracture mice model, following AMD3100 treatment, the MSC migration was inhibited, and new bone formation was significantly reduced by 47% at the fracture sites in comparison with the controls treated with PBS [87]. However, when comparing to CG, UAG showed slightly better fracture healing, as shown by the early callus bridging (started from week 2), less chondroid tissues at fracture site and higher ultimate load. Taken together, these data suggest that SDF-1/CXCR4 signaling plays a very important role in fracture healing. Disrupting the SDF/CXCR4 pathway during LIPUS treatment could partially reduce, but not fully abolish LIPUS-induced fracture healing, indicating that other factors or signaling pathways might also be involved in this accelerated fracture repair process, such as increased blood circulation [88], upregulated osteogenic genes [89], and gap junctional cell-to-cell intercellular communication in rat MSCs [69].

The limitation of our study is that we used transplanted GFP-labeled allogenic MSCs to explore the migration activities of endogenous MSCs in vivo, which might not fully reflect the exact pattern of MSCs recruitment during facture healing. Further research may be necessary to understand the possible roles of endogenous MSCs in participating in tissue repair process.

Conclusion

In conclusion, our study demonstrated that the application of LIPUS treatment could enhance SDF-1 signaling pathway, promote MSCs migration towards the fracture site, and accelerate fracture healing. This is the first evidence showing the micromechanical stimulation produced by LIPUS's pressure waves can regulate MSCs migration through SDF-1/CXCR4 pathway in

fracture healing. It provides novel insights into comprehensive mechanisms, through which LIPUS promotes fracture healing, and will potentially lead to the development of LIPUS enhanced MSC therapies for improving bone regeneration in a wide range of orthopaedic conditions.

Author Contributions

Conceived and designed the experiments: FYW KSL GL LQ WHC. Performed the experiments: FYW SH JHQ SKHC MHS. Analyzed the data: FYW WHC. Wrote the paper: FYW WHC.

References

1. Claes L, Recknagel S, Ignatius A (2012) Fracture healing under healthy and inflammatory conditions. Nat Rev Rheumatol 8: 133–143.
2. Caplan AI (1991) Mesenchymal stem cells. Journal of orthopaedic research : official publication of the Orthopaedic Research Society 9: 641–650.
3. Minguell JJ, Erices A, Conget P (2001) Mesenchymal stem cells. Exp Biol Med (Maywood) 226: 507–520.
4. Short B, Brouard N, Occhiodoro-Scott T, Ramakrishnan A, Simmons PJ (2003) Mesenchymal stem cells. Arch Med Res 34: 565–571.
5. Schindeler A, McDonald MM, Bokko P, Little DG (2008) Bone remodeling during fracture repair: The cellular picture. Semin Cell Dev Biol 19: 459–466.
6. Kolar P, Gaber T, Perka C, Duda GN, Buttgereit F (2011) Human early fracture hematoma is characterized by inflammation and hypoxia. Clin Orthop Relat Res 469: 3118–3126.
7. Bastian O, Pillay J, Alblas J, Leenen L, Koenderman L, et al. (2011) Systemic inflammation and fracture healing. J Leukoc Biol 89: 669–673.
8. Kitaori T, Ito H, Schwarz EM, Tsutsumi R, Yoshitomi H, et al. (2009) Stromal cell-derived factor 1/CXCR4 signaling is critical for the recruitment of mesenchymal stem cells to the fracture site during skeletal repair in a mouse model. Arthritis Rheum 60: 813–823.
9. Ponomaryov T, Peled A, Petit I, Taichman RS, Habler L, et al. (2000) Induction of the chemokine stromal-derived factor-1 following DNA damage improves human stem cell function. J Clin Invest 106: 1331–1339.
10. Ma M, Ye JY, Deng R, Dee CM, Chan GC (2011) Mesenchymal stromal cells may enhance metastasis of neuroblastoma via SDF-1/CXCR4 and SDF-1/CXCR7 signaling. Cancer Lett 312: 1–10.
11. Houthuijzen JM, Daenen LG, Roodhart JM, Voest EE (2012) The role of mesenchymal stem cells in anti-cancer drug resistance and tumour progression. Br J Cancer 106: 1901–1906.
12. Granero-Molto F, Weis JA, Miga MI, Landis B, Myers TJ, et al. (2009) Regenerative effects of transplanted mesenchymal stem cells in fracture healing. Stem Cells 27: 1887–1898.
13. Duarte LR (1983) The stimulation of bone growth by ultrasound. Arch Orthop Trauma Surg 101: 153–159.
14. Kristiansen TK, Ryaby JP, McCabe J, Frey JJ, Roe LR (1997) Accelerated healing of distal radial fractures with the use of specific, low-intensity ultrasound. A multicenter, prospective, randomized, double-blind, placebo-controlled study. The Journal of bone and joint surgery American volume 79: 961–973.
15. Leung KS, Lee WS, Tsui HF, Liu PP, Cheung WH (2004) Complex tibial fracture outcomes following treatment with low-intensity pulsed ultrasound. Ultrasound Med Biol 30: 389–395.
16. Fung CH, Cheung WH, Pounder NM, de Ana FJ, Harrison A, et al. (2012) Effects of different therapeutic ultrasound intensities on fracture healing in rats. Ultrasound in medicine & biology 38: 745–752.
17. Cheung WH, Chow SK, Sun MH, Qin L, Leung KS (2011) Low-intensity pulsed ultrasound accelerated callus formation, angiogenesis and callus remodeling in osteoporotic fracture healing. Ultrasound Med Biol 37: 231–238.
18. Khan Y, Laurencin CT (2008) Fracture repair with ultrasound: clinical and cell-based evaluation. J Bone Joint Surg Am 90 Suppl 1: 138–144.
19. Tam KF, Cheung WH, Lee KM, Qin L, Leung KS (2008) Osteogenic effects of low-intensity pulsed ultrasound, extracorporeal shockwaves and their combination - an in vitro comparative study on human periosteal cells. Ultrasound Med Biol 34: 1957–1965.
20. Wang YX, Leung KC, Cheung WH, Wang HH, Shi L, et al. (2010) Low-intensity pulsed ultrasound increases cellular uptake of superparamagnetic iron oxide nanomaterial: results from human osteosarcoma cell line U2OS. J Magn Reson Imaging 31: 1508–1513.
21. Arnsdorf EJ, Tummala P, Kwon RY, Jacobs CR (2009) Mechanically induced osteogenic differentiation-the role of RhoA, ROCKII and cytoskeletal dynamics. J Cell Sci 122: 546–553.
22. Li YJ, Batra NN, You L, Meier SC, Coe IA, et al. (2004) Oscillatory fluid flow affects human marrow stromal cell proliferation and differentiation. J Orthop Res 22: 1283–1289.
23. Kasper G, Glaeser JD, Geissler S, Ode A, Tuischer J, et al. (2007) Matrix metalloprotease activity is an essential link between mechanical stimulus and mesenchymal stem cell behavior. Stem Cells 25: 1985–1994.
24. Kumagai K, Takeuchi R, Ishikawa H, Yamaguchi Y, Fujisawa T, et al. (2012) Low-intensity pulsed ultrasound accelerates fracture healing by stimulation of recruitment of both local and circulating osteogenic progenitors. J Orthop Res 30: 1516–1521.
25. Xu L, Song C, Ni M, Meng F, Xie H, et al. (2012) Cellular retinol-binding protein 1 (CRBP-1) regulates osteogenenesis and adipogenesis of mesenchymal stem cells through inhibiting RXRalpha-induced beta-catenin degradation. Int J Biochem Cell Biol 44: 612–619.
26. Dominici M, Le Blanc K, Mueller I, Slaper-Cortenbach I, Marini F, et al. (2006) Minimal criteria for defining multipotent mesenchymal stromal cells. The International Society for Cellular Therapy position statement. Cytotherapy 8: 315–317.
27. Rui YF, Lui PP, Li G, Fu SC, Lee YW, et al. (2010) Isolation and characterization of multipotent rat tendon-derived stem cells. Tissue Eng Part A 16: 1549–1558.
28. Leung KS, Cheung WH, Zhang C, Lee KM, Lo HK (2004) Low intensity pulsed ultrasound stimulates osteogenic activity of human periosteal cells. Clin Orthop Relat Res: 253–259.
29. Wynn RF, Hart CA, Corradi-Perini C, O'Neill L, Evans CA, et al. (2004) A small proportion of mesenchymal stem cells strongly expresses functionally active CXCR4 receptor capable of promoting migration to bone marrow. Blood 104: 2643–2645.
30. Fricker SP, Anastassov V, Cox J, Darkes MC, Grujic O, et al. (2006) Characterization of the molecular pharmacology of AMD3100: a specific antagonist of the G-protein coupled chemokine receptor, CXCR4. Biochem Pharmacol 72: 588–596.
31. Guo Y, Hangoc G, Bian H, Pelus LM, Broxmeyer HE (2005) SDF-1/CXCL12 enhances survival and chemotaxis of murine embryonic stem cells and production of primitive and definitive hematopoietic progenitor cells. Stem Cells 23: 1324–1332.
32. Leung KS, Shi HF, Cheung WH, Qin L, Ng WK, et al. (2009) Low-magnitude high-frequency vibration accelerates callus formation, mineralization, and fracture healing in rats. J Orthop Res 27: 458–465.
33. Sun MH, Leung KS, Zheng YP, Huang YP, Wang LK, et al. (2012) Three-dimensional high frequency power Doppler ultrasonography for the assessment of microvasculature during fracture healing in a rat model. J Orthop Res 30: 137–143.
34. Furlani D, Li W, Pittermann E, Klopsch C, Wang L, et al. (2009) A transformed cell population derived from cultured mesenchymal stem cells has no functional effect after transplantation into the injured heart. Cell Transplant 18: 319–331.
35. Cheung WH, Chin WC, Wei FY, Li G, Leung KS (2013) Applications of exogenous mesenchymal stem cells and low intensity pulsed ultrasound enhance fracture healing in rat model. Ultrasound Med Biol 39: 117–125.
36. De Clercq E (2003) The bicyclam AMD3100 story. Nat Rev Drug Discov 2: 581–587.
37. Shi HF, Cheung WH, Qin L, Leung AH, Leung KS (2010) Low-magnitude high-frequency vibration treatment augments fracture healing in ovariectomy-induced osteoporotic bone. Bone 46: 1299–1305.
38. Cheung WH, Chin WC, Qin L, Leung KS (2012) Low intensity pulsed ultrasound enhances fracture healing in both ovariectomy-induced osteoporotic and age-matched normal bones. J Orthop Res 30: 129–136.
39. Ekeland A, Engesoeter LB, Langeland N (1982) Influence of age on mechanical properties of healing fractures and intact bones in rats. Acta Orthop Scand 53: 527–534.
40. Toupadakis CA, Wong A, Genetos DC, Chung DJ, Murugesh D, et al. (2012) Long-term administration of AMD3100, an antagonist of SDF-1/CXCR4 signaling, alters fracture repair. J Orthop Res 30: 1853–1859.
41. Carter DR, Van Der Meulen MC, Beaupre GS (1996) Mechanical factors in bone growth and development. Bone 18: 5S–10S.
42. Luu YK, Capilla E, Rosen CJ, Gilsanz V, Pessin JE, et al. (2009) Mechanical stimulation of mesenchymal stem cell proliferation and differentiation promotes osteogenesis while preventing dietary-induced obesity. J Bone Miner Res 24: 50–61.
43. Lai CH, Chen SC, Chiu LH, Yang CB, Tsai YH, et al. (2010) Effects of low-intensity pulsed ultrasound, dexamethasone/TGF-beta1 and/or BMP-2 on the

transcriptional expression of genes in human mesenchymal stem cells: chondrogenic vs. osteogenic differentiation. Ultrasound Med Biol 36: 1022–1033.

44. Ponte AL, Marais E, Gallay N, Langonne A, Delorme B, et al. (2007) The in vitro migration capacity of human bone marrow mesenchymal stem cells: comparison of chemokine and growth factor chemotactic activities. Stem Cells 25: 1737–1745.

45. Nagaya N, Kangawa K, Itoh T, Iwase T, Murakami S, et al. (2005) Transplantation of mesenchymal stem cells improves cardiac function in a rat model of dilated cardiomyopathy. Circulation 112: 1128–1135.

46. Caplan AI, Dennis JE (2006) Mesenchymal stem cells as trophic mediators. J Cell Biochem 98: 1076–1084.

47. Walter MN, Wright KT, Fuller HR, MacNeil S, Johnson WE (2010) Mesenchymal stem cell-conditioned medium accelerates skin wound healing: an in vitro study of fibroblast and keratinocyte scratch assays. Exp Cell Res 316: 1271–1281.

48. Meirelles Lda S, Fontes AM, Covas DT, Caplan AI (2009) Mechanisms involved in the therapeutic properties of mesenchymal stem cells. Cytokine Growth Factor Rev 20: 419–427.

49. Iwai T, Harada Y, Imura K, Iwabuchi S, Murai J, et al. (2007) Low-intensity pulsed ultrasound increases bone ingrowth into porous hydroxyapatite ceramic. J Bone Miner Metab 25: 392–399.

50. Bhakta S, Hong P, Koc O (2006) The surface adhesion molecule CXCR4 stimulates mesenchymal stem cell migration to stromal cell-derived factor-1 in vitro but does not decrease apoptosis under serum deprivation. Cardiovasc Revasc Med 7: 19–24.

51. Son BR, Marquez-Curtis LA, Kucia M, Wysoczynski M, Turner AR, et al. (2006) Migration of bone marrow and cord blood mesenchymal stem cells in vitro is regulated by stromal-derived factor-1-CXCR4 and hepatocyte growth factor-c-met axes and involves matrix metalloproteinases. Stem Cells 24: 1254–1264.

52. Garlet TP, Coelho U, Repeke CE, Silva JS, Cunha Fde Q, et al. (2008) Differential expression of osteoblast and osteoclast chemmoatractants in compression and tension sides during orthodontic movement. Cytokine 42: 330–335.

53. Li F, Guo WY, Li WJ, Zhang DX, Lv AL, et al. (2009) Cyclic stretch upregulates SDF-1alpha/CXCR4 axis in human saphenous vein smooth muscle cells. Biochem Biophys Res Commun 386: 247–251.

54. Ernstrom GG, Chalfie M (2002) Genetics of sensory mechanotransduction. Annu Rev Genet 36: 411–453.

55. Huang S, Ingber DE (1999) The structural and mechanical complexity of cell-growth control. Nat Cell Biol 1: E131–138.

56. Schwartz MA, Assoian RK (2001) Integrins and cell proliferation: regulation of cyclin-dependent kinases via cytoplasmic signaling pathways. J Cell Sci 114: 2553–2560.

57. Katsumi A, Orr AW, Tzima E, Schwartz MA (2004) Integrins in mechanotransduction. J Biol Chem 279: 12001–12004.

58. Peled A, Kollet O, Ponomaryov T, Petit I, Franitza S, et al. (2000) The chemokine SDF-1 activates the integrins LFA-1, VLA-4, and VLA-5 on immature human CD34(+) cells: role in transendothelial/stromal migration and engraftment of NOD/SCID mice. Blood 95: 3289–3296.

59. Jing D, Fonseca AV, Alakel N, Fierro FA, Muller K, et al. (2010) Hematopoietic stem cells in co-culture with mesenchymal stromal cells–modeling the niche compartments in vitro. Haematologica 95: 542–550.

60. Cencioni C, Capogrossi MC, Napolitano M (2012) The SDF-1/CXCR4 axis in stem cell preconditioning. Cardiovasc Res 94: 400–407.

61. Cheng M, Qin G (2012) Progenitor cell mobilization and recruitment: SDF-1, CXCR4, alpha4-integrin, and c-kit. Prog Mol Biol Transl Sci 111: 243–264.

62. Yamaguchi J, Kusano KF, Masuo O, Kawamoto A, Silver M, et al. (2003) Stromal cell-derived factor-1 effects on ex vivo expanded endothelial progenitor cell recruitment for ischemic neovascularization. Circulation 107: 1322–1328.

63. Dar A, Goichberg P, Shinder V, Kalinkovich A, Kollet O, et al. (2005) Chemokine receptor CXCR4-dependent internalization and resecretion of functional chemokine SDF-1 by bone marrow endothelial and stromal cells. Nat Immunol 6: 1038–1046.

64. Kitaori T, Ito H, Schwarz EM, Tsutsumi R, Yoshitomi H, et al. (2009) Stromal cell-derived factor 1/CXCR4 signaling is critical for the recruitment of mesenchymal stem cells to the fracture site during skeletal repair in a mouse model. Arthritis and rheumatism 60: 813–823.

65. Murata K, Kitaori T, Oishi S, Watanabe N, Yoshitomi H, et al. (2012) Stromal cell-derived factor 1 regulates the actin organization of chondrocytes and chondrocyte hypertrophy. PLoS One 7: e37163.

66. Jung Y, Wang J, Schneider A, Sun YX, Koh-Paige AJ, et al. (2006) Regulation of SDF-1 (CXCL12) production by osteoblasts; a possible mechanism for stem cell homing. Bone 38: 497–508.

67. Katayama Y, Battista M, Kao WM, Hidalgo A, Peired AJ, et al. (2006) Signals from the sympathetic nervous system regulate hematopoietic stem cell egress from bone marrow. Cell 124: 407–421.

68. Azuma Y, Ito M, Harada Y, Takagi H, Ohta T, et al. (2001) Low-intensity pulsed ultrasound accelerates rat femoral fracture healing by acting on the various cellular reactions in the fracture callus. J Bone Miner Res 16: 671–680.

69. Sena K, Angle SR, Kanaji A, Aher C, Karwo DG, et al. (2011) Low-intensity pulsed ultrasound (LIPUS) and cell-to-cell communication in bone marrow stromal cells. Ultrasonics 51: 639–644.

70. Sun JS, Hong RC, Chang WH, Chen LT, Lin FH, et al. (2001) In vitro effects of low-intensity ultrasound stimulation on the bone cells. J Biomed Mater Res 57: 449–456.

71. Li JK, Chang WH, Lin JC, Ruaan RC, Liu HC, et al. (2003) Cytokine release from osteoblasts in response to ultrasound stimulation. Biomaterials 24: 2379–2385.

72. Naruse K, Miyauchi A, Itoman M, Mikuni-Takagaki Y (2003) Distinct anabolic response of osteoblast to low-intensity pulsed ultrasound. J Bone Miner Res 18: 360–369.

73. Sant'Anna EF, Leven RM, Virdi AS, Sumner DR (2005) Effect of low intensity pulsed ultrasound and BMP-2 on rat bone marrow stromal cell gene expression. J Orthop Res 23: 646–652.

74. Fung CH, Cheung WH, Pounder NM, de Ana FJ, Harrison A, et al. (2012) Effects of different therapeutic ultrasound intensities on fracture healing in rats. Ultrasound Med Biol 38: 745–752.

75. Mehta S, Antich P (1997) Measurement of shear-wave velocity by ultrasound critical-angle reflectometry (UCR). Ultrasound Med Biol 23: 1123–1126.

76. Hsu HC, Fong YC, Chang CS, Hsu CJ, Hsu SF, et al. (2007) Ultrasound induces cyclooxygenase-2 expression through integrin, integrin-linked kinase, Akt, NF-kappaB and p300 pathway in human chondrocytes. Cell Signal 19: 2317–2328.

77. Pounder NM, Harrison AJ (2008) Low intensity pulsed ultrasound for fracture healing: a review of the clinical evidence and the associated biological mechanism of action. Ultrasonics 48: 330–338.

78. Yang R-S, Lin W-L, Chen Y-Z, Tang C-H, Huang T-H, et al. (2005) Regulation by ultrasound treatment on the integrin expression and differentiation of osteoblasts. Bone 36: 276–283.

79. Whitney NP, Lamb AC, Louw TM, Subramanian A (2012) Integrin-mediated mechanotransduction pathway of low-intensity continuous ultrasound in human chondrocytes. Ultrasound in medicine & biology.

80. Dyson M, Brookes M (1983) Stimulation of bone repair by ultrasound. Ultrasound Med Biol Suppl 2: 61–66.

81. Wang SJ, Lewallen DG, Bolander ME, Chao EY, Ilstrup DM, et al. (1994) Low intensity ultrasound treatment increases strength in a rat femoral fracture model. J Orthop Res 12: 40–47.

82. Yang KH, Parvizi J, Wang SJ, Lewallen DG, Kinnick RR, et al. (1996) Exposure to low-intensity ultrasound increases aggrecan gene expression in a rat femur fracture model. J Orthop Res 14: 802–809.

83. Hadjiargyrou M, McLeod K, Ryaby JP, Rubin C (1998) Enhancement of fracture healing by low intensity ultrasound. Clinical orthopaedics and related research 355: S216–S229.

84. Augat P, Merk J, Genant HK, Claes L (1997) Quantitative assessment of experimental fracture repair by peripheral computed tomography. Calcif Tissue Int 60: 194–199.

85. Dai K-R, Hao Y-Q (2007) Quality of healing compared between osteoporotic fracture and normal traumatic fracture. Advanced Bioimaging Technologies in Assessment of the Quality of Bone and Scaffold Materials: Springer. 531–541.

86. Nyman JS, Munoz S, Jadhav S, Mansour A, Yoshii T, et al. (2009) Quantitative measures of femoral fracture repair in rats derived by micro-computed tomography. Journal of biomechanics 42: 891–897.

87. Liu X, Zhou C, Li Y, Ji Y, Xu G, et al. (2013) SDF-1 promotes endochondral bone repair during fracture healing at the traumatic brain injury condition. PLoS One 8: e54077.

88. Rawool NM, Goldberg BB, Forsberg F, Winder AA, Hume E (2003) Power Doppler assessment of vascular changes during fracture treatment with low-intensity ultrasound. J Ultrasound Med 22: 145–153.

89. Favaro-Pipi E, Bossini P, de Oliveira P, Ribeiro JU, Tim C, et al. (2010) Low-intensity pulsed ultrasound produced an increase of osteogenic genes expression during the process of bone healing in rats. Ultrasound Med Biol 36: 2057–2064.

21

The Effect of Naturally Occurring Chronic Kidney Disease on the Micro-Structural and Mechanical Properties of Bone

Anna Shipov[1]*[᠑], **Gilad Segev**[1᠑], **Hagar Meltzer**[1], **Moran Milrad**[1], **Ori Brenner**[2], **Ayelet Atkins**[1], **Ron Shahar**[1]

1 Koret School of Veterinary Medicine, Hebrew University of Jerusalem, Rehovot, Israel, 2 Department of Veterinary Resources, Weizmann Institute, Rehovot, Israel

Abstract

Chronic kidney disease (CKD) is a growing public health concern worldwide, and is associated with marked increase of bone fragility. Previous studies assessing the effect of CKD on bone quality were based on biopsies from human patients or on laboratory animal models. Such studies provide information of limited relevance due to the small size of the samples (biopsies) or the non-physiologic CKD syndrome studied (rodent models with artificially induced CKD). Furthermore, the type, architecture, structure and biology of the bone of rodents are remarkably different from human bones; therefore similar clinicopathologic circumstances may affect their bones differently. We describe the effects of naturally occurring CKD with features resembling human CKD on the skeleton of cats, whose bone biology, structure and composition are remarkably similar to those of humans. We show that CKD causes significant increase of resorption cavity density compared with healthy controls, as well as significantly lower cortical mineral density, cortical cross-sectional area and cortical cross-sectional thickness. Young's modulus, yield stress, and ultimate stress of the cortical bone material were all significantly decreased in the skeleton of CKD cats. Cancellous bone was also affected, having significantly lower trabecular thickness and bone volume over total volume in CKD cats compared with controls. This study shows that naturally occurring CKD has deleterious effects on bone quality and strength. Since many similarities exist between human and feline CKD patients, including the clinicopathologic features of the syndrome and bone microarchitecture and biology, these results contribute to better understanding of bone abnormalities associated with CKD.

Editor: Luc Malaval, Université Jean Monnet, France

Funding: This work was supported by the Israel Science Foundation (Grant No 151/08) to RS. The funders had no role in study design, data collection and analysis, decision to publish, or preparation of the manuscript.

Competing Interests: The authors have declared that no competing interests exist.

* Email: anna.shipov@mail.huji.ac.il

᠑ These authors contributed equally to this work.

Introduction

Chronic kidney disease (CKD) is a growing public health concern worldwide, with increasing incidence in all age groups. The prevalence of moderate to severe CKD in the general population is reported to be as high as 8.5% [1,2]. The disease is irreversible and progressive in nature, and as it progresses, metabolic derangements worsen. This is particularly true in the ageing population, where CKD has become a major cause of morbidity and mortality.

CKD-associated bone diseases include several different types of bone pathologies, such as adynamic bone disease and osteomalacia which are characterized by low bone turnover, osteitis fibrosa cystica which is characterized by high bone turnover (due to secondary hyperparathyroidism) and mixed uremic osteodystrophy which is characterized by either high or low turnover and abnormal mineralization [3].

One of the inevitable metabolic consequences of CKD is secondary renal hyperparathyroidism (SRH) [4]. The pathophysiology of SRH is complex and involves phosphorus retention leading to hyperphosphatemia, ionized hypocalcemia, decreased circulating 1,25-dihydroxyvitamin D (calcitriol) concentration and increased concentrations of parathyroid hormone (PTH) and fibroblast growth factor 23 [FGF23, [5]]. FGF23, which has been shown to have a pivotal role in mineral homeostasis, is produced mainly by osteocytes and osteoblasts [6]. Serum levels of FGF23 increase already in the early stages of CKD, when patients are still normo-phosphatemic and have normal PTH levels [7–9]. When PTH levels increase, they promote bone resorption, and persistently high PTH concentrations, as documented in CKD patients, eventually lead to, osteopenia, and increased risk of pathological fractures [10].

It is widely recognized that bone fragility increases markedly in patients with CKD, and that fracture risk increases with progression of the disease [11–13]. The risk of pathological fractures has been reported to increase by 9% with each 200-pg/mL increase in PTH concentration, and by 72% with PTH concentrations above 900 pg/mL (reference range, 150–300 pg/mL), [10], Furthermore, the United States Renal Data System

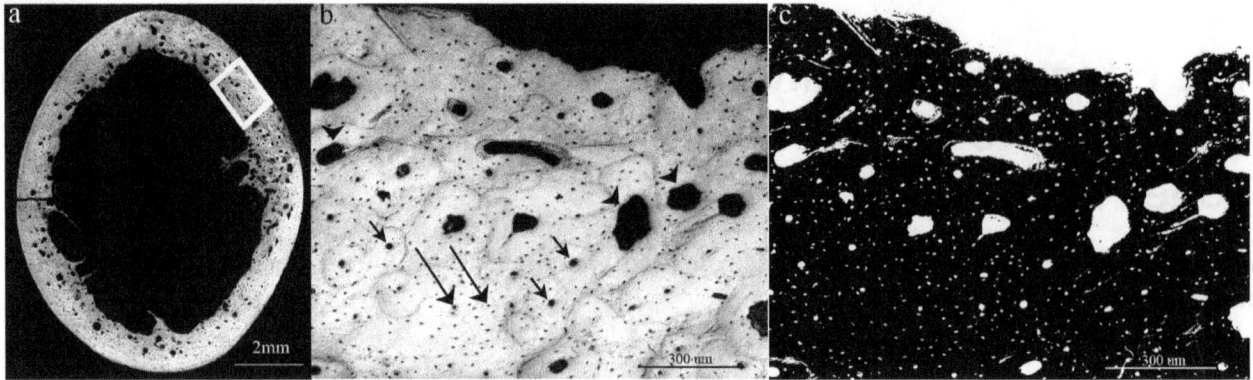

Figure 1. Bone cavity analysis. (a) Light microscopy image of a typical transverse cross section of the bone created by stitching together of many individual images. Classification of voids was performed on based on their size. (b) An individual image from the cross sectional image (marked by a white rectangle in image a). (c) Each image was first binarized, separating it into 'bone' (white) and 'void' (black) (right image). Cavities within the range of 9–50 μm^2 were considered to be lacunae (long arrows), cavities within the range of 151–2000 μm^2 were considered to be Haversian canal (short arrows) and cavities larger than 2000 μm^2 were considered resorptive lesions (arrowheads).

identified a 4-fold greater risk of hip fractures in human dialysis patients as compared to the general population [11].

The precise structural and compositional changes in the skeleton that occur in CKD patients, are not entirely clear. Data regarding the nature of these changes is crucial to the understanding of the skeletal consequences of CKD, since they determine the quality of bone material and the quantitative deterioration of material's mechanical properties. Most previous studies were based on data collected from human patients or from rodent models. For obvious reasons, studies conducted on human patients are subjected to severe inherent limitations, primarily the need to rely on non-invasive or minimally invasive (biopsy) methods. Biopsies, while a valuable diagnostic tool, by their nature provide data relevant to a very small region in a bone, and therefore can provide only limited information. Noninvasive methods used in human patient studies, such as determination of bone mass (or apparent mineral density) by dual x-ray absorptiometry (DEXA) provide imprecise information because of technical limitations, as described eloquently by Parfitt [14]. The main limitations of DEXA include its reliance on a 2-D proxy of mineral density (g/cm^2) measurement rather than true 3-D density, and the inability to obtain separate data for cortical and trabecular bone. Peripheral quantitative computed tomography (pQCT) provides true 3-D information and is therefore a valuable tool, however its resolution is in the mm range.

On the other hand, use of model animals (almost exclusively mice and rats), while allowing the use of a wide array of testing methods, is hindered by the fact that in most studies, CKD is induced by non-physiologic means, mostly partial nephrectomy [15]. This obviously does not mimic with precision the disease in human patients, and may affect the skeleton in ways which are subtly (or even substantially) different from those caused by the natural course of the disease in humans. Moreover, the structure and architecture of rat and mouse cortical bone differs greatly from that of human cortical bone, as shown recently by Shipov *et al* and Bach-Gansmo *et al* [16,17]. Therefore, the ability to directly extend the observed effects of artificially-induced CKD in the rat skeleton to the effects of naturally-occurring CKD in the human skeleton is limited.

The course, pathology and pathophysiology, diagnosis and treatment of feline CKD mirror those of the human disease very closely, and the disease is very prevalent in the feline population

[18]. Another major advantage of studying the effects of CKD in cats is that the bones of mature cats are structurally and compositionally very similar to those of humans, both consisting mostly of remodeled secondary osteons [19].

Here we present a detailed study of the skeletal changes, both structural and mechanical, in cats with naturally occurring CKD. In this study we compare the femora and vertebra of cats diagnosed with CKD and those of age-matched cats with normally functioning kidneys.

Materials and Methods

2.1 Animals and data collection

The study was prospective, based on the patient population of the Veterinary Teaching Hospital of the Hebrew University of Jerusalem, and was approved by the institutional animal care and use committee. Cats considered for the study either died or were euthanized at their owners' request after medical management had failed. Euthanasia was performed using 200 mg/kg pentobarbital (CTS chemical industries LTD, Israel) administered intravenously. Cats were enrolled only after their owners had signed an informed consent form and donated the body to science. The study group consisted of 13 cats diagnosed with Stage III or IV CKD, based on the classification scheme of the International Renal Interest Society guidelines [20], for at least 6 months prior to death or euthanasia. These criteria included documentation of persistent azotemia (3 occasions, at least 2 weeks interval, serum creatinine concentration >2.8 mg/dL), urine specific gravity <1.020 and ultrasonographic changes consistent with CKD. CKD was additionally confirmed in all cats by histopathological examination showing moderate to severe interstitial nephritis accompanied by moderate to severe fibrosis.

The control group included 13 healthy cats without any clinicopathologic signs of CKD (e.g., normal creatinine, concentrated urine) that died or were euthanized in the Veterinary Medical Teaching Hospital due to reasons unrelated to diseases of the urinary system. Cats with concurrent metabolic diseases that could potentially affect the skeleton were excluded, as were cats that were treated for more than 2 weeks during the 6 months prior to their death with medications that could alter bone metabolism (e.g., vitamin D derivatives, corticosteroids).

Table 1. Morphometric characteristics of cortical bone of the distal femur in CKD and healthy controls by light microscopy.

Light microscopy morphometry	CKD (mean ±SD)	Controls (mean ±SD)	P value
Oseteocytic lacunae			
Size [μm^2]	33.9±3.1	33.1±3.9	0.60
Density [mm^{-2}]	510±55	524±106	0.70
Haversian canals			
Size [μm^2]	481±115	411±48	0.10
Density [mm^{-2}]	22.3±4.3	23.3±4.6	0.58
Resorption cavities			
Size [μm^2]	10,342±11,888	12,406±25,258	0.78
Density [mm^{-2}]	2.2±2.4	0.4±0.1	**0.04**

2.2 Sample collection and preparation

Blood and serum samples from all cats were collected *antemortem* for complete blood count and biochemical analysis. Sera were stored at −80°C for determination of PTH and vitamin D levels.

Tissue sample collection was performed within 12 hours of death. The right femora and lumbar vertebrae were carefully removed and cleaned of all soft tissue, wrapped in saline-soaked gauze, placed in a sealed plastic bag and stored at −20°C until testing. Kidney samples were harvested and stored in 10% formalin for histologic evaluation.

2.3 Light microscopy

Thin transverse slices (400 microns thick) of the mid-diaphyseal region of all right femora were cut by a water-cooled slow-speed diamond saw (Buehler Isomet low Speed saw, USA). The slices were then polished by increasingly fine grit (Buehlet Minimet Polisher, USA), from 320 grit to 1 μm diamond paste. Transverse cross-sections of all cortical samples were viewed by reflective light microscopy (Olympus BX-51) and their detailed architecture characterized by analysis of images captured by a dedicated high-resolution camera (Olympus DP 71, 12 MegaPixels).

Quantitative analysis of the transverse cross-sectional images, particularly quantification of voids and their classification, was performed with a public domain image processing software (ImageJ, NIH, v. 1.44p). Several microstructural parameters were measured, such as the number, size and density of the osteocytic lacunae, Haversian canals and resorption cavities within each cross section [21]. Specifically, each cross-sectional image was first binarized by selection of an appropriate threshold, separating it into 'bone' (white) and 'void' (black) entities [22], (Fig. 1 a, b). Next, the ImagJ 'analyze particles' command was applied to each cross-section to identify all individual voids within it. This command analyzes each void by its size and reports the results in a tabular form (see Figure 1) [23]. Based on sizes of osteocytic

lacuna and Haversian canals reported in the literature [24–27] voids with an area in the range of 9–150 μm^2 were considered to be osteocytic lacunae, voids with an area in the range of 151–2000 μm^2 were considered to be Haversian canals, while voids larger than 2000 μm^2 were considered resorption cavities which are in the process of remodeling [21]. Voids smaller than 9 μm^2 were considered to be artifacts. Images were also visually examined by two of the authors (AS and HM) and the results of thresholding and void categorization were manually corrected if indicated. Overall porosity was calculated as the ratio of total void area (i.e. resorption cavities, lacunae and blood vessels) to total bone area.

2.4 Mechanical testing

Mechanical properties of cortical bone were evaluated using four-point bending tests performed on bone beams prepared from the cranial aspect of the mid-diaphyseal cortical region of the right femora. Beam sizes were 20 mm×1.5 mm×1 mm (long dimension along the bone axis).

Mechanical testing was performed with the samples immersed in saline, using a custom-built micromechanical-testing device as previously described [28]. All samples were thawed immediately before testing for one hour at room temperature. The beams were placed within a saline-containing testing chamber that had a stationary anvil attached to its wall [28]. This anvil consisted of two supports which were 15 mm apart. A movable double-pronged loading anvil was attached to a load-cell (model 31, Honeywell Sensotec, Colombus, OH, USA), which was in turn attached to a high-precision linear motor (PI GmbH, Karlsruhe, Germany). The loading anvil had a span of 5 mm between its two prongs, which were centered between the two supports of the stationary anvil.

The upper prongs were brought into contact with the tested beams at a predetermined preload (2N), the chamber was filled with physiologic saline solution at room temperature until the

Table 2. Morphometry of cortical bone of the mid-diaphyseal femur in CKD and healthy controls.

Microtomography	CKD (mean ±SD)	Controls (mean ±SD)	P value
Cortical cross sectional area (mm^2)	28.5±3.6	32.2±5.5	**0.04**
Cross-sectional thickness (mm)	1.1±0.2	1.2±0.2	**0.03**
Mean polar area of inertia (μm^4)	432±95	520 ±154	0.08

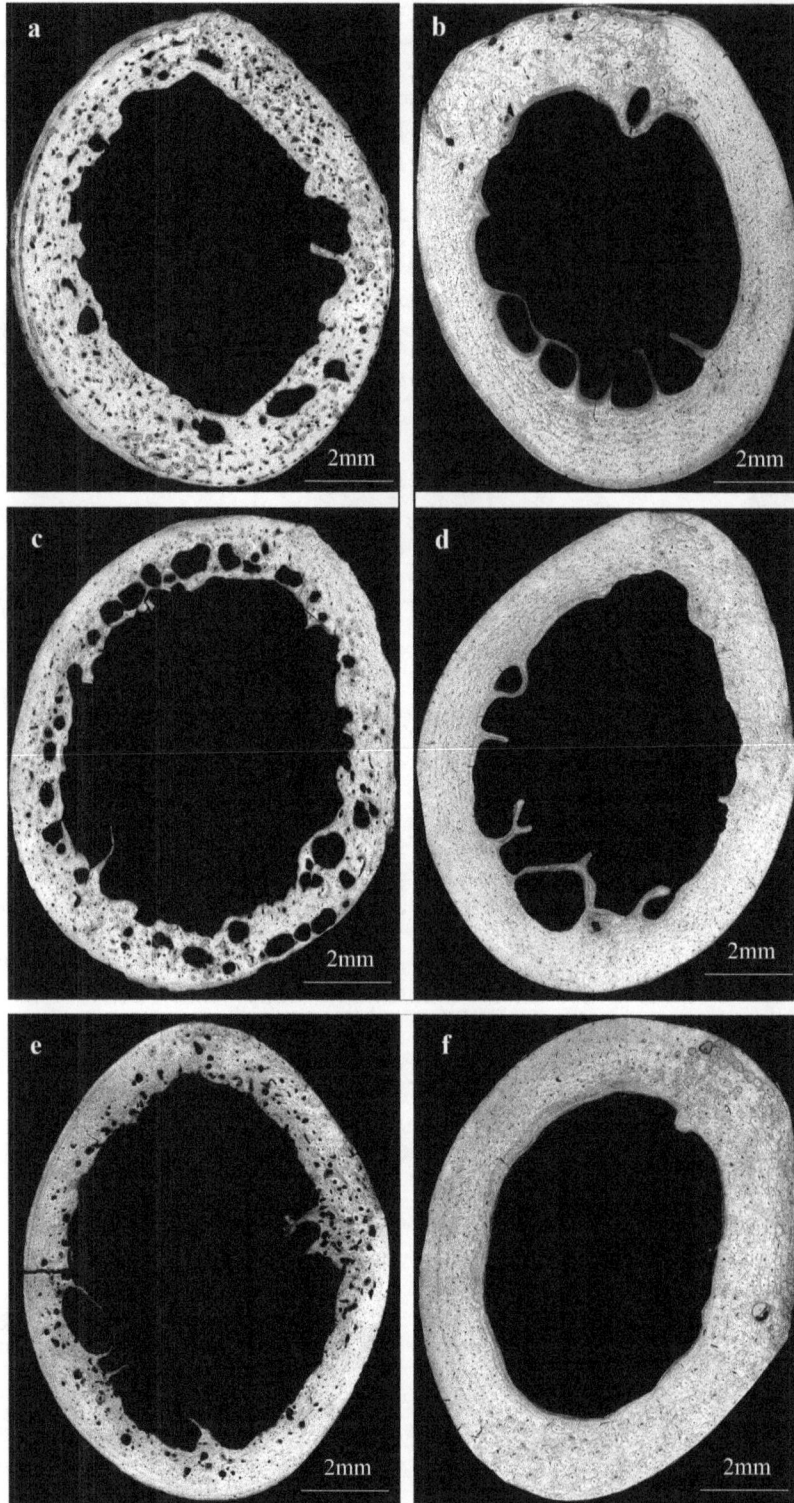

Figure 2. Light microscopy images of three transverse cross-sections of the femoral mid-diaphysis of (a, c, e) cats with CKD and (b, d, f) healthy cats. Note dramatic increase in unfilled resorption cavities in CKD cats compared to the healthy cats.

samples were fully immersed, and bending tests were conducted under displacement control at a rate of 500 μm/180 seconds up to failure. Force-displacement data were collected by custom-written software (LabView, National Instruments, Texas, USA) at 50 Hz.

Load and displacement values were converted to stress and strain, respectively, based on beam theory [29]. The stress-strain curves were used to estimate Young's modulus of the beam material, as well as yield and failure stresses and strains. It should be noted that

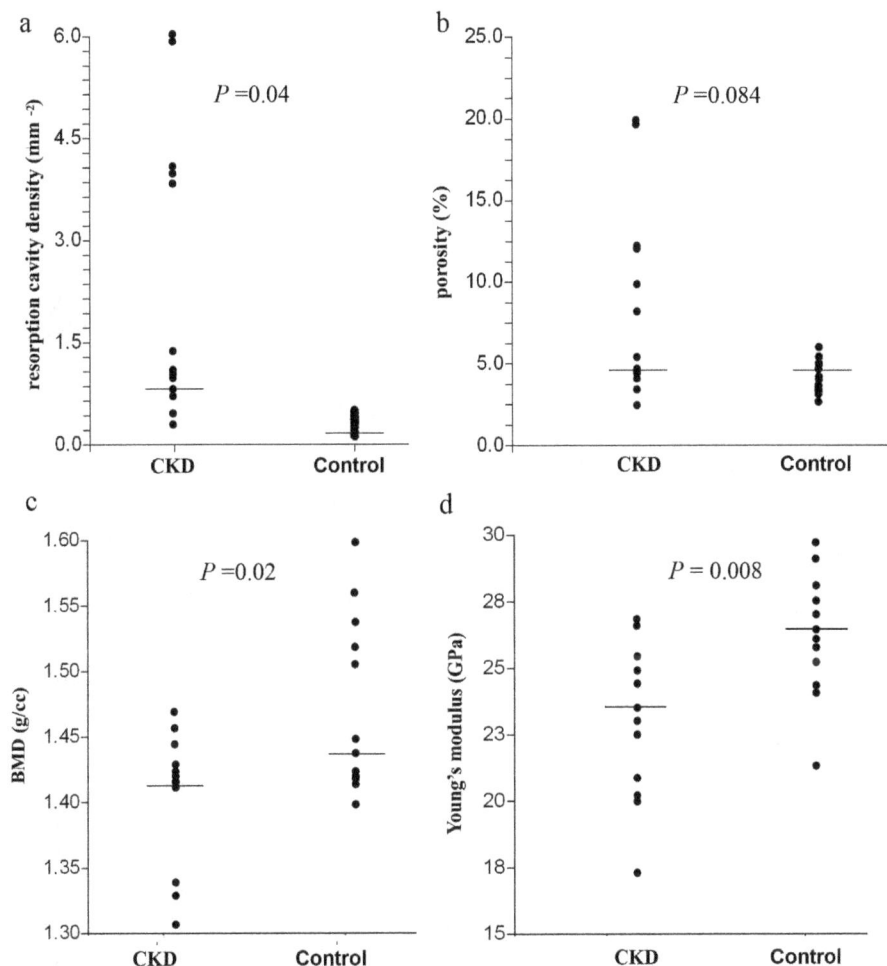

Figure 3. Dot plots depicting the data of CKD and control cats; the horizontal line represents the median. (a) resorption cavity density (b) porosity, (c) BMD and (d) Young's modulus.

care was taken to minimize shear deformation at the supports by maintaining a ratio of distance between supports/beam depth of 15:1 [30,31]. Yield point was determined for each beam as the point at which a line parallel with the linear portion of the stress-strain curve and offset by 0.03% strain intersected with this curve [32].

2.5 Microstructural characterization by Micro-CT

All cortical beams, the right femur and the 6th and 7th lumbar vertebrae were scanned by microCT (Skyscan 1174 compact micro-CT scanner, Belgium), with the beams scanned prior to mechanical testing. Analyses were performed on the entire beam, the mid-diaphyseal femoral cortex (cortical bone analysis), and in the distal femoral metaphyses and vertebral bodies (cancellous bone analysis).

The X-ray source was set at 50 kVp and 800 µA. A total of 450 projections were acquired over an angular range of 180°. The samples were scanned with an isotropic voxel size of 11.1 µm for the cortical bone beams and 19.6 µm for the femoral cortex and cancellous bone of both the femora and vertebrae. Integration time for all scans was 4500 ms, and a 0.25 mm aluminum filter was used. Scans were reconstructed and analyzed using commercial software (NRecon Skyscan software, version 1.6.1.2 and CT

analyser Skyscan software, version 1.9.3.2, respectively). Cortical bone mineral density (BMD) of the beams was determined based on calibration with 2 phantoms of known mineral density (0.25 g/cm^3 and 0.75 g/cm^3) supplied by SkyScan, which were scanned under exactly the same condition as were the bone specimens.

2.6 Statistical analysis

The distribution of continuous parameters (normal vs. non-normal) was assessed using the Shapiro-Wilk's test. Normally and non-normally distributed continuous parameters were compared between the study and the control group using Student's t-test and Mann-Whitney U test, respectively. Gender proportion between the study group and the control groups was compared using the Fischer Exact test. Correlations between continuous parameters (e.g., biomechanical parameters and PTH concentration) were performed using the Pearson or the Spearman Rank correlation test, according to data distribution. For all tests $P<0.05$ was considered statistically significant. All calculations were performed using a statistical software (SPSS 17.0 for Windows, SPSS Inc; Chicago, IL, USA).

Results

3.1 Animals

The study population included 26 cats, of which 13 were diagnosed with CKD and 13 were healthy controls. There were eight males and five females in the study group and six males and seven females in the control group, with no gender proportion differences between the study groups. Mean body weight was significantly lower in cats with CKD compared with healthy controls (2.8 ± 0.6 kg vs. 3.7 ± 0.9 kg, respectively; $P = 0.01$). There was no statistically significant difference in mean age between the study and control groups (10.5 ± 5.6 years compared to 9.7 ± 3.9 years, respectively; $P = 0.7$).

3.2 Clinical pathology

Median serum creatinine concentration within the CKD group was 8.2 mg/dL (range 3.5–16.0 mg/dL) compared with a median of 0.9 mg/dL (range 0.6–1.2 mg/dL) [reference interval (RI), 0.5–1.6 mg/dL] of the control group. Three cats in the study group (23%) were classified as Stage III CKD, and the rest (77%) were classified as Stage IV CKD. Median phosphorous concentration in the study group was 8.8 mg/dL (range 5.3–21.7 mg/dL; RI, 3.0–6.2 mg/dL). Median concentration of ionized calcium in the study group was 0.80 mmol/L (range 0.65–1.01 mmol/L; RI, 0.9–1.4 mmol/L). PTH concentration, available for five cats of the study group, had a median concentration of 15.70 pmol/L (range, 0.9–32.9 pmol/L; RI, 0.4–2.5 pmol/L). Vitamin D concentration was below normal in five out of the seven cats in which it was measured (median 63 nmol/L, range 35–143 nmol/L; RI 65–170 nmol/L).

3.3 Cortical bone architecture

The results of the architectural analysis are presented in Table 1 (microscopy) and Table 2 (microCT). CKD-affected cats had significantly higher density of resorption cavities compared to healthy controls (Table 1, Figures 2, 3a). Other structural parameters were not significantly different between the groups. Porosity tended to be higher in the CKD group, however the difference between groups did not reach statistical significance ($P = 0.084$, Figure 3b).

Micro-CT analysis of cortical bone of the femoral diaphysis showed significantly lower cortical cross-sectional area and cross-sectional thickness in CKD cats (Table 2). Mean polar area moment of inertia tended to be lower in the CKD group, but the differences did not reach statistical significance ($P = 0.083$) (Table 2).

3.4 Cortical mineral density

Cortical mineral density of CKD cats was lower by 4.8% compared to controls, ($P = 0.02$, as shown in Figure 3c).

3.5 Mechanical properties of cortical bone

Table 3 and Figure 3 present a comparison of several mechanical properties of cortical bone material between the CKD and control groups. Bones from the CKD group were shown to have inferior mechanical properties compared to the control group, in particular lower stiffness (Young's modulus), lower yield stress, and lower ultimate stress (Table 3, Figure 3).

Correlation could not be demonstrated between PTH levels and any of the mechanical properties of the cortical bone.

3.6 Architecture of trabecular bone

Analysis of cancellous bone in the 6th and 7th lumbar vertebrae and in the distal femur revealed significantly lower trabecular thickness and bone volume over total volume (BV/TV) in bones belonging to the CKD group, compared with control cats (Figure 4).

Discussion

This study demonstrates that advanced CKD in cats results in deterioration of bone quality, in particular a dramatic increase of resorption cavities and decreased bone mineral density. These results provide insight into skeletal changes occurring in human CKD due to the similarity between cats and humans in terms of the pathophysiology of the syndrome and the type of bone architecture.

To the best of our knowledge, this is the first study to measure the detailed mechanical and structural effects of CKD on the skeleton of an animal model with *naturally occurring* CKD. Feline and human CKD have very similar clinicopathologic features and progression, therefore cats are superior models compared to rodents in which pharmaceutical and surgical interventions are usually employed to induce kidney disease [33–36]. Furthermore, the cat skeleton shows great similarities to human bone, in particular because it remodels continuously throughout life and therefore consists mostly of secondary osteons. Rodent bones on the other hand are remarkably different from those of human bones in terms of type, architecture, structure and biology. Most dramatically, rodent cortical bone does not remodel [16,17]. Therefore similar clinicopathologic circumstances, such as those occurring due to CKD, may affect rodent bones differently from human (or cat) bones.

We found that naturally occurring CKD results in several alterations to the architecture and morphology of the bones of the skeleton. Cortical thickness was found to be decreased by approximately 17% in CKD cats compared with controls. This change compromises the mechanical performance of long bones, by reducing their flexural stiffness and is likely to lead to increased fracture risk. Previous studies conducted in human patients with CKD showed similar tendencies, however with much smaller changes. For instance, a recent study documented 4.2% increase

Table 3. Mechanical properties of the cortical bone of CKD and control groups.

Parameter	CKD (mean ±SD)	Controls (mean ±SD)	P value
Young's modulus (GPa)	23.5±2.9	27.1±2.8	0.01
Yield stress (MPa)	151±18	166±14	0.04
Yield strain (millistrain)	5.6±0.8	5.6±0.4	.95
Ultimate stress (MPa)	185±16	205±15	0.01
Ultimate strain (millistrain)	8.6±1.0	8.1±0.4	0.37

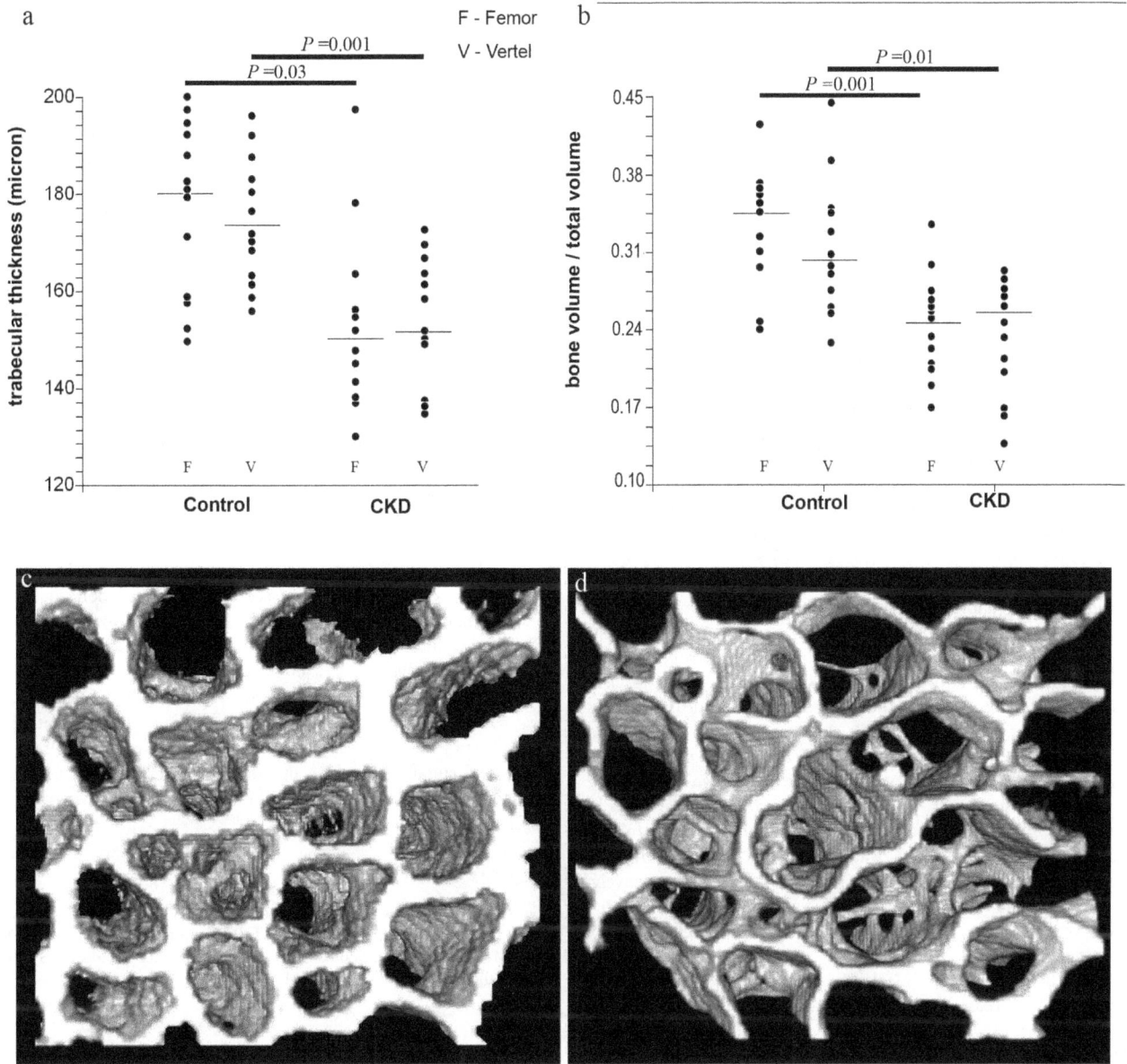

Figure 4. Trabecular bone analysis: Trabecular thickness (a) and bone volume/total volume (BV/TV, b) in the cancellous bone of CKD and control cats for both the femora (F) and the vertebrae (V). Data are presented as dot plots. The horizontal line represents the median. Data from the femur and vertebra are similar within the study groups; however, there is a significant difference for both parameters between the study and the control groups. Micro-CT scans of cancellous bone of a control (c) and CKD (d) cat.

in porosity, 2.9% decrease in cortical area and 2.8% decrease in cortical thickness, indicating progressive loss of cortical bone [37]. These results, like those of other human studies, were based on DEXA and high resolution peripheral quantitative CT. These methods have been shown to be limited in precision in terms of bone volume quantification due to inability to separate cortical from cancellous bone (DEXA), and low-resolution volumetric measurements, compared to microCT and whole bone sampling [38].

Overall cortical porosity in cats with advanced CKD tended to be somewhat higher compared with controls (Figure 3b), but this difference did not reach statistical significance, most likely due to small group size and biological variation. However, the density (number per unit area) of resorption cavities in CKD patients is greatly increased (5-fold, $2.22/mm^2$ vs. $0.41/mm^2$, $P = 0.04$). Such a difference is expected to affect mechanical behavior of long bones radically, and is likely to play an important role in the increased fragility of CKD patients. Previous studies in a rat model demonstrated that persistently elevated PTH concentrations result in high bone turnover, exhibited as elevated numbers and size of osteoclasts, increased osteoblastic activity and enhanced bone resorption [33]. Consequently, these studies found extensive endocortical, intracortical and periosteal resorption, resulting in a dramatic increase in cortical porosity (9.75% compared to 0%). It should be noted however that these results were observed in rat bone, which normally does not remodel, as oppose to cat (and human) bone.

Bone mineral density is another major determinant of bone quality, and the current clinical standard for prediction of fracture risk in osteoporotic patients [39]. Therefore, determining the influence of CKD on BMD was a major objective of this study. The cortical mineral density of cats with CKD was significantly lower compared to controls. It should be noted that despite appearing small, this decrease (4.8%) is clinically significant, as even a small decrease in BMD substantially decreases the stiffness of the bone and increases fracture risk [40], since the relationship between them is exponential [41]. A decrease in BMD was also reported in various studies in humans CKD patients, with a wide range of values [1.3% to 17.5%; [37,42,43]]. However, these studies were based on areal BMD (g/cm^2, using DEXA), histomorphometry or pQCT, while the current study measured volumetric BMD at high resolution and precision using microCT [38,44].

Reliable measurement of material properties requires precise and accurate mechanical testing. Such testing is difficult to achieve in rodents due to the small size of their bones, which are often tested by bending tests of whole bones, using the 3-point bending technique. Results are dependent upon the geometry of the bones and mechanical properties of the material, often leading to underestimation of Young's modulus [45,46]. The size of cat bones allowed us to prepare cortical bone beams, enabling accurate and reliable assessment of the material properties using four-point bending testing. Furthermore, four-point bending tests of beams allowed us also to measure other material properties, like yield and strength, and to document significant decrease in yield stress and failure stress in the cortical bone of the CKD cats compared to controls.

This study demonstrates that the cortical bone material of CKD cats is less stiff and more prone to micro-damage which occurs at lower loads, and will result in increased bone fragility. In particular, Young's modulus of the cortical bone, which reflects the stiffness of the material, was shown here to be significantly lower in the CKD group (23.5 GPa vs 27.1 GPa, respectively, $P = 0.008$). The lower Young's modulus in CKD cats compared with controls is most likely due to a combination of higher porosity and lower BMD.

Cancellous bone is also significantly affected by CKD, as reflected by decreased trabecular thickness and lower bone volume (BV/TV). Furthermore, the effect on cancellous bone was multi-site, shown to occur both in the vertebral bodies as well as in the long bones. Both changes (trabecular thickness and BV/TV) negatively affect bone quality and were shown to be associated with increased risk for fracture [47].

Since this study was based on cats with naturally occurring kidney disease, some variability existed among them in terms of the severity and the chronicity of the disease. Furthermore, the number of cats included in this study was relatively small, though comparable to numbers typically seen in published rodent model studies. It should also be noted that cats with advanced CKD often exhibit decreased appetite and thus might fail to consume enough food to meet their caloric requirements. Such nutritional deficiencies, as in human patients, might contribute to their decreased bone quality. Additionally, all of the cats in this study had a single etiology (interstitial nephritis, which is the etiology in more than 70% of cases with CKD in cats [18]), whereas human patients with CKD have multiple etiologies (e.g., diabetic nephropathy, transplant patients, etc.). Nevertheless, slowly progressive natural disease represents the human syndrome and its consequences, including renal osteopathy, much better than artificially-induced rodent models.

In conclusion, the current study demonstrates the deleterious effects of CKD on remodeled (secondary osteonal) bone quality and strength, including increased bone resorption, decreased BMD, and inferior mechanical and structural properties. Due to these changes, the bones of CKD patients become more fragile. Since many similarities have been demonstrated between human and feline CKD patients, in terms of the clinic-pathologic features of the syndrome, as well as bone-associated effects, cats are an extremely suitable and relevant animal model for studying the development of bone abnormalities in humans suffering from CKD.

Author Contributions

Conceived and designed the experiments: AS GS HM MM RS. Performed the experiments: AS GS HM MM AA OB RS. Analyzed the data: AS GS HM MM AA RS. Contributed reagents/materials/analysis tools: RS OB. Wrote the paper: AS HM GS MM OB AA RS.

References

1. Lora CM, Daviglus ML, Kusek JW, Porter A, Ricardo AC, et al. (2009) Chronic kidney disease in United States Hispanics: a growing public health problem. Ethn Dis 19: 466–472.

2. McClellan WM, Plantinga LC (2013) A public health perspective on CKD and obesity. Nephrol Dial Transplant 28 Suppl 4: iv37–iv42.

3. Spasovski GB, Bervoets AR, Behets GJ, Ivanovski N, Sikole A, et al. (2003) Spectrum of renal bone disease in end-stage renal failure patients not yet on dialysis. Nephrol Dial Transplant 18: 1159–1166.

4. Levin A, Bakris GL, Molitch M, Smulders M, Tian J, et al. (2007) Prevalence of abnormal serum vitamin D, PTH, calcium, and phosphorus in patients with chronic kidney disease: results of the study to evaluate early kidney disease. Kidney Int 71: 31–38.

5. Cozzolino M, Ciceri P, Volpi EM, Olivi L, Messa PG (2009) Pathophysiology of calcium and phosphate metabolism impairment in chronic kidney disease. Blood Purif 27: 338–344.

6. Nabeshima Y (2008) The discovery of alpha-Klotho and FGF23 unveiled new insight into calcium and phosphate homeostasis. Cell Mol Life Sci 65: 3218–3230.

7. Saito H, Kusano K, Kinosaki M, Ito H, Hirata M, et al. (2003) Human fibroblast growth factor-23 mutants suppress Na+ -dependent phosphate co-transport activity and 1 alpha,25-dihydroxyvitamin D-3 production. J Biol Chem 278: 2206–2211.

8. Shimada T, Hasegawa H, Yamazaki Y, Muto T, Hino R, et al. (2004) FGF-23 is a potent regulator of vitamin D metabolism and phosphate homeostasis. J Bone Miner Res 19: 429–435.

9. Ketteler M, Biggar PH, Liangos O (2013) FGF23 antagonism: the thin line between adaptation and maladaptation in chronic kidney disease. Nephrol Dial Transplant 28: 821–825.

10. Danese MD, Kim J, Doan QV, Dylan M, Griffiths R, et al. (2006) PTH and the risks for hip, vertebral, and pelvic fractures among patients on dialysis. Am J Kidney Dis 47: 149–156.

11. Alem AM, Sherrard DJ, Gillen DL, Weiss NS, Beresford SA, et al. (2000) Increased risk of hip fracture among patients with end-stage renal disease. Kidney Int 58: 396–399.

12. Coco M, Rush H (2000) Increased incidence of hip fractures in dialysis patients with low serum parathyroid hormone. Am J Kidney Dis 36: 1115–1121.

13. Stehman-Breen CO, Sherrard DJ, Alem AM, Gillen DL, Heckbert SR, et al. (2000) Risk factors for hip fracture among patients with end-stage renal disease. Kidney Int 58: 2200–2205.

14. Parfitt AM (1998) A structural approach to renal bone disease. J Bone Miner Res 13: 1213–1220.

15. Iwasaki Y, Kazama JJ, Yamato H, Shimoda H, Fukagawa M (2013) Accumulated uremic toxins attenuate bone mechanical properties in rats with chronic kidney disease. Bone 57: 477–483.

16. Bach-Gansmo FL, Irvine SC, Bruel A, Thomsen JS, Birkedal H (2013) Calcified Cartilage Islands in Rat Cortical Bone. Calcif Tissue Int 92: 330–338.

17. Shipov A, Zaslansky P, Riesemeier H, Segev G, Atkins A, et al. (2013) Unremodeled endochondral bone is a major architectural component of the cortical bone of the rat (Rattus norvegicus). J Struct Biol 183: 132–140.

18. Polzin DJ (2010) Chronic kidney disease. In: Ettinger SJ, Feldman EC, editors.Textbook of Veterinary Internal Medicine. 7th ed. St. Louis: Saunders. pp. 1990–2020.

19. Hillier ML, Bell LS (2007) Differentiating human bone from animal bone: A review of histological methods. J Forensic Sci 52: 249–263.
20. Segev G, Palm C, LeRoy B, Cowgill LD, Westropp JL (2013) Evaluation of neutrophil gelatinase-associated lipocalin as a marker of kidney injury in dogs. J Vet Intern Med 27: 1362–1367.
21. Zebaze RM, Ghasem-Zadeh A, Bohte A, Iuliano-Burns S, Mirams M, et al. (2010) Intracortical remodelling and porosity in the distal radius and post-mortem femurs of women: a cross-sectional study. Lancet 375: 1729–1736.
22. Bousson V, Meunier A, Bergot C, Vicaut E, Rocha MA, et al. (2001) Distribution of intracortical porosity in human midfemoral cortex by age and gender. J Bone Miner Res 16: 1308–1317.
23. Montanari S, Brusatte SL, De Wolf W, Norell MA (2011) Variation of osteocyte lacunae size within the tetrapod skeleton: implications for palaeogenomics. Biol Lett 7: 751–754.
24. Urbanova P, Novotny V (2005) Distinguishing between human and non-human bones: Histometric method for forensic anthropology. Anthropologie 43: 77–85.
25. Currey JD, Shahar R (2013) Cavities in the compact bone in tetrapods and fish and their effect on mechanical properties. J Struct Biol 183: 107–122.
26. Kuchler U, Pfingstner G, Busenlechner D, Dobsak T, Reich K, et al. (2013) Osteocyte lacunar density and area in newly formed bone of the augmented sinus. Clin Oral Implants Res 24: 285–289.
27. Tazawa K, Hoshi K, Kawamoto S, Tanaka M, Ejiri S, et al. (2004) Osteocytic osteolysis observed in rats to which parathyroid hormone was continuously administered. J Bone Miner Metab 22: 524–529.
28. Cohen L, Dean M, Shipov A, Atkins A, Monsonego-Ornan E, et al. (2012) Comparison of structural, architectural and mechanical aspects of cellular and acellular bone in two teleost fish. J Exp Biol 215: 1983–1993.
29. Sharir A, Barak MM, Shahar R (2008) Whole bone mechanics and mechanical testing. Vet J 177: 8–17.
30. Spatz HC, O'Leary EJ, Vincent JF (1996) Young's moduli and shear moduli in cortical bone. Proc Biol Sci 263: 287–294.
31. Draper ER, Goodship AE (2003) A novel technique for four-point bending of small bone samples with semi-automatic analysis. J Biomech 36: 1497–1502.
32. Turner CH, Burr DB (1993) Basic biomechanical measurements of bone: a tutorial. Bone 14: 595–608.
33. Miller MA, Chin J, Miller SC, Fox J (1998) Disparate effects of mild, moderate, and severe secondary hyperparathyroidism on cancellous and cortical bone in rats with chronic renal insufficiency. Bone 23: 257–266.
34. Cao HH, Nazarian A, Ackerman JL, Snyder BD, Rosenberg AE, et al. (2010) Quantitative P-31 NMR spectroscopy and H-1 MRI measurements of bone mineral and matrix density differentiate metabolic bone diseases in rat models. Bone 46: 1582–1590.
35. Iwasaki Y, Kazama JJ, Yamato H, Fukagawa M (2011) Changes in chemical composition of cortical bone associated with bone fragility in rat model with chronic kidney disease. Bone 48: 1260–1267.
36. Jokihaara J, Jarvinen TLN, Jolma P, Koobi P, Kalliovalkama J, et al. (2006) Renal insufficiency-induced bone loss is associated with an increase in bone size and preservation of strength in rat proximal femur. Bone 39: 353–360.
37. Nickolas TL, Stein EM, Dworakowski E, Nishiyama KK, Komandah-Kosseh M, et al. (2013) Rapid cortical bone loss in patients with chronic kidney disease. J Bone Miner Res 28: 1811–1820.
38. Barou O, Valentin D, Vico L, Tirode C, Barbier A, et al. (2002) High-resolution three-dimensional micro-computed tomography detects bone loss and changes in trabecular architecture early: comparison with DEXA and bone histomorphometry in a rat model of disuse osteoporosis. Invest Radiol 37: 40–46.
39. Nickolas TL, Leonard MB, Shane E (2008) Chronic kidney disease and bone fracture: a growing concern. Kidney Int 74: 721–731.
40. Yenchek RH, Ix JH, Shlipak MG, Bauer DC, Rianon NJ, et al. (2012) Bone mineral density and fracture risk in older individuals with CKD. Clin J Am Soc Nephrol 7: 1130–1136.
41. Wasnich RD, Ross PD, Davis JW, Vogel JM (1989) A Comparison of Single and Multi-Site Bmc Measurements for Assessment of Spine Fracture Probability. J Nucl Med 30: 1166–1171.
42. Balon BP, Hojs R, Zavratnik A, Kos M (2002) Bone mineral density in patients beginning hemodialysis treatment. Am J Nephrol 22: 14–17.
43. Rix M, Andreassen H, Eskildsen P, Langdahl B, Olgaard K (1999) Bone mineral density and biochemical markers of bone turnover in patients with predialysis chronic renal failure. Kidney Int 56: 1084–1093.
44. Leonard MB (2009) A Structural Approach to Skeletal Fragility in Chronic Kidney Disease. Semin Nephrol 29: 133–143.
45. van Lenthe GH, Voide R, Boyd SK, Muller R (2008) Tissue modulus calculated from beam theory is biased by bone size and geometry: implications for the use of three-point bending tests to determine bone tissue modulus. Bone 43: 717–723.
46. Torcasio A, Van Oosterwyck H, van Lenthe GH (2008) The systematic errors in tissue modulus of murine bones when estimated from three-point bending. J Biomech 41: S14.
47. Nickolas TL, Stein E, Cohen A, Thomas V, Staron RB, et al. (2010) Bone mass and microarchitecture in CKD patients with fracture. J Am Soc Nephrol 21: 1371–1380.

22

The Roles of P2Y$_2$ Purinergic Receptors in Osteoblasts and Mechanotransduction

Yanghui Xing, Yan Gu, James J. Bresnahan, Emmanuel M. Paul, Henry J. Donahue, Jun You*

Division of Musculoskeletal Sciences, Department of Orthopaedics and Rehabilitation, The Pennsylvania State University College of Medicine, Hershey, Pennsylvania, United States of America

Abstract

We previously demonstrated, using osteoblastic MC3T3-E1 cells, that P2Y$_2$ purinergic receptors are involved in osteoblast mechanotransduction. In this study, our objective was to further investigate, using a knockout mouse model, the roles of P2Y$_2$ receptors in bone mechanobiology. We first examined bone structure with micro-CT and measured bone mechanical properties with three point bending experiments in both wild type mice and P2Y$_2$ knockout mice. We found that bones from P2Y$_2$ knockout mice have significantly decreased bone volume, bone thickness, bone stiffness and bone ultimate breaking force at 17 week old age. In order to elucidate the mechanisms by which P2Y$_2$ receptors contribute to bone biology, we examined differentiation and mineralization of bone marrow cells from wild type and P2Y$_2$ knockout mice. We found that P2Y$_2$ receptor deficiency reduces the differentiation and mineralization of bone marrow cells. Next, we compared the response of primary osteoblasts, from both wild type and P2Y$_2$ knockout mice, to ATP and mechanical stimulation (oscillatory fluid flow), and found that osteoblasts from wild type mice have a stronger response, in terms of ERK1/2 phosphorylation, to both ATP and fluid flow, relative to P2Y$_2$ knockout mice. However, we did not detect any difference in ATP release in response to fluid flow between wild type and P2Y$_2$ knock out osteoblasts. Our findings suggest that P2Y$_2$ receptors play important roles in bone marrow cell differentiation and mineralization as well as in bone cell mechanotransduction, leading to an osteopenic phenotype in P2Y$_2$ knockout mice.

Editor: Damian Christopher Genetos, University of California Davis, United States of America

Funding: This work was supported by National Institutes of Health Grant AR054851 (to J.Y.). The funder had no role in study design, data collection and analysis, decision to publish, or preparation of the manuscript.

Competing Interests: The authors have declared that no competing interests exist.

* Email: jxy118@psu.edu

Introduction

It is well known that mechanical loading of bone results in a variety of biophysical signals that affect bone cell metabolism and differentiation [1–3]. Oscillatory fluid flow, one of these biophysical signals, has been demonstrated to be a potent stimulator to osteoblast activity, differentiation, extracellular matrix protein production, and gene expression [4–8]. Previously, we reported that oscillatory fluid flow induced MC3T3-E1 osteoblastic cell intracellular calcium mobilization via the inositol 1, 4, 5-trisphosphate pathway and increased steady state mRNA levels of osteopontin, a bone extracellular matrix protein [4]. Others have demonstrated that oscillatory fluid flow can enhance COX-2 protein levels, activate ERK1/2 phosphorylation pathways and cause the release of ATP into extracellular space [9–11]. However, the mechanisms by which oscillatory fluid flow initiates osteoblast mechanotransduction pathways are still unclear. Previous studies suggest that various membrane bound structures, such as channels, integrins, and primary cilium, are involved [12–14]. Additionally, our studies suggested that the G-protein coupled P2Y$_2$ purinergic receptors may also play an important role in osteoblast mechanotransduction [5].

Accumulating evidence suggests mechanical signals induce the release of nucleotides, such as ATP, from bone cells, which then activate nucleotide P2 purinergic receptors in adjacent cells to initiate cellular biological responses [15–17]. For example, fluid flow induces ATP release in vascular endothelial cells and smooth muscle cells, and ATP acts on P2 receptors to initiate cellular responses which ultimately regulate vascular tone and blood flow [18,19]. In addition, ATP released by chondrocytes in response to mechanical loading or inflammation contributes to cartilage destructing processes by activating signaling pathways involved in articular pathology, especially in the early stage of arthritic diseases [20]. More importantly, fluid flow induces ATP release in osteoblastic and osteocytic cells [10]. Recently our observations suggest the extracellular nucleotide ATP is involved in fluid flow induced intracellular calcium mobilization in MC3T3-E1 cells in that oscillatory fluid flow failed to increase intracellular calcium concentration in the presence of apyrase, an enzyme that hydrolyses ATP to AMP [4,5]. This suggests that extracellular nucleotide ATP and its receptors, P2 receptors, may be critical to initiating cellular responses to biophysical signals in osteoblastic cells.

Consistent with the concept that P2 receptors play an important role in bone remodeling, it has been reported that decreased cancellous and cortical bone mass are found in mice lacking the P2X$_7$ receptor [21]. However, our previous studies demonstrated that it is P2Y receptors, not P2X receptors, which are responsible for fluid flow induced intracellular calcium mobilization [5]. In addition, Bowler et al. reported that extracellular ATP, acting via the P2Y receptors, can potentiate the response of human osteoblasts to systemic growth and differentiation factors [22]. Moreover, Orriss et al. reported that ATP and UTP at low concentration strongly inhibit in vitro bone formation by osteoblasts via P2Y2 receptors. They also found 9–17% increases in bone mineral content of hindlimbs of P2Y2 deficient mice. However, the roles of P2Y2 receptor in bone mechanotransduction are still unclear.

In addition to P2Y receptor activation by ATP, Bagchi et al. and Yu et al. have reported that P2Y receptors interact with actin-binding proteins (filamin A) and integrins which are widely believed to play an important role in cell mechanotransduction and adhesion, suggesting that mechanical force may activate P2Y receptors through integrins [23,24]. Furthermore, $P2Y_2$ has been reported to play an important role in cell migration [25]. Take together this evidence suggests a pivotal role of P2Y receptors in bone cell mechanotransduction. Thus, we hypothesize that the $P2Y_2$ purinergic signaling pathway, activated by ATP release due to mechanical stimulation, is essential to regulation of bone cell response to fluid flow.

In this study, we employed $P2Y_2$ knockout (KO) mice models, to elucidate the role of $P2Y_2$ in bone. We first examined bone structure and measured bone mechanical properties in both wild type mice (WT) and $P2Y_2$ KO mice. Then, we tested the differentiation and mineralization of bone marrow cells. Next, we compared the response of primary osteoblasts from both WT and $P2Y_2$ KO mice to ATP stimulation and oscillatory fluid flow. Finally, ATP releases from both WT and $P2Y_2$ KO primary osteoblasts in response to oscillatory fluid flow were quantified. Our observations suggest that $P2Y_2$ receptors play an important role in bone metabolism.

Methods

Wild Type and Knockout Mice

$P2Y_2$ KO mice were generated as previously described [23] and were provided by Dr. Beverly H. Koller (University of North Carolina, Chapel Hill, NC). The genetic backgrounds of the male littermate WT and KO mice used in this study were SV129 (Taconic Farms). The full-time veterinary staff at Pennsylvanian State University College of Medicine took care of the experimental mice fed by regular rodent diet with a normal light-dark cycle. The mouse pups were routinely weaned at 21 days of age, and then small tissues (2–3 mm^2) from mouse ears were collected for genotyping following the genotype procedures provided by Taconic Farms. This study was carried out in strict accordance with the recommendations in the Guide for the Care and Use of Laboratory Animals of the National Institutes of Health. The protocol was approved by the Committee on the Ethics of Animal Experiments of the Pennsylvanian State University College of Medicine (IACUC No.: 42925). Due to different ages of mice in this study, only male mice were used for bone structure phenotype study to eliminate the effects of female hormones on bone.

Micro-CT analysis of bone structure

Femurs from the right side of 8 week and 17 week old mice were harvested for micro-CT analysis. The diaphyses were scanned starting at the midpoint of the bone and acquiring 76 slices distally using a Scanco vivaCT 40 (Scanco Medical AG) with scan settings of 55 KVp, 145 µA, and 200 ms integration time. Images were reconstructed as a matrix of 2048×2048×76 isotropic voxels measuring 10.5 µm. Images were gaussian filtered (sigma = 0.8, support = 1) and a threshold (24% of full scale) was applied to

remove the surrounding soft tissue. The periosteal and endosteal boundaries of the cortical bone were segmented using the Scanco semi-automated edge detection algorithm. Periosteal volume, endosteal volume, bone porosity, cortical bone thickness, and BMD were calculated for the diaphysis of each femur using the Scanco Image Processing Library routines.

Mechanical Testing for Bone Mechanical Properties

Femurs from 8 week and 17 week old mice were stored in PBS at −80°C before being mechanically tested to failure in three-point bending experiments using a Bose Electroforce Load Frame (EnduraTec MN, USA). The flexural support spans were 8 mm while a loading rate of 1 mm/minute was applied. Femurs were consistently oriented so that loading occurred in the medial to lateral direction. All testing was executed with the bones hydrated and at ambient temperature.

Cell Culture

Mice at age of 6 to 12 weeks were sacrificed and their femurs and tibias removed. Marrow cells were flushed out with a syringe needle system until the bones appeared white. Marrow cells were collected and cultured in cell differentiation medium (α–MEM with 10% FBS, 50 µg/ml L-Ascorbic Acid, 10 nM Dexamethasone, 10 mM Beta-Glycerol-phosphate, 1% P/S) in six well plates. At day 7, 14 and 21, cells were stained for alkaline phosphatase (AP) and extracellular deposited calcium. While bone marrow cells were widely employed as an ideal cell model to examine cell differentiation and mineralization, osteoblasts/osteocytes are major cells in bone to sense mechanical signals and maintain dynamic functions. Thus, primary osteoblastic cells were used to examine bone cell responses induced by mechanical loading. To isolate osteoblastic cells, bones were incubated with 0.01 percent collagenese for 70 minutes and then chopped into small chips with a size approximately 1–2 mm in diameter. Bone chips were then cultured with anti-contamination maintenance medium (α-MEM with 15%FBS, 1% Antibiotic-Antimycotic, 50 µg/ml L-Ascorbic Acid) for 5 days. The medium was switched to normal maintenance media for another 21 days. Cells migrating out of the chips were trypsinized and removed to another larger culture plate for further growth. Finally, cells were subcultured onto glass slides and subjected to oscillatory fluid flow.

Alkaline phosphatase staining and quantification

For AP staining, a commercially available kit (Sigma) was used. Cells were fixed and stained following manufactures instructions. For quantification, AP activity was determined by the colorimetric conversion of p-nitrophenol phosphate to p-nitrophenol (Sigma) and normalized to total protein (Pierce).

Calcium Staining and quantification

Extracellular calcium was identified using the von Kossa method. Briefly, cells were fixed with 4% formaldehyde, then incubated with 5% sliver nitrate for 20 minutes, and rinsed with distilled water three times. To quantify calcium, the o-cresolphthalein method was used following instructions from the calcium assay kit (Cayman Chemicals).

RT–PCR analysis

Cells were lysed and homogenized with a QIAshredder mini column (QIAGEN). Total RNA was extracted with Qiagen RNeasy mini kit. cDNA was prepared from 1 µg total RNA using the iScript Kit (Bio-rad). PCR amplification was performed in a 30 µl reaction with 1 µl cDNA reaction using Qiagen PCR kit according to the manufacturer's protocol. For comparison between the cells from WT and P2Y$_2$ KO mice, RT-PCR was performed and the products were analyzed by agarose gel electrophoresis. The primers for mouse osteocalcin were forward 5'-CAG GAG GGC AAT AAG GTA GT-3' and reverse 5'-GAG GAC AGG GAG GAT CAA G-3'. The primers for mouse β-actin as controls were forward 5'-AGA GGG AAA TCG TGC GTG AC-3' and reverse 5'-CAA GAA GGA AGG CTG GAA AA-3'.

Fluid Flow Experiments

To expose cells to oscillatory fluid flow, slides were positioned in parallel plate flow chambers and connected to a servopneumatic materials testing device (EnduraTec), oscillating at 1Hz, via glass Hamilton syringes and rigid wall tubing. Flow rate was monitored in real time with an ultrasonic flowmeter (Transonic Systems). We utilized a flow regime that facilitated the oscillatory movement of a defined volume of fluid across the cell monolayer. Oscillatory fluid flow mimics the shear stresses associated with the loading and unloading of long bones during normal gait and, as such, was implemented at a physiological frequency of 1 Hz (i.e., 1 step/s). For all experiments cells were exposed to fluid flow sufficient to induce 10 dynes/cm^2 shear stress.

Western immunoblotting

To examine the effect of fluid flow on ERK1/2 phosphorylation, osteoblasts were exposed to oscillatory fluid flow inducing a shear stress at 10 dynes/cm^2 for 5 min. Immediately after fluid flow, slides with cells were frozen in $-80°C$ and cells lysed with radio-immunoprecipitation assay buffer (40 mM Hepes (pH 7.4), 1% Triton X-100, 0.5% Na-deoxylcholate, 0.1% SDS, 100 mM NaCl, 1 mM EDTA and 25 mM β-glycerolphosphate) supplemented with 0.2 mM Na$_3$VO$_4$ and a protease inhibitor cocktail (Calbiochem). The total protein concentrations in cell lysates was quantified with a BCA protein assay kit (Pierce). 15 μg protein from each sample was resolved by SDS-PAGE and transferred to PVDF membranes. The membrane was then probed with a Phospho-ERK1/2 antibody. Total ERK1/2 was used as a control. Visualization of immunoreactive proteins was achieved employing an ECL detection system and membrane exposure to film. Densitometric analysis was carried out with Quality One image analysis software (Bio-Rad).

ATP release

After oscillatory fluid flow exposure, samples of conditioned media were collected and immediately stored at $-80°C$ until analyzed. ATP concentration in samples of conditioned media was determined using a commercially available ATP bioluminescence kit (Roche). ATP in 50 μl of each sample serves as a co-factor for luciferase, to convert D-luciferin, in 50 μl of a luciferin-luciferase assay buffer, into oxyluciferin and light. The luminescence from each reaction, as measured by a Monolight 3010 luminometer (BD Pharmingen), was compared with a standard curve created by serially diluting an ATP standard. Duplicate measurements were taken from each sample. Control experiments were performed with each pharmacological inhibitor to ensure that they had no detrimental effect on the reaction. Results were normalized to cellular protein concentration using bicinchoninic acid assay (Pierce).

Data Analysis

The results were analyzed using the statistical package MINITAB (Minitab Inc) and were expressed as the mean ± standard error (SE). To compare observations, the non-parametric Mann-Whitney test was used in which sample variance was not assumed to be equal. $p < 0.05$ was considered statistically significant.

Results

Micro-CT Analysis of WT and P2Y$_2$ KO mice bone structure

To examine the bone phenotype of P2Y$_2$ KO mice, we used micro-CT to analyze femoral structure of WT and P2Y$_2$ KO mice at age of 8 and 17 weeks. We found that trabecular bone volume and thickness are significantly decreased in P2Y$_2$ KO mice relative to WT mice at both 8 and 17 weeks of age (Fig 1A). For cortical bone, there were no significant differences between WT and P2Y$_2$ KO mice at 8 weeks. However, at 17 weeks, the cortical bone volume and mean cortical thickness are significantly decreased in P2Y$_2$ KO mice (Fig 1B).

Three-Point Bending Test of Bone Mechanical Properties

To further confirm bone structure difference between WT and P2Y$_2$ KO mice, we employed three-point bending experiments to measure the mechanical properties of mouse femurs. At 8 weeks the ultimate force and bone stiffness were not different between bones from WT and P2Y$_2$ KO mice. However, at 17 weeks, P2Y$_2$ KO mice showed significantly lower bone ultimate force and stiffness than WT mice (Fig 2).

Alkaline phosphatase, calcium staining and osteocalcin mRNA of bone marrow cells from WT and P2Y$_2$ KO mice

The same amount of bone marrow cells from WT and P2Y$_2$ KO mice were cultured in 6-well plates for 7, 14 and 21 days in cell differentiation media. Subsequently, cells were stained for AP. There was no difference in AP activity at day 7 between WT and P2Y$_2$ KO cells. However, the AP activity in WT cells was significantly higher than that in P2Y$_2$ KO mice at days 14 and 21 (Fig 3A). Similarly, we used von Kossa staining to examine calcium deposition during marrow cell differentiation. Cells from WT mice deposited significantly more calcium at 14 and 21 days than did cells from P2Y$_2$ KO mice (Fig 3B). Additionally, we examined another osteoblast-related gene osteocalcin changes and found that there was no difference in the mRNA levels of osteocalcin at day 7 between WT and P2Y$_2$ KO cells. However, the osteocalcin mRNA levels in WT cells were significantly higher than those in P2Y$_2$ KO cells at day 14 (Fig 3C). The results suggest that bone marrow cells from P2Y$_2$ KO mice, relative to those form WT mice, display a decreased ability to differentiate into osteoblastic cells. This may at least partially explain their osteopenic phenotype.

When measuring AP activities, we also determined total protein contents in each well. We found the total proteins from wells of WT cells were significantly higher than those from P2Y$_2$ KO cells at 7 days (345±41 μg/well vs. 272±24 μg/well) and 14 days (838±64 μg/well vs. 617±49 μg/well), respectively, suggesting proliferation deceases in P2Y$_2$ KO cells.

Response of primary osteoblastic cells ATP and mechanical stimulation

To examine the mechanism underlying the osteopenic phenotype of P2Y$_2$ KO mice, we examined the response of osteoblastic cells from these mice to both ATP and mechanical stimulation. Osteoblastic cells from WT and P2Y$_2$ KO mice were subcultured into 6-well plates. After 3 days, we added ATP at concentrations of 5 μM or 20 μM directly to plates containing osteoblastic cells from

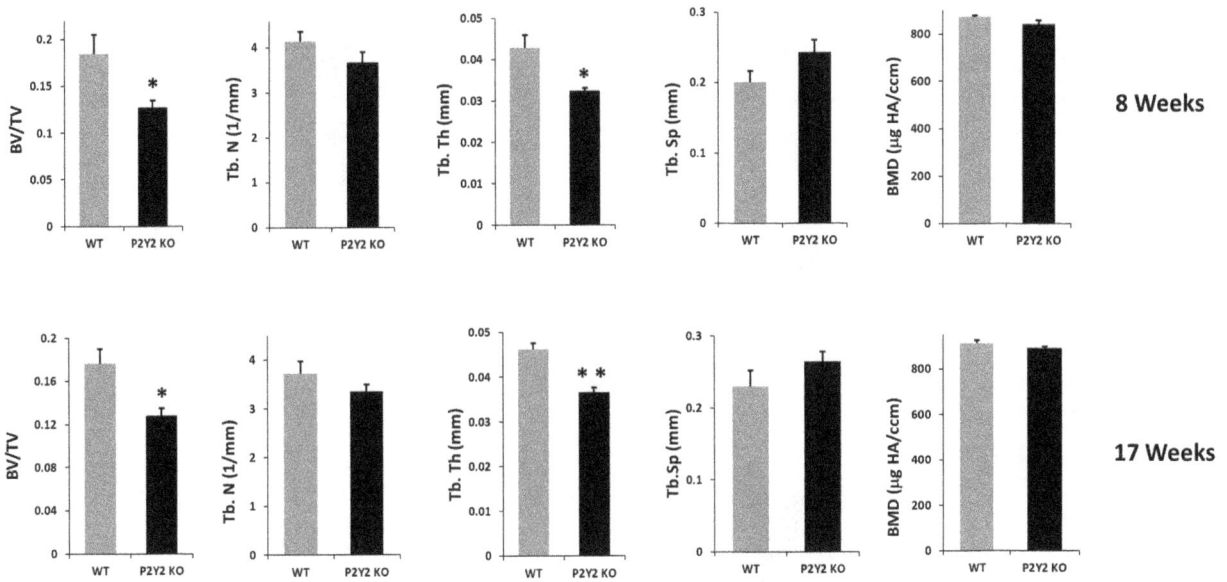

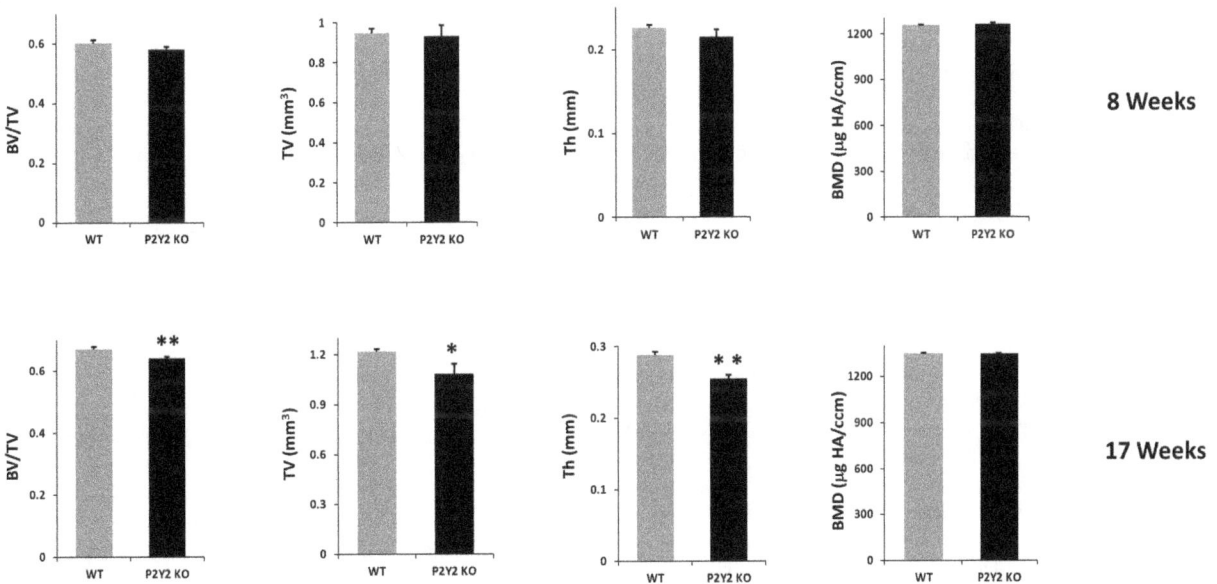

Figure 1. Bone loss in P2Y$_2$ KO age-matched WT and P2Y$_2$ KO male mice (8 and 17 weeks of age) were subjected to micro-CT analysis. (A) Parameters of trabecular bone mass, including bone volume fraction (BV/TV), trabecular number (Tb. N), trabecular thickness (Tb. Th), trabecular separation (Tb. Sp), and bone mineral density (BMD) were quantified; (B) Parameters of cortical bone quantified included bone volume fraction (BV/TV), total volume (TV), cortical bone thickness (Th), and bone mineral density (BMD). (n = 4–6, *p<0.05, **p<0.01) Error bars represent SEM.

WT and P2Y$_2$ KO mice. ATP stimulated ERK1/2 phosphorylation was significantly greater in WT cells relative to P2Y$_2$ KO cells (Fig 4A).

Additionally osteoblasts from WT and P2Y$_2$ KO mice were subcultured on glass slides as described before. After 3 days, the slides were exposed to oscillatory fluid flow at 10 dynes/cm^2 for 5 minutes. Western blot analysis revealed that fluid flow stimulated ERK1/2 phosphorylation was significantly greater in WT cells relative to P2Y$_2$ KO cells (Fig 4B). The results suggest that osteoblastic cells from P2Y$_2$ KO mice are less sensitive to mechanical stimulation, which may at least partially explain their bone phenotype.

ATP releases in response to mechanical stimulation

To further investigate role of P2Y$_2$ in bone mechanotransduction, we examined ATP releases from osteoblastic cells from WT and P2Y$_2$ KO mice exposed to fluid flow. ATP release from WT and P2Y$_2$ KO cells was not statistically different at the first minute

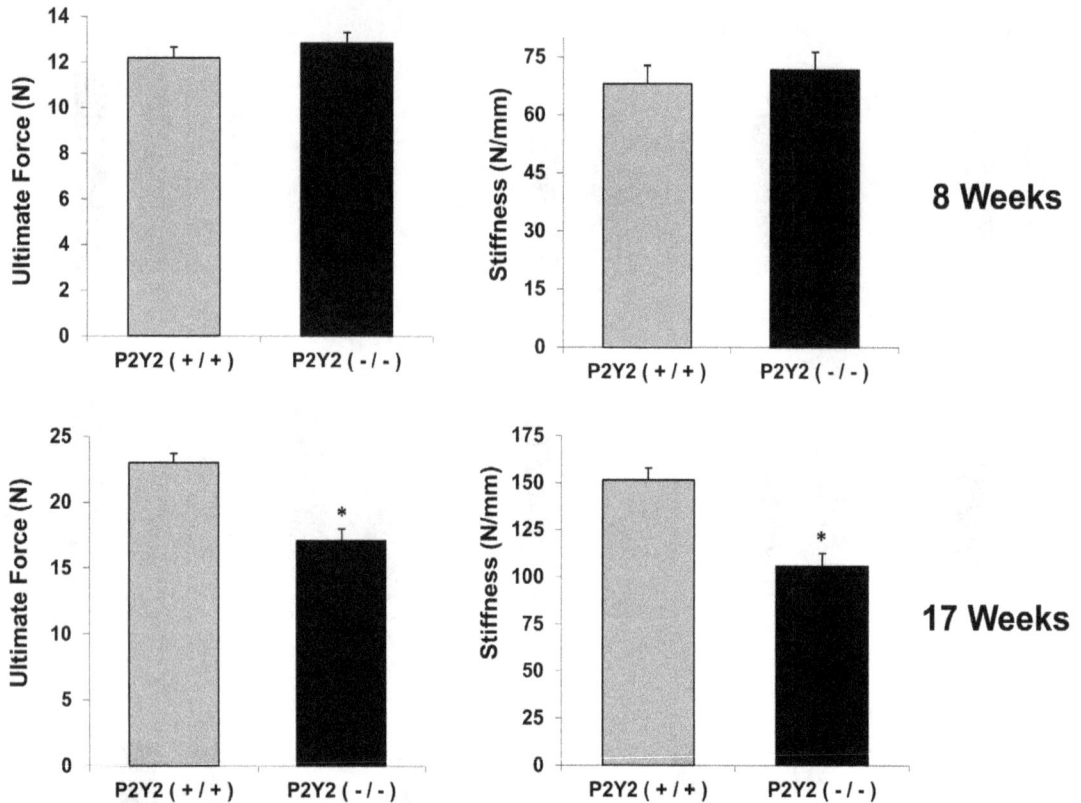

Figure 2. Mechanical properties of cortical bone were decreased in P2Y$_2$ KO mice. Femurs of age-matched WT and P2Y$_2$ KO male mice (8 and 17 weeks of age) were subjected to three point bending. Ultimate force and stiffness were recorded. (n = 6–8, *p<0.05) Error bars represent SEM.

when the highest ATP concentration was measured (12.65±3.16 nM and 11.97±3.47 nM, respectively).

Discussion

Extensive studies have sought to discover the mechanotransduction pathways in bone and bone cells, but the exact underlying mechanisms are still elusive. To date, researchers have focused on the roles of integrin, membrane channels, primary cilium etc to study bone mechanobiology [12–14]. Given the complexity of mechanical transduction, it is no surprise that several pathways are involved in this process. It is well known that mechanical stimulation is able to induce the release of ATP molecules from bone cells, which subsequently activate P2 purinergic receptors [26]. Accumulating evidence suggests these P2 purinergic receptors play an important role in bone cell mechanotransduction [15,17].

We have previously shown that P2Y$_2$ purinergic receptors are responsible for oscillatory fluid flow induced intracellular calcium mobilization in MC3T3-E1 osteoblast cells [4,5]. P2Y$_2$ receptors also have an important role in regulating ERK1/2 phosphorylation in bone and chondrocyte cell lines in response to mechanical stimulation [27,28]. In this study, we showed that osteoblastic cells from P2Y$_2$ KO mice have a weaker response to ATP stimulation than those from WT mice. This result was not surprising as ATP is able to initiate ERK1/2 phosphorylation and P2Y$_2$ receptors are one of the major ATP receptors in osteoblasts. Previously, researchers have demonstrated that flow induced ATP release in endothelial cells and smooth muscle cells is able to act on P2 receptors to initiate cellular responses which ultimately regulate

vascular tone and blood flow [18,19]. In addition, it has been reported that ATP released by chondrocytes in response to mechanical loading or inflammation contributes to cartilage tissue destruction by activating signaling pathways involved in articular pathology [20]. These pieces of evidence suggested the importance of P2 receptors in mechanotransduction of different cell types.

Our studies suggest that bone marrow stromal cells from P2Y$_2$ KO mice display decreased proliferation and osteogenic differentiation rate relative to those from WT mice. The data may explain the importance of ATP release in response to fluid flow for bone marrow stromal cell proliferation as shown in a previous study [29]. According to some earlier studies, P2Y$_2$ receptors play an important role in differentiation of several other cell types, including neurons [30,31]. P2Y$_2$ may also be involved in bone marrow cell differentiation into osteoblasts, although there is a possibility that the strong differentiation ability in WT cells is enhanced by high cell density due to their higher proliferation rate.

Our *in vitro* results of differentiation and mineralization were in contrast to what Orriss et al. have previously reported [32]. This may be due to the different cells used in their study, in which rat primary calvarial osteoblasts were treated with ATP or UTP, suggesting the role of P2Y$_2$ receptors in mineralized bone formation. It is well known that ATP and UTP can activate different subtypes of P2X and P2Y receptors, thus the responses of ATP and UTP may not accurately reflect the role of only P2Y$_2$ receptors. However, our approach was to employ P2Y$_2$ deficient bone marrow stromal cells to elucidate the role of P2Y$_2$ receptors in bone mineralization. Moreover, heterogeneous cell populations

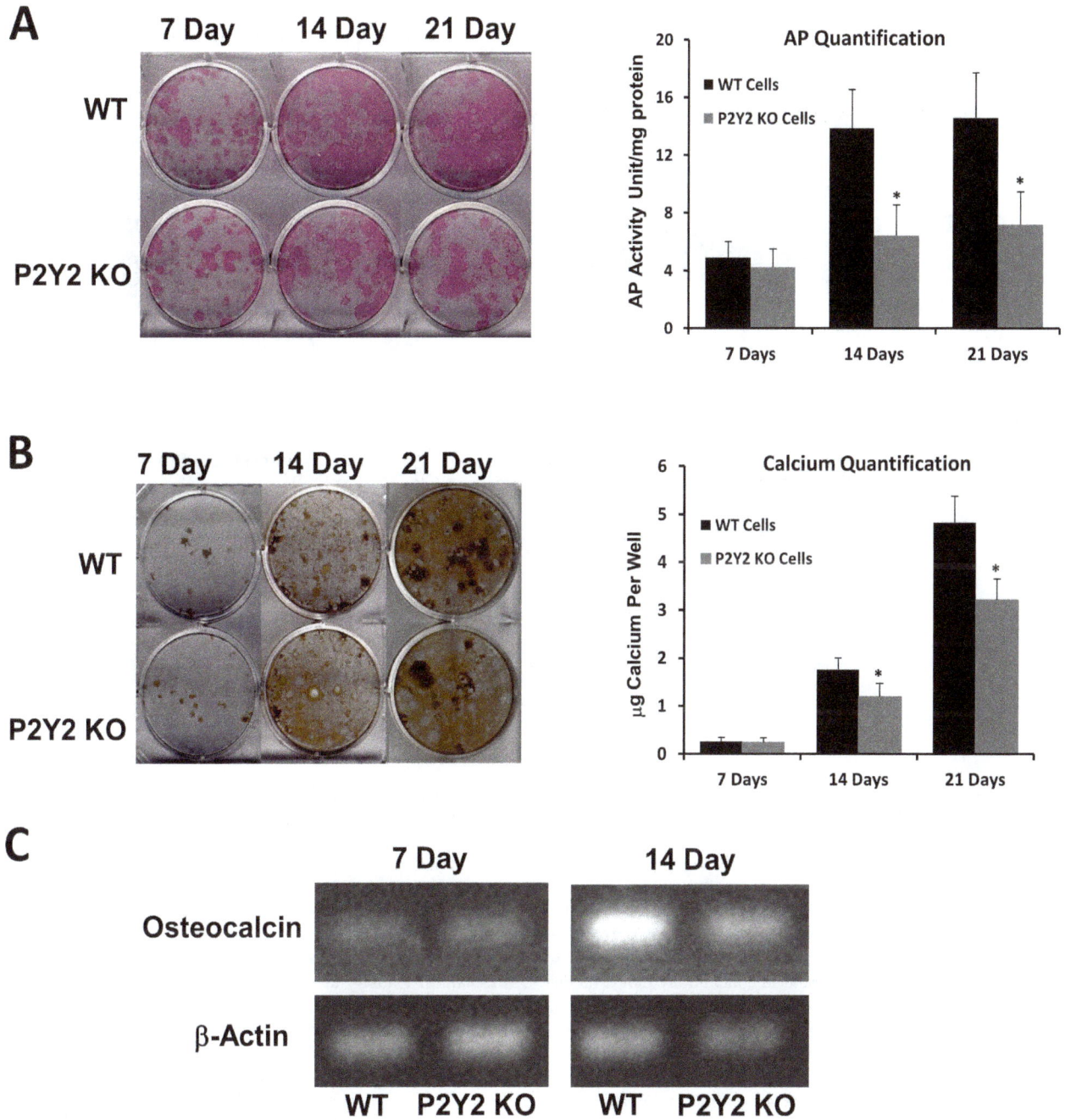

Figure 3. Differentiation (A, C) and mineralization (B) of bone marrow cells was decreased in P2Y$_2$ KO mice. (A) Left: Images of AP staining from bone marrow cell cultures; Right: Bar graph representation of AP quantified by colorimetric conversion of p-nitrophenol phosphate to p-nitrophenol and normalized to total protein. (B). Left: Images of von Kossa staining from bone marrow cell cultures; Right: Bar graph representation of calcium qualification by o-cresolphthalein method in each well. (C). Represented images of 35 cycles of RT-PRC show the osteocalcin mRNA levels in bone marrow cells from both WT and P2Y$_2$ KO mice (β-actin as controls). (n = 4, *p<0.05) Error bars represent SEM.

of bone marrow stromal cells were more likely to mimic *in vivo* environment for studying bone mineralization.

In addition to cell based studies *in vitro*, we also investigated the difference in bone structure between WT and P2Y$_2$ KO mice by micro-CT analysis. We found there are no significant differences between these mice in terms of cortical bone total bone volume and mean cortical bone thickness when the mice were 8 weeks old. But interestingly, cortical bone from 17 week old P2Y$_2$ KO mice displayed significantly less total bone volume and bone thickness

than did WT mice. The results suggest that the result of P2Y$_2$ receptor deficiency does not manifest until a more advanced age. This may be because it takes some time for the decreased responsiveness to mechanical load to affect bone phenotype. Interestingly, we did not found any significant difference in BMD between wild type and P2Y$_2$ KO mice, which conflicts with previous report [33]. This may be due to different genetic background of the inbred strains used. However, our results of reduced bone volume in P2Y$_2$ KO mice are consistent with other

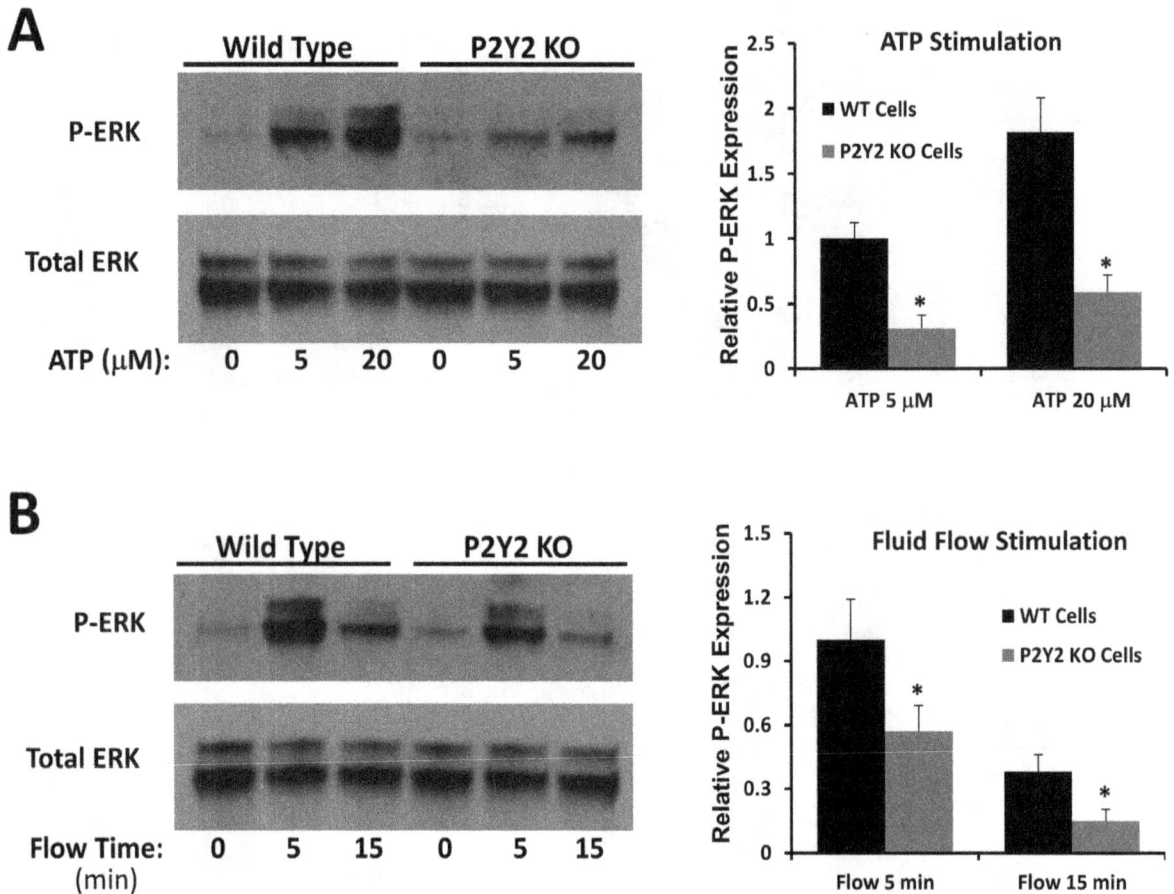

Figure 4. Response to ATP (A) and oscillatory fluid flow (B) was decreased in P2Y₂ KO mice. (Left: western blot analysis of ERK1/2 phosphorylation in primary osteoblastic cells. Right: Bar graph representation of ERK1/2 phosphorylation quantified by scanning densitometry normalized to total ERK1/2). (n=4–6, *p<0.05) Each bar represents SEM.

in vivo studied in P2Y₁ and P2X₇ [21,34], other subtype P2 receptors. More importantly, both P2Y₂ KO mice were P2Y₂ deficient in all cell types. Any environment or other tissue changes may complicate experimental results in bone. Future mouse model study of P2Y₂ deficient only in osteoblast/osteocytes will elucidate the roles of P2Y₂ receptor in bone biology.

We further examined ultimate force and stiffness of bones, through 3-point bending experiments, from these two types of mice. Our results showed that there is no significant difference in either ultimate force or stiffness between WT and P2Y₂ KO mice at 8 weeks, which is consistent with our micro-CT data (no significant differences in cortical parameters between WT and KO mice at 8 weeks). But at 17 weeks, both ultimate force and stiffness of femurs from P2Y₂ KO mice were dramatically decreased compared to those in WT mice. This may have resulted from the loss of bone volume and bone thickness in 17 week old P2Y₂ KO mice as revealed by micro-CT. The difference between 8 and 17 weeks of cortical bones is that it takes some time for the decreased responsiveness to mechanical load to affect bone phenotype. To fully understand the molecular mechanisms of P2Y₂ receptor in bone mechanobiology, our ongoing and future approaches will examine in vivo bone formation rates at 8 and 17 weeks, and investigate mechanical loading induced bone formation at 17 weeks. Due to the fact that it takes a long time to observe a difference in bone structure changes between WT and KO mice, unlike in vitro cell models we may not see a difference in bone

formation changes in vivo by using dynamic histomorphometric measurements in a few days. Thus, ideal approaches will be to examine in vivo bone formation under mechanical challenging (mechanical loading) in mice which are P2Y₂ deficient only in osteoblast/osteocytes eliminating other tissue effects.

ATP is a well-known local mediator in bone and can initiate a series of other cellular activities, such as intracellular calcium mobilization and ERK1/2 phosphorylation as shown in our previous studies [4,5]. ATP works through P2 receptors, which are divided into two subclasses, P2X and P2Y. Earlier studies have demonstrated that P2X₇ is important for maintaining bone sensitivity to mechanical stimulation through PGE2 signaling pathways [21] and P2Y₁ is able to enhance bone resorption through the RANKL pathway [35]. In addition, our results suggest that P2Y₂ receptors are one of the major contributors to osteoblast mechanotransduction through ERK1/2 pathways. Our AP, osteocalcin and total protein results suggest the absence of P2Y₂ receptors in bone tissue results in decreased sensitivity to loading as well as slower bone cell differentiation and proliferation.

We found that fluid flow-induced ATP release from WT or P2Y₂ KO osteoblastic cells was the same, strongly suggesting that the effects observed in this study are directly due to the lack of P2Y₂ receptor, instead of ATP release. Our previous studies suggest that membrane structural components related to lipid rafts, cholesterol and GPI-anchored proteins may play an important role in mechanically induced ATP releases in osteoblastic cells [27]. Thus, our results suggest that $P2Y_2$ receptors may not be involved in mechanically induced ATP releases, but involved in subsequent responses after released ATP stimulation.

In conclusion, we demonstrated a reduction in differentiation and mineralization in bone marrow cells and decrease in response to ATP and fluid flow in primary osteoblastic cells from $P2Y_2$ KO mice, relative to those from WT mice. These changes may account for the bone phenotype observed in 17-week old $P2Y_2$ KO mice and our *in vitro* results may imply an important role of $P2Y_2$ in bone mechanotransduction.

References

1. Klein-Nulend J, Bacabac RG, Mullender MG (2005) Mechanobiology of bone tissue. Pathol Biol (Paris) 53: 576–580.
2. Robling AG, Castillo AB, Turner CH (2006) Biomechanical and molecular regulation of bone remodeling. Annu Rev Biomed Eng 8: 455–498.
3. Bonewald LF (2007) Osteocytes as dynamic multifunctional cells. Ann N Y Acad Sci 1116: 281–290.
4. You J, Reilly GC, Zhen X, Yellowley CE, Chen Q, et al. (2001) Osteopontin Gene Regulation by Oscillatory Fluid Flow via Intracellular Calcium Mobilization and Activation of Mitogen-activated Protein Kinase in MC3T3-E1 Osteoblasts. J Biol Chem 276: 13365–13371.
5. You J, Jacobs CR, Steinberg TH, Donahue HJ (2002) P2Y purinoceptors are responsible for oscillatory fluid flow-induced intracellular calcium mobilization in osteoblastic cells. J Biol Chem 277: 48724–48729.
6. Li YJ, Batra NN, You L, Meier SC, Coe IA, et al. (2004) Oscillatory fluid flow affects human marrow stromal cell proliferation and differentiation. J Orthop Res 22: 1283–1289.
7. Ponik SM, Triplett JW, Pavalko FM (2007) Osteoblasts and osteocytes respond differently to oscillatory and unidirectional fluid flow profiles. J Cell Biochem 100: 794–807.
8. Jaasma MJ, O'Brien FJ (2008) Mechanical stimulation of osteoblasts using steady and dynamic fluid flow. Tissue Eng Part A 14: 1213–1223.
9. Wadhwa S, Godwin SL, Peterson DR, Epstein MA, Raisz LG, et al. (2002) Fluid flow induction of cyclo-oxygenase 2 gene expression in osteoblasts is dependent on an extracellular signal-regulated kinase signaling pathway. J Bone Miner Res 17: 266–274.
10. Okumura H, Shiba D, Kubo T, Yokoyama T (2008) P2X7 receptor as sensitive flow sensor for ERK activation in osteoblasts. Biochem Biophys Res Commun 372: 486–490.
11. Genetos DC, Kephart CJ, Zhang Y, Yellowley CE, Donahue HJ (2007) Oscillating fluid flow activation of gap junction hemichannels induces atp release from MLO-Y4 osteocytes. J Cell Physiol.
12. Jekir MG, Donahue HJ (2009) Gap junctions and osteoblast-like cell gene expression in response to fluid flow. J Biomech Eng 131: 011005.
13. Scott A, Khan KM, Duronio V, Hart DA (2008) Mechanotransduction in human bone: in vitro cellular physiology that underpins bone changes with exercise. Sports Med 38: 139–160.
14. Malone AM, Anderson CT, Tummala P, Kwon RY, Johnston TR, et al. (2007) Primary cilia mediate mechanosensing in bone cells by a calcium-independent mechanism. Proc Natl Acad Sci U S A 104: 13325–13330.
15. Jorgensen NR, Geist ST, Civitelli R, Steinberg TH (1997) ATP- and gap junction-dependent intercellular calcium signaling in osteoblastic cells. Journal of Cell Biology 139: 497–506.
16. Jorgensen NR, Henriksen Z, Brot C, Eriksen EF, Sorensen OH, et al. (2000) Human osteoblastic cells propagate intercellular calcium signals by two different mechanisms [In Process Citation]. J Bone Miner Res 15: 1024–1032.
17. Yamazaki S, Weinhold PS, Graff RD, Tsuzaki M, Kawakami M, et al. (2003) Annulus cells release ATP in response to vibratory loading in vitro. J Cell Biochem 90: 812–818.
18. Yamamoto K, Sokabe T, Ohura N, Nakatsuka H, Kamiya A, et al. (2003) Endogenously released ATP mediates shear stress-induced Ca2+ influx into pulmonary artery endothelial cells. Am J Physiol Heart Circ Physiol 285: H793–803.
19. Liu C, Mather S, Huang Y, Garland CJ, Yao X (2004) Extracellular ATP facilitates flow-induced vasodilatation in rat small mesenteric arteries. Am J Physiol Heart Circ Physiol 286: H1688–1695.
20. Berenbaum F, Humbert L, Bereziat G, Thirion S (2003) Concomitant recruitment of ERK1/2 and p38 MAPK signalling pathway is required for activation of cytoplasmic phospholipase A2 via ATP in articular chondrocytes. J Biol Chem 278: 13680–13687.
21. Li J, Liu D, Ke HZ, Duncan RL, Turner CH (2005) The P2X7 nucleotide receptor mediates skeletal mechanotransduction. J Biol Chem 280: 42952–42959.
22. Bowler WB, Dixon CJ, Halleux C, Maier R, Bilbe G, et al. (1999) Signaling in human osteoblasts by extracellular nucleotides. Their weak induction of the c-fos proto-oncogene via Ca2+ mobilization is strongly potentiated by a parathyroid hormone/cAMP-dependent protein kinase pathway independently of mitogen-activated protein kinase. J Biol Chem 274: 14315–14324.
23. Bagchi S, Liao Z, Gonzalez FA, Chorna NE, Seye CI, et al. (2005) The P2Y2 nucleotide receptor interacts with alphav integrins to activate Go and induce cell migration. J Biol Chem 280: 39050–39057.
24. Yu N, Erb L, Shivaji R, Weisman GA, Seye CI (2008) Binding of the P2Y2 nucleotide receptor to filamin A regulates migration of vascular smooth muscle cells. Circ Res 102: 581–588.
25. Chen Y, Corriden R, Inoue Y, Yip L, Hashiguchi N, et al. (2006) ATP release guides neutrophil chemotaxis via P2Y2 and A3 receptors. Science 314: 1792–1795.
26. Grol MW, Pereverzev A, Sims SM, Dixon SJ (2013) P2 receptor networks regulate signaling duration over a wide dynamic range of ATP concentrations. J Cell Sci 126: 3615–3626.
27. Xing Y, Gu Y, Xu LC, Siedlecki CA, Donahue HJ, et al. (2011) Effects of membrane cholesterol depletion and GPI-anchored protein reduction on osteoblastic mechanotransduction. J Cell Physiol 226: 2350–2359.
28. Xing Y, Gu Y, Gomes RR, Jr., You J (2011) P2Y(2) receptors and GRK2 are involved in oscillatory fluid flow induced ERK1/2 responses in chondrocytes. J Orthop Res 29: 828–833.
29. Riddle RC, Taylor AF, Rogers JR, Donahue HJ (2007) ATP Release Mediates Fluid Flow-Induced Proliferation of Human Bone Marrow Stromal Cells. J Bone Miner Res 22: 589–600.
30. Glaser T, Resende RR, Ulrich H (2013) Implications of purinergic receptor-mediated intracellular calcium transients in neural differentiation. Cell Commun Signal 11: 12.
31. Chen X, Molliver DC, Gebhart GF (2010) The P2Y2 receptor sensitizes mouse bladder sensory neurons and facilitates purinergic currents. J Neurosci 30: 2365–2372.
32. Orriss IR, Knight GE, Ranasinghe S, Burnstock G, Arnett TR (2006) Osteoblast responses to nucleotides increase during differentiation. Bone 39: 300–309.
33. Orriss IR, Utting JC, Brandao-Burch A, Colston K, Grubb BR, et al. (2007) Extracellular nucleotides block bone mineralization in vitro: evidence for dual inhibitory mechanisms involving both P2Y2 receptors and pyrophosphate. Endocrinology 148: 4208–4216.
34. Orriss I, Syberg S, Wang N, Robaye B, Gartland A, et al. (2011) Bone phenotypes of P2 receptor knockout mice. Front Biosci (Schol Ed) 3: 1038–1046.
35. Takasaki J, Kamohara M, Saito T, Matsumoto M, Matsumoto S, et al. (2001) Molecular cloning of the platelet P2T(AC) ADP receptor: pharmacological comparison with another ADP receptor, the P2Y(1) receptor. Mol Pharmacol 60: 432–439.

Acknowledgments

We are grateful for technical supports from Drs. Yue Zhang and Feng Li.

Author Contributions

Conceived and designed the experiments: YX JY. Performed the experiments: YX YG JJB. Analyzed the data: YX EMP JY. Contributed reagents/materials/analysis tools: HJD JY. Wrote the paper: YX HJD JY. Manuscript editing: HJD.

23

Lubricin Protects the Temporomandibular Joint Surfaces from Degeneration

Adele Hill[1◑]**, Juanita Duran**[2◑]**, Patricia Purcell**[2]*

1 Department of Orthopaedic Surgery, Boston Children's Hospital, Boston, Massachusetts, United States of America; Department of Genetics, Harvard Medical School, Boston, Massachusetts, United States of America, **2** Department of Plastic and Oral Surgery, Boston Children's Hospital and Harvard Medical School, Boston, Massachusetts, United States of America

Abstract

The temporomandibular joint (TMJ) is a specialized synovial joint essential for the mobility and function of the mammalian jaw. The TMJ is composed of the mandibular condyle, the glenoid fossa of the temporal bone, and a fibrocartilagenous disc interposed between these bones. A fibrous capsule, lined on the luminal surface by the synovial membrane, links these bones and retains synovial fluid within the cavity. The major component of synovial fluid is lubricin, a glycoprotein encoded by the gene proteoglycan 4 (Prg4), which is synthesized by chondrocytes at the surface of the articular cartilage and by synovial lining cells. We previously showed that in the knee joint, Prg4 is crucial for maintenance of cartilage surfaces and for regulating proliferation of the intimal cells in the synovium. Consequently, the objective of this study was to determine the role of lubricin in the maintenance of the TMJ. We found that mice lacking lubricin have a normal TMJ at birth, but develop degeneration resembling TMJ osteoarthritis by 2 months, increasing in severity over time. Disease progression in $Prg4^{-/-}$ mice results in synovial hyperplasia, deterioration of cartilage in the condyle, disc and fossa with an increase in chondrocyte number and their redistribution in clusters with loss of superficial zone chondrocytes. All articular surfaces of the joint had a prominent layer of protein deposition. Compared to the knee joint, the osteoarthritis-like phenotype was more severe and manifested earlier in the TMJ. Taken together, the lack of lubricin in the TMJ causes osteoarthritis-like degeneration that affects the articular cartilage as well as the integrity of multiple joint tissues. Our results provide the first molecular evidence of the role of lubricin in the TMJ and suggest that $Prg4^{-/-}$ mice might provide a valuable new animal model for the study of the early events of TMJ osteoarthritis.

Editor: Yann Herault, IGBMC/ICS, France

Funding: AH was supported by the National Institute of Arthritis and Musculoskeletal and Skin Diseases R01AR050180 (Drs. M. Warman and G. Jay; co-PIs) and The Boston Plastic and Oral Surgery Foundation, Inc supported PP and JD. The funders had no role in study design, data collection and analysis, decision to publish, or preparation of the manuscript.

Competing Interests: The authors have declared that no competing interests exist.

* Email: ppurcell@post.harvard.edu

◑ These authors contributed equally to this work.

Introduction

The temporomandibular joint (TMJ) is a specialized synovial joint essential for the mobility and function of the mammalian jaw, including nutritional intake and communication. The TMJ is composed of the mandibular condyle, the glenoid fossa of the temporal bone, and an intra-articular fibrocartilagenous disc that lies between the two bones. A fibrous capsule, lined on the luminal surface by the synovial membrane, connects these two bones and encapsulates the synovial fluid within the joint cavity [1].

The epiphyses of the condyle and fossa are covered by articular cartilage, which together with the joint disc and the synovial fluid permit a smooth movement of the jaw. With age, the articular surfaces are prone to degeneration, due to intense daily use, frequently resulting in osteoarthritis (OA), a common but debilitating disorder that affects all synovial joints including the TMJ. In osteoarthritis chondrocytes respond to biomechanical and biologic stresses, resulting in breakdown of the matrix and structural changes in the underlying bone [2,3]. OA affects the integrity of the multiple tissues that form the joint, including synovium, bone, ligaments, supporting musculature and fibrocartilaginous structures [4].

Although disorders of the TMJ, including OA, have been previously documented [5,6,7], no studies have focused on the role of lubricin, a major component of the synovial fluid. Lubricin studies have been aimed at its mechanical role in the TMJ [8,9]. Synovial fluid lubricates the joint and protects the articular cartilage surfaces from erosion and protein deposition. Lubricin is a large proteoglycan encoded by the gene proteoglycan 4 (Prg4) and is essential in the boundary lubrication of the knee articular surfaces to maintain joint integrity [10]; however its specific role in the TMJ has not been determined.

Patients with the autosomal recessive disorder camptodactyly-arthropathy-coxa vara-pericarditis syndrome (CACP) caused by mutations in PRG4 have joints that appear normal at birth, but over time develop severe degeneration of the cartilage surface and synoviocyte hyperplasia leading to precocious joint failure [11]. This indicates that lubricin has a dual role in synovial joints: cartilage protection and inhibition of synovial cell outgrowth [12].

In this study we report the phenotype of the TMJ in mice lacking the Prg4 gene. The TMJ in these animals presents changes comparable to those previously described for the knee joint [10], and are analogous to histopathological findings described for human TMJ-OA, the most common degenerative joint disease of the TMJ [5,13]. The etiology of TMJ-OA is unknown, although host adaptive factors (i.e age, systemic illness, and hormones) and mechanical factors (i.e trauma, parafunction, malocclusion, overloading and increased joint friction) may all play a role [14]. Due to the difficulty in studying this disease in humans, degeneration of the TMJ has been poorly characterized. Recently several mouse models have been reported which are attributable to mutations in components of extracellular matrix; however no studies have described deletion of synovial fluid components [15,16,17,18,19,20].

The studies described herein provide the first molecular evidence that lubricin, a major component of the synovial fluid, is essential in the maintenance of the TMJ by protecting the articular cartilage surfaces and regulating synovial cell growth. Furthermore, the $Prg4^{-/-}$ mouse may provide a new animal model for the study of early events of OA-like degeneration in the TMJ.

Materials and Methods

Animals

All mice were housed and all experiments were conducted in compliance with protocols approved by the Institutional Animal Care and Use Committee of Boston Children's Hospital. Mice were sacrificed by CO_2 inhalation. Homozygous $Prg4^{-/-}$ and age-matched heterozygous $Prg4^{+/-}$ mice were generated as reported [10]. Jaws from heterozygous mice were indistinguishable from *wild-type* mice and were used as controls. For each genotype and age, heads were hemi-sected and TMJs were isolated from mice at 2 $Prg4^{-/-}$ (n = 16), $Prg4^{+/-}$ (n = 8); 4 $Prg4^{-/-}$ (n = 8), $Prg4^{+/-}$ (n = 10); 6 $Prg4^{-/-}$ (n = 8), $Prg4^{+/-}$ (n = 8); and 9 months of age $Prg4^{-/-}$ (n = 6), $Prg4^{+/-}$ (n = 6).

Histology

Skin and brain tissue were removed and heads were fixed in 4% paraformaldehyde overnight, decalcified in 14% EDTA in PBS, pH 7.5 for 14 days and embedded in paraffin or O.C.T. compound. Sections (8–10 µm) were stained with hematoxylin/eosin, and adjacent sections were stained with Safranin O or tartrate resistant acid phosphate (TRAP), following standard procedures.

Osteoclast quantitation

For quantitation, an image of the TMJ stained for TRAP (20X) was divided into 120 equal units; multinucleated TRAP+ cells (>2 nuclei) were considered osteoclasts.

Immunohistochemistry

Paraffin-embedded sections of the TMJ at 2, 6, and 9 months of age were stained with rabbit polyclonal antibody against aggrecan neopeptide at 1:200 dilution (NB-100-74350, Santa Cruz Biotechnology). Sections were reacted with biotinylated secondary antibody (Jackson Laboratories, West Grove, PA) and color was developed with ImmPact NovaRED peroxidase substrate (Vector Labs, Burlingame, CA).

In situ hybridization

In situ hybridization was performed on frozen or paraffin sections using digoxigenin (DIG)-labeled probes and color detected

with BM purple (Roche, IN, USA), as previously described [21,22]. The Prg4 mouse probe used was generated by PCR (nucleotides 2370–3070). Images were captured on a Nikon Eclipse 80i microscope with a Spot RT3 camera.

Results

Loss of *Prg4* causes age-related degeneration of the TMJ

To determine the role of lubricin in the development and maintenance of the mouse TMJ, we examined sections of the TMJ of $Prg4^{-/-}$ and age-matched control $Prg4^{+/-}$ mice at embryonic day 16, birth, and 2, 4, 6 and 9 months postnatally. Embryonic and newborn $Prg4^{-/-}$ mice had joints of normal appearance, comparable to those of control mice (data not shown). However, by 2 months of age the joints of $Prg4^{-/-}$ mice displayed signs of joint degeneration when compared to control ($Prg4^{+/-}$) mice (Fig. 1, A vs D). In controls, the surfaces of the mandibular condyle and the glenoid fossa were smooth with flattened chondrocytes distributed across the articular surface (Fig. 1A, inset). The disc exhibited no signs of degeneration and the characteristic biconcave shape was preserved (Fig. 1A). In contrast in $Prg4^{-/-}$ mice the articular surfaces were irregular, the superficial flattened chondrocytes were located at a distance from the surface and fewer flattened cells were observed (Fig. 1D, inset). Additionally, a weakly stained material that appeared to represent protein deposition was observed on all articular surfaces including the condyle, the fossa and both sides of the disc (Fig. 1D brackets).

Four-month-old $Prg4^{-/-}$ mice had a very similar phenotype, and thus we did not study this time point further. In mutant mice of 6 and 9 months of age, fewer superficial chondrocytes were seen at the cartilage surface, and clusters of chondrocytes could be distinguished, characteristic of joint degeneration in osteoarthritis (Fig 1. E and F insets). As early as 6 months, *wild-type* mice presented early signs of naturally occurring TMJ-OA with areas of acellularity and some chondrocyte clustering [23]. However, flat superficial chondrocytes were still observed across the surface of the condyle and in general the joint appeared normal with smooth surfaces and no evidence of protein deposition (Fig. 1B).

At 9 months, $Prg4^{-/-}$ mice joints showed evidence of disease progression. Chondrocytes in the superficial layer of the condyle were scarce and the disc had a markedly increased thickness compared to controls. Cartilage tears were observed at the surfaces (Fig.1 C vs F). In addition, there was marked hyperplasia of the synovium (Fig. 2C, 2D). The articular surface of the condyle was almost completely absent, and the area of synoviocyte infiltration extended to the layer of columnar chondrocytes, especially in the center of the cartilage surface, likely due to synovial infiltration from both sides of the joint (Fig.1F). Tears in the cartilage were observed at the surface and there was almost complete loss of superficial zone chondrocytes.

Prg4 regulates synovium growth

We studied RNA expression of *Prg4* by *in situ* hybridization at embryonic day 18 (E18), postnatal day 1 (P1) and adult TMJ (2, 6 and 9 months). We found that at all three stages of the growing TMJ, *Prg4* was strongly expressed in the synovial lining cells of the upper joint cavity and in the synovium of both edges of the TMJ, suggesting a major role for lubricin at these sites (Fig. 1 top panel) [24]. In control $Prg4^{+/-}$ mice, a thin single cell synovial lining, similar to that found in *wild-type* mice, was observed (Fig. 2 A, C, E and G). In contrast, a thickened synovium was seen in $Prg4^{-/-}$ mice (Fig. 2 B, D, F and H), characteristic of synovitis. Over time, a remarkable increase in the severity of synovial hypertrophy was apparent, as observed by comparing 2-month-old to 9-month-old

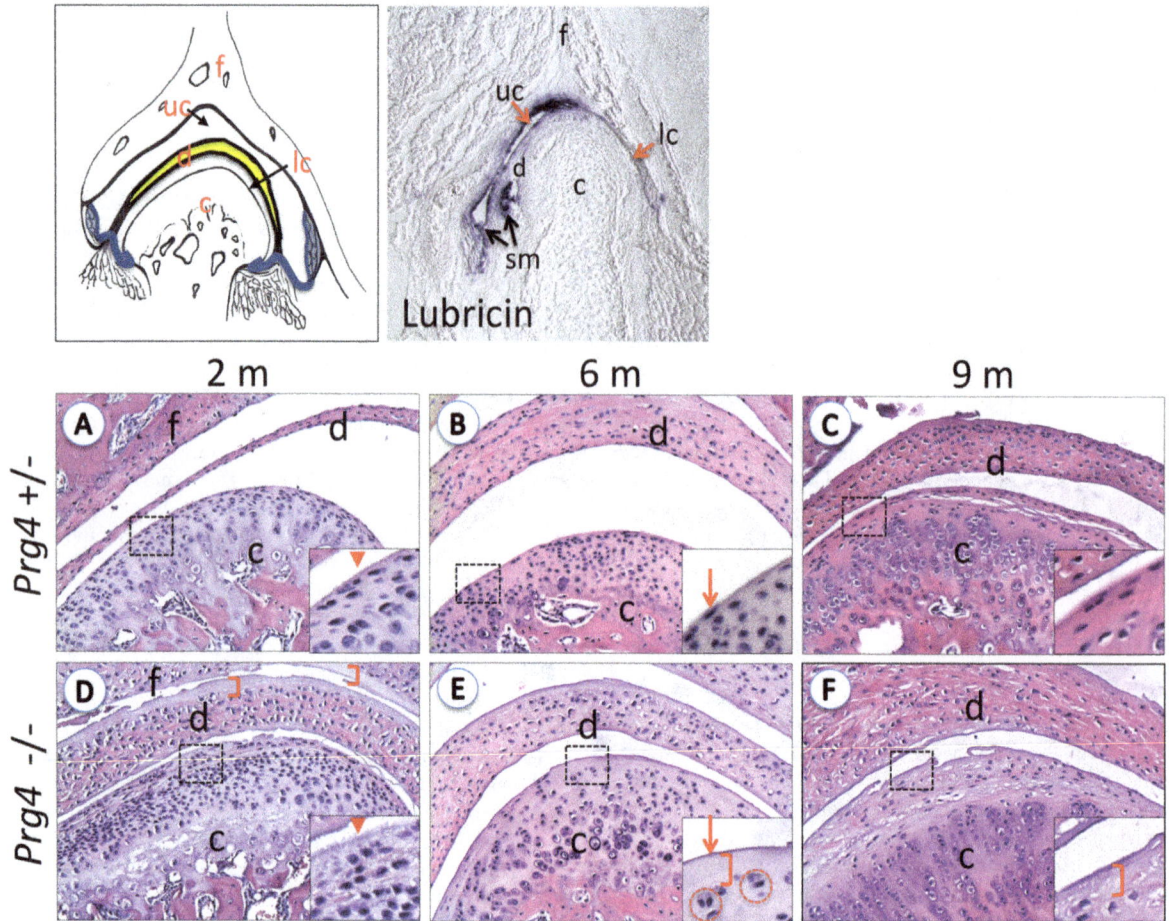

Figure 1. Morphological changes of the TMJ resembling osteoarthritis-like characteristics in Lubricin null (Prg4$^{-/-}$) mice. Top Panels: TMJ diagram indicating the different components of the TMJ. C: condyle, f: fossa, d: disc, uc: upper joint cavity, lc: lower joint cavity. Highlighted in blue the synovial membrane (sm). On the right, lubricin *in situ* hybridization shows localization of the *Prg4* gene primarily in the synovial membrane and upper joint cavity in mice at postnatal day 1 (10× magnification). **(A–F)** Representative coronal sections of the TMJ of lubricin control (*Prg4$^{+/-}$*) (A–C) and null (*Prg4$^{-/-}$*) (D–F) mice at 2-, 6- and 9-months of age, stained with H&E. *Prg4$^{-/-}$* articular surfaces are irregular (arrowheads), there is an increased number of chondrocytes (dark purple dots) in the condyle (c), increased chondrocyte clusters (red circles in E), decreased number of flat cells in the most superficial layer (arrows) of the condyle, disc (d) and fossa (f), and deposit of cartilage and cellular detritus over the different structures of the joint (brackets). Thickness and cellularity in the disc progressively increases in the *Prg4$^{-/-}$* mice compared to *Prg4$^{+/-}$* control mice. Pictures are taken with 20× magnification and insets are 40X.

Prg4$^{-/-}$ mice (Fig. 2 F and H). In addition, 9-month mutants presented with villous digitations (Fig. 2 H arrowhead), cartilage debris surrounded by synovial membrane (open arrowhead) and detritus rich zones (*). These inflammatory changes were seen primarily in the upper joint cavity, between the disc and the fossa where lubricin was normally expressed most abundantly (Fig. 1).

TMJs of *Prg4$^{-/-}$* mice show osteoarthritis-like changes

The glycosaminoglycan content and bone resorptive activity are two parameters used to evaluate joint osteoarthritis. For proteoglycans we analyzed sections from 6-month-old *Prg4$^{-/-}$* mice stained with Safranin O (SO), which binds to negatively charged glycosaminoglycans. *Prg4$^{-/-}$* mice showed reduced pericellular SO staining compared to control mice (Fig. 3, A and B). In addition, we evaluated the expression of aggrecan neopeptide, which reflects the degraded extracellular matrix proteins aggrecan and collagen [25], and found a dramatic increase in products of degradation in mutant mice (Fig 3, C and D), consistent with a reduction in the proteoglycan content of the cartilage. It is relevant

to note that chondrocytes that stained positive for SO were negative for aggrecan neopeptide and vice-versa (Fig. 3 A–D).

Another characteristic of osteoarthritis is bone resorption as a result of increased osteoclast activity [2]; therefore we evaluated changes in bone resorption in *Prg4$^{-/-}$* mice with respect to their aged-matched controls. We used the TRAP assay to measure osteoclast activity. At all time points studied, TRAP positive multinucleated cells were significantly increased in the subchondral bone region of the mutant *Prg4$^{-/-}$* mice as compared to their respective age-matched control *Prg4$^{+/-}$* mice, and these differences increased with age, (Fig. 3 E–J). TRAP activity in the mutants was elevated 39%, 41% and 46% in 2-, 6- and 9-month old mice, respectively, relative to age-matched controls (Fig. 3K). Interestingly, TRAP staining peaked at 2 months in both mutant mice and age-matched controls (Fig. 3 F, J and K), indicating that this peak activity was unrelated to the *Prg4* mutation and that greater bone resorption occurred at this earlier time point. Thus, we observed an increased osteoclast activity in the absence of the *Prg4* gene that augments with age. Taken together these results

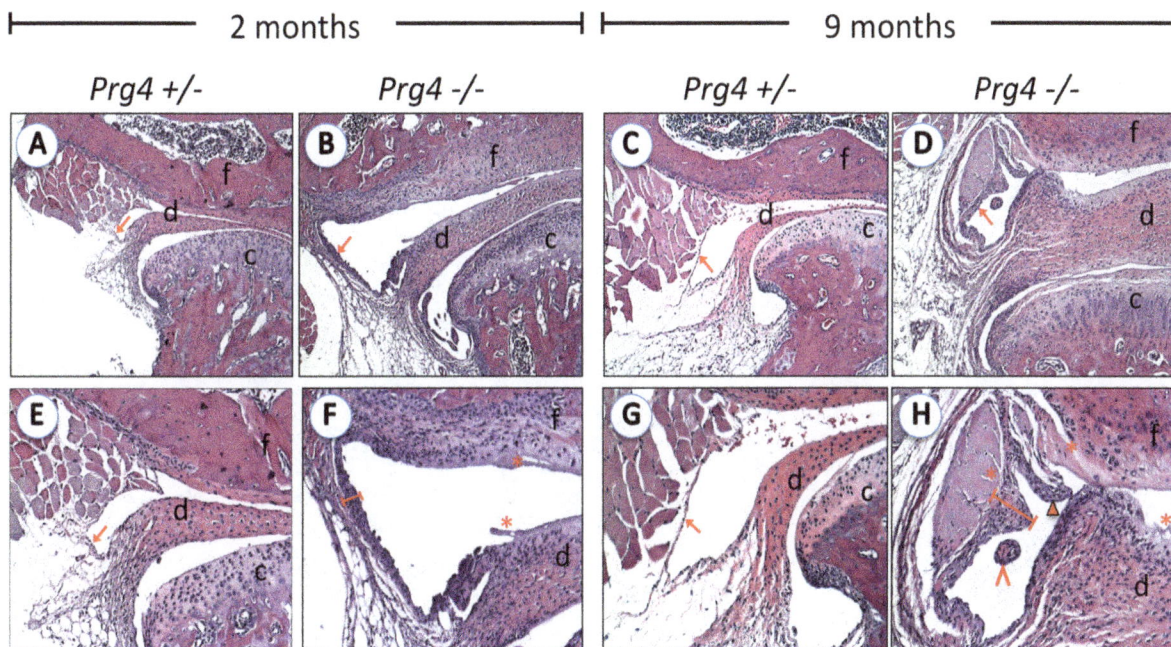

Figure 2. Synovial hypertrophy in *Prg4*$^{-/-}$ mice TMJs. TMJ sections stained for H&E at 2- and 9-months old mice in control (*Prg4*$^{+/-}$) and lubricin null mice (*Prg4*$^{-/-}$). Synovial membranes in the upper joint cavity are indicated by arrows. Control *Prg4*$^{+/-}$ mice exhibit a thin synovial lining (A, C, E, and G) compared to the hypertrophied synovium (red bar) observed in *Prg4*$^{-/-}$ mice (B, D, F, and H) and characteristic of synovitis. An increase in the severity of synovitis is observed over time, as shown in 2-month-old *Prg4*$^{-/-}$ (B, F) versus 9-month-old *Prg4*$^{-/-}$ (F, H) mice. Villous digitations (arrowhead), cartilage debris surrounded by synovial membrane (open arrowhead) and detritus rich zones (*) are observed in 9-month-old *Prg4*$^{-/-}$ mice. These inflammatory changes were seen primarily in the upper joint cavity, located between the disc (d) and fossa (f). (A–D: 10X, E–H: 20X).

strongly suggest that lubricin protects TMJ synovial joints from degeneration, and in the absence of lubricin, mice present a premature osteoarthritis-like phenotype in the TMJ.

TMJ degeneration is more severe than knee joint degeneration in *Prg4*$^{-/-}$ mice

We previously reported that knee joints from lubricin-deficient mice appear normal at birth, but progressively degenerate over time [10]. We therefore examined the evolution of pathology in the TMJ of *Prg4*$^{-/-}$ mice compared to the knee joints of 2 and 9 month-old mice. By 2 months of age, the articular surfaces of knee joints in *Prg4*$^{-/-}$ mice showed loss of superficial zone chondrocytes, particularly from the tibial plateau where areas of acellularity were observed (Fig. 4B arrowheads). There was also evidence of protein deposition across the entire surface of both the femoral condyle and the tibial plateau (Fig. 4B brackets). Interestingly, TMJs of 2 month-old *Prg4*$^{-/-}$ mice had a much greater loss of superficial zone chondrocytes, especially on the surface of the glenoid fossa (Fig. 4A brackets). As in the articular surfaces of the knee, protein had accumulated on the surface of both the fossa and the mandibular condyle. By 9 months of age, the cartilage surfaces of both joints had worsened, superficial zone chondrocytes were not observed at the articular surface of either the knee or the TMJ, and protein deposition on the cartilage surfaces had increased (Figs. 4 C and D). In the TMJ, the entire upper layer of articular cartilage was lost and was replaced by an infiltration of synoviocytes (Fig. 4C *). In the knee, some cartilage was still observed, though there was evidence of cartilage clefts (Fig. 4D arrow), suggesting that the osteoarthritis-like phenotype observed in the TMJ was more severe than that observed for the knee joint in age-matched mice.

Discussion

The *Prg4*$^{-/-}$ mouse is an established animal model for CACP syndrome, in which the absence of lubricin causes degenerative changes in the knee joint, resembling the fundamental features of osteoarthritis [10]. Synovial joints are the most common joint type in mammals and are characterized by the presence of synovial fluid. The TMJ is a unique synovial joint both in terms of its structure, as well as in the developmental genetic pathways that govern its formation [22]. The knee articular surfaces are covered with hyaline cartilage, whereas the TMJ articular surfaces are covered with fibrocartilage. Although TMJ-OA can be associated with arthropathies affecting other joints, few patients presenting with TMJ-OA have generalized osteoarthritis [5].

We therefore examined *Prg4*$^{-/-}$ mice to determine whether the TMJ also presents an OA-like phenotype that would, if comparable to that seen in the knee, be indicative of early changes and would allow the investigation of the first stages of TMJ-OA. We found that in adult *Prg4*$^{-/-}$ mice, the TMJ displayed significant signs of joint degeneration that increased in severity over time. Embryonic and newborn mice showed no apparent abnormalities suggesting that lubricin is not required for normal development of the TMJ, but is essential for its maintenance.

Early osteoarthritis is characterized by fibrillation and erosion of the articular cartilage surface, associated with loss of cells in the superficial zone, and protein deposition from the synovial fluid. We observed these characteristics in the TMJ of *Prg4*$^{-/-}$ mice, suggesting that a lack of lubricin induces joint degeneration that mimics human TMJ OA. Osteoarthritis is characterized by early loss of proteoglycans, leading to reduction in compressive strength of the cartilage, and subsequent joint failure [25,26]. *Prg4*$^{-/-}$

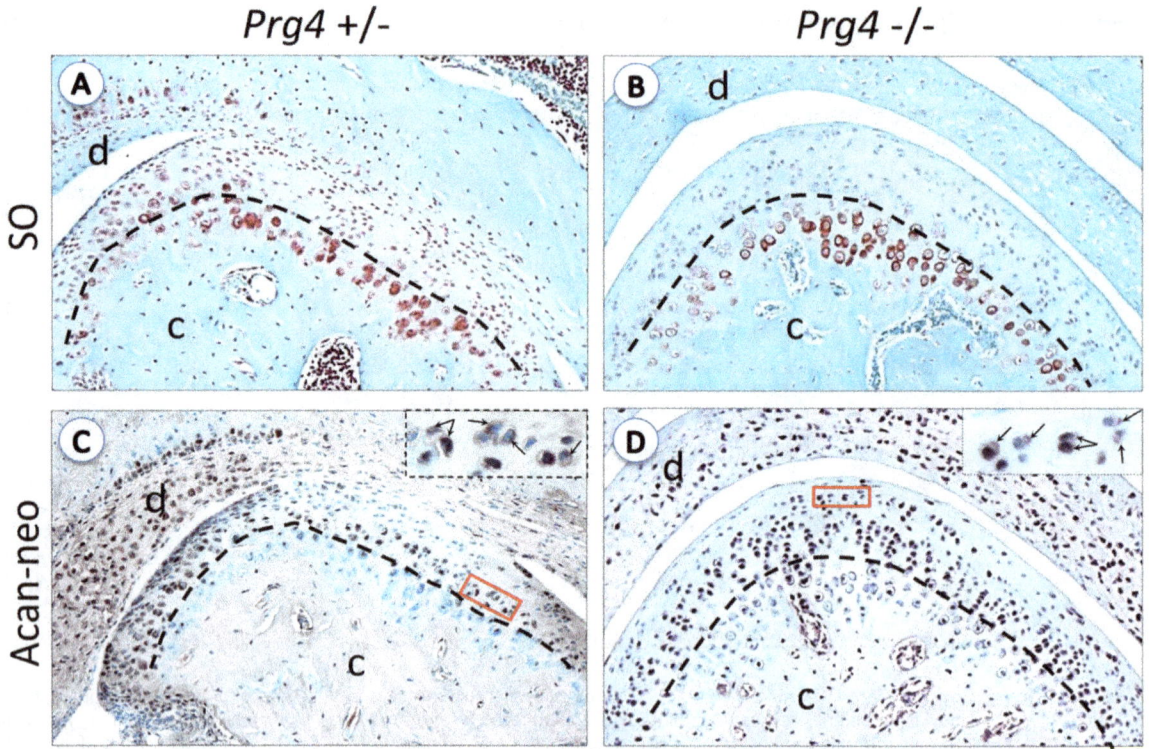

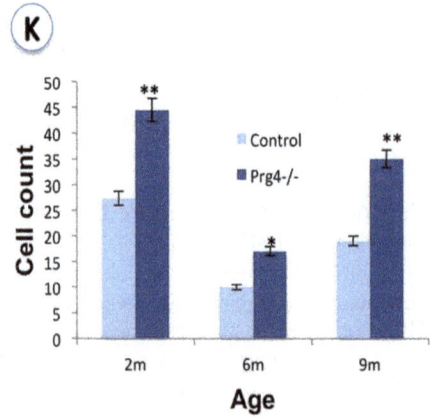

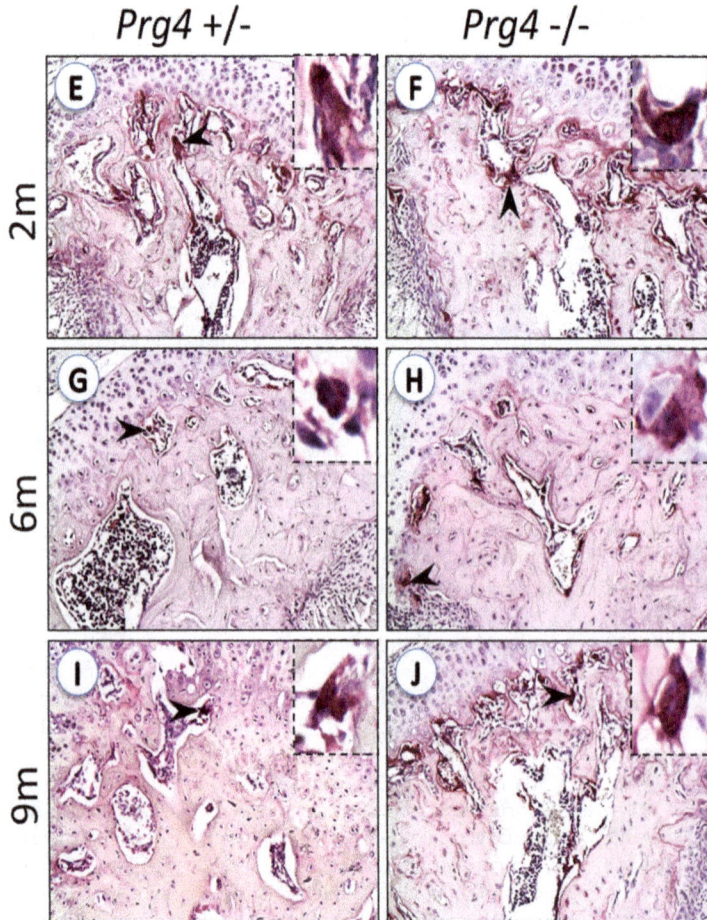

Figure 3. Increased TMJ Osteoarthritis-like characteristics in *Prg4⁻/⁻* mice. (A–D) Decreased extracellular matrix components in *Prg4⁻/⁻* mice. Paraffin embedded, formalin fixed TMJ sections at 6-months, histologically stained for Safranin O (SO) (A, B) and immunostained for aggrecan neopeptide (C, D) in *Prg4⁺/⁻* and *Prg4⁻/⁻* mice. A black dotted line separates the region of columnar chondrocytes in the condyle (below) from the proliferating chondrocytes (above). Note the abundant SO positive staining (red dots) in the proliferating zone in control *Prg4⁺/⁻* mice and the few positive SO staining in *Prg4⁻/⁻* null mice. As expected, the opposite is observed for aggrecan neopeptide immuno localization, which shows abundant aggregan neopeptide in proliferating chondrocytes, indicating that in *Prg4⁻/⁻* mice aggrecan degradation is increased. Insets in C and D correspond to a magnification of the field indicated in the red dotted line and highlight the extracellular localization of aggrecan neopeptide staining (arrows). **(E–J)** TRAP staining in 2-, 6- and 9- month TMJ sections. Black arrowheads show TRAP+ multinucleated cells (MNC) in resorptive areas of the condyles stained in red (magnified panel on right). **(K)** Quantitation of TRAP+ MNC at 2-, 6-, and 9- month-old mice showed an increase in osteoclastogenesis of 38.65%, 41.2%, and 45.8% respectively in *Prg4⁻/⁻ mice*, compared to age-matched control mice. Student t-test: *, p<0.05; **, p<0.01. Dotted lines demarcate the surfaces of the condyle and fossa. c: condyle, d: disc, f: fossa.

mice exhibited loss of proteoglycan staining at all ages analyzed, and an increase in osteoclast activity (Fig. 3), hallmarks of both OA and active joint remodeling.

In addition, we observed a thickening of the disc in the TMJs of *Prg4⁻/⁻* mice with a loss of the characteristic bi-concave shape (Fig.1), a phenotype that has also previously been shown in human TMJ disorders [27]. This thickening may represent a defense mechanism to maintain the smoothness of the joint by compensating for the degeneration of the cartilage surface. The integrity of the disc is vital in maintaining the homeostasis of the joint, degenerative changes in the disc, including perforation, lead to disruption of the joint [28]. Disc displacements are also associated with TMJ OA [29]. Thus, our results suggest that there is inter-

dependency between the integrity of the cartilage and the integrity of the disc, and that changes in either structure will have an effect on the health of the joint.

Lubricin has been shown to prevent adhesion and regulate synoviocyte cell proliferation in the knee [10]. We observed both synovial overgrowth and severe synoviocyte infiltration in the TMJ of *Prg4⁻/⁻* mice, most strikingly at 9 months of age (Fig. 2D, 4C), suggesting that synovial growth is also controlled by lubricin in the TMJ. The increased thickness of the disc may also be due in part to synovial infiltration. The cellular organization of the growth plate of the TMJ is not linear as it is in the knee [22], allowing for mutli- rather than uni-directional growth [23]. In addition, the articular cartilage components are different between the two joints,

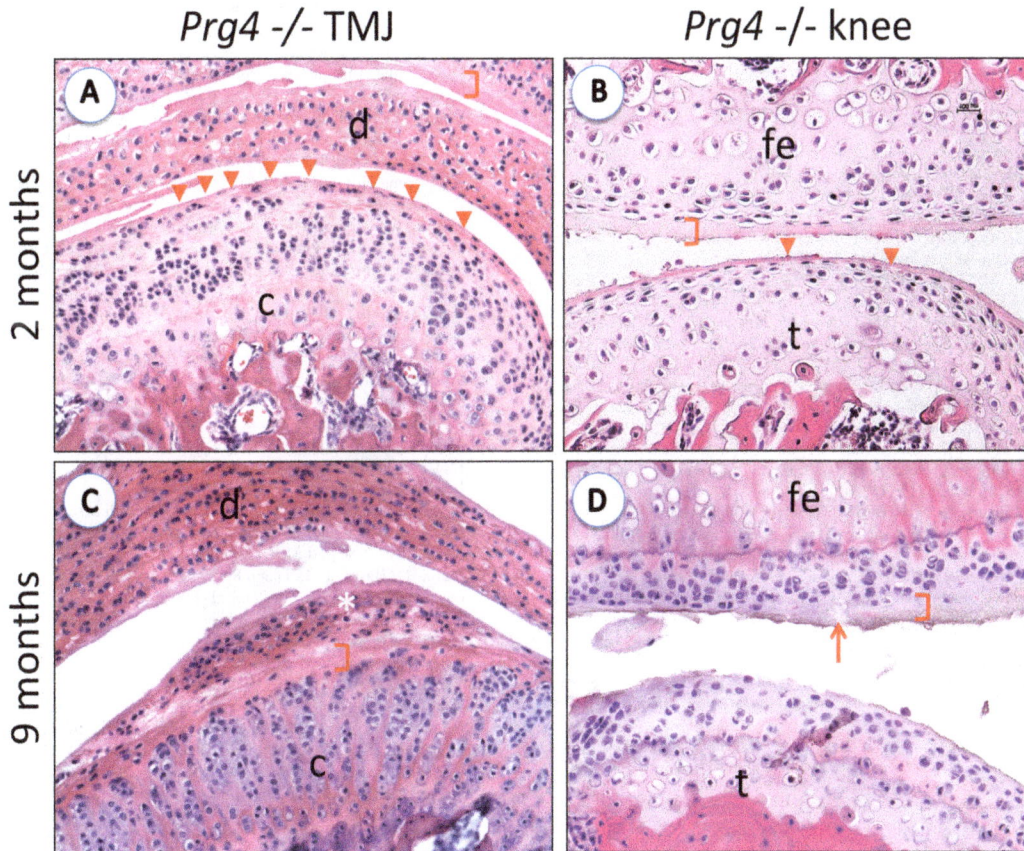

Figure 4. Osteoarthritis-like characteristics in TMJ and knee joints in *Prg4⁻/⁻* mice. (A–D) Representative coronal sections of the TMJ and knee joints of lubricin null (*Prg4⁻/⁻*) mice at 2- and 9- months of age, stained with H&E. At 2 months, the articular surface of the TMJ condyle displays very few superficial flat chondrocytes, whereas, several superficial zone chondrocytes can still be observed at the cartilage surface in the knee joint. (A and B, arrowheads). In both joints, evidence of lightly stained protein deposition across the entire joint surfaces can be observed (Fig. 4 brackets). At 9 months of age, the cartilage surfaces of both joints display disease progression, chondrocytes are absent from the articular surfaces of both the knee and the TMJ, and the protein layer deposited on all surfaces is enlarged and disrupted (Figs. 4 C and D). In addition, in the TMJ there is a large infiltration of synoviocytes deposited on the surface of the condyle (Fig. 4C *). c: condyle, d: disc, fe: femur, t: tibia.

predominated by hyaline cartilage in the knee and fibrocartilage in the TMJ. This suggests that lubricin is able to prevent adhesion and excessive proliferation of the synovium in many joints on multiple surfaces. This information could be of particular relevance for the lubrication of engineered joint replacements.

In the mouse, the knee develops earlier than the TMJ, with joint cavitation seen at E14.5, whereas the TMJ does not develop until E16. Nevertheless there was a more severe OA-like phenotype in the TMJ than in the knee at both 2 months and 9 months in $Prg4^{-/-}$ mice (Fig.4), indicating that the TMJ has an earlier onset of OA-like pathology in $Prg4^{-/-}$ mice than the knee joint. Other TMJ-OA mouse models have been reported, for example mice double deficient in biglycan and fibromodulin, which exhibit OA of both the knees and TMJ similar to that observed in $Prg4^{-/-}$ mice [23,30]. However, in these mice OA is observed in the knee at 1 month of age, whereas degeneration of the TMJ was not seen until 6 months.

In summary our results present the first structural and molecular evidence of the essential role of lubricin, the main component of the synovial fluid, in the maintenance of the integrity of the TMJ. In addition, these data also suggest that TMJ

degeneration in $Prg4^{-/-}$ mice occurs much earlier than in other reported models [23,30]. The $Prg4^{-/-}$ mouse therefore offers a distinct and valuable model for investigating an earlier onset of TMJ-OA and precocious joint failure, which in addition to the existing models will lead to a better understanding of TMJ disorders.

Acknowledgments

We are grateful to Dr. Matthew Warman (Boston Children's Hospital, Boston, MA) for providing the $Prg4$ knockout mice and for valuable guidance throughout this project. To Drs. Philip Stashenko (The Forsyth Institute, Cambridge, MA), Yefu Li (Harvard School of Dental Medicine, Boston, MA) and Pamela Tran (University of Kansas, Kansas city, KS) for helpful suggestions and discussion. To Michela Grunebaum for technical support. The authors declare that they have no conflicts of interest.

Author Contributions

Conceived and designed the experiments: PP AH JD. Performed the experiments: PP AH JD. Analyzed the data: PP AH JD. Contributed reagents/materials/analysis tools: PP AH JD. Contributed to the writing of the manuscript: PP AH JD.

References

1. Avery JK (2001) Oral development and histology: Thieme Medical Publishers. 3rd Ed p. 435.
2. Embree M, Ono M, Kilts T, Walker D, Langguth J, et al. (2011) Role of subchondral bone during early-stage experimental TMJ osteoarthritis. J Dent Res 90: 1331–1338.
3. Kumar D, Schooler J, Zuo J, McCulloch CE, Nardo L, et al. (2013) Trabecular bone structure and spatial differences in articular cartilage MR relaxation times in individuals with posterior horn medial meniscal tears. Osteoarthritis Cartilage 21: 86–93.
4. Sellam J, Berenbaum F (2010) The role of synovitis in pathophysiology and clinical symptoms of osteoarthritis. Nat Rev Rheumatol 6: 625–635.
5. Scrivani SJ, Keith DA, Kaban LB (2008) Temporomandibular disorders. N Engl J Med 359: 2693–2705.
6. Ingawale S, Goswami T (2009) Temporomandibular joint: disorders, treatments, and biomechanics. Ann Biomed Eng 37: 976–996.
7. Romero-Reyes M, Uyanik JM (2014) Orofacial pain management: current perspectives. J Pain Res 7: 99–115.
8. Jay GD, Torres JR, Warman ML, Laderer MC, Breuer KS (2007) The role of lubricin in the mechanical behavior of synovial fluid. Proc Natl Acad Sci U S A 104: 6194–6199.
9. Kure-Hattori I, Watari I, Takei M, Ishida Y, Yonemitsu I, et al. (2012) Effect of functional shift of the mandible on lubrication of the temporomandibular joint. Arch Oral Biol 57: 987–994.
10. Rhee DK, Marcelino J, Baker M, Gong Y, Smits P, et al. (2005) The secreted glycoprotein lubricin protects cartilage surfaces and inhibits synovial cell overgrowth. J Clin Invest 115: 622–631.
11. Marcelino J, Carpten JD, Suwairi WM, Gutierrez OM, Schwartz S, et al. (1999) CACP, encoding a secreted proteoglycan, is mutated in camptodactyly-arthropathy-coxa vara-pericarditis syndrome. Nat Genet 23: 319–322.
12. Bahabri SA, Suwairi WM, Laxer RM, Polinkovsky A, Dalaan AA, et al. (1998) The camptodactyly-arthropathy-coxa vara-pericarditis syndrome: clinical features and genetic mapping to human chromosome 1. Arthritis Rheum 41: 730–735.
13. Gynther GW, Holmlund AB, Reinholt FP, Lindblad S (1997) Temporomandibular joint involvement in generalized osteoarthritis and rheumatoid arthritis: a clinical, arthroscopic, histologic, and immunohistochemical study. Int J Oral Maxillofac Surg 26: 10–16.
14. Tanaka E, Detamore MS, Mercuri LG (2008) Degenerative disorders of the temporomandibular joint: etiology, diagnosis, and treatment. J Dent Res 87: 296–307.
15. Ameye LG, Young MF (2006) Animal models of osteoarthritis: lessons learned while seeking the "Holy Grail". Curr Opin Rheumatol 18: 537–547.
16. Rintala M, Metsaranta M, Saamanen AM, Vuorio E, Ronning O (1997) Abnormal craniofacial growth and early mandibular osteoarthritis in mice harbouring a mutant type II collagen transgene. J Anat 190 (Pt 2): 201–208.
17. Wadhwa S, Embree MC, Kilts T, Young MF, Ameye LG (2005) Accelerated osteoarthritis in the temporomandibular joint of biglycan/fibromodulin double-deficient mice. Osteoarthritis Cartilage 13: 817–827.
18. Xu L, Flahiff CM, Waldman BA, Wu D, Olsen BR, et al. (2003) Osteoarthritis-like changes and decreased mechanical function of articular cartilage in the joints of mice with the chondrodysplasia gene (cho). Arthritis Rheum 48: 2509–2518.
19. Chen J, Gupta T, Barasz JA, Kalajzic Z, Yeh WC, et al. (2009) Analysis of microarchitectural changes in a mouse temporomandibular joint osteoarthritis model. Arch Oral Biol 54: 1091–1098.
20. Xu L, Peng H, Glasson S, Lee PL, Hu K, et al. (2007) Increased expression of the collagen receptor discoidin domain receptor 2 in articular cartilage as a key event in the pathogenesis of osteoarthritis. Arthritis Rheum 56: 2663–2673.
21. Purcell P, Jheon A, Vivero MP, Rahimi H, Joo A, et al. (2012) Spry1 and spry2 are essential for development of the temporomandibular joint. J Dent Res 91: 387–393.
22. Purcell P, Joo BW, Hu JK, Tran PV, Calicchio ML, et al. (2009) Temporomandibular joint formation requires two distinct hedgehog-dependent steps. Proc Natl Acad Sci U S A 106: 18297–18302.
23. Wadhwa S, Embree M, Ameye L, Young MF (2005) Mice deficient in biglycan and fibromodulin as a model for temporomandibular joint osteoarthritis. Cells Tissues Organs 181: 136–143.
24. Ochiai T, Shibukawa Y, Nagayama M, Mundy C, Yasuda T, et al. (2010) Indian hedgehog roles in post-natal TMJ development and organization. J Dent Res 89: 349–354.
25. Lotz M, Martel-Pelletier J, Christiansen C, Brandi ML, Bruyere O, et al. (2014) Republished: Value of biomarkers in osteoarthritis: current status and perspectives. Postgrad Med J 90: 171–178.
26. Franz T, Hasler EM, Hagg R, Weiler C, Jakob RP, et al. (2001) In situ compressive stiffness, biochemical composition, and structural integrity of articular cartilage of the human knee joint. Osteoarthritis Cartilage 9: 582–592.
27. Wang XD, Kou XX, Mao JJ, Gan YH, Zhou YH (2012) Sustained inflammation induces degeneration of the temporomandibular joint. J Dent Res 91: 499–505.
28. Lang TC, Zimny ML, Vijayagopal P (1993) Experimental temporomandibular joint disc perforation in the rabbit: a gross morphologic, biochemical, and ultrastructural analysis. J Oral Maxillofac Surg 51: 1115–1128.
29. Cortes D, Exss E, Marholz C, Millas R, Moncada G (2011) Association between disk position and degenerative bone changes of the temporomandibular joints: an imaging study in subjects with TMD. Cranio 29: 117–126.
30. Ameye L, Aria D, Jepsen K, Oldberg A, Xu T, et al. (2002) Abnormal collagen fibrils in tendons of biglycan/fibromodulin-deficient mice lead to gait impairment, ectopic ossification, and osteoarthritis. FASEB J 16: 673–680.

Permissions

The contributors of this book come from diverse backgrounds, making this book a truly international effort. This book will bring forth new frontiers with its revolutionizing research information and detailed analysis of the nascent developments around the world.

We would like to thank all the contributing authors for lending their expertise to make the book truly unique. They have played a crucial role in the development of this book. Without their invaluable contributions this book wouldn't have been possible. They have made vital efforts to compile up to date information on the varied aspects of this subject to make this book a valuable addition to the collection of many professionals and students.

This book was conceptualized with the vision of imparting up-to-date information and advanced data in this field. To ensure the same, a matchless editorial board was set up. Every individual on the board went through rigorous rounds of assessment to prove their worth. After which they invested a large part of their time researching and compiling the most relevant data for our readers.

The editorial board has been involved in producing this book since its inception. They have spent rigorous hours researching and exploring the diverse topics which have resulted in the successful publishing of this book. They have passed on their knowledge of decades through this book. To expedite this challenging task, the publisher supported the team at every step. A small team of assistant editors was also appointed to further simplify the editing procedure and attain best results for the readers.

Apart from the editorial board, the designing team has also invested a significant amount of their time in understanding the subject and creating the most relevant covers. They scrutinized every image to scout for the most suitable representation of the subject and create an appropriate cover for the book.

The publishing team has been an ardent support to the editorial, designing and production team. Their endless efforts to recruit the best for this project, has resulted in the accomplishment of this book. They are a veteran in the field of academics and their pool of knowledge is as vast as their experience in printing. Their expertise and guidance has proved useful at every step. Their uncompromising quality standards have made this book an exceptional effort. Their encouragement from time to time has been an inspiration for everyone.

The publisher and the editorial board hope that this book will prove to be a valuable piece of knowledge for researchers, students, practitioners and scholars across the globe.

List of Contributors

R. Adam Horch
Department of Biomedical Engineering, Vanderbilt University, Nashville, Tennessee, United States of America
Institute of Imaging Science, Vanderbilt University, Nashville, Tennessee, United States of America

Daniel F. Gochberg
Institute of Imaging Science, Vanderbilt University, Nashville, Tennessee, United States of America
Department of Radiology and Radiological Sciences, Vanderbilt University, Nashville, Tennessee, United States of America

Jeffry S. Nyman
Department of Biomedical Engineering, Vanderbilt University, Nashville, Tennessee, United States of VA Tennessee Valley Healthcare System, Nashville, Tennessee, United States of America
Department of Orthopaedics and Rehabilitation, Vanderbilt University Medical Center, Nashville, Tennessee, United States of America
Center for Bone Biology, Vanderbilt University Medical Center, Nashville, Tennessee, United States of America

Mark D. Does
Department of Biomedical Engineering, Vanderbilt University, Nashville, Tennessee, United States of Institute of Imaging Science, Vanderbilt University, Nashville, Tennessee, United States of America
Department of Radiology and Radiological Sciences, Vanderbilt University, Nashville, Tennessee, United States of America
Department of Electrical Engineering, Vanderbilt University, Nashville, Tennessee, United States of America

Rishi R. Rampersad, Teresa K. Tarrant, Christopher T. Vallanat, Tatiana Quintero-Matthews, Michael F. Weeks, Alan M. Fong and Peng Liu
Department of Medicine, Thurston Arthritis Research Center, University of North Carolina at Chapel Hill, Chapel Hill, North Carolina, United States of America

Denise A. Esserman
Division of General Medicine and Clinical Epidemiology, Department of Medicine, University of North Carolina at Chapel Hill, Chapel Hill, North Carolina, United States of America
Department of Biostatistics, University of North Carolina at Chapel Hill, Chapel Hill, North Carolina, United States of America

Jennifer Clark
Department of Biostatistics, University of North Carolina at Chapel Hill, Chapel Hill, North Carolina, United States of America

Dhavalkumar D. Patel
Department of Medicine, Thurston Arthritis Research Center, University of North Carolina at Chapel Hill, Chapel Hill, North Carolina, United States of America
Novartis Institutes for BioMedical Research, Basel, Switzerland

Franco Di Padova
Novartis Institutes for BioMedical Research, Basel, Switzerland

Yifang Fan, Changsheng Lv and Zhang Bo
Center for Scientific Research, Guangzhou Institute of Physical Education, Guangzhou, People's Republic of China

Yubo Fan
Key Laboratory for Biomechanics and Mechanobiology of Ministry of Education, School of Biological Science and Medical Engineering, Beihang University, Beijing, People's Republic of China

Zhiyu Li
College of Foreign Studies, Jinan University, Guangzhou, People's Republic of China

Mushtaq Loan
International School, Jinan University, Guangzhou, People's Republic of China

Susannah J. Sample, Zhengling Hao and Aliya P. Wilson, Peter Muir
Comparative Orthopaedic Research Laboratory, School of Veterinary Medicine, University of Wisconsin-Madison, Madison, Wisconsin, United States of America

Friederike A. Schulte, Davide Ruffoni, Floor M. Lambers, David Christen, Duncan J. Webster, Gisela Kuhn and Ralph Müller
Institute for Biomechanics, ETH Zurich, Zurich, Switzerland

Grac₌a Cardadeiro, Fátima Baptista and Luís B. Sardinha
Exercise and Health Laboratory, Faculty of Human Movement, Technical University of Lisbon, Lisbon, Portugal

Rui Ornelas
Centre of Social Sciences, Department of Physical Education and Sport, University of Madeira, Funchal, Portugal

Kathleen F. Janz
Department of Health and Human Physiology, Department of Epidemiology, University of Iowa, Iowa City, Iowa, United States of America

Jangwoo Lee and David M. Gardiner
Department of Developmental and Cell Biology, University of California Irvine, Irvine, California, United States of America
The Developmental Biology Center, University of California Irvine, Irvine, California, United States of America

Neil Curtis, Junfen Shi and Michael J. Fagan
Medical and Biological Engineering Research Group, Department of Engineering, University of Hull, Hull, United Kingdom

Marc E. H. Jones and Susan E. Evans
Research Department of Cell and Developmental Biology, University College London, London, United Kingdom

Paul O'Higgins
Hull-York Medical School, University of York, York, United Kingdom

Lovisa Hessle, Christina Wenglén, Dick Heinegård and Patrik Önnerfjord
Sections of Molecular Skeletal Biology and Rheumatology, Department of Clinical Sciences Lund, Lund University, Lund, Sweden

Gunhild A. Stordalen, Sverre-Henning Brorson, Finn P. Reinholt and Espen S. Baekkevold
Department of Pathology, University of Oslo, and Oslo University Hospital, Rikshospitalet, Oslo, Norway

Christiane Petzold
Faculty of Odontology, University of Oslo, Oslo, Norway

Elizabeth K. Tanner
School of Engineering, University of Glasgow, Glasgow, United Kingdom
Section of Orthopaedics, Department of Clinical Sciences Lund, Lund University, Lund, Sweden

Xiang Li and Cheng-Tao Wang
School of Mechanical Engineering, Shanghai Jiao Tong University, State Key Laboratory of Mechanical System and Vibration, Shanghai, China

Ya-Fei Feng, Wei Lei and Lin Wang
Department of Orthopaedics, Xijing Hospital, The Fourth Military Medical University, Xi'an, China

Guo-Chen Li
Department of Orthopaedics, Tangdu Hospital, The Fourth Military Medical University, Xi'an China

Zhi-Yong Zhang
Department of Orthopaedics, Xijing Hospital, The Fourth Military Medical University, Xi'an, China
Department of Plastic and Reconstructive Surgery, Shanghai 9th People's Hospital, Shanghai Key Laboratory of Tissue Engineering, School of Medicine, Shanghai Jiao Tong University, Shanghai, China

Kim Henriksen, Christian S. Thudium, Anita V. Neutzsky-Wulff and Morten A. Karsdal
Nordic Bioscience A/S, Herlev, Denmark

Carmen Flores, Maria Askmyr and Johan Richter
Molecular Medicine and Gene Therapy, Lund University, Lund, Sweden

Jesper S. Thomsen and Anne-Marie Brüel
Institute of Anatomy, University of Aarhus, Aarhus, Denmark

Geerling E. J. Langenbach
Department of Functional Anatomy, Academic Centre of Dentistry Amsterdam (ACTA), University of Amsterdam and VU University Amsterdam, Research Institute MOVE, Amsterdam, The Netherlands

Natalie Sims and Thomas J. Martin
St. Vincent's Institute for Medical Research, Melbourne, Australia

Vincent Everts
Department of Oral Cell Biology, Academic Centre of Dentistry Amsterdam (ACTA), University of Amsterdam and VU University Amsterdam Research Institute MOVE, Amsterdam, The Netherlands

Susan A. Novotny, Tara L. Mader, Angela G. Greising and Dawn A. Lowe
Program in Physical Therapy and Rehabilitation Sciences, University of Minnesota, Minneapolis, Minnesota, United States of America

Angela S. Lin and Robert E. Guldberg
Institute for Bioengineering and Bioscience, Georgia Institute of Technology, Atlanta, Georgia, United States of America

Gordon L. Warren
Department of Physical Therapy, Georgia State University, Atlanta, Georgia, United States of America

Danielle E. Green, Benjamin J. Adler, Meilin Ete Chan, James J. Lennon and Clinton T. Rubin
Department of Biomedical Engineering, Stony Brook University, Stony Brook, New York, United States of America

Alvin S. Acerbo and Lisa M. Miller
Department of Biomedical Engineering, Stony Brook University, Stony Brook, New York, United States of America
Photon Sciences Directorate, Brookhaven National Laboratory, Upton, New York, United States of America

Wagner S. Vicente, Luciene M. dos Reis, Rafael G. Graciolli, Fabiana G. Graciolli, Wagner V. Dominguez and Vanda Jorgetti
Nephrology Division, Medical School, University of São Paulo, São Paulo, Brazil

Hamilton Roschel and Bruno Gualano
Department of Sports, School of Physical Education and Sport, University of Sao Paulo, Sao Paulo, Brazil

Tatiana L. Fonseca
Department of Anatomy, Institute of Biomedical Sciences, University of Sao Paulo, São Paulo, Brazil

Ana P. Velosa and Walcy R. Teodoro
Rheumatology Division, Medical School, University of São Paulo, São Paulo, Brazil

Charles C. Wang
Department of Physiological Sciences, Federal University of São Carlos, São Paulo, Brazil

Cesar A. Olguin
Animal Science Tech, Applied to Wildlife Management Res.Group, IREC Sec. Albacete, IREC (UCLM-CSIC-JCCM), Campus UCLM, Albacete, Spain
Grupo de Recursos Cinegéticos, Instituto de Desarrollo Regional (IDR), Universidad de Castilla-La Mancha (UCLM), Albacete, Spain

Tomas Landete-Castillejos, Andrés J. García and Laureano Gallego
Animal Science Tech, Applied to Wildlife Management Res.Group, IREC Sec. Albacete, IREC (UCLM-CSIC-JCCM), Campus UCLM, Albacete, Spain
Grupo de Recursos Cinegéticos, Instituto de Desarrollo Regional (IDR), Universidad de Castilla-La Mancha (UCLM), Albacete, Spain
Departamento de Ciencia y Tecnología Agroforestal y Genética, ETSIA, Universidad de Castilla-La Mancha (UCLM), Albacete, Spain

Francisco Ceacero
Department of Animal Science and Food Processing in Tropics and Subtropics, Faculty of Tropical AgriSciences – Czech University of Life Sciences, Praha– Suchdol, Czech Republic

Karen A. Roddy and Paula Murphy
Department of Zoology, School of Natural Sciences, Trinity College Dublin, Dublin, Ireland
Trinity Centre for Bioengineering, School of Engineering, Trinity College Dublin, Dublin, Ireland

Patrick J. Prendergast
Trinity Centre for Bioengineering, School of Engineering, Trinity College Dublin, Dublin, Ireland

Jan-Erik Ode
Julius Wolff Institute, Charité - Universitätsmedizin, Berlin, Germany

Andrea Ode and Sven Geissler
Julius Wolff Institute, Charité - Universitätsmedizin, Berlin, Germany
Berlin-Brandenburg Center for Regenerative Therapies, Berlin, Germany

Georg N. Duda and Carsten Perka
Julius Wolff Institute, Charité - Universitätsmedizin, Berlin, Germany
Berlin-Brandenburg Center for Regenerative Therapies, Berlin, Germany
Klinik für Orthopädie, Centrum für Muskuloskeletale Chirurgie, Charité - Universitätsmedizin, Berlin, Germany

Patrick Strube and Stephan Pauly
Julius Wolff Institute, Charité - Universitätsmedizin, Berlin, Germany
Klinik für Orthopädie, Centrum für Muskuloskeletale Chirurgie, Charité - Universitätsmedizin, Berlin, Germany

Xinhua Qu and Zhifeng Yu
Shanghai Key Laboratory of Orthopaedic Implants, Department of Orthopaedic Surgery, Shanghai Ninth People's Hospital, Shanghai Jiaotong University School of Medicine, Shanghai, China

Gang Wang
Shanghai Key Laboratory of Orthopaedic Implants, Department of Orthopaedic Surgery, Shanghai Ninth People's Hospital, Shanghai Jiaotong University School of Medicine, Shanghai, China
Department of Orthopedic Surgery, the Second Affiliated Hospital of Nanjing Medical University, Nanjing, China

Maxime A. Gallant, Drew M. Brown, Amy Y. Sato, Justin N. Williams and David B. Burr
Department of Anatomy and Cell Biology, Indiana University School of Medicine, Indianapolis, Indiana, United States of America

Kathleen M. Hill Gallant
Department of Anatomy and Cell Biology, Indiana University School of Medicine, Indianapolis, Indiana, United States of America
Department of Nutrition Science, Purdue University, West Lafayette, Indiana, United States of America

Fang-Yuan Wei, Gang Li, Jianghui Qin, Simon Kwoon-Ho Chow, Shuo Huang and Ming-Hui Sun
Department of Orthopaedics and Traumatology, Clinical Sciences Building, The Chinese University of Hong Kong, Shatin, New Territories, Hong Kong SAR, China

Kwok-Sui Leung, Ling Qin and Wing-Hoi Cheung
Department of Orthopaedics and Traumatology, Clinical Sciences Building, The Chinese University of Hong Kong, Shatin, New Territories, Hong Kong SAR, China
Translational Medicine Research & Development Center, Institute of Biomedical and Health Engineering, Shenzhen Institutes of Advanced Technology, Chinese Academy of Sciences, Shenzhen, China

Anna Shipov, Gilad Segev, Hagar Meltzer, Moran Milrad, Ayelet Atkins and Ron Shahar
Koret School of Veterinary Medicine, Hebrew University of Jerusalem, Rehovot, Israel

Ori Brenner
Department of Veterinary Resources, Weizmann Institute, Rehovot, Israel

Yanghui Xing, Yan Gu, James J. Bresnahan, Emmanuel M. Paul, Henry J. Donahue and Jun You
Division of Musculoskeletal Sciences, Department of Orthopaedics and Rehabilitation, The Pennsylvania State University College of Medicine, Hershey, Pennsylvania, United States of America

Adele Hill
Department of Orthopaedic Surgery, Boston
Children's Hospital, Boston, Massachusetts, United
States of America; Department of Genetics, Harvard
Medical School, Boston, Massachusetts, United
States of America

Juanita Duran and Patricia Purcell
Department of Plastic and Oral Surgery, Boston
Children's Hospital and Harvard Medical School,
Boston, Massachusetts, United States of America

Index

www.ingramcontent.com/pod-product-compliance
Lightning Source LLC
Chambersburg PA
CBHW061245190326
41458CB00011B/3581